Nuclear and Particle Physics

Nuclear and Particle Physics

W. S. C. Williams

Department of Physics, University of Oxford
and St Edmund Hall, Oxford

CLARENDON PRESS · OXFORD
1991

Oxford University Press, Walton Street, Oxford OX2 6DP

Oxford New York Toronto
Delhi Bombay Calcutta Madras Karachi
Petaling Jaya Singapore Hong Kong Tokyo
Nairobi Dar es Salaam Cape Town
Melbourne Auckland

and associated companies in
Berlin Ibadan

Oxford is a trade mark of Oxford University Press

Published in the United States
by Oxford University Press, New York

British Library Cataloguing in Publication Data
Williams, W. S. C.
Nuclear and particle physics.
I. Title
539.7
ISBN 0–19–851999–0
ISBN 0–19–852046–8 (Pbk)

Library of Congress Cataloging in Publication Data
Williams, W. S. C. (William S. C.)
Nuclear and particle physics / W. S. C. Williams.
Includes bibliographical references.
1. Nuclear physics. 2. Particles (Nuclear physics) I. Title.
QC776.W55 1990 539.7′2—dc20 90–7110
ISBN 0–19–851999–0
ISBN 0–19–852046–8 (pbk.)

Typeset by Latimer Trend & Company Ltd, Plymouth
Printed and bound in Great Britain by Butler & Tanner Ltd, Frome, Somerset

For

Renée, Claire, and Matthieu

Preface

My intention in preparing this book was to provide a text on nuclear and particle physics for use by university undergraduates in the penultimate year of their studies for a first degree in physics. I have assumed that at this point they will have completed courses in electromagnetism, quantum mechanics, and atomic physics. I have chosen the level to which the subject has been treated in a way which I trust will satisfy two groups. The first is those who need an introduction to a subject that they will study in greater depth and detail in their final year using other, more advanced texts. The second group contains those who do not propose such studies but who need a complete and, as far as possible, up-to-date coverage suitable for a less specialized course. This policy is exemplified in the problems; many have been adapted from questions set in past Final Honours School Examinations in Physics at the University of Oxford. They have been selected from among the examination papers set each year that all students in physics take but not from the advanced paper in nuclear and particle physics that some choose to take.

I am indebted to E.H. Bellamy, J. H. Cobb, J. Gulley, J. E. Paton, and J.V. Peach for reading parts of the typescript and for many helpful comments. I am also grateful to other colleagues for useful discussions about physics. Naturally I am responsible for the final result and for any mistakes it may contain.

I am grateful to my daughter, Claire Williams, for help with Figs 2.8 and 2.9 and to Marilyn Coles for producing word-processed disk files from a draft manuscript; to Irmgard Smith and Alan Holmes who worked on the drawings, and to Cyril Band and his staff, particularly Jane Burrage, who did the photography. Thanks also to the staff at OUP for their help.

Oxford W.S.C.W.
December 1990

Acknowledgements

For photographs of distinguished physicists I am indebted to the following: The American Institute of Physics Niels Bohr Library for photographs of Rutherford (courtesy of Otto Hahn and Lawrence Badash), Becquerel, Planck, Heisenberg and Dirac (gift of Mrs Mark Zemansky); Professor Aage Bohr for the photograph of his father, Niels Bohr; Ruth Braunzier for the photograph of her father Erwin Schrödinger (courtesy of Zentralbibliothek für Physik, Wien); The Cavendish Laboratory, Cambridge, for the photograph of J.J. Thomson; The Deutsches Museum, München, for the photographs of Röntgen and of Yukawa; The Hebrew University of Jerusalem, Israel, for the photograph of Einstein (courtesy of the AIP Niels Bohr Library); Professor Walter Moore for the photograph of de Broglie, originally published in *Erlebte Physik* by Franz von Krebk, Berlin, 1942.

I am also pleased to thank, for material and permission to reproduce: Sidney Harris © for the cartoon of Einstein (Fig. 1.3) and for Figs 9.23 and 14.6; Bob Thaves © 1984 and United Media, New York, for the cartoon of Ernest and Frank (Fig. 9.9); Professor J.J. Simpson, University of Guelph, for the spectrum and Kurie plot for tritium beta decay (Fig. 12.10); Gallerie dell'Accademia, Venezia, for Leonardo da Vinci's Vitruvian Man (Fig. 2.1); The European Centre for Nuclear Research (CERN Photo) for bubble-chamber photographs (Figs 10.3, 10.4, 10.5, 12.17, and 12.18) and for computer reconstruction of a W and of a Z event (9.19 and 9.20) taken by the UA1 Collaboration (with permission); The Lawrence Berkeley Laboratory, California, for a bubble-chamber photograph (Fig. 10.8); The European Southern Observatory © for the before and after photographs of SN1987A; Deutsches Electronen-Synchrotron (DESY) for computer reconstructions of two events from the TASSO detector (Figs 10.17 and 10.19).

My thanks are due to publishers and authors (when possible) for permission to reprint or to adapt from the publications named: The American Physical Society, from *Physical Review* and *Physical Review Letters* (Figs 2.7, 3.4, 5.3, 7.13 and 11.2b); The American Institute of Physics, from *The Journal of Chemical Physics* (Fig. 11.2c); Cambridge University Press from *Essays in Nuclear Astrophysics* (Fig. 14.3) and from *Radiations from radioactive substances* by Rutherford, Chadwick, and Ellis (Table 1.3); Elsevier Science Publications, SV, from *Nuclear Physics* and *Physics Letters* (Figs 3.5, 3.6, 7.8, 7.13, 9.19 and 9.20); The Institute of Physics Publishing Ltd, from *Reports on Progress in Physics* (Fig. 9.14); MacMillan Magazines Ltd © 1947, Fig. 11.1 reprinted by permission from *Nature* vol. 159, p.126; MacMillan Publishers Ltd, for *Alice Through the Looking Glass* by Tenniel (Fig. 12.13); Oxford University Press, for examination questions; Pergamon Press, Oxford, from *The Journal of the Franklin Institute* (Fig. 11.8); The Royal Society, London, from *The Proceedings of the Royal Society* (Figs 7.1 and 11.4); Springer-Verlag, Berlin, from *Naturwissenschaft* (Fig. 6.6).

Contents

1 Introduction

2 Some Quantitative Formalities

3 The Size and Shape of Nuclei

4 The Masses of Nuclei

5 Nuclear Instability

6 Alpha Decay

7 Nuclear Collisions and Reactions

8 Nuclear Models

9 Forces and Interactions

10 Hadrons and the Quark–Parton Model

11 The Electromagnetic Interaction

12 The Weak Interaction

13 Particles: Summary and Outlook

14 Nuclear and Particle Astrophysics

Fig. 1.1 Ernest Rutherford (1871–1937), the father of nuclear physics. This photograph was taken about 1906, five years before his discovery of the atomic nucleus. In 1908 he received the Nobel Prize in chemistry 'for his investigations into the disintegration of the elements and the chemistry of radioactive substances'

1

Introduction

1.1 Historical perspective

In their classic work, *Radiations from radioactive substances* published in 1930, Rutherford, Chadwick, and Ellis refer to three discoveries as marking the beginning of an epoch in which rapid progress became possible in attacking the fundamental problems of physics on the nature of electricity and the constitution and relation of the atoms of the elements:

1895 The discovery of X-rays by Röntgen;

1896 The discovery of the radioactivity of uranium by Becquerel;

1897 The discovery of the electron by J. J. Thomson.

As vital to the interpretation of the investigations that followed they mention the role played by two other events:

1900 The discovery of the black body radiation formula by Planck;

1913 Bohr's theory of the hydrogen atom.

It is our belief that, insofar as the development of nuclear physics is concerned, other developments must be added:

1905 The development of the theory of special relativity by Einstein;

1911 The discovery of the atomic nucleus by Rutherford;

1926 and later
The development of non-relativistic and relativistic quantum mechanics by de Broglie, Schrödinger, Heisenberg, Dirac and others.

Much that followed from these developments is now a part of our culture and has had a profound impact on our technology and science. However, we need to distinguish three areas of physics.

Atomic physics is the physics of the behaviour of atoms and in particular of their electronic structure. That layer of the structure of matter depends for its existence on the nuclear atom but it is not markedly affected by the properties of the nucleus, apart from its charge and, to a lesser extent, by its spin and magnetic moment. Similarly the behaviour of the nucleus is little affected by the electronic structure surrounding it so that **nuclear physics** can, as a subject, be almost completely isolated from atomic physics. However, all methods of detecting nuclear and particle radiation depend on atomic physics.

From nuclear physics came **Elementary particle physics**. Two events marked its beginning:

1935 Yukawa's meson hypothesis (see Section 9.10);

DEFINITIONS AND KEYWORDS

Atomic physics The physics of the electronic, extra-nuclear structure of atoms.

Nuclear physics The physics of the atomic nucleus, believed to be constituted of neutrons and protons.

Elementary particle physics The physics of quarks and gluons, believed to be the constituents of protons and neutrons, and of leptons and gauge bosons and ... who knows what else!

Notation Names given to particles emitted in radioactive changes and what they are:

α-particle Nucleus of the helium atom.

β-particle Electron.

γ-ray Photon.

Next alphabetically but not emitted in radioactive change:

δ-ray Electron knocked out of an atom by the passage of a fast charged particle. Sometimes called a knock-on electron.

1 2 3

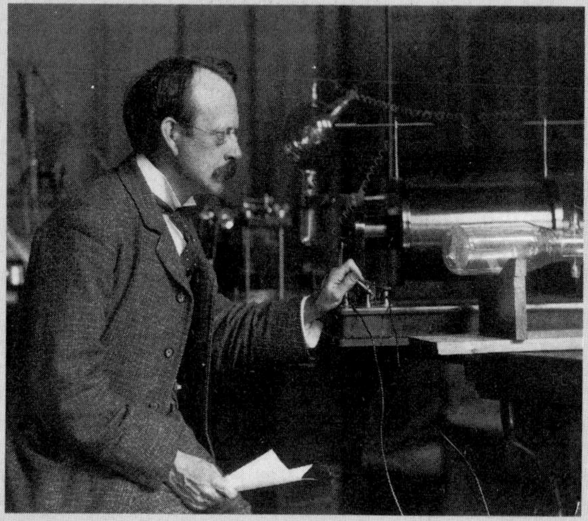

4

1 Wilhelm Röntgen (1845–1923). Discovered X-rays (1895) while investigating the properties of cathode rays. (The passage of electricity through gases at low pressures was known to cause the emission from the cathode of what was called cathode rays.)

2 Henri Becquerel (1852–1908). Discovered the radioactivity of uranium (1896) while investigating the fluorescence properties of uranium salts.

3 Max Planck (1858–1947). Discovered (1900) the formula correctly describing the spectrum of the electromagnetic radiation emitted by a black body and thereby introduced the idea of the quantum to physics.

4 J. J. Thomson (1856–1940). Discovered the electron in his investigations into the nature of cathode rays (1897).

Fig. 1.2 Four of the scientists whose discoveries initiated the exploration of the structure of atoms, nuclei, and matter.

1946 The discovery of pi-mesons by Powell and co-workers (see Section 9.10).

This subject is concerned with the neutron and proton and related particles, and in particular with their constituents, the quarks and gluons. It is also concerned with other entities such as the photon and the leptons, that appear, like quarks and gluons, to have no substructure. This level of the structure of matter is so far removed from normal experience that it has made little impact on our culture; none the less it exists. It is not clear to what extent this layer dominates the behaviour of the one above, the nuclear physics level. Whether or not there is a layer of new structure below the one of quarks, gluons, and leptons, which we call elementary particles, is presently a matter for speculation.

From 1913 onwards atomic physics and nuclear physics advanced in parallel. The flowering of quantum mechanics brought order to atomic physics and is

Fig. 1.3 Albert Einstein (1879–1955). In the cartoon the artist, Sidney Harris, has drawn a figure instantly recognizable as that of the mature Einstein. But, of course, he was a young man when he derived $E=mc^2$. The photograph was taken near the time (1905) when he published his *Special Theory of Relativity*. This was one of his many notable achievements in the field of theoretical physics, the greatest being the *General Theory of Relativity* published in 1915.

essential to the interpretation of nuclear physics. In this chapter we shall trace with some historical emphasis the discovery of the nucleus and the early stages in the development of the subject. In later chapters we shall not use that approach but attempt to gain an up-to-date and immediate view and understanding of the subject at a simple level appropriate to an introductory course. The latter part of this volume will be concerned with a description of the physics of the elementary particles.

Following the discovery of the radioactivity of uranium, there was a successful search for other naturally occurring radioactive elements and the recognition by Rutherford and Soddy that radioactivity involved a change in the mass and chemical nature of an element. The new element produced in any radioactive change was itself frequently radioactive. In fact, three radioactive series were recognized in which heavy elements lost mass and changed their atomic number in successive changes, the changes ending only when the element became an isotope of lead. In all these spontaneous changes three types of radiation were recognized. They are:

(1) α-rays: these were found to be positively charged particles with a ratio of charge to mass about one half that of a singly charged hydrogen atom. It became clear that these rays were energetic nuclei of helium.

1

2

3

1 Niels Bohr (1885–1962) developed the first successful model of the hydrogen atom (1913) and made many important contributions to the interpretation of quantum mechanics and to the development of atomic and nuclear physics. This photograph was taken in about 1917.

2 Louis de Broglie (1892–1987) who triggered the development of quantum mechanics by associating a wave-like behaviour with matter (1923). This informal photograph was taken in about 1924.

3 Erwin Schrödinger (1887–1961) derived the equation which bears his name (1926). It is the basis of the wave mechanics approach to quantum phenomena and allowed the solution of many problems in solid, atomic, and nuclear physics. This photograph was taken in about 1927.

4 Werner Heisenberg (1901–1976) formulated the matrix mechanics approach to quantum phenomena (1925) and the uncertainty relations that bear his name (1927). The photograph was taken near the time of these advances.

5 Paul Dirac (1902–1984) discovered the relativistic wave equation for the electron (1928). He gave, in Einstein's opinion, '. . . the logically most perfect presentation . . .' of quantum mechanics. The date of this photograph is not known but it was probably taken shortly after 1928.

4

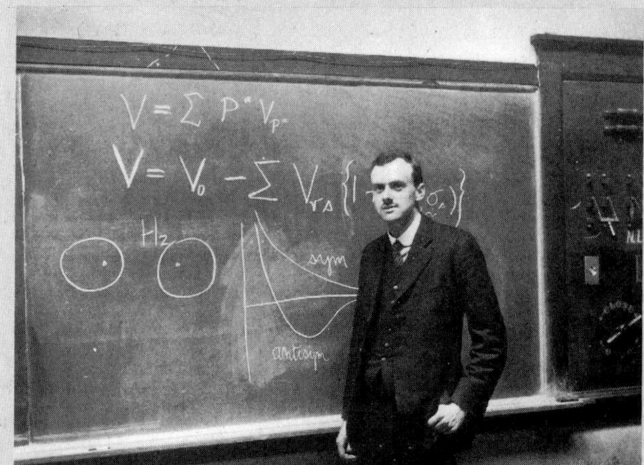

5

Fig. 1.4 A few of the famous names associated with the development of non-relativistic and relativistic quantum mechanics. The achievements mentioned in the captions do not do full justice to the many other contributions made by these scientists. In addition, the choice of these five must not lead us to neglect the many others who, at the same time and since, have put their mark on the progress of quantum mechanics. Yet, in spite of its formidable successes, the significance of the interpretation placed upon it remains a topic of debate.

(2) β-rays: these are negatively charged particles which were found to be identical to the electrons found by J. J. Thomson, although in general much more energetic than those generated in electrical discharges.

(3) γ-rays: these are electrically neutral particles with properties which identified them as energetic photons.

These radiations were present to lesser or greater extent in the emissions from each radioactive element, the differences being in the relative intensities and the energy of the radiation, not in its nature. Thus every radioactive change involving the emission of an **α-particle**, for example, might be expected to cause a change in the **atomic mass** and **number**. The **displacement laws** were enunciated by Russell and by Soddy and Fajans. We give them in a modern form:

1. The emission of an α-particle reduces the atomic mass number by 4 and the atomic number by 2.
2. The emission of a **β-particle** increases the atomic number by 1 and leaves the mass number unchanged.

Displacement laws

1. The consequence of the emission of an α-particle is to reduce the atomic mass number by 4 and the atomic number by 2.
2. The consequence of the emission of a β-particle is to increase the atomic number by one and to leave the mass number unchanged.

It was through the investigation of the scattering of α-particles that the most important advances were made. Rutherford in 1906 observed that a beam of α-particles became spread slightly on traversing a thin layer of material. A layer of material insufficient to stop α-particles (e.g. gold 4 μm thick) would scatter α-particles an average of about 9 degrees. However, in 1909 Rutherford's colleagues, Geiger and Marsden, observed that one in a few thousand α-particles suffered a scattering of greater than 90°. This was an astonishing result. At that time there was an atomic model due to J. J. Thomson which envisaged the atom as a sphere of uniformly distributed positive charge in which was embedded a number of negatively charged particles (electrons): in this model the scattering of posively charged particles would be by repulsion by the positive charge and by the attraction of the electrons. Such an atom can only give small deflections to an α-particle that traverses it: the deflections observed after traversing a layer of material were taken to be the effect of many such collisions acting randomly on the particle's trajectory. Such an atom cannot give the rare, large-angle scatters that were observed, and Rutherford showed that the observed rate of large-angle scattering could not be explained as the sum of multiple collisions with atoms of this kind. He proposed the nuclear model of atoms and showed that such a model can explain both small deflections of a beam of α-particles traversing a thin foil and the rare large-angle deflections. The former was explained as the result of many very small angle deflections (**multiple scattering**) and the latter as the result of rare single encounters of α-particles with atoms in which there is a large deflection (**single scattering**). Rutherford used his model to calculate a cross-section for the large-angle scattering that was in accord with observation. This agreement established his nuclear model. We shall derive the Rutherford scattering formula in the next section.

DEFINITIONS AND KEYWORDS

Single scattering The deflection of the path of a particle crossing a layer of material is the result of a significant deflection in one, and only one, encounter with an atom.

Multiple scattering The deflection of the path is the result of the sum of many very small deflections in many atoms, all uncorrelated.

Multiple scattering is inevitable for charged particles and a significant single scatter will have superimposed on it the uncorrelated effect of the multiple scattering. Plural scattering, a phrase which is rarely used, means several single scatters. The magnitude of these processes, both relative and absolute, is determined by the thickness and nature of the material, and by the charge and energy of the particle.

1.2 The Rutherford scattering formula

Although Rutherford derived this formula with α-particles as the incident, to-be-scattered particle, we shall be slightly more general and assume that incident particle carries positive charge ze, where e is the magnitude of the electronic charge. Firstly the model:

Rutherford scattering cross section formula

Assumptions

1 The nuclear model
2 Target nucleus fixed (no recoil)
3 Point-like charges
4 Coulomb force only
5 Elastic scattering
6 Classical mechanics

1. The atom contains a nucleus with positive charge Ze and almost the entire mass of the atom.
2. The electrically neutral atom contains Z electrons moving around the nucleus.

It is easy to show that the electrons cannot cause a single scattering with significant deflection of α-particles of the kinetic energy that Rutherford considered, so we shall now neglect them. Our other assumptions are:

3. That the target nucleus is very much more massive than the incident particle and therefore does not recoil significantly in the collision.
4. That classical mechanics can be used to describe the collision. (And, of course, we thereby include the conservation of momentum, angular momentum and energy.)
5. That the target nucleus and the incident particle have point-like charge distributions so that the Coulomb potential $V(r) = Zze^2/4\pi\varepsilon_0 r$ acts between them, where r is the distance separating their centres. We will be treating the orbit of the incident particle classically and will work out the case of like charges (as is the case in α-particle-nucleus scattering) and therefore of a repulsive force.
6. That there is no other force acting other than that due to the Coulomb potential.
7. That there is no excitation of incident or target particle: each remains unchanged. This is elastic scattering.

The symbols we shall use are defined in Table 1.1. Figure 1.5 shows an orbit. The incident particle, if undeflected, would pass the centre (at O) of the target nucleus at a distance b, the **impact parameter**. In fact, the orbit is hyperbolic and at D the incident particle is at its distance of closest approach, d. The orbit is

Table 1.1 The notation for quantities used in deriving Rutherford's formula for the differential scattering cross-section for the elastic scattering of one charged particle by a fixed charged target particle.

m	mass	Incident particle
v	velocity	
T	kinetic energy	
ze	electric charge	
Ze	charge of target nucleus (at O)	See Fig. 1.5.
b	impact parameter	
d	distance of closest approach (at D)	
u	velocity of incident particle at D	
θ	angle of scatter	
r, φ	polar coordinates with respect to OD of point (X) on the trajectory of particle.	
p	distance of closest approach for b=0.	See Fig. 1.6.

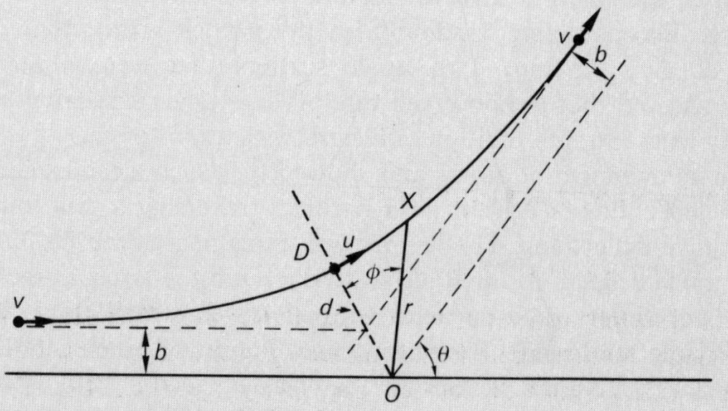

Fig. 1.5 The classical orbit of the incident particle in Rutherford scattering for non-zero impact parameter b.

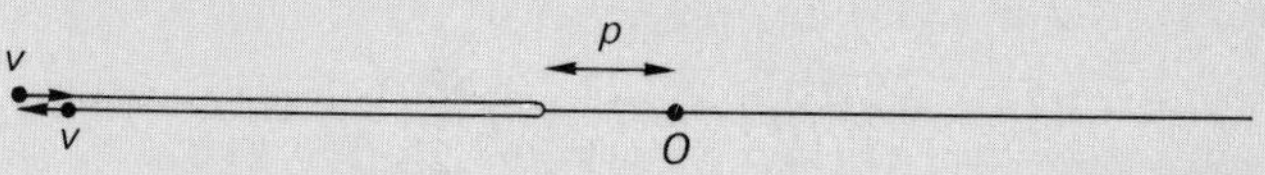

Fig. 1.6 The classical orbit in Rutherford scattering for zero impact parameter. Conservation of energy requires that the incident particle's distance of closest approach p, is given by

$$p = Zze^2/4\pi\varepsilon_0 T.$$

clearly symmetric about the line OD. If b was zero the incident particles would approach to a distance p (see Fig. 1.6). At this point the incident kinetic energy is transformed into mechanical potential energy in the Coulomb field, therefore:

$$\tfrac{1}{2}mv^2 p = Zze^2/4\pi\varepsilon_0. \tag{1.1}$$

Step 1 To find the connection between b and θ.

We use the conservation of angular momentum about O to connect the incident velocity to the component of the velocity transverse to OX at X:

$$mvb = mr^2\frac{d\varphi}{dt}, \tag{1.2}$$

hence

$$\frac{dt}{r^2} = \frac{d\varphi}{vb}. \tag{1.3}$$

Consider now the component of the linear momentum in the direction OD. This changes from $-mv\sin(\theta/2)$ to $+mv\sin(\theta/2)$. At X the rate of change of this momentum is the component of the Coulomb repulsion in the direction OD. Hence

$$2mv\sin\frac{\theta}{2} = \int_{-\infty}^{+\infty}(Zze^2/4\pi\varepsilon_0 r^2)\cos\varphi\, dt.$$

We use equation (1.3) to change the variable of integration from time, t, to φ, obtaining

$$2mv\sin\frac{\theta}{2} = \frac{Zze^2}{4\pi\varepsilon_0 vb}\int_{\varphi=-(\pi-\theta)/2}^{\varphi=(\pi-\theta)/2}\cos\varphi\, d\varphi = \frac{p}{2}\frac{mv}{b}\Big[\sin\varphi\Big]_{-(\pi-\theta)/2}^{(\pi-\theta)/2}$$

which gives

$$\tan\frac{\theta}{2} = \frac{p}{2b}. \tag{1.4}$$

This is the relation required from Step 1.

Step 2 To derive a first cross-section.

The relation (1.4) tells us that as b decreases θ increases. Therefore to suffer an angle of scatter greater than Θ the impact parameter b must be less than $(p/2)\cot(\Theta/2)$. That means the incident particle must strike a disc of this radius centred at O and perpendicular to v. The area, σ, presented by the nucleus for scattering through an angle greater than Θ is the area of this disc. That is

$$\sigma(\theta>\Theta) = \frac{\pi p^2}{4}\cot^2\frac{\Theta}{2}, \tag{1.5}$$

or in its full glory:

$$\sigma(\theta>\Theta) = \frac{\pi}{4}\left(\frac{Zze^2}{4\pi\varepsilon_0 T}\right)^2\cot^2\frac{\Theta}{2}.$$

DEFINITIONS AND KEYWORDS

Impact parameter This is the perpendicular distance from the centre of the target to the extrapolation of the trajectory of the incident particle before it suffers any deflection.

Momentum transfer This is the momentum transferred to the target in a single scatter (symbol q).

Comment These quantities are not specific to Rutherford scattering which is our present subject of discussion. However, impact parameter is a classical concept.

The area σ is called a cross-section: if the reader is concerned about the meaning and use of this term we suggest reading Section 2.9, where a fuller description of the concept is given, before proceeding.

Step 3 To obtain the angular differential cross-section.

What we want is $d\sigma/d\Omega$, which is the cross-section per unit solid angle located at an angle θ. The element of solid angle $d\Omega$ between θ and $\theta + d\theta$ is given by

$$d\Omega = 2\pi \sin\theta \, d\theta.$$

Therefore

$$\frac{d\sigma}{d\Omega} = \frac{1}{2\pi \sin\theta} \frac{d\sigma}{d\theta}.$$

The $d\sigma/d\theta$ we need is $(d/d\Theta)\sigma(\theta > \Theta)$ from Equation (1.5) and hence we obtain

$$\frac{d\sigma}{d\Omega} = \left(\frac{Zze^2}{16\pi\varepsilon_0 T}\right)^2 \operatorname{cosec}^4\frac{\theta}{2}. \tag{1.6}$$

Table 1.2 A summary of results obtained for Rutherford scattering.

The impact parameter b, the angle of scatter θ, and p are related by

$$p = 2b \tan(\theta/2),$$

where

$$p = \frac{Zze^2}{4\pi\varepsilon_0 T}.$$

The cross-section for scattering through an angle greater than Θ is

$$\sigma(\theta > \Theta) = 4\pi \left(\frac{Zze^2}{16\pi\varepsilon_0 T}\right)^2 \cot^2\frac{\Theta}{2}.$$

The angular differential cross-section for scattering through an angle θ is

$$\frac{d\sigma(\theta)}{d\Omega} = \left(\frac{Zze^2}{16\pi\varepsilon_0 T}\right)^2 \operatorname{cosec}^4\frac{\theta}{2}.$$

Putting in some members, use

$$\sigma(\theta > \Theta) = \frac{\pi}{4} Z^2 z^2 \left(\frac{e^2}{4\pi\varepsilon_0 \hbar c}\right)^2 \left(\frac{\hbar c}{T}\right)^2 \cot^2\frac{\Theta}{2},$$

$$= \frac{\pi Z^2 z^2}{4} \left(\frac{1}{137}\right)^2 \left(\frac{197}{(T\,\text{MeV})}\right)^2 \cot^2\frac{\Theta}{2}\ \text{fm}^2,$$

and

$$\frac{d\sigma(\theta)}{d\Omega} = \frac{Z^2 z^2}{16} \left(\frac{197}{137(T\,\text{MeV})}\right)^2 \operatorname{cosec}^4\frac{\theta}{2}\ \text{fm}^2\,\text{sr}^{-1}.$$

For an explanation of the units, see Section 2.2.

This is the famous Rutherford formula for the differential cross-section in Coulomb scattering.

We assumed a repulsive Coulomb force. If the force is attractive the orbit is changed so that $\theta \to -\theta$, but the differential cross-section is unchanged.

PROBLEMS

Try the following problems, keeping an eye on Sections 1.2 and 1.3, in order to take the unfamiliarity out of the Rutherford scattering formula.

1.1 Calculate the cross-section for the scattering of a 10 MeV α-particle by a gold nucleus (Z=79, A=197) through an angle greater than (a) 10°, (b) 20°, (c) 30°. Neglect nuclear recoil.

1.2 For the same circumstance as in Problem 1.1 calculate the differential scattering cross-section $d\sigma/d\Omega$ fm^2 sr^{-1} for scattering at 10°.

1.3 The Rutherford scattering formula has the property that as the angle of scatter, $\theta \to 0°$, $d\sigma/d\Omega \to \infty$. The practical circumstances are that the incident particle will be scattered by the whole atom. Suggest what happens to $d\sigma/d\Omega$ as $\theta \to 0$ in this case.

1.4 Show that the distance of closest approach d, in Rutherford scattering leading to an angle of deflection θ, is given by

$$d = \frac{p}{2}(1 + \text{cosec}\,\theta/2).$$

where p is defined in Fig. 1.6.
[Use the convservation of energy and angular momentum.]

1.3 The properties of the Rutherford differential cross-section

The cross-section

(1) decreases rapidly with increasing angle, θ,
(2) becomes infinite at $\theta = 0$,
(3) is inversely proportional to the square of the incident particles kinetic energy, T,
(4) is proportional to the square of the charge of the incident particle and of the target nucleus.

Let us examine these results to get a feel for what they mean physically. As θ increases the **momentum transfer** (q, see Fig. 1.7) increases. The greater it is the larger the electric force the particles must experience to bring it about. Large electric fields mean close collisions; the closer, the rarer they are and the cross section decreases with increasing θ. It is now possible to understand property 3: at a fixed angle the required momentum transfer will increase as does T. Thus the cross-section must decrease with increasing T.

The difficulty implied in property 2 is really non-existent. We ask the reader to consider this difficulty in Problem 1.3.

Property 4 is typical of electromagnetic interactions and we shall meet this idea again. Essentially these two particles interact through the electromagnetic field. The quantum-mechanical amplitude for charge Z to interact with the field is proportional to Ze. Thus the Rutherford scattering amplitude is proportional to zZe^2 and the final intensity (cross-section) is proportional to $(zZe^2)^2$.

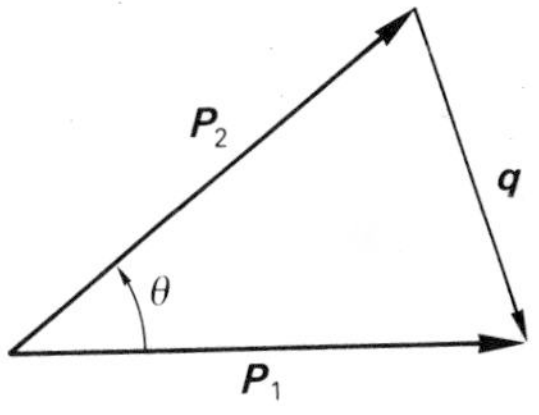

Fig. 1.7 The momentum transfer $\boldsymbol{q}$ in elastic scattering at a fixed target. The vectors $\boldsymbol{P}_1$ and $\boldsymbol{P}_2$ represent the incident and scattered particle momenta respectively ($|\boldsymbol{P}_1| = |\boldsymbol{P}_2| = P$). If the angle of scatter is θ, the geometry gives

$$q = 2P\sin(\theta/2).$$

1.4 The experiments of Rutherford and his colleagues

Geiger, in 1911, verified that the angular distribution of α-particles scattered between 30° and 50° by a thin gold foil was in agreement with the theory. Later Geiger and Marsden in a more extensive series of experiments verified that

(1) the angular distribution was varying as $\text{cosec}^4\theta/2$, for $5° < \theta < 150°$,
(2) the yield in this angular range was proportional to the thickness of the foil, and
(3) the cross-section for scattering in Al, Cu, Ag, Sn, and Au varied approximately as the square of the atomic weight.

Result 1 was a verification of Rutherford's formula and of the essential correctness of his nuclear model of the atom. Result 2 verified that the large-angle scattering was due to single encounters and not due to the sum of many small-angle scatters (an essential fact in establishing the correctness of Rutherford's interpretation; see Problem 1.7). The last result we now know follows from the fact that Z for a nucleus is nearly proportional to A. The identification of the nuclear charge in units of e came from Moseley's investigations of X-ray spectra. Chadwick, in 1920, made direct measurements of the nuclear Z using α-particle scattering. In Table 1.3 we reproduce some pages from Rutherford, Chadwick and Ellis describing the apparatus used by Geiger and Marsden and the results obtained.

1.5 Examination of the assumptions

Rutherford assumed a nuclear model: that assumption was vindicated by the experimental results. Let us look at the other assumptions:

1. Neglect of nuclear recoil: this can be avoided by transforming the centre-of-mass of the collision. The formula is the same but the effective T is now the total kinetic energy in that frame and the angle of scatter and differential cross-section apply in that frame. These latter quantities must be transformed back to the laboratory frame (target at rest) for a comparison with experimental results.
2. The classical approach to the orbit: a simple quantum-mechanical approach using the Born approximation gives the same answer. The value of the classical approach is its transparency and that it gives the right answer! We shall use some of the ideas in the classical approach later but what we say must not distract us from knowing that the quantum-mechanical approach is 'more correct'.
3. The point-like charges: Rutherford and Chadwick could find no deviation from the formula in the scattering of α-particles from radium B and C, using gold, silver and copper targets: knowing the classical distance of closest approach they concluded that these nuclei, if they had size, had radii which were less than 3.2, 2.0 and 1.2×10^{-14} m respectively. Later, using targets of several light elements deviations from the Rutherford formula were found: for example, in aluminium deviations occurred for a classical distance of closest approach of about 8×10^{-15} m. We know that this represented the effect of nuclear interaction, as distinct from the Coulomb interaction, coming into play at short distances of approach. We shall discuss this when we discuss nuclear size in Chapter 3.
4. Absence of other forces: Of course, we have just introduced the nuclear

Table 1.3 An extract from *Radiations from Radioactive Substances* by Sir Ernest Rutherford, James Chadwick, and C. D. Ellis, published by Cambridge University Press, 1930.

§ 43. Experimental test of the nuclear theory. The first point, the angular distribution of the scattered particles, was tested by Geiger* in some preliminary experiments. He found that the distribution between 30° and 150° of the particles scattered by a thin gold foil was in agreement with this theory. Later, a beautiful series of experiments was carried out by Geiger and Marsden†, in which the above conclusions drawn by Rutherford from his theory were

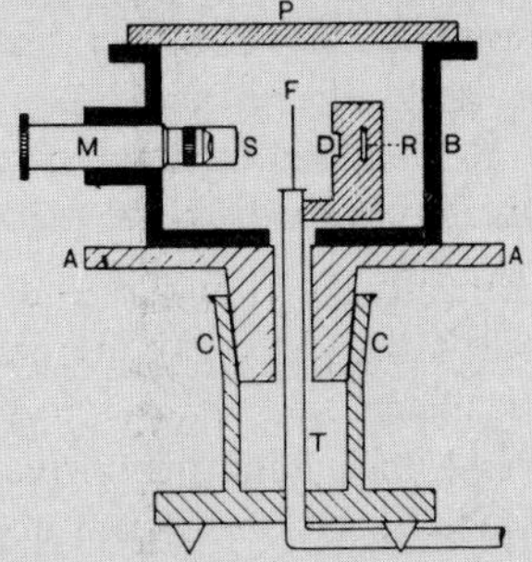

tested point by point. They first investigated the variation of scattering of α particles over a wide range of angles, using the apparatus shown in Fig. 47. The apparatus consisted in essentials of a strong cylindrical metal box B which contained the source of α particles R, the scattering foil F, and a zinc sulphide screen S attached rigidly to a microscope M. The box was fixed to a graduated circular platform A, which could be rotated in the airtight joint C. The microscope and zinc sulphide screen rotated with the box, while the scattering foil and source remained fixed. The box was

* Geiger, *Proc. Manch. Lit. Phil. Soc.* **55**, 20, 1911.
† Geiger and Marsden, *Phil. Mag.* **25**, 604, 1913.

closed by a ground glass plate P and could be exhausted through the tube T.

The source of α particles was an α ray tube filled with radon. The beam of α particles was therefore not homogeneous, for it contained particles from the products radium A and radium C in addition to those from the radon. This fact did not interfere with the investigation of the law of scattering with angle, since each group of α particles is scattered according to the same law. A narrow pencil of α particles from the source R was directed through the diaphragm D to fall normally on the scattering foil F.

By rotating the platform A the α particles scattered in different directions could be observed on the zinc sulphide screen. Observations were taken for angles of scattering between 5° and 150° and both silver and gold foils were used as scattering material. Two sets of experiments were carried out, the first comparing angles from 15° to 150° and the second angles from 5° to 30°.

For the smaller angles of scattering, the aperture of the diaphragm D was reduced considerably in order to ensure that the angle at which the scattered particles were counted was large compared with the angular width of the incident pencil. The number of particles scattered to the zinc sulphide screen was found to decrease very rapidly as the angle of scattering increased. Since it was not feasible to count more than about 90 scintillations per minute or less than 5, observations were made only over a relatively small range of angles at the same time. The scattering was first measured at the larger angles, and as the amount of radon decreased the measurements were extended to smaller and smaller angles, due correction being made for the decay.

Even when no scattering foil was in position at F a few scintillations were observed. These were due to particles scattered from the edge of the diaphragm limiting the incident pencil and from the walls of the vessel. The number of these extraneous particles was reduced by lining the vessel with paper and by using aluminium for the material of the diaphragm, for the scattering, as will be seen, is least for substances of low atomic weight. This extraneous effect was determined at different angles and allowed for in the subsequent measurements of the true scattered particles.

The collected results of these experiments are given in the following table. The first column gives the values of the angles ϕ between the direction of the incident pencil of α particles and the direction

in which the scattered particles were counted, and the second column gives the corresponding values of $\operatorname{cosec}^4 \frac{1}{2}\phi$. Columns III and V give the observed numbers N of scintillations for silver and gold respectively. Columns IV and VI show the ratio of N to $\operatorname{cosec}^4 \frac{1}{2}\phi$. This ratio is seen to be approximately constant for both sets of experiments.

Variation of scattering with angle.

I	II	III Silver	IV Silver	V Gold	VI Gold
Angle of deflection, ϕ	$\operatorname{cosec}^4 \frac{1}{2}\phi$	Number of scintillations, N	$\frac{N}{\operatorname{cosec}^4 \frac{1}{2}\phi}$	Number of scintillations, N	$\frac{N}{\operatorname{cosec}^4 \frac{1}{2}\phi}$
150°	1·15	22·2	19·3	33·1	28·8
135	1·38	27·4	19·8	43·0	31·2
120	1·79	33·0	18·4	51·9	29·0
105	2·53	47·3	18·7	69·5	27·5
75	7·25	136	18·8	211	29·1
60	16·0	320	20·0	477	29·8
45	46·6	989	21·2	1435	30·8
37·5	93·7	1760	18·8	3300	35·3
30	223	5260	23·6	7800	35·0
22·5	690	20300	29·4	27300	39·6
15	3445	105400	30·6	132000	38·4
30	223	5·3	0·024	3·1	0·014
22·5	690	16·6	0·024	8·4	0·012
15	3445	93·0	0·027	48·2	0·014
10	17330	508	0·029	200	0·0115
7·5	54650	1710	0·031	607	0·011
5	276300	—	—	3320	0·012

Fitting the experiments at the smaller angles to those at larger angles, the numbers of scattered particles are proportional to $\operatorname{cosec}^4 \frac{1}{2}\phi$ over the whole range investigated, where $\operatorname{cosec}^4 \frac{1}{2}\phi$ varies from 1 to 250,000. These experiments thus afford abundant proof of the law of scattering with angle deduced by Rutherford from the nuclear theory of the atom.

Geiger and Marsden next examined the variation of scattering with the thickness of the scattering material. In these and most subsequent experiments it was necessary to use a source of homogeneous α particles, for it was found—and it was predicted by the theory—that the scattering increased very rapidly as the velocity of the α particle was reduced. In this series of experiments the angle of scattering was kept constant, while the thickness of the scattering foil was varied. The α particles from a source R of radium (B + C) forces but if any of the colliding particles have spin (and therefore magnetic moment) then there will be magnetic effects, i.e. an interaction between the spins and between each spin and the current represented by the passage of the charge of the other particle. Later we illustrate the effect of one particle having spin (electron scattering, see Section 3.4).

5. Elastic scattering: The α-particles used by Rutherford were not sufficiently energetic to cause a significant number of inelastic collisions. By inelastic collision we mean that one or both of the particles involved become excited or disintegrate. After Rutherford, artificial sources of more energetic α-particles became available and these certainly can cause inelastic collisions.

Another implicit assumption which we must be prepared to abandon is that there are no relativistic effects: it is clear that there is no problem with Rutherford's α-particle sources: the velocities were about $\frac{1}{15}$ of the velocity of light. However, the scattering of electrons must be treated relativistically, since electrons of sufficient energy for nuclear scattering investigations are relativistic.

PROBLEMS

1.5 Define the differential cross-section $d\sigma/d\Omega$ in a scattering process. Derive Rutherford's formula for the scattering cross-section of α-particles incident with kinetic energy T on a heavy nucleus with charge Ze:

$$\frac{d\sigma}{d\Omega} = \left(\frac{Ze^2 \operatorname{cosec}^2(\theta/2)}{8\pi\varepsilon_0 T}\right)^2$$

where e is the electronic charge and θ is the scattering angle. Under what conditions do deviations from this formula occur? Illustrate your answers by discussing the scattering of 15 MeV α-particles by thin targets of aluminium (atomic number Z=13, atomic weight A=27) and gold (Z=79, A=197). Assume the nuclear radius R is given approximately by the formula

$$R = 1.2A^{1/3} \times 10^{-15}\,\text{m}.$$

(Adapted from the 1968 examination in the Honours School of Natural Science, Physics, University of Oxford)

1.6 Show that the Rutherford differential scattering cross-section formula can be written in terms of the squared momentum transfer q^2 as

$$\frac{d\sigma}{dq^2} = \frac{4\pi Z^2 z^2 \alpha^2 (\hbar c)^2}{q^4 v^2},$$

where α is the fine-structure constant and v is the velocity of the deflected particle.

1.7 If the large angle scatters observed by Rutherford had not been the effect of single scattering but of multiple scattering (see p. 5), how would the yield at fixed angle vary with the thickness of the target material? In the former case, neglecting attenuation of the incident beam, the yield will increase linearly with thickness. What would be the dependence for multiple scattering?

[Consider a random walk problem: each small scatter is like a hop on the surface of a sphere. Starting at a pole, n randomly directed small hops will take you a distance from the pole. That distance will vary with each trial of n hops but it has an average over many such trials. How is the magnitude of that average distance expected to depend on n? The connected ideas are those of *diffusion* and *Brownian motion*.]

1.6 The nuclear constituents

The unravelling of the details of atomic structure and of the relation between the chemical elements established that the charge (positive) on the nucleus of an atom was Ze, where the integer Z is the **atomic number**, the number which gives the number of atomic electrons and hence the ordered position of that species of atom in the periodic table, and e is the magnitude of the charge on the electron. The measurement of atomic masses by the techniques of mass spectroscopy had established that atoms of a given chemical could occur with different masses (**isotopes**) and that throughout the periodic table each mass is always close to an integer multiple of the mass (M_H) of a hydrogen atom. For each isotope the nearest integer is called A, the **atomic mass number**. The notation we shall use to represent a given **nuclide** is

$$^{A}_{Z}\text{Chemical symbol}$$

Table 1.4 A tribute to Dmitri Mendeleev (1834–1907), the discoverer of the periodic nature of the chemical properties of the elements.

Key: Z / Chemical symbol / † Atomic Weight

The Isotope $^{12}_{6}C$ is assigned an atomic mass of exactly 12 unified atomic mass units.

1 **H** 1.008																	2 **He** 4.003
3 **Li** 6.941	4 **Be** 9.012											5 **B** 10.811	6 **C** 12.011	7 **N** 14.007	8 **O** 15.999	9 **F** 18.998	10 **Ne** 20.180
11 **Na** 22.990	12 **Mg** 24.305											13 **Al** 26.981	14 **Si** 28.086	15 **P** 30.974	16 **S** 32.066	17 **Cl** 35.453	18 **Ar** 39.948
19 **K** 39.098	20 **Ca** 40.078	21 **Sc** 44.956	22 **Ti** 47.88	23 **V** 50.942	24 **Cr** 51.996	25 **Mn** 54.938	26 **Fe** 55.847	27 **Co** 58.933	28 **Ni** 58.69	29 **Cu** 63.546	30 **Zn** 65.39	31 **Ga** 69.723	32 **Ge** 72.61	33 **As** 74.922	34 **Se** 78.96	35 **Br** 79.904	36 **Kr** 83.80
37 **Rb** 85.468	38 **Sr** 87.62	39 **Y** 88.906	40 **Zr** 91.224	41 **Nb** 92.906	42 **Mo** 95.94	43 **Tc**	44 **Ru** 101.07	45 **Rh** 102.91	46 **Pd** 106.42	47 **Ag** 107.87	48 **Cd** 112.41	49 **In** 114.82	50 **Sn** 118.71	51 **Sb** 121.75	52 **Te** 127.60	53 **I** 126.90	54 **Xe** 131.29
55 **Cs** 132.91	56 **Ba** 137.33	57–71 **La series**	72 **Hf** 178.49	73 **Ta** 180.95	74 **W** 183.85	75 **Re** 186.21	76 **Os** 190.2	77 **Ir** 192.22	78 **Pt** 195.08	79 **Au** 196.97	80 **Hg** 200.59	81 **Tl** 204.38	82 **Pb** 207.21	83 **Bi** 208.98	84 **Po**	85 **At**	86 **Rn**
87 **Fr**	88 **Ra**	89–103 **Ac series**	(104)	(105)	(106)	(107)	(108)	(109)									

Lanthanide series	57 **La** 138.91	58 **Ce** 140.12	59 **Pr** 140.91	60 **Nd** 144.24	61 **Pm**	62 **Sm** 150.36	63 **Eu** 151.97	64 **Gd** 157.25	65 **Tb** 158.93	66 **Dy** 162.50	67 **Ho** 164.93	68 **Er** 167.26	69 **Tm** 168.93	70 **Yb** 173.04	71 **Lu** 174.97
Actinide series	89 **Ac**	90 **Th** 232.04	91 **Pa**	92 **U** 238.03	93 **Np**	94 **Pu**	95 **Am**	96 **Cm**	97 **Bk**	98 **Cf**	99 **Es**	100 **Fm**	101 **Md**	102 **No**	103 **Lw**

† The **atomic weight** is the average of the atomic mass for the isotopic mixture found terrestrially (although that can vary depending on the source of the element). Elements not given an atomic weight have no stable isotopes; exceptions to this in the table are uranium and thorium which both have isotopes with mean lives sufficiently long to have ensured their survival in terrestrial material. Look at Appendix A to find the full name for each symbol.

(104–109) These elements may have been produced but have not been named.

Thus, for example, the stable nucleus of the element beryllium is $^9_4$Be. The Periodic Table is shown in Table 1.4. The symbols may be turned into names using the appendix Table A.1.

The simplest atom is hydrogen and its nucleus was assumed to be one of the simplest constituents of nuclei: it was given the name the **proton**. Since the hydrogen atom is proton plus electron and since the mass of the proton is about 1840 times that of the electron, the mass of atoms has to be explained almost entirely in terms of nuclear mass. These facts lead originally to the following nuclear model:

> An atom of atomic number Z and atomic mass number A has a nucleus containing A protons plus $(A-Z)$ electrons.

The model predicts, correctly, a mass of $\sim AM_H$, a nuclear charge Ze and has the appealing feature that it might explain the source of electrons emitted in β-radioactivity. But it is wrong!

The argument which shows that it is wrong is simple. Consider the nucleus of the helium atom, for example. In the model 4 protons and 2 electrons occupy a volume of half linear dimension of about 2 fermis: the uncertainty principle immediately says that the uncertainty in the electron momentum is ~ 50 MeV/c (see Section 2.2). To have that kind of momentum the electrons must have a kinetic energy of the order 50 MeV and such electrons can remain bound only if there is a potential well at least that deep. Such a potential well would have effects on the optical spectroscopy of helium which do not exist. The conclusion

DEFINITIONS AND KEYWORDS

Isotopes Atoms of the same chemical properties but different mass.

Atomic number, Z The position of a chemical element in the periodic table of the elements.

Atomic mass number, *A* This is the nearest integer to the ratio of the mass of an atom of a given isotope to the mass of an atom of the lightest isotope of hydrogen.

Atomic mass This is the ratio of the mass of an atom of a given isotope to $\frac{1}{12}$ of the mass of an atom of the isotope of carbon with $A = 12$.

Atomic weight The atomic mass averaged over the terrestrial isotopic abundances for an element with more than one stable isotope.

Proton Mass = 938.27 MeVc/c^2 (= 1.007276 u, see Section 2.2).
Electric charge = +1 (in units of the magnitude of the electronic charge).
Spin = $\frac{1}{2}\hbar$.

Neutron Mass = 939.57 MeV/c^2 (= 1.008665 u.)
Electric charge = 0.
Spin = $\frac{1}{2}\hbar$.

Atom (*Z,A*) Z = number of protons in the nucleus.
$A - Z \equiv N$ = number of neutrons in the nucleus.

Nucleon The generic name for proton and neutron.

Nuclide A specific nuclear species of fixed A and Z.

Isobars Atoms of the same atomic mass number. Their nuclei have the same number of nucleons A but different proton number.

Isotones Atoms of the same $A-Z$ but different Z. Their nuclei have the same neutron number but differing proton number.

is that there can be no electrons in the nucleus. There is also another experimental result supporting this conclusion. The molecular spectroscopy of nitrogen (N_2 = two atoms with nucleus $^{14}_{7}N$) has features that indicate that each nucleus acts as a boson, that is, it has integer spin. Now in this nuclear model each nitrogen nucleus contains 14 protons and 7 electrons—21 fermions in all. An assembly of an odd number of fermions also has half-odd integer spin and is a fermion. So this result indicates that there cannot be an odd number of fermions in a nitrogen nucleus and that the model must be wrong.

The way forward was indicated by the discovery of the neutron by Chadwick in 1932. We shall describe this discovery in Section 7.3. The **neutron** is a neutral counterpart to the proton and has a mass slightly greater than that of the proton (see opposite). The more correct nuclear model is now

> An **atom** of atomic number Z and mass number A has a nucleus of Z protons and $A-Z$ ($\equiv N$) neutrons.

This model is consistent with the absence of electrons from the nucleus and with the molecular spectroscopy of nitrogen. Although even this model is certainly an over-simplification, it is a good basis for constructing more detailed nuclear models and understanding nuclear processes; in addition, the problem of the source of electrons in β-decay is not really a problem, as we shall find in Section 5.3.

The word **nucleon** is used as the generic name for neutron and proton. Nuclei of the same N but different Z are called **isotones**.

Why is this model an oversimplification? Consider a hydrogen atom which we believe contains an electron moving around a proton. The Coulomb potential that keeps the electron bound is electromagnetic in origin and the modern view requires the Coulomb force to be the result of the exchange of quanta of the electromagnetic field between the proton and the electron. Photons are such quanta but in this case the photons exchanged are not real but **virtual**, in that they carry momentum but no energy, when **real** photons carry energy equal to

DEFINITIONS AND KEYWORDS

Real photons These are photons which have energy equal to their momentum multiplied by c, the velocity of light.

Virtual photons These are photons for which the energy is not equal to the momentum times c.

Comment Only real photons can become independent of their source. They satisfy Einstein's relation between total energy E and momentum P:

$$E^2 = M^2c^4 + P^2c^2,$$

with the rest mass M equal to zero. Virtual photons do not satisfy this relation and can be time-like, $E > Pc$, or space-like, $E < Pc$. They cannot become free and exist only for a time allowed by the uncertainty principle: they are the carriers of the electromagnetic interaction between electrically charged particles.

Positron This is the positive version of the negative electron. The electron and positron have identical mass and spin. They have opposite electric charge and magnetic moment.

Comment The positron is said to be the **antiparticle** to the electron (and vice-versa). Later we shall meet other examples and explore their role in elementary particle physics.

their momentum ($\times c$, the velocity of light). So the hydrogen atoms contain photons, albeit transitory, as well as electrons and protons. Now the electric field of photons can polarize the vacuum, that is create an electron plus **positron** pair which exists for a very brief period then recombines to recreate the photon from which the pair materialised. That all these processes are occurring is the view taken from the standpoint of modern quantum mechanics. Only by taking their effects into account can that theory correctly predict the energy levels of the hydrogen atom. Thus we can see that the hydrogen atom is more than electron and proton. So what are the extras to the neutrons and protons of the nucleus? That part of the story will begin to appear in Chapter 9.

1.7 What is coming?

The atomic nucleus is a very complicated object and the subject of nuclear physics a very large one. In this volume, as befits an introductory course, we shall consider the more obvious properties of nuclei only. These properties divide, roughly, into static and dynamic. The ones we shall discuss, in varying degrees of detail, are listed in Table 1.5. Our objectives are

1. To provide a broad basis of knowledge of the properties of nuclei and of the simpler models which contribute to explaining those properties.
2. To give the reader some idea about the nature of nuclear forces and the role played by other interactions in nuclear physics.

Table 1.5 Properties of nuclei

Static	Dynamic
1 charge	1 radioactivity
2 size and shape	2 excited states
3 mass and binding energy	3 nuclear reactions
4 spin and parity	
5 electromagnetic moments	

The last objective leads us to a study of elementary particle physics, as defined at the beginning of this chapter. There our objectives will duplicate those we enumerated for nuclear physics with the addition that we aim to introduce the reader to the most fundamental parts of matter known at the present time. In addition we shall look at the role of nuclear and particle physics in astrophysics and cosmology.

Reference

Rutherford, Sir Ernest, Chadwick, J., and Ellis, C. D. (1930). *Radiations from radioactive substances*. Cambridge University Press.

2

Some Quantitative Formalities

2.1 Introduction

The greater part of the experimental investigation of nuclear and particle physics involves the study of collisions or of spontaneous change. The first of these requires a knowledge of the concept of collision cross-section and its use. Spontaneous change has to be understood in terms of a transition rate. These quantities and their properties can be interpreted in terms of the basic physics involved. However, these interpretations are not rigorous unless given a quantitative foundation. To that end we must start by considering suitable units and we shall encourage the reader to take a quantitative approach to understanding the physics of the nucleus and of the elementary particles.

2.2 The scale of nuclear physics and suitable units

The work of Rutherford and his colleagues established the existence of the atomic nucleus and their measurements showed its linear dimension to be of the order of 2 to 7×10^{-15} m. These discoveries extended the scale of human knowledge down from the atomic scale by approximately four orders of magnitude. On a logarithmic scale of the universe man stands below midway in the 41 decades from the radius of the known universe down to the size of the constituents of the nucleus (Fig. 2.1).

We need some convenient units with which to do nuclear calculations. Clearly the results must be independent of those chosen but the familiar SI units based on the metre, kilogram, etc., are hardly the most straightforward when dealing with nuclear dimensions of the order of 10^{-15} m and nuclear masses of the order of 10^{-27} kg.

To start we need to find a convenient energy unit. On the atomic scale that unit is the electron volt ($1\ \text{eV} = 1.6 \times 10^{-19}$ J). On the nuclear scale we shall find that it is usually the megaelectron (MeV). When we come to the particle scale it is the gigaelectron ($1\ \text{GeV} = 10^9$ eV). However, we shall not stick rigorously to this division, so be prepared to meet all the other standard multiples of the electron volt.

The next step is to use the relation between total energy (E), momentum (P) and rest mass (M) of special relativity:

$$E^2 = P^2c^2 + M^2c^4. \tag{2.1}$$

Fig. 2.1 The scale of the known universe and the position of humans within it.

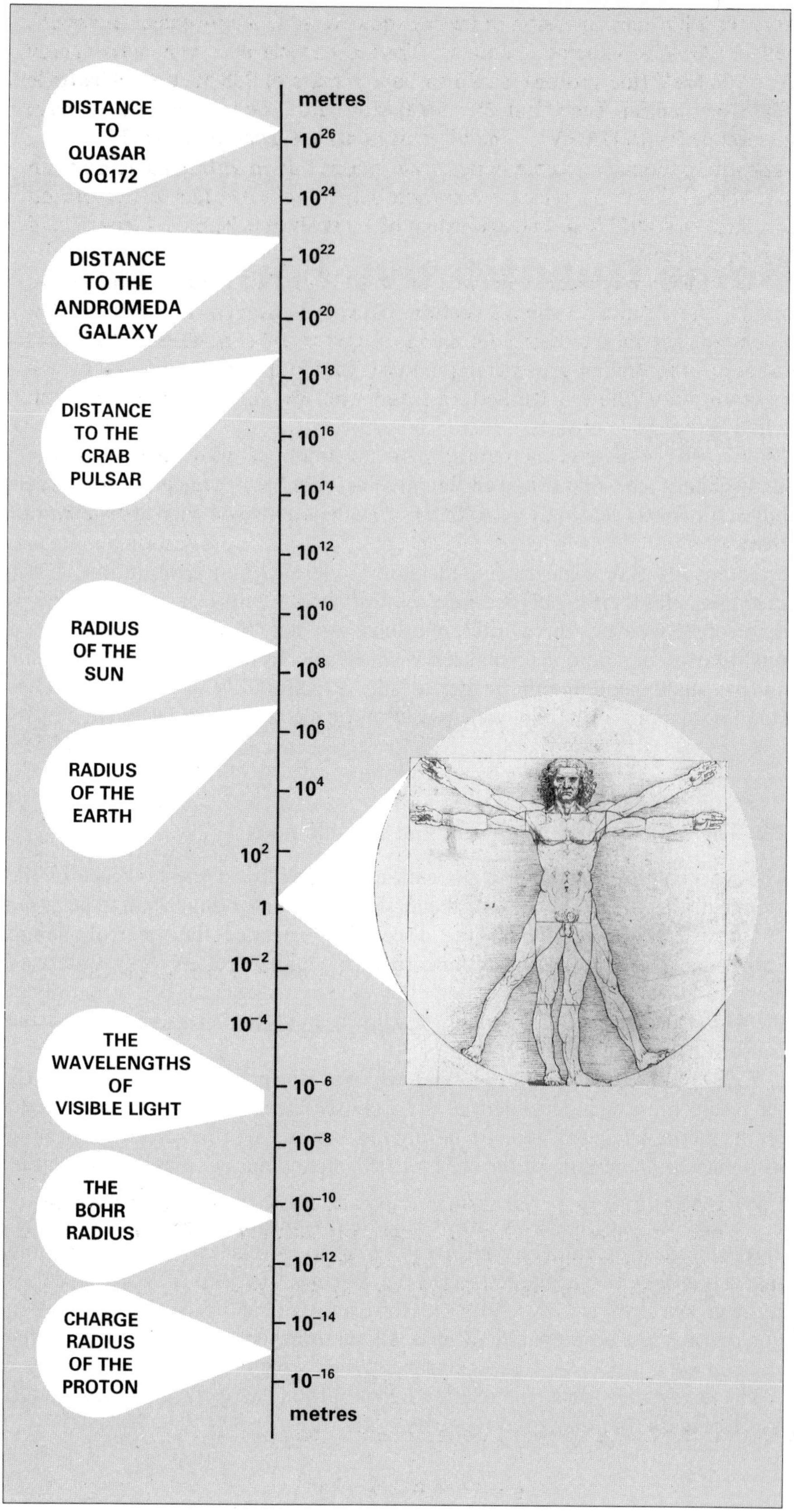

Every term has the dimension of energy squared, so it is convenient to use eV, MeV, or GeV to express P and M. Thus a particle with rest mass energy $Mc^2 = 938$ MeV (the proton) is said to have a mass of 938 MeV/c^2. A particle with a momentum such that $Pc = 1000$ MeV, for example, is said to have momentum $P = 1000$ MeV/c. The SI equivalents are given in Table 2.1.

The internationally accepted mass scale is based upon setting the mass of an atom of the ^{12}C isotope to have a mass precisely 12 units. On this scale, a proton has a mass of 1.007276 u. The equivalent of 1 u is given in Table 2.1 (u = unified atomic mass unit).

Now for length: nuclear radii are 2 to 7×10^{-15} m, that is 2 to 7 femtometres (fm). The last is the SI name convention, but we will deviate from this to call a femtometre a fermi although continuing to use the abbreviation fm. This is in honour of the Italian–American physicist Enrico Fermi, who was of the generation that followed Rutherford's and who was famous for his work in nuclear and particle physics.

Now some useful hints and numbers for calculating things in nuclear physics. Clearly Planck's constant ($\hbar$) and the velocity of light (c) are going to appear frequently. The author has found that a very convenient and easy-to-remember quantity is

$$\hbar c = 197.3 \text{ MeV fm} = 0.1973 \text{ GeV fm}. \tag{2.2}$$

Therefore

$$\hbar = 197.3 \frac{\text{MeV}}{c} \text{ fm} \tag{2.3}$$

or, since

$$c = 2.998 \times 10^{23} \text{ fm s}^{-1}, \tag{2.4}$$

$$\hbar = 6.582 \times 10^{-22} \text{ MeV s}. \tag{2.5}$$

The relation (2.3) is used for application of the uncertainty principle in the form

$$\Delta p\, \Delta x \approx \hbar/2,$$

whereas the relation (2.5) is useful for

$$\Delta E\, \Delta t \approx \hbar/2.$$

Thus, if the uncertainty on the position of a particle is 1 fm, then the uncertainty in its momentum is about 99 MeV/c. If the average lifetime of a state is 10^{-22} s then its energy uncertainty is ~3.3 MeV (see Section 2.9).

The fine structure constant, α, is given in the SI system by

$$\alpha = \frac{e^2}{4\pi\varepsilon_0 \hbar c},$$

where e is the charge on the electron. In any system.

$$\alpha = 1/137.036 \approx 1/137.$$

Table 2.1 Useful units and their SI equivalents

		SI Values
Energy	1 eV	$=1.602\times10^{-19}$ J
	1 MeV$=10^6$ eV	$=1.602\times10^{-13}$ J
	1 GeV$=1000$ MeV	$=1.602\times10^{-10}$ J
Momentum	1 MeV/c	$=5.344\times10^{-22}$ kg m s^{-1}
Mass	1 MeV/c^2	$=1.783\times10^{-30}$ kg
The unified atomic mass unit ($^{12}_{6}$C scale)	1 u$=931.5$ MeV/c^2	$=1.661\times10^{-27}$ kg
Length	1 fermi (fm)	$=1.0\times10^{-15}$ m
Other quantities	$\hbar c=197.3$ MeV fm	$=3.162\times10^{-26}$ J m
	$c=2.998\times10^{23}$ fm s^{-1}	$=2.998\times10^{8}$ m s^{-1}
	$\hbar=6.588\times10^{-22}$ MeV s $=197.3$ MeV/c fm	$=1.055\times10^{-34}$ J s

The fine-structure constant

$$\frac{e^2}{4\pi\varepsilon_0\hbar c}=\frac{1}{137.04}$$

Natural units

$\hbar=c=1$

1 unit of mass = 1 GeV

1 unit of length = 1 GeV^{-1} = 0.1975 fm

1 unit of time = 1 GeV^{-1} = 6.588×10^{-25} s

For more accurate numbers see Cohen and Taylor (1987).

The use of α and of $\hbar c=197.3$ MeV fm makes many calculations simple: see, for example, Problem 2.1.

There is a system called natural units which we will not use but which it is useful to know about and is found in many books. In this system we set $\hbar=c=1$ and express mass, length and time in units of GeV. Clearly 1 unit of mass is 1 GeV. Length has the dimensions of $\hbar c$/energy so is GeV^{-1}: 1 GeV^{-1} is, from equation (2.2), equal to 0.1975 fm. Time has the dimensions of $\hbar$/energy so is also GeV^{-1}: 1 GeV^{-1} is 7×10^{-25} s. Although convenient, dimensional homogeneity is not so evident in this system. The SI equivalents are given in Table 2.1.

2.3 The radioactive decay law

A familiar situation in atomic and subatomic physics is that of a state X, energetically able to make a spontaneous transition to a state Y. Examples are many: a hydrogen atom in an excited state (X) may be free to change to the ground state (Y) and a photon; one nucleus (Z, A) may be able to change to the nucleus ($Z+1$, A) and a β^--particle, or to a nucleus ($Z-2$, $A-4$) and an α-particle. In Fig 2.2 we represent the states involved by a term diagram in which the vertical distance between the terms represents the energy release. The **transition rate**, ω, is the probability that the state X makes the transition to state

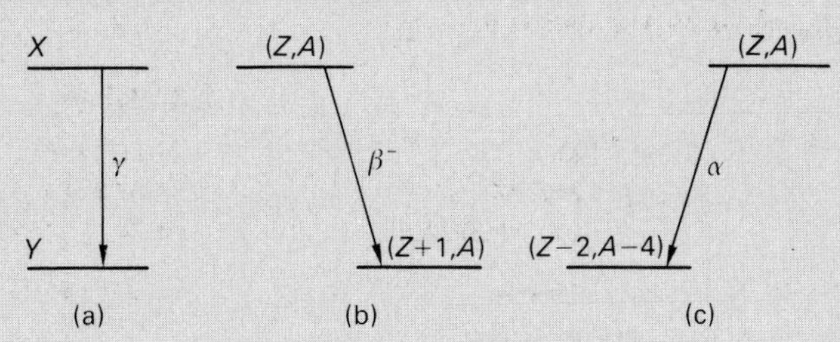

Fig. 2.2 This figure shows typical representations of spontaneous transitions from one state to another state. The vertical distance represents the energy release. (a) Photon emission by an excited state X of to another state Y of hydrogen. (b) Emission of a β^--particle by the nucleus Z,A. (c) Emission of an α-particle by the nucleus Z,A. The horizontal displacement represents a change in Z and/or A.

Y in one second. Suppose at time $t=0$ we have an assembly of a number $N(0)$ of X and at time t a number $N(t)$ survive. Then, from the definition of ω,

$$\mathrm{d}N(t) = -\omega N(t)\,\mathrm{d}t. \qquad (2.6)$$

Hence

$$N(t) = N(0)\,\mathrm{e}^{-\omega t}. \qquad (2.7)$$

This is the **radioactive decay law**. It is a survival equation giving the number of the state X remaining unchanged until the time t. It follows that **mean lifetime**, τ, of the state X is given by

$$\tau = \omega^{-1}. \qquad (2.8)$$

We shall find that τ can range over very many decades in different decays (see Table 2.3), but of course it is a constant for one particular decay.

In early studies of radioactivity another quantity was used, called the **half-life**. At the end of one half-life, half the original nuclei survive without decaying. The relation between this quantity and τ is given alongside and shown in Fig. 2.3.

The detection of radioactivity normally requires the detection of radiation

DEFINITIONS AND KEYWORDS

Transition rate The probability that a state of a system will make a transition to another state in a period of one second. In the context of radioactivity the transition rate is sometimes called the decay constant. Symbol ω.

Radioactive decay law or **Survival equation** This is the equation which gives the number of states N expected to survive to a time t given the number N_0 present at time $t=0$:

$$N = N_0 \exp(-\omega t).$$

Mean lifetime or **mean life** The average lifetime of an unstable state in a large sample of such states. It is also the time at which the number of states surviving is e^{-1} of the number present initially. Symbol τ.

Half-life The time at which the number of states surviving is one half the number present initially. Symbol $t_{1/2}$.

$$(t_{1/2} = \tau \ln 2 = 0.693\tau.)$$

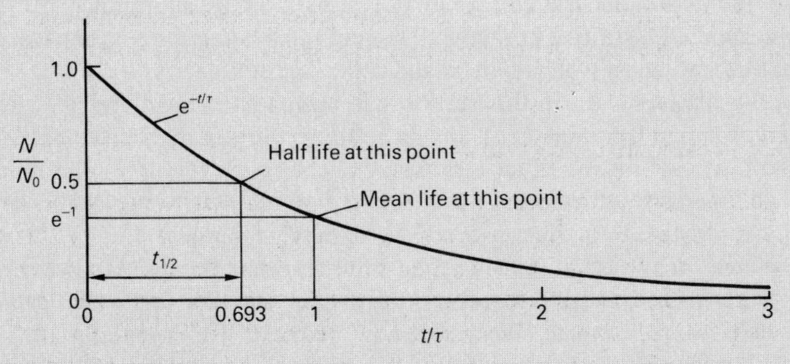

Fig. 2.3 The graphical form of the radioactive decay law showing the relation between mean life (τ) and the half-life ($t_{1/2}$).

emitted when a nucleus decays. Thus the total intensity detectable at time t could be given by

$$I(t) = -\frac{\mathrm{d}N(t)}{\mathrm{d}t} = N(0)\omega \mathrm{e}^{-\omega t},$$

or

$$I(t) = I(0)\,\mathrm{e}^{-\omega t}. \tag{2.9}$$

Thus the radioactivity detected by the intensity of particles emitted itself decays at the same rate as does the number of surviving nuclei.

Reader beware! The equations (2.6) to (2.9) are strictly probability relations. For example, equation (2.7) states that the probability one nucleus survives a time t is $\mathrm{e}^{-\omega t}$. This is clearly not the case because, for a finite number of nuclei, $\mathrm{d}N$ and N are integers, cannot vary smoothly, and are subject to statistical fluctuations. Only in the limit of very large numbers do the statistical fluctuations become relatively small. Thus an understanding of the statistics of these fluctuations is required if measurements are to be made on small numbers or low intensities with a minimum of error. This situation arises from the random nature of the decay process. It is implicit in everything that follows in this chapter but we will not repeat the warning until Section 2.11.

The radioactive decay law is independent of the real time at $t=0$. By this, from the nature of an exponential, it is impossible to determine the age of a pure radioactive sample with stable products from a measurement either of the intensity of radiation emitted or of the amount of surviving active material alone. However, there are many circumstances where ages can be determined, and Problems 2.8–2.11 illustrate some possibilities.

The units of radioactivity are the curie (Ci) and the becquerel (Bq) (see Table 2.2). The former is the historical unit and is still very widely used. It honours Marie Curie-Sklodowska (1867–1934), the discoverer (1898) of radium and polonium. The becquerel is the SI unit.

There is a subtle dual use of the curie and becquerel. They can designate the instantaneous activity of a radioactive material or material containing radioactivity, which need not be pure. They can also be used to designate the amount of a pure radioactive material that is present: thus a millicurie of pure radium is a milligram of radium.

An important subject is the natural and artificial radiation dose received by people and its biological effect. We do not embark on this subject here but Table 2.2 gives some illustrative examples. The real safety situation is complicated by the nature and energy of radiation and other factors.

The vocabulary used in this section has been somewhat ill-defined. Radioactivity was first investigated in the naturally occurring active elements. Thus the word 'decay' means either a radioactive change or the decline in numbers as in the intensity of detected activity, but not necessarily nuclear in source. We have written as of nuclei decaying, or having a mean life. In fact these and related ideas apply whenever one state can make a spontaneous change to another. The initial researchers uncovered activity for which the half lives were hours to years. We now know of mean lives spanning 10^{10} years to 10^{-25} seconds. For some from the nuclear and particle domain, see Table 2.3.

The decay situation which we have presented in this section is the simplest

Table 2.2 Units of radioactivity

Curie One curie (symbol Ci) is the amount of a radioactive material in which the number of disintegrations in one second is the same as that of one gram of pure radium. This number is $3.7 \times 10^{10}\ s^{-1}$.

Becquerel One becquerel (symbol Bq) is the amount of a radioactive material in which the average number of disintegrations in one second is one.

Examples of activities

1. The activity of one person weighing 70 kg is about 10^{-7} Ci$=3.7 \times 10^3$ Bq mainly due to potassium ($^{40}_{19}K$) and carbon 14 ($^{14}_{6}C$).
2. The activity of one cubic metre of air in a dwelling house depends on the nature of the building materials and of the ground below, and on the ventilation. It therefore varies, from below 100 Bq to 1000 Bq or higher, mainly due to a radon isotope ($^{222}_{86}Rn$) and its decay products.

How close can you remain to a source of cobalt 60 (emits γ-rays of about 1 MeV) and receive a radiation dose not greater than the United States maximum permissible occupational whole body-dose of 50 millisieverts per year? (The sievert (Sv) is defined in a way which takes account of the susceptibility of human tissue to long-term risk from radiation; approximately, 1 Sv is 1 J of energy deposited per kilogram of absorbing material in the case of electrons and γ-rays. Most of us receive about 1 mSv per year from naturally occurring background.)

Source of $^{60}_{27}Co$	Minimum safe distance for one year
1 Ci ($=3.7 \times 10^{10}$ Bq)	approximately 40 m

Therefore this strength of radioactive source of γ-rays needs local shielding.

Table 2.3 Examples of decay processess and their mean lives.

Decay	Mean life
$^{238}_{92}U \rightarrow ^{234}_{90}Th + \alpha$	6.5×10^9 years
$^{215}_{84}Po \rightarrow ^{211}_{82}Pb + \alpha$	1.9×10^3 s
$\mu^+ \rightarrow e^+ + \nu_e + \bar{\nu}_\mu$	2.2×10^{-6} s
$\Lambda \rightarrow p + \pi^-$	2.6×10^{-10} s
$\pi^0 \rightarrow \gamma + \gamma$	8.3×10^{-17} s
$\Delta^{++} \rightarrow p + \pi^+$	6×10^{-24} s

These new symbols are for particles that we shall meet in later chapters. The important point is the huge range of mean lifetimes that exist and that we wish to understand.

and one which illustrates the concepts. However, there are more complicated situations which we must now consider.

2.4 Multimodal decays

A particular decay process is called a **decay mode**. Let us now consider a radioactive material for which two modes of decay are possible. An example is

PROBLEMS

2.1 Calculate the potential energy in MeV due to the Coulomb repulsion of

(a) two protons separated by a distance of 1 fm;

(b) a gold nucleus (Z=79) and an α-particle (Z=2) with their centres 10 fm apart;

(c) two nuclei Z=46, A=115, radius $R=1.2\times A^{1/3}$ fm, just not touching.

2.2 An investigation of the mean life of the K^+-meson yielded 20 events of the decay mode

$$K^+ \rightarrow \mu^+ + \nu_\mu,$$

in which the time interval between the arrival of the K^+-meson in a detector and the decay was determined. Estimate the mean life of the K^+-meson from these measurements (time in ns).

Measurement number	1	2	3	4	5	6	7	8	9	10
Time	31.1	2.1	2.1	0.3	38.4	1.1	1.7	7.1	2.3	4.4
Measurement number	11	12	13	14	15	16	17	18	19	20
Time	13.1	15.2	19.0	11.2	5.3	19.7	5.1	13.0	3.0	3.8

2.3 In a sample of one litre of carbon dioxide at STP an average of 5 disintegrations

$${}^{14}_{6}\mathrm{C} \rightarrow {}^{14}_{7}\mathrm{N} + e^- + \bar{\nu}_e$$

are observed per minute. Calculate the atomic fraction of ${}^{14}_{6}\mathrm{C}$ present if the mean life of this nucleus is 8267 years.

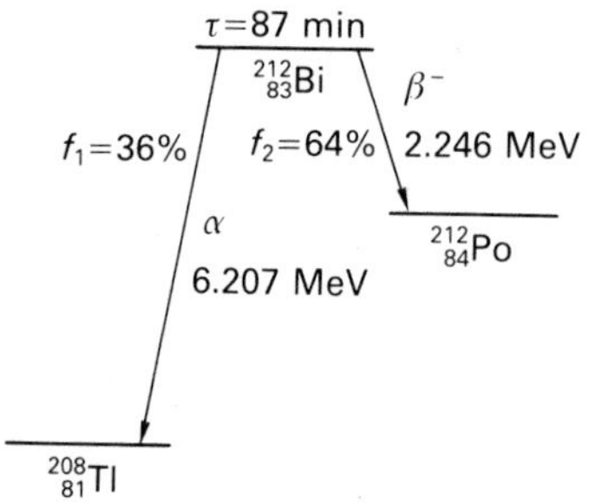

Fig. 2.4 An example of a bimodal decay. The branching fractions f_1 and f_2 are given in per cent.

${}^{212}_{83}$Bi which can decay either by α-particle emission or by β^--particle emission (see Fig. 2.4). The fraction decaying by each mode are the **branching fractions**, f_1 and f_2. There are two **partial transition rates**, ω_1 and ω_2, for the two decay modes separately. They are independent so that the total transition rate for decay of the **parent nucleus** is given by

$$\frac{dN}{dt} = -\omega_1 N - \omega_2 N, \tag{2.10}$$

and therefore

$$N(t) = N(0)\, e^{-(\omega_1+\omega_2)t}. \tag{2.11}$$

The parent therefore decays as if the transition rate were $\omega=\omega_1+\omega_2$ and its mean life is therefore $\omega^{-1}=(\omega_1+\omega_2)^{-1}$. In any period of time the ratio of the probability of decay by mode 1 to decay by mode 2 is ω_1/ω_2, so that

$$f_1=\omega_1/\omega \quad \text{and} \quad f_2=\omega_2/\omega. \tag{2.12}$$

This situation is easily extended to **multimodal decays**: the total transition rate from the parent state is the sum of the partial transition rates:

$$\omega = \sum_i \omega_i, \tag{2.13}$$

with the sum taken over all decay modes. The parent mean life is ω^{-1}. An example is the K^+-meson, which has six principal decay modes (others are very rare). In Problem 2.4 there is a table giving these decay modes with their branching fractions and the mean life of the K^+-meson. There is a blank column headed 'partial transition rate'. Calculate the partial rate for each mode and fill in this column. (Don't worry about what each mode is or means: this unfamiliar material will come later. At the moment this is just a simple problem in radioactivity.)

PROBLEM

2.4 Fill in the last column of this table:

The decay of the K⁺ meson

Decay mode	Branching fraction	Partial transition rate (s^{-1})
$K^+ \rightarrow \mu^+ + \nu_\mu$	0.635	
$K^+ \rightarrow \pi^+ + \pi^0$	0.212	
$K^+ \rightarrow \pi^+ + \pi^+ + \pi^0$	0.056	
$K^+ \rightarrow \pi^+ + \pi^0 + \pi^0$	0.017	
$K^+ \rightarrow \pi^+ + \mu^+ + \nu_\mu$	0.032	
$K^+ \rightarrow \pi^+ + e^+ + \nu_e$	0.048	

The mean life of the K^+-meson is 1.237×10^{-8} s.

2.5 The production of radioactive material

An important method of producing radioactive materials is to expose some target materials to the flux of neutrons in a nuclear reactor (see Section 7.12). Many nuclei can absorb a neutron and become β-active. For example,

$$^{23}_{11}\text{Na} + \text{n} \rightarrow {}^{24}_{11}\text{Na} \rightarrow {}^{24}_{12}\text{Mg} + \text{e}^- + \bar{\nu}_e.$$

In a normal sample of sodium, the depletion of the target material is negligible, so that the rate of production of $^{24}_{11}$Na is constant if the reactor operates at constant power. However, as soon as any $^{24}_{11}$Na is produced it is subject to decay. If the rate of production is p and the decay constant is ω, then the number N of nuclei of $^{24}_{11}$Na satisfies the differential equation

$$\frac{\mathrm{d}N}{\mathrm{d}t} = p - \omega N.$$

If $N=0$ at $t=0$, the beginning of the exposure, the solution is

$$N(t) = \frac{p}{\omega}(1 - \mathrm{e}^{-\omega t}).$$

This is shown in Fig. 2.5. Clearly there is a saturation of the yield which occurs when the rate of production p is matched by the rate of loss ωN due to decay. Try Problem 2.5.

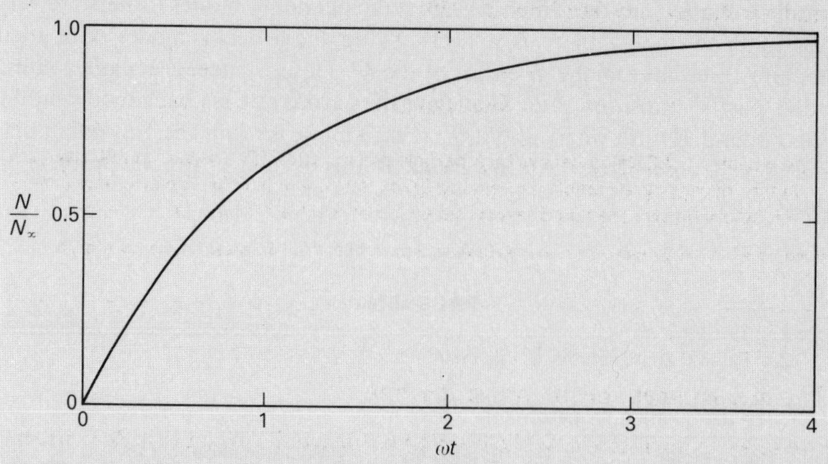

Fig. 2.5 The accumulated yield of radioactive parent material as a function of time for a constant rate of production. The decay rate of this parent is ω and the accumulated yield is N at time t. The quantity N_∞ is the yield if the production is continued indefinitely.

PROBLEM

2.5 A sample of gold is exposed to a neutron beam of constant intensity such that 10^{10} neutrons per second are absorbed in the reaction

$$^{197}_{79}\text{Au}+\text{n}\rightarrow{}^{198}_{79}\text{Au}+\gamma.$$

The nuclide $^{198}_{79}$Au undergoes β-decay to $^{198}_{80}$Hg with a mean life of 3.89 days. How many atoms of $^{198}_{79}$Au will be present after 6 days of irradiation? How many atoms of $^{198}_{80}$Hg will be present at that time, assuming that the $^{198}_{80}$Hg is unaffected by the neutron beam? What is the equilibrium number of $^{198}_{79}$Au atoms?

(Adapted from the 1974 examination in the Honours School of Natural Science, Physics, University of Oxford.)

DEFINITIONS AND KEYWORDS

Decay mode The general term for a particular process that can occur when a state, for example a nucleus, decays.

Multimodal decay The situation when a given state has more than one possible mode of decay.

Partial transition rate The transition rate for one mode in a multimodal decay.

Branching fraction The ratio of the probability of decay by a given decay mode to the probability of decay by any mode. The branching fraction is often known as the *branching ratio.*

Parent nucleus A nucleus which is subject to radioactivity decay.

Daughter nucleus The nucleus which is the product of the decay of a parent nucleus.

2.6 Sequential decays

If a parent decays to give a **daughter nucleus** which is also active, then the equations governing growths and decays are:

$$\frac{dN_1}{dt}=-\omega_1N_1,$$

$$\frac{dN_2}{dt}=-\omega_2N_2+\frac{dN_1}{dt},$$

where N_1 and N_2 are the numbers of nuclei of parent and daughter, ω_1 and ω_2 their respective decay constants. If $N_2=0$ at $t=0$, clearly N_2 will grow as it is fed by decay of the parent nucleus 1, but later as the rate of supply decreases the growth of N_2 will be halted and its own decay will become dominant and will set up a decrease in N_2. The details of such a situation depend on the relative values of ω_1 and ω_2. This and more complicated situations were recognized and analysed by Rutherford (see Fermi, 1950).

PROBLEM

2.6 $^{210}_{83}Bi$ (mean life 7.2 days) decays by β-particle emission to $^{210}_{84}Po$ (mean life 200 days), which in turn decays by α-particle emission to $^{206}_{82}Pb$. If a source initially contains only pure $^{210}_{83}Bi$, after how long will the rate of α-particle emission reach a maximum?

(Adapted from the 1960 examination in the Honours School of Natural Science, Physics, University of Oxford.)

2.7 The measurement of the transition rate

Here we investigate briefly how the transition rate can be measured. Remember the two equations for the decay of a pure source:

Survival equation: $$N(t)=N(0)\,e^{-\omega t}. \tag{2.7}$$

Intensity of emitted radiation: $$I(t)=I(0)\,e^{-\omega t} \tag{2.9}$$

where

$$I=-\frac{dN}{dt},$$

so that

$$I=\omega N. \tag{2.14}$$

(a) Single-mode decay

1. If $e^{-\omega t}$ changes significantly in a reasonable time, clearly ω can be derived from measurements of N or I as a function of t. The former is difficult because it could entail destruction of the source and thus prevent a determination of the time variation. However, if radioactive nuclei or particles can be generated singly at a known time and the time of decay measured, then a sample should conform to equation (2.7) (with fluctuations) and give ω. This technique can apply only to activities of short mean life, so that there is time to repeat many individual lifetimes. A measurement based on equation (2.9) is as simple as a measurement of the intensity of emitted radiation as a function of time. Even if the detection efficiency is not 100%, that is irrelevant, as long as it is constant in time.
2. If ω is small and $e^{-\omega t}$ is not changing detectably or significantly, then equation (2.14) must be used. A measurement of I and N yields ω. This has many difficulties. The intensity of emitted radiation, I, must be measured with known efficiency—not always easy—and the measurement of N, that is the number of nuclei present in the source, again is not normally an easy measurement.

(b) Multimode decay

The same equations apply but now

$$\omega=\omega_1+\omega_2+\cdots,$$

where $\omega_1, \omega_2, \ldots$ are the transition rates for the individual modes. The source survival equation applies as does the equation for the total intensity of all radiations emitted as a function of time. However, if we have I_i as the intensity of radiation associated with mode i ($I = \sum_i I_i$), then

$$I_i(t) = \omega_i \, N(t)$$
$$= \omega_i \, N(0) \, \mathrm{e}^{-\omega t}.$$

So the individual intensities fall off with same decay rate, given by ω.

1. If $\mathrm{e}^{-\omega t}$ changes significantly in the time available, then ω can be found in the same way as for single-mode decay. Every intensity declines with the same exponential. To find the individual ω_i, the relative partial intensities $I_1 : I_2 : I_3 : \ldots$ and hence the branching fractions must be found; then $\omega_i = \omega f_i$. This is a technique that will depend on knowing the relative efficiency for detecting the radiation from each mode.
2. If $\mathrm{e}^{-\omega t}$ does not change significantly, then again we must use the equations

$$I = \omega N,$$

and

$$I_i = \omega_i N, \qquad i = 1, 2, 3, \ldots$$

The measurement of total I, or partial intensities I_i and of N, have all the difficulties recognized for the case of a single-mode decay.

These methods are summarized in Table 2.4.

Table 2.4 Measurement of transition rates: summary.

Since $\quad N(t) = N(0)\exp(-\omega t)$

and $\quad I(t) = I(0)\exp(-\omega t)$,

(1) measure $N(t)$ as a function of t, or
(2) measure $I(t)$ as a function of t.

If ω is small it will be impossible to detect the decline in activity in a reasonable measuring time. In this case since

$$I(t) = \omega N(t),$$

(3) measure $I(t)$ and $N(t)$.

For multimodal decays

$$\omega = \omega_1 + \omega_2 + \omega_3 + \omega_4 + \ldots$$
$$I = I_2 + I_2 + I_3 + I_4 + \ldots$$

(1)–(3) still apply to total transition rates and intensities.

Now

$$I_1 : I_2 : I_3 : I_4 : \ldots = \omega_1 : \omega_2 : \omega_3 : \omega_4 : \ldots,$$

(4) measure I_i and I to find ω_i from ω.

or since

$$I_i(t) = \omega_i N(t)$$

(5) measure $I_i(t)$ and $N(t)$ to find ω_i.

And, of course, the mean life of the state $\tau = \omega^{-1}$.

PROBLEM

2.7 Natural potassium has an atomic weight of 39.089 and contains 0.0118 atomic percent of the isotope $^{40}_{19}K$, which has two decay modes:

$$^{40}_{19}K \rightarrow {}^{40}_{20}Ca + \beta^- + \bar{\nu}_e \qquad (\beta \text{ decay}),$$

$$^{40}_{19}K + e^- \rightarrow {}^{40}_{18}Ar^* + \nu_e \quad \text{(electron capture, see Section 5.3)}$$

$$\qquad \hookrightarrow {}^{40}_{18}Ar + \gamma.$$

where $^{40}_{18}Ar^*$ means an excited state of $^{40}_{18}Ar$. In this case this excited state decays to the ground state by emitting a single γ-ray. The total intensity of β-particles emitted is $2.7 \times 10^4\ kg^{-1}$ of natural potassium and on the average there are 12 γ-rays emitted to every 100 β-particles emitted. Estimate the mean life of $^{40}_{19}K$.

2.8 Radioactive dating

Many beautiful techniques now exist for the dating materials of archaeological, geological or cosmological interest. Many are applications of the simple ideas we have discussed in the previous sections. Although of great interest we cannot pursue them in detail here. Problems 2.8–2.11 are simple examples and we hope the reader will solve them in order to consolidate comprehension of the decay laws.

One simple example is that of carbon dating. Biological carbon comes from atmospheric CO_2 which contains the active isotope $^{14}_{6}C$ (decays to $^{14}_{7}N$) in the ratio of 1 atomic part in about 10^{12} of the stable isotope $^{12}_{6}C$. Once fixed biologically, the decay of the $^{14}_{6}C$ causes a decline in this ratio with a half-life of 5730 years. The $^{14}_{6}C$ is produced in the atmosphere by the action of cosmic rays and if we believe their intensity has not changed significantly (probably not true) then the ratio $^{14}_{6}C/^{12}_{6}C$ at the time of biological fixing is the same as it was before 1945 (when atmospheric nuclear weapon testing injected unnatural $^{14}_{6}C$ into the atmosphere). Thus an assumed ratio at biological fixing and a measured ratio at investigation will give the age of a biological specimen. That ratio is easily measured by determining the actual number of $^{14}_{6}C$ disintegrations per second in the specimen. Clearly this method has, at least, the uncertainty of the assumed initial ratio. This uncertainty has been removed by calibrating against other methods of dating when applicable.

There are now improved methods of measuring the $^{14}_{6}C/^{12}_{6}C$ ratio using mass spectroscopy. In these methods the residual $^{14}_{6}C$ is detected directly rather than by its decay, a potentially more sensitive method. (This is like measuring N rather than I, see Section 2.7.)

2.9 Decay and the uncertainty principle

A decaying state is a system with an uncertainty of lifetime equal to the mean life (ω^{-1}) of that state. It follows that there is an uncertainty in the total energy given by

$$\Delta E\, \Delta t \simeq \hbar/2.$$

PROBLEMS

2.8 Natural uranium found in the Earth's crust contains the isotopes $^{235}_{92}U$ and $^{239}_{92}U$ in the atomic ratio 7.3×10^{-3} to 1. Assuming that at the time of formation these two isotopes were produced equally, estimate the time since formation given that the mean lives are 1.03×10^9 and 6.49×10^9 years respectively.

2.9 Chemical analysis of a meteorite shows that it contains 1 g of potassium and 10^{-5} g of argon formed from the decay of $^{40}_{19}K$. Using the results of Problem 2.7 and assuming no argon has escaped, find a value for the age of the meteorite.

(Adapted from the 1966 examination in the Honours School of Natural Science, Physics, University of Oxford.)

2.10 Write down the law of radioactivity decay. Define the half-life and mean life of a radioactive nucleus and obtain the relation between them.

The nucleus $^{87}_{37}Rb$ decays into the ground state of $^{87}_{38}Sr$, with a half-life of 4.7×10^{10} years and a maximum β-energy of 272 keV. Discuss briefly the difficulties you might encounter in attempting to measure this half-life.

Five different samples of chondritic meteorites are found to have the following proportions of $^{87}_{37}Rb$, $^{87}_{38}Sr$ and $^{86}_{38}Sr$.

Meteorites	$^{87}_{37}Rb/^{86}_{38}Sr$	$^{87}_{38}Sr/^{86}_{38}Sr$
Modoc	0.86	0.757
Homestead	0.8	0.751
Bruderheim	0.72	0.747
Kyushu	0.6	0.739
Bath Furnace	0.09	0.706

Given that the nucleus $^{86}_{38}Sr$ is not a daughter product of any long-lived radioactive nucleus, show that these data are consistent with a common primordial ratio $^{87}_{38}Sr/^{86}_{38}Sr$ and a common age for all these meteorites and find that age.

(Adapted from the 1979 examination in the Honours School in Natural Science, Physics, University of Oxford.)

2.11 Given that the carbon dioxide of problem 2.3 is from a sample of carbon which was fixed in a biological specimen when the $^{14}_{6}C/^{12}_{6}C$ ratio was 10^{-12}, calculate the age of the specimen.

This situation is identical to that in atomic physics and is, of course, general. The uncertainty in energy of an excited state is reflected in the line shape of the radiation emitted in decay to the ground state. That shape is Lorentzian; it is given by

$$\frac{1}{(E-E_0)^2+\Gamma^2/4},$$

which is shown in Fig. 2.6. E_0 is the central energy and the expression gives the relative probability of finding an energy E. Γ is the full width at half height, so that $\Gamma/2$ is the energy uncertainty. The Lorentzian shape is the Fourier transform of the exponential time decay, yielding

$$\Gamma = \hbar\omega,$$

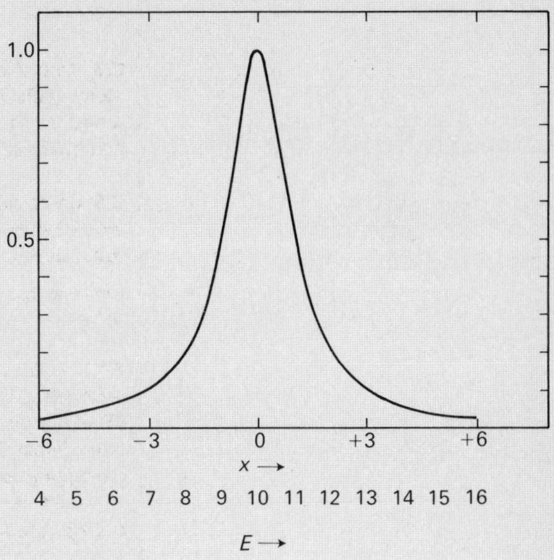

Fig. 2.6 The Lorentzian line shape. This is the shape of the energy spectrum of an isolated state of finite lifetime. The mathematical form of the curve drawn is

$$\frac{\Gamma^2/4}{(E-E_0)^2+\Gamma^2/4}=\frac{1}{1+x^2}\quad \text{where } x=\frac{E-E_0}{\Gamma/2}.$$

The abscissa has two scales: one has x from -6 to $+6$, the other has E from 4 to 16 where $E_0=10$ and $\Gamma=2$ energy units. The full width at half height (Γ) corresponds to x from -1 to $+1$.

as expected from the uncertainty principle. The ω is the sum of all partial transition rates for a multimodal decay.

Significant Γ are not detectable in nuclear or particle physics unless the mean life is very short or if the energy resolution is exceptionally good. This is exemplified in Table 2.5, which gives values of Γ for a range of mean lives. Fig. 2.7 shows a typical experimentally observed line shape for the decay of the rho-meson into two pi-mesons

$$\rho^{\mathrm{o}} \rightarrow \pi^{+}+\pi^{-},$$

superimposed on a background. The uncertainty analysis of the line width of 153 MeV gives a mean life of about 4×10^{-24} s. Situations with line widths and mean lives at the opposite extreme (typically 10^{-8} eV and 10^{-7} s) are to be found in the techniques and application of Mössbauer spectroscopy (see Section 11.11).

Table 2.5 Some line widths Γ and mean lives.

Mean life (s)	Line width
10^{-24}	660 MeV
10^{-18}	660 eV
10^{-12}	6.6×10^{-4} eV
10^{-6}	6.6×10^{-10} eV
1	6.6×10^{-16} eV

2.10 Collisions and cross-sections

The study of collisions between particles, nuclear or atomic, is an essential part of experimental physics. The fact that the discovery of the nucleus came from a study of collisions of α-particles with atoms emphasizes this. We cannot, by their nature, aim one atom or subatomic particle to hit another such 'target' particle; instead we have to rely on the random collisions that occur when an incident beam of projectile particles crosses a space occupied by target particles. See Figs. 2.8 and 2.9.

It is a **cross-section** which quantifies the probability of a collision occurring between two particles and in this section we shall discuss the concept and its partial and differential forms. Quantitatively these quantities are related to the nature of the colliding particles and the forces between them and so their measurement can give important information. Later it will become one theme in this book that the value and behaviour of a cross section are intimately connected to the basic physics and that their interpretation is an important skill.

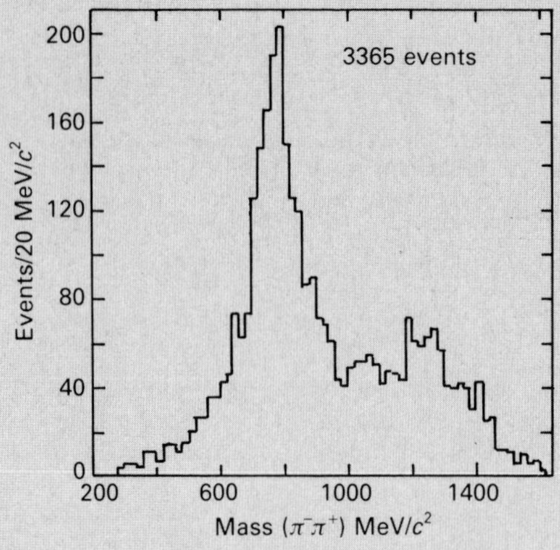

Fig. 2.7 A histogram of the number of events against the mass of $\pi^+\pi^-$ system in the final state of the reaction

$$\pi^- + p \rightarrow \pi^+ + \pi^- + n,$$

at incident π^- momenta of 2.75 and 3.00 GeV/c. The peak at a mass of 770 MeV/c^2 is due to the reaction

$$\pi^- + p \rightarrow \rho^0 + n,$$

followed by

$$\rho^0 \rightarrow \pi^+ + \pi^-.$$

A smaller peak at a mass of 1250 MeV/c^2 is due to the production of a spin 2 particle, the f_2, which decays in the same way as the ρ^0. The peak due to the ρ^0 has a width of about 150 MeV, corresponding to a mean life of approximately 9×10^{-24} s. Both peaks are superimposed on smooth background of events in which the two pions do not come from the decay of a heavier bosons like the ρ^0 and the f_2. (Data from Hagopian *et al.*, 1966.)

(How is the mass of the $\pi^+\pi^-$ system calculated? If a particle of mass M is to be observed, then its total energy E and momentum P satisfy

$$Mc^2 = \sqrt{(E^2 - (\boldsymbol{P}.\boldsymbol{P})c^2)}.$$

If it decays before direct observation into particle 1 and particle 2, then

$$E = E_1 + E_2\,,$$

$$\boldsymbol{P} = \boldsymbol{P}_1 + \boldsymbol{P}_2\,.$$

Therefore

$$Mc^2 = \sqrt{((E_1 + E_2)^2 - (\boldsymbol{P}_1 + \boldsymbol{P}_2).(\boldsymbol{P}_1 + \boldsymbol{P}_2)c^2)}.$$

Thus if the hypothesis is that two particles are the product of the decay of another, calculation of the right-hand side, using measurements of their energies and momenta, will give the mass of this parent. It is this mass which determines the abscissa for each event entered into the histogram. The prominent peak is consistent with the production of a particle of mass 770 MeV/c^2 and width 153 MeV decaying into $\pi^+\pi^-$.)

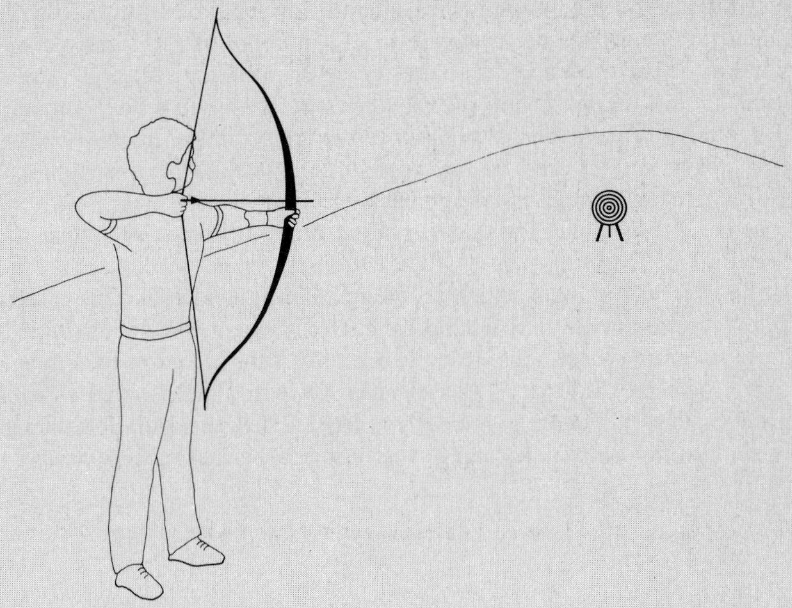

Fig. 2.8 Collisions in atomic and subatomic physics cannot be arranged in the same way as an archer tries to score a bull's-eye on the target. Instead the situation is more like that in Fig. 2.9.

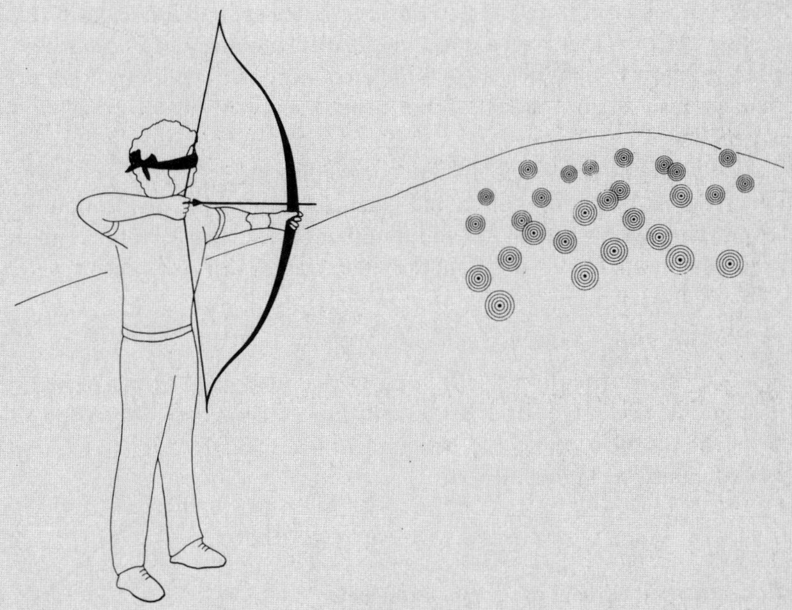

Fig. 2.9 Clearly the chance that the blindfolded archer hits any target is proportional to the density of targets and to the area (cross-section) presented by each, and to the number of arrows that he fires if he has more than one attempt, provided, of course, that he is firing into the region of the assembly of targets. When an arrow does strike the target the archer will score depending on which ring of the target is struck. The probability of a given single score for the blind archer will be proportional to that partial area of the whole target which yields that score. In atomic and nuclear collisions the total cross-section gives the probability that a collision will occur and a partial cross-section gives the probability that the collision has a given outcome.

A knowledge of a cross-section allows a prediction of the number of collisions expected under given conditions of incident beam flux and of target thickness and density. Conversely the observation of a number of collisions allows the calculation of the cross-section. Note, however, that the collision process is random, so that the prediction is really an expectation and an experiment will yield a number of collisions, a quantity subject to statistical fluctuation. In the reverse calculation, the error on the cross-section must include a part due to the statistical error on the observed number of collisions.

A good starting point for understanding cross-section is to return to the elementary kinetic theory of gases. Readers will remember the idea of mean free path of a molecule of a gas which can be related to the idea of a cross-section. This cross-section is the area around the centre of one molecule into which the path of the centre of a second molecule must fall if the second is to collide with the first (assumed stationary). If molecules are hard, sharp-edged spheres of radius R that cross-section, σ, is clearly $\pi(2R)^2$. A simple calculation then gives that the mean free path (m.f.p., λ,) of a molecule in a gas of n molecules per unit volume is

$$\lambda = \frac{1}{n\sigma}. \tag{2.15}$$

The reader will remember that this formula is not correct for a gas where all molecules are moving. However, redefining σ, it does give the m.f.p. for a fast foreign particle moving through molecules which are moving relatively slowly—for example a 10 eV electron moving through a gas—and, by extension, for a fast nuclear particle moving through the material of a target in which the collisions are collisions with the target nuclei assumed to be separated by distances that are large in relation to their radii. So we shall use equation (2.15). The next problem is the cross-section: molecular, atomic, and nuclear particles do not have sharp boundaries, so that the cross-section may not be conceived in a geometric fashion, but reflects other properties involved—a point we have already made.

What about the collisions? Consider a collision which effectively removes one particle from a beam of collimated particles either by absorption or by scattering. The beam which starts with number of particles N_0 when it impinges on the target has a number, N, of unaffected particles after a distance x, given by

$$N = N_0\, e^{-x/\lambda}, \tag{2.16}$$

where λ is given by equation (2.15), n being the number of target particles per unit volume of the target and σ the collision cross-section. Equation (2.16) describes the attenuation of the beam as it traverses the target. Clearly the number of collisions in distance x is

$$C \equiv N_0 - N = N_0(1 - e^{-x/\lambda}). \tag{2.17}$$

For thin targets (that is $x \ll \lambda$) this reduces to

$$C = \frac{N_0 x}{\lambda} = N_0 n\sigma x. \tag{2.18}$$

Expressed in another way: the probability an incident particle undergoes a collision in a distance x of the target material is $(1 - e^{-x/\lambda})$ or, for thin targets, it is x/λ, which is equal to $n\sigma x$.

Let us consider now the outcome of a collision. We will classify the possibilities into very broad categories. The incident particle could suffer either

(1) elastic scattering by the target (subscript e), or

(2) inelastic scattering (subscript i), or

(3) absorption by the target (subscript a).

The collision cross-section, σ, can be divided into three parts σ_e, σ_i, σ_a: each is a measure of the probability the collision occurs and leads to the outcome designated by the subscript. If no other process is possible,

$$\sigma = \sigma_e + \sigma_i + \sigma_a . \tag{2.19}$$

We now call σ the **total cross-section** and the others **partial cross-sections.** Clearly it is σ which determines the attenuation of the beam, as in equation (2.16). The number of expected collisions is given by equation (2.17). These collisions are expected to divide in the ratios

$$\sigma_e : \sigma_i : \sigma_a ,$$

among elastic scattering, inelastic scattering and absorption respectively. Thus the probability of inelastic scattering is

$$\frac{\sigma_i}{\sigma}(1 - e^{-n\sigma x}). \tag{2.20}$$

For thin target ($x \ll \lambda$) this reduces to $\sigma_i nx$.

Let us consider elastic scattering. The incident particle can in general scatter at any angle with respect to the incident direction. If we consider a finite element of solid angle $\Delta\Omega$ in direction θ, then there is a part of σ_e, which we call $\Delta\sigma_e$, that corresponds to the probability that the particle scatters into the solid angle $\Delta\Omega$. Then the **differential cross-section** for elastic scattering at angle θ is defined by

$$\frac{d\sigma_e}{d\Omega} = \lim_{\Delta\Omega \to 0} \frac{\Delta\sigma_e}{\Delta\Omega} . \tag{2.21}$$

Clearly we have

$$\int_{\text{all directions}} d\Omega = 4\pi \text{ sr},$$

and

$$\int \frac{d\sigma_e}{d\Omega} d\Omega = \sigma_e .$$

This differential form has to be expanded in the case of inelastic scattering: the scattered particle can not only appear at a certain angle but it can have an energy E depending on the inelasticity of the collision. Therefore the relevant derivative which descibes the probability is $d^2\sigma_i/d\Omega dE$. These two derivatives will in general be functions of the angle of scatter, and in the latter case, of the energy of the particle also.

DEFINITIONS AND KEYWORDS

Cross-section The area that is a quantitative measure of the probability of collision between two particles.

Total cross-section The cross section for a collision to occur with any possible outcome.

Partial cross-section The cross-section for a collision to occur with a particular outcome.

Differential cross-section When a kinematic variable for one of the products of a collision can be continuous, the partial cross section per unit range of that variable at a specified value of the variable is called a differential cross section.

Barn A unit of cross-section equal to 10^{-24} cm^2.

Cross-section is an area and we are free to express it in any suitable units. However, these nuclear cross-sections are very small on the human scale and although often expressed in centimetres squared are more conveniently expressed in fermis squared ($1\ \text{fm}^2 = 10^{-30}\ \text{m}^2 = 10^{-26}\ \text{cm}^2$) or in barns ($1\ \text{b} = 10^{+2}\ \text{fm}^2 = 10^{-24}\ \text{cm}^2$). How this unit came to be called a barn is given in Table 2.6. In the natural units the cross-section is GeV^{-2} and $1\ \text{GeV}^{-2}$ is equal to $0.039\ \text{fm}^2$.

There is one habit that nuclear physicists have and to which the reader should become accustomed, and it is that of expressing thickness of target presented to a beam of particles not in metres (or in centimetres) but in units which give the mass per unit area of that thickness. This is convenient as it removes the density of the target material from some equations. However, there is one complication: it is customary to use grams per square centimetre (g cm^{-2}) instead of the more correct SI unit of kilograms per square metre (kg m^{-2}). We may use either, so beware! Therefore

$$\text{distance in mass/unit area} = \text{distance in length units} \times \text{density}.$$

Another irritating inconsistency appears in calculating the n of equation (2.15). The number density of nuclei n is given by

$$n = \rho N_A / A',$$

where ρ is the density, N_A is Avogadro's number and A' is the atomic weight of the target material (not the atomic mass number). In SI units N_A is given in number per gram mole whereas ρ is properly kg m^{-3} and n is number per cubic metre. Thus if you work in SI units, N_A is 6.022×10^{23} per mole, so A' must be the mass in kilograms of one gram mole. Again beware!

If λ is to be found in kg m^{-2}, equation (2.15) becomes

$$\lambda = A'/(N_A \sigma),$$

which is a simplification if you can keep control of the units.

The reader should do the cross-section problems (2.12–2.14) to make the concepts familiar and to extend knowledge of the units used. Table 2.7 gives operational definitions which may help in calculations.

Table 2.6 Units of cross-section. The cross-section is an area so for nuclear and particle collisions a convenient unit is the fermi squared (fm^2). However, usage favours the **barn** and multiples thereof (milli-, micro-, nano-, and picobarn):

$$1\ \text{b} = 10^{-24}\ \text{cm}^2 = 10^{-28}\ \text{m}^2 = 10^{+2}\ \text{fm}^2,$$
$$1\ \text{mb} = 10^{-27}\ \text{cm}^2 = 10^{-31}\ \text{m}^2 = 10^{-1}\ \text{fm}^2,$$
$$1\ \mu\text{b} = 10^{-30}\ \text{cm}^2 = 10^{-34}\ \text{m}^2 = 10^{-4}\ \text{fm}^2,$$
$$1\ \text{nb} = 10^{-33}\ \text{cm}^2 = 10^{-37}\ \text{m}^2 = 10^{-7}\ \text{fm}^2,$$
$$1\ \text{pb} = 10^{-36}\ \text{cm}^2 = 10^{-40}\ \text{m}^2 = 10^{-10}\ \text{fm}^2.$$

The natural unit ($\hbar = c = 1$) of cross-section has

$$1\ \text{GeV}^{-2} = 0.039\ \text{fm}^2 = 0.39\ \text{mb}.$$

Comment The unit of a barn was invented in 1942 when physicists were measuring the cross-sections for nuclear reactions involving certain light nuclei. Where they expected a cross-section of order $10^{-26}\ \text{cm}^2$, they found one of order $10^{-24}\ \text{cm}^2$: As big as a (farm) barn!

Table 2.7 Operational definitions of cross-section, or how to calculate expected yields.

Suppose σ is the relevant total or partial cross-section of the defined collision for which the expected yield has to be calculated.

Case 1 Target immersed in a beam of incident particles. The cross-section is the probability one target particle suffers the defined collision when the incident beam is one incident particle per unit area. Or, if you prefer: The cross-section is the probability that one target particle suffers one defined collision per unit time when the incident beam flux is one particle per unit time per unit area.

Case 2 Thin target, thickness dx, containing n target particles per unit volume. The probability one incident particle, crossing the target normally, causes or undergoes the defined collision is $n\sigma\,dx$. In this context 'thin' means that the probability of any collision for an incident particle is small compared with one.

PROBLEMS

2.12 A beam of neutrons of kinetic energy 0.29 eV, intensity $10^5\ s^{-1}$ traverses normally a foil of $^{235}_{92}U$, thickness $10^{-1}\ kg\ m^{-2}$. Any neutron–nucleus collision can have one of three possible results:

(1) elastic scattering of neutrons: $\sigma_e = 2\times10^{-30}\ m^2$,
(2) capture of the neutron followed by the emission of a γ-ray by the nucleus: $\sigma_c = 7\times10^{-27}\ m^2$,
(3) capture of the neutron followed by splitting of nucleus in two almost equal parts (fission): $\sigma_f = 2\times10^{-26}\ m^2$.

Calculate:

(a) the attenuation of the neutron beam by the foil;
(b) the number of fission reactions occurring per second in the foil, caused by the incident beam;
(c) the flux of elastically scattered neutrons at a point 10 m from the foil and out of the incident beam, assuming isotropic distribution of the scattered neutrons.

(Adapted from the 1970 examination of the Honours School of Natural Science, Physics, University of Oxford.)

2.13 Define what is meant by the term 'cross-section' applied to a nuclear reaction.

A liquid hydrogen target of volume $10^{-4}\ m^3$ and density $60\ kg\ m^{-3}$ is immersed in a broad, uniform, monoenergetic beam of negative pions, of intensity 10^7 particles $m^{-2}\ s^{-1}$. At a beam momentum of 300 MeV/c the only reaction, apart from elastic scattering, which takes place is

$$\pi^- + p \rightarrow \pi^0 + n$$

with a cross-section of 45 mb ($4.5\times10^{-30}\ m^2$). (The neutral pion decays in the mode $\pi^0 \rightarrow 2\gamma$, with negligibly short lifetime.)

Calculate the number of γ-rays emitted per second from the target.

(Adapted from the 1974 examination of the Honours School of Natural Science, Physics, University of Oxford.)

2.14 A target of natural boron (atomic composition 20% $^{10}_{5}B$, 80% $^{11}_{5}B$) in the form of a thin foil which has a mass per unit area of $1\ kg\ m^{-2}$ is traversed normally by a beam of neutrons of kinetic energy 1 keV. The only significant neutron absorption is by the $A=10$ isotope for which the cross-section has the value of 19.3 barns at 1 keV. This absorption leads to the emission of an α-particle. The residual nucleus is left in its ground state in 30% of the reactions and for the remainder in an excited state at about 500 keV from which it decays by single γ-ray emission to the ground state. State the Z and A of the residual nucleus and calculate the yield of gamma rays when the beam has an intensity of 10^5 per second. The effect of elastic scattering of neutrons in the target may be neglected.

(Adapted from the 1988 examination of the Honours School of Natural Science, Physics, University of Oxford.)

2.11 Probabilities, expectations, and fluctuations

Collision cross-section and transition rates can be used to calculate expected yields under given conditions. However, these are random processes so that the actual number of events observed in repeated measurements will fluctuate. As the experiment is repeated the mean number of events will converge to that expected (if our calculation is correct) and we can observe the probability with which the observation of n events occurs where m are expected and is the mean for an infinite number of observations. In the limit this probability is that of the Poisson distribution and is given by

$$P(n,m) = \frac{m^n e^{-m}}{n!} .$$

(see Table 2.8). This distribution has a mean of m (as required) and a variance of m also. A part of the distribution for $m=3.15$ is shown in Fig. 2.10. Except for small m such distributions approximate to Gaussians with means of m and

Table 2.8 A mnemonic for remembering the Poisson distribution, $P(n,m)$.

Now $1=\exp(-m)\times\exp(+m)$.

Expand $\exp(+m)$:

$$1=\exp(-m)+m\exp(-m)+\frac{m^2}{2!}\exp(-m)+\cdots+\frac{m^n}{n!}\exp(-m)+\cdots$$

which is, term by term,

$$1=P(0,m)+P(1,m)+P(2,m)+\cdots+P(n,m)+\cdots.$$

Thus $P(n,m)$ is $\exp(-m)$ times the $(n+1)$th term of the expansion of $\exp(+m)$.

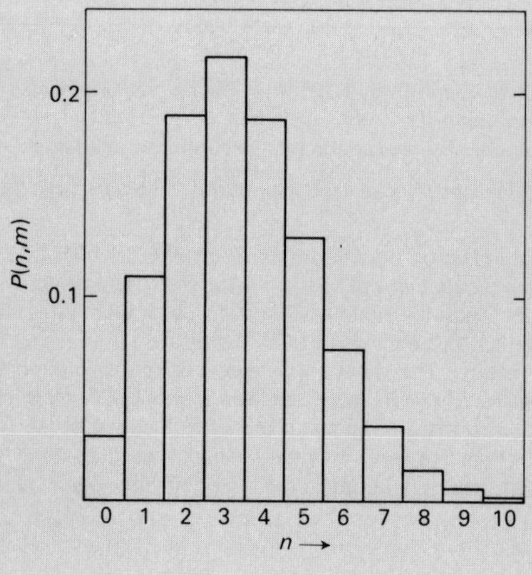

Fig. 2.10 The Poisson distribution, $P(n,m)$, for $m=3.45$ and $n=0$–10.

standard deviations of $\sqrt{m}$. Thus if m is 100, for example, one measurement has an approximately 30% chance of being outside the range 90 to 110.

Conversely, if we do not know m and make one measurement yielding n events, then our best estimate of m is n and our best estimate of the error is $\sqrt{n}$. As already mentioned, this error must be propagated to the cross-section or transition rate if that is what is to be derived from n.

Problem 2.15 is an application of the Poisson statistics to nuclear counting. Problem 2.16 also involves these statistics.

PROBLEMS

2.15 A particle counter registers an average of 0.453 counts per second. What is the probability that it registers 2 in any one second?

A larger counter registers 1296 in ten minutes so that the best estimate of its rate is 2.16 per second: estimate the error on this measured rate and the probability that this measurement is low by more than the error. Assume that the Poisson distribution $P(n,m)$ for large m is approximated by a Gaussian distribution with mean m and standard deviation $m^{1/2}$.

2.16 An α-particle of mass $4M$ and speed v passes near the nucleus of an atom of atomic number Z and atomic weight A ($\gg 4$). The α-particle has an impact parameter b and is scattered through an angle θ. Derive the approximate relation, for small θ,

$$\theta = (Ze^2)/(4\pi\varepsilon_0 M v^2 b).$$

A beam of α-particles with speed $v = 2\times10^7\ \mathrm{m\,s^{-1}}$ is incident normally on a foil, thickness 10^{-5} m, of gold ($Z=79$, $A=197$, density $1.9\times10^4\ \mathrm{kg\,m^{-3}}$). Estimate the proportion of α-particles suffering double scattering, where each scatter is of at least 10°, in traversing the foil.

(Adapted from the 1970 Examination in the Honours School of Natural Science, Physics, University of Oxford.)

References

Cohen, E. R. and Taylor, B. N. (1987). The 1986 adjustment of the fundamental constants. *Reviews of Modern Physics*, **59**, 1121–48.

Fermi, E. (1950). *Nuclear Physics*. University of Chicago Press. (Notes compiled by J. Orear, A. H. Rosenfeld and R. A. Schuter.)

Hagopian, V., Selove, W., Alitti, J., Baton, J. P. and Neveu-René, M. (1966). *Physical Review*, **145**, 1128–35.

3

The Size and Shape of Nuclei

3.1 The size of nuclei

In fact this chapter is mostly about the size of nuclei and how the nuclear matter is distributed inside a sphere, if the nucleus is spherical. The important matter of whether it is spherical we discuss undeservedly briefly in Section 3.9.

We must start by considering what we mean by the size of a nucleus. In atomic physics, the boundary of an atom is not sharp since the wavefunction of the outer electrons decreases monotonically: therefore we must keep in mind that we should expect a similar situation to occur in nuclear physics. As we have seen, when Rutherford was investigating the scattering of α-particles by gold, his scattering measurements were in agreement with the predictions found assuming that the interaction was the Coulomb potential between two point charges. Thus for gold at the classical distance of closest approach that could be reached with the α-particle energies that Rutherford had available, he was unable to detect any deviations which might indicate that the charges on the α-particle and the gold nucleus were not point-like. However, using light nuclei as targets, Rutherford found deviations which indicated a breakdown in the Coulomb law when the classical distance of closest approach was about 10^{-14} m. These deviations occur at short distances of approach, firstly because the finite charge density distributions overlap and secondly because the strong **nuclear forces** which exist between the α-particle and a nucleus come into play. This situation raises a question: what size are we measuring if we investigate these effects? It is necessary to distinguish between the distribution of the sources of electric fields (charge) and the sources of nuclear fields (which we will call matter), although these sources are probably strongly linked. To make progress we should use a probe for one which does not feel the other. The electron is coupled to an electric field due to charge sources but does not experience the nuclear force. Conversely, the neutron is electrically neutral but does experience the nuclear force. Thus separate scattering experiments with these two particles should, in principle, go some way to measuring these two distributions. In the following sections we will discuss firstly the measurement of electric charge distributions and secondly the matter distributions.

There is one significant conclusion we can draw from Rutherford's results. Beyond distances of separation of the order of 10^{-14} m the scattering of α-particles showed the potential is Coulomb and, of course, is varying as r^{-1}. At smaller separations the strong nuclear forces take over and swamp the Coulomb force. This meant that not only were the nuclear forces stronger at those distances but that their potential decreased more rapidly than r^{-1}: we call such forces **short range**. We shall find that the range of nuclear forces is 1 to 2 fermi. Of course the range may not be precisely defined but if the potential were to vary as e^{-ar}/r then a^{-1} would be the range.

3.2 The scattering of electrons by nuclei

Our discussion in the last section indicated that a study of electron scattering would be a method of measuring the distribution of the sources of the static electromagnetic field of the nucleus. The scattering of β-particles emitted by naturally occurring radioactive materials was observed early in the investigations of radioactivity. However, these β-particles are not monoenergetic, they have low momentum and suffer, in addition to large single scatters, severe multiple scattering effects. This property prevented clear results being obtained from early measurements of electron scattering. However, the scattering of more energetic electrons (> 100 MeV) is a very important tool in the investigation of the nuclear size. The formula for the differential cross-section for the elastic scattering of relativistic electrons, which is analogous to Rutherford's for α-particle scattering, was derived by Mott using relativistic quantum mechanics. In Table 3.1 we have rewritten Rutherford's formula and below it Mott's formula. Note that the latter still neglects nuclear size and nuclear recoil but, because it was derived using Dirac's relativistic theory of the electron, it contains the effect of electron spin.

3.3 The nuclear electric charge distribution

We need a model. The methods used to measure charge distribution find the time average so we can define a time-independent charge density. For the present we shall assume spherically symmetric nuclei so that we can define a radial charge density, $\rho(r)$. Two models are given in Fig. 3.1. Model I with its sharp-edged charge distribution is very unlikely but can be tested. Model II softens the hard edges by assuming a charge distribution with a mathematical form normally associated with the Fermi–Dirac statistics but which, applied to nuclei, is called the **Saxon–Woods** form. So the challenge to experiments is to see if either model makes predictions which fit the data and if so, to determine the parameters ρ_0, a and d, and if not, to find a better model and its parameters.

Fig. 3.1 Two models of the radial electric charge distribution of nuclei.

(a) Model I: $\rho(r) = \rho_0,\ r < a,$
$\rho(r) = 0,\ r > a.$

(b) Model II: $$\rho(r) = \frac{\rho_0}{1 + \exp\left(\frac{r-a}{d}\right)}.$$

The $r = a \pm d/2$ points have densities 62.2% and 37.8% of the central density respectively. The 90% to 10% thickness is $4.39d$. This shape is called the Saxon–Woods form. The charge density, ρ_0, is fixed by normalizing to the total nuclear charge $Z|e|$.

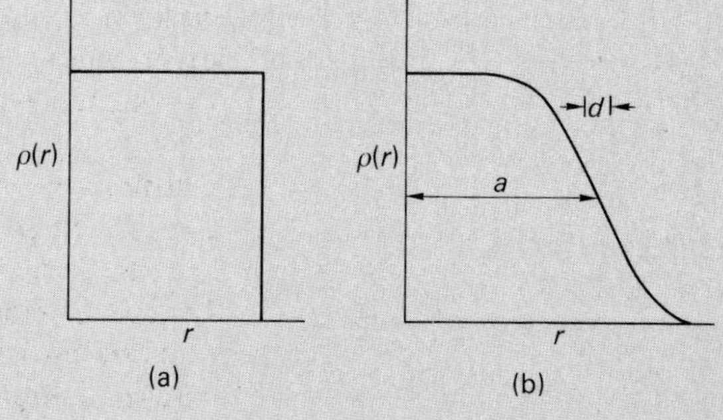

DEFINITIONS AND KEYWORDS

Nuclear force This is the force that exists between the nuclear constituents and is responsible for binding these constituents. At the separations typical of neutrons and protons in nuclei, the nuclear force is much stronger than the Coulomb force between two protons at the same separation and it is therefore said to be 'strong'.

Short-range forces Forces are said to be short range if the corresponding potential decreases with distance from the source faster than $1/r$. If the potential varies as $\exp(-ar)/r$, then $1/a$ is the range. Thus the Coulomb potential is sometimes said to have infinite range.

Saxon–Woods form The form of mathematical expression frequently used to describe the distribution of nuclear charge or the shape of the strong nuclear potential. It is

$$\frac{1}{\left(1+\exp\left[\frac{r-a}{d}\right]\right)},$$

where a is the nuclear radius, d describes the density fall-off at the surface and r is the radial distance from the centre.

Form factor This is the factor in a formula for the scattering cross-section that accounts for the effect of the extension of the charge distribution of a non-point-like object scattering photons or charged particles.

3.4 The nuclear electric form-factor

How do we deal with the effect of an extended nuclear charge on the Mott scattering of electrons? The answer is to do as in classical optics where we derive the Fraunhofer diffraction pattern of an aperture in a screen by taking the Fourier transform of that aperture. For electron scattering the aperture is replaced by a spherical distribution of charge. We take the nucleus to have charge Ze where e is the charge on the proton. If that charge was point-like at $r=0$ we can imagine that it gives rise to a scattered wave amplitude $Zef(\theta)$ at large distances at an angle θ, defined so that

$$\left.\frac{\mathrm{d}\sigma}{\mathrm{d}\Omega}\right|_{\mathrm{Mott}} = Z^2e^2|f(\theta)|^2. \tag{3.1}$$

If that charge is spread out then an element of charge $\mathrm{d}(Ze)$ at a point $\mathbf{r}$ will give rise to a contribution to the amplitude of $f(\theta)\mathrm{d}(Ze)$ where δ is the extra 'optical' path length introduced by wave scattering by the element of charge at the point $\boldsymbol{r}$ compared to zero optical path for scattering at $r=0$.

Consider now Fig. 3.2. The incident and scattered electron have momentum $\boldsymbol{p}$ and $\boldsymbol{p}'$ with $p=|\boldsymbol{p}|=|\boldsymbol{p}'|$. The momentum transfer $\boldsymbol{q}(|q|=2p\sin(\theta/2)$, see Fig. 1.7) is along OZ, O being the nuclear centre. The 'optical ray' P_1OP_1' is taken to have zero relative path length. The ray P_2SP_2' has equal angles of incidence and reflection at the plane AXA' which is perpendicular to OZ. Therefore the path length d is the same for all rays parallel to P_1OP_1' and reflected at any point in AXA'. This path length is given by $d=2OX\sin(\theta/2)$. The 'optical' path length, δ, is $2\pi d/\lambda$, where λ is the de Broglie wavelength. Now the reduced wavelength $\lambda/2\pi=\hbar/p$ so

$$\delta=pd/\hbar=\frac{2p\sin(\theta/2)}{\hbar}OX=\frac{q}{\hbar}OX.$$

Return to the point S: let the charge density be $\rho(r)$ when $r=OS$ (we assume spherical symmetry). If we have polar coordinate r, α, β, where the polar axis is along Z and $S\hat{O}Z=\alpha$, then a volume element $\mathrm{d}V$ at S is $r^2\sin\alpha\,\mathrm{d}r\,\mathrm{d}\alpha\,\mathrm{d}\beta$ and the charge is $\rho(r)\mathrm{d}V$. Then this charge element gives rise to an amplitude

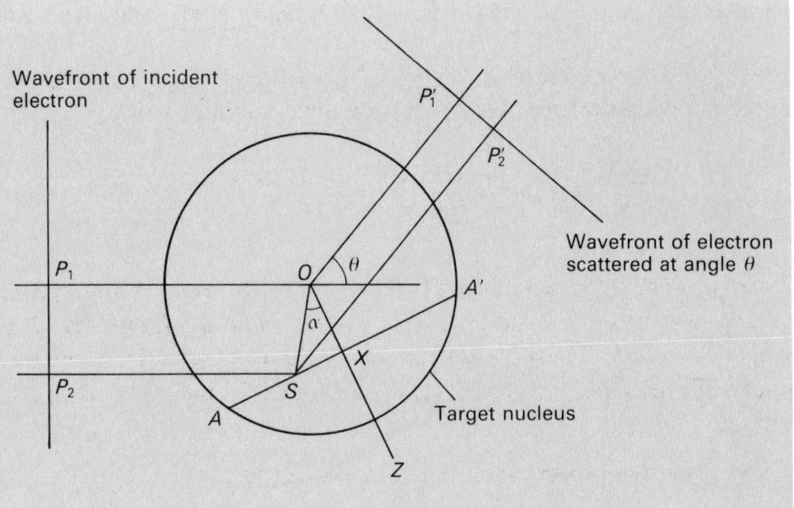

Fig. 3.2 The optical picture of electron scattering at an extended nucleus showing the geometrical construction required to calculate the Fraunhofer diffraction scattering pattern.

$$\text{charge} \times f(\theta) \times e^{i\delta}$$

$$= \rho(r)r^2 \sin\alpha \, dr \, d\alpha \, d\beta \, f(\theta) e^{i\boldsymbol{q}.\boldsymbol{r}/\hbar}.$$

The exponent contains $\boldsymbol{q}.\boldsymbol{r}$ because $q(OX) = qr\cos\alpha = \boldsymbol{q}.\boldsymbol{r}$. Then the total scattered amplitude

$$A(\theta) = f(\theta) \int_0^{2\pi} \int_0^{\pi} \int_0 \rho(r)r^2 \sin\alpha \, dr \, d\alpha \, d\beta \, e^{i\boldsymbol{q}.\boldsymbol{r}/\hbar}, \tag{3.2}$$

the integration being over the whole nucleus. The integration over the azimuthal angle β around OZ is trivial because the point S just traverses the plane AXA' for which $\boldsymbol{q}.\boldsymbol{r}$ is constant.

$$A(\theta) = f(\theta) \int_0^{\pi} \int_0^{\infty} 2\pi\rho(r)r^2 \sin\alpha \, dr \, d\alpha \, e^{i\boldsymbol{q}.\boldsymbol{r}/\hbar}.$$

Now we have for the total charge

$$Ze = \int_0^{\pi} \int_0^{\infty} 2\pi\rho(r)r^2 \sin\alpha \, dr \, d\alpha,$$

so we see that we can write

$$A(\theta) = Zef(\theta) \frac{\int\int \rho(r)r^2 \sin\alpha \, dr \, d\alpha \, e^{i\boldsymbol{q}.\boldsymbol{r}/\hbar}}{\int\int \rho(r)r^2 \sin\alpha \, dr \, d\alpha} = Zef(\theta)F(\theta).$$

Thus the Mott (or Rutherford) scattering amplitude $Zef(\theta)$ is changed by a factor $F(\theta)$ and the scattering cross-section becomes

$$\frac{d\sigma}{d\Omega} = |F(\theta)|^2 \frac{d\sigma}{d\Omega}\bigg|_{\text{Mott}}.$$

$F(\theta)$ is called a **form factor**. (This is a name inherited from the description of atomic X-ray scattering.) Somewhat formally it could be written

$$F(\theta) = \frac{\int \rho(\boldsymbol{r}) e^{i\boldsymbol{q}.\boldsymbol{r}/\hbar} dV}{\int \rho(\boldsymbol{r}) dV} = \frac{1}{Ze} \int \rho(\boldsymbol{r}) e^{i\boldsymbol{q}.\boldsymbol{r}/\hbar} dV.$$

Clearly as $\theta \to 0$, $q \to 0$ and $F(0) = 1$. Since the form factor is more properly a function of q than of θ it has become usual to write $F(q^2)$ rather than $F(\theta)$ and we note $F(q^2)$ is the Fourier transform of the charge distribution (remember the optics of Fraunhofer diffraction). These ideas and formulae are summarized in Table 3.1.

Now let us look to see what this formula means:

1. When $q \to 0$, $F(q^2) \to 1$ and the scattering is no different from that for a

Table 3.1 The form factor

The effect of an extended nuclear charge is to reduce the differential cross-section for elastic electron scattering from that for a point-like nucleus by a factor which is the square of the form factor:

$$\frac{d\sigma}{d\Omega} \rightarrow \left|F(q^2)\right|^2 \frac{d\sigma}{d\Omega},$$

where the form factor

$$F(q^2) = \frac{1}{Ze}\iiint \rho(\boldsymbol{r})\, e^{i\boldsymbol{q}.\boldsymbol{r}/\hbar}\, dV,$$

and q is the momentum transfer, $q = |\boldsymbol{q}|$ and $\rho(\boldsymbol{r})$ is the charge density. The volume integral is to be taken over the entire nucleus. If the nucleus is spherically symmetric, then

$$F(q^2) = \frac{4\pi\hbar}{Zeq}\int \rho(r) r \sin\left(\frac{qr}{\hbar}\right) dr.$$

point-like nucleus; as q increases the oscillatory nature of the exponential in eq. 3.2 for an extended nucleus reduces $|F(q^2)|$ from 1 and the scattering is reduced. This is not unexpected: an extended electric charge has greater difficulty in taking up the momentum transfer than does the point-like arrangement of the same total charge.

2. Since we know the rough size of nuclei, $R < 10^{-14}$ m ($\simeq 10$ fm), we can now estimate the q needed to see a significant reduction in scattering intensity due to size. We would want $\boldsymbol{q}.\boldsymbol{R}/\hbar$ to be of order 1, hence $q \simeq \hbar/R$. Remember $\hbar c = 197$ MeV fm, so we see that $q > 20$ MeV$/c$. To reach this at 30° scattering requires incident electrons of 40 MeV. In fact this is hardly adequate since nuclear radii are somewhat less than 10 fm and we want to see detail with a resolution of better than 1 fm. Therefore we should be aiming for $q \simeq \hbar/(1\text{ fm}) \simeq 200$ MeV$/c$. The first detailed measurements were made with electrons of 150 MeV but later work has increased the energies used to 500 MeV.

Now what does $F(q^2)$ look like? As an exercise you are asked to show (Problem 3.1) that the form factor for model I is

$$F(q^2) = \frac{3\{\sin x - x\cos x\}}{x^3}, \qquad x = qa/\hbar.$$

This looks rather unmanageable but in fact is the spherical Bessel function $j_1(x)$. The square of this function is the factor by which the point-like Mott differential cross-section is reduced. This factor is plotted in Fig. 3.3 for the case of $a = 4.1$ fm ($^{58}_{28}$Ni nucleus) and the abscissa is marked in units of q MeV$/c$ and of θ degrees for 450 MeV incident electrons. We notice immediately the diffraction zeros near 27°, 48°, 69°, and 95°. This is typical of an object with sharp edges. In Fig. 3.4a we give the measured differential cross-section: it shows diffraction minima at q values expected from model I. On the same figure is a fitted curve using a model close to our model II. The softer edges of the distribution fill in the diffraction minima and give results closer to the data than the model I prediction.

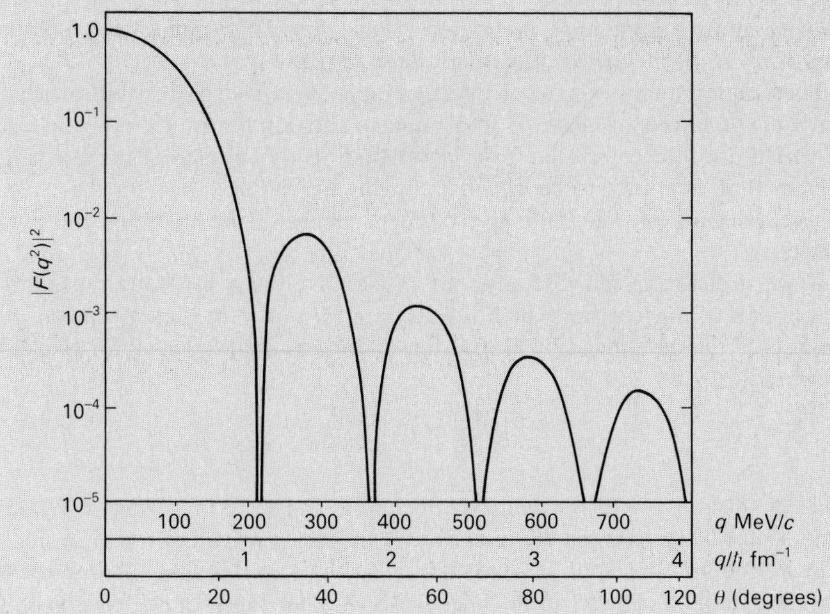

Fig. 3.3 The square of the form factor $|F(q^2)|^2$ as a function of q for a model I nucleus having $a = 4.1$ fm. The abscissa is also marked in inverse fermis ($q/\hbar$) and in degrees for an angle of scatter at a fixed nucleus for incident electrons of 450 MeV. Note that the ordinate is logarithmic.

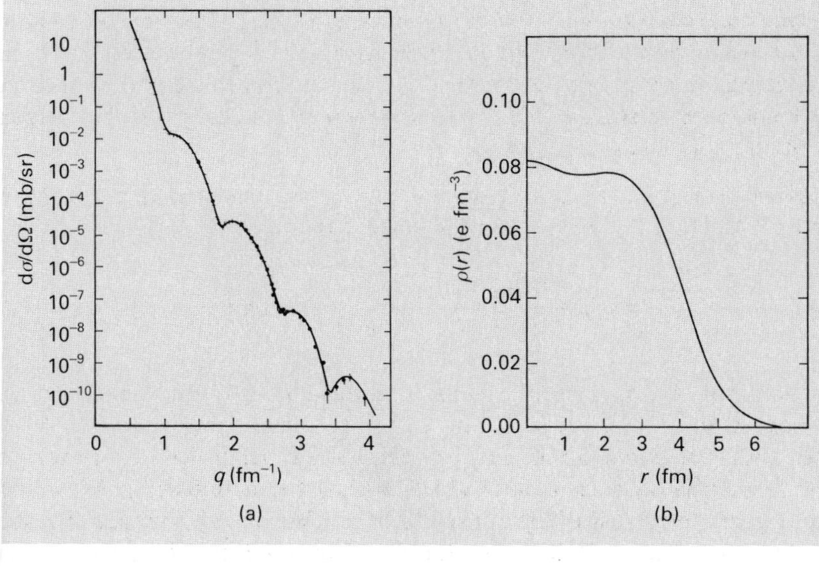

Fig. 3.4 (a) The measured differential scattering cross-section $d\sigma/d\Omega$ for the scattering of 450 MeV electrons by $^{58}_{28}$Ni. The positions of the diffraction minima should be matched against those of Fig. 3.3. The solid line is the prediction for the charge distribution shown in (b) which is close to that of our model II (Sick *et al.* 1975).

The actual experimental work means measuring differential scattering cross-sections which are varying so rapidly with angle that they can change by five or six decades in 60°: this situation presents many problems if sufficiently accurate results are to be obtained. The analysis of the data to obtain the charge distributions can be done in several ways. Here we make one remark and describe the conceptually simplest method. The remark is that since $F(q^2)$ is the Fourier transform of $\rho(r)$, if this form factor is known for all q^2, the inverse Fourier transform might be computed to give $\rho(r)$. This is impossible because the form can be measured only in a limited range of q^2.

The simplest analysis is to construct a model of the charge distribution. Such a model will have parameters (for example, a and d in our model II) which can be varied until the calculated form factor best fits the measurements available. This method is known to be imperfect, giving unreliable values for the central charge densities and misleadingly confident values of the errors on the parameters.

If we neglect the finer details, our model II gives a modestly reasonable description of nuclear size which is adequate for our immediate purposes. An analysis of the data made by Hofstadter and Collard (1967) gave for the half-point radius.

$$a = 1.18A^{1/3} - 0.48 \text{ fm},$$

and a 90% to 10% density thickness 2.4 ± 0.3 fm (Saxon–Woods: $d = 0.55 \pm 0.07$ fm), where the $\pm$ indicates the range of values found in nuclei with $A > 40$, not the error, which is normally less than this value. Below $A = 40$ there are marked changes in thickness with A. Since the volume depends on A but the charge on Z, we could expect the central charge density to decrease through the periodic table as $2Z/A$ from the value in light nuclei where $Z = N$. However, in light nuclei the surface thickness is a large part of the whole and any region of uniform density is very limited. In addition, the charge density

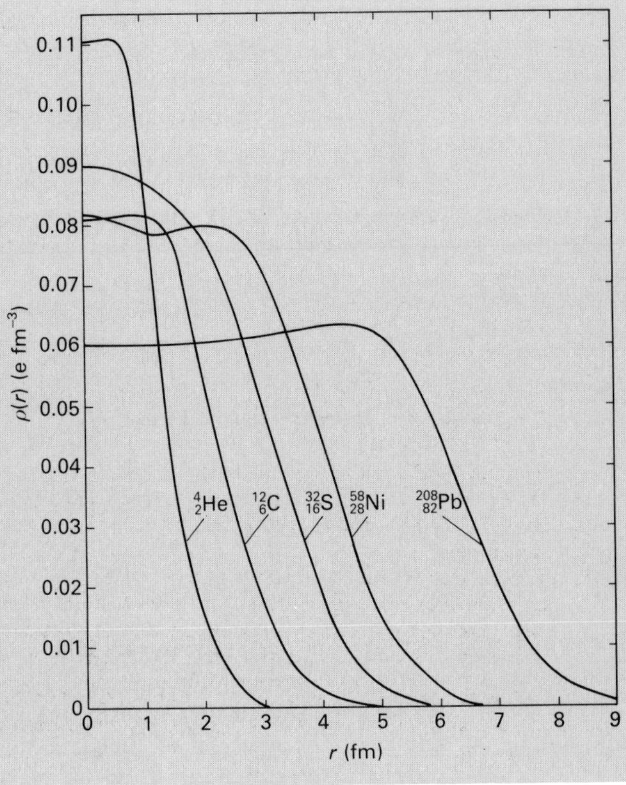

Fig. 3.5 The charge density against radial distance for several nuclei as derived from electron scattering measurements. The resolution of these results is such that there is typically an uncertainty of about 10% on the central density and a decreasing uncertainty as the radial distance increases. Clearly our model II, a Saxon–Woods shape, is only an approximation. The curves in this figure are adapted from the results of analyses by Dreher *et al.* (1974), Sick *et al.* (1976) and by Sick (1974).

could show effects which reflect either the shell model structure of nuclei or other properties of nuclear matter. Measurements have been made at high momentum transfers (up to 800 MeV/c) and with particular nuclei; some results are shown in Fig. 3.5. It is clear that our model II is an over-simplification.

The assumptions implicit in our use of the Mott formula are listed in Table 3.2. A brief discussion follows on each of the points: where we have simplified matters the more sophisticated treatment may improve the determination of the charge distributions. However, they do not radically affect the results that we need.

3.5 The isotope shift

The energy of optical transitions in atomic physics can be measured with quite extraordinary accuracy. If it were possible to calculate the wavefunctions of the electrons, allowing for the finite size of the nuclear charge distribution, then the parameters of the latter could be varied to bring the calculations into line with observation. Unfortunately the calculations can not be done because the theory

Table 3.2 The Rutherford and Mott scattering-cross-sections

The Rutherford formula for the differential angular cross-section for the elastic scattering of a non-relativistic, spin-less particles of unit charge, momentum P, and velocity v, at a fixed (no recoil) target nucleus of atomic number Z, is

$$\frac{d\sigma}{d\Omega} = \frac{Z^2\alpha^2\hbar^2c^2}{4P^2v^2}\operatorname{cosec}^4(\theta/2).$$

If the incident particle is an electron (or positron, see Section 1.6) which is unpolarized and can be relativistic ($v \Rightarrow c$), the formula becomes

$$\frac{d\sigma}{d\Omega} = \frac{Z^2\alpha^2\hbar^2c^2}{4P^2v^2}\operatorname{cosec}^4(\theta/2)\,[1-\frac{v^2}{c^2}\sin^2(\theta/2)].$$

This is the Mott formula. It assumes

(1) relativistic quantum mechanics for the electron;
(2) that first-order perturbation theory is adequate to calculate the scattering cross section;
(3) that there is no nuclear recoil;
(4) that the nuclear electric charge is point-like;
(5) that the nuclear spin is zero.

Commentary on these assumptions

(1) This is clearly the right thing to do; it takes care of the effect of relativistic velocities and of electron spin.
(2) Only true for Z not too large ($Z\alpha \ll 1$).
(3) Allowing for nuclear recoil means including an extra kinematic factor.
(4) The inclusion of a form factor covers the effect of the charge distribution of a spin-less nucleus.
(5) If the nucleus has spin j, then in principle it has $2j+1$ form factors each of which will appear with different kinematic factors in the formula for the cross-section and can vary independently with momentum transfer. However, the effect of all but the electric form factor is small and they need not concern us here.

It is clear that most of these points are only a complication to the discussion about form factors in the text. We list them here to remind readers that we are sometimes using a very simplified approach.

PROBLEMS

3.1 Show that, for a spherically symmetric charge distribution,

$$\left.\frac{\mathrm{d}F(q^2)}{\mathrm{d}q^2}\right|_{q^2=0} = -\frac{\langle r^2\rangle}{6\hbar^2},$$

where $\langle r^2\rangle$ is the mean square of the electric charge distribution. [*Hint*: expand sin $(qr/\hbar)$ in the formula in Table 3.1]

3.2 Show that the form factor for the charge distribution of model I is

$$F(q^2) = \frac{3\{\sin(qa/\hbar) - (qa/\hbar)\cos(qa/\hbar)\}}{(qa/\hbar)^3}.$$

3.3 Find the form factor for a charge distribution

$$\rho(r) = \rho_0 e^{-r/a}/r.$$

3.4 An electron of momentum 330 MeV/c is scattered at an angle of 10° by a calcium nucleus. Assuming no recoil, find the momentum transfer and its reduced de Broglie wavelength. Also calculate the Mott differential cross-section (point-like nucleus), and by what factor it is reduced if the calcium nucleus (A=40) can be assumed to be represented by model I with $a = 1.2A^{1/3}$ fm.

of many-electron systems is not sufficiently refined. (However, in the next section we shall meet a circumstance in which this unrewarding situation is reversed.) All that can be done is observe the optical spectrum of a given element (constant Z) from isotope to isotope. As A increases at constant Z the nuclear charge distribution will change: the changes in the wavelengths of spectral lines are consistent with the model II that we have favoured.

3.6 X-ray spectroscopy of mu-mesic atoms

Among the particles that exist in nature there are heavy versions of the electron and positron called the negative and positive mu-meson, or μ^- and μ^+ for short (see Section 9.10). They have 207 times the mass of the electron. The μ^- behaves just like a heavy version of the electron with one exception: it decays radioactively if free. However, its mean lifetime is long enough (2×10^{-6} s) to allow it to be used in many useful experiments, and we know how to make a moderately copious supply in the form of beams of momentum ~ 28 MeV/c and higher. So we can arrange to intercept the beam with a suitable material. What happens? The μ^- mesons enter the material, slow down by collisions with electrons and finally are attracted by a positive nuclear charge. Each μ^- then jumps down a series of energy levels emitting photons (or causing other atomic changes: Auger electron emission). By the time it has reached levels with principal quantum number (n) about 4 the μ^- orbit is well inside the Bohr radius of the lowest filled electronic levels and the photons emitted are in the X-ray region. We now have a situation which can be calculated: the μ^--nucleus system is almost completely isolated from the effects of the many electrons and the spectrum of photons emitted as the μ^- makes the last steps into the 1s level can be calculated sufficiently precisely. The 1s level can bring the μ^- close to the nucleus, in fact in lead the μ^- in that orbit is predominantly to be found inside the nucleus. Thus the nuclear charge distribution has a profound effect on the energy levels and the measurement of the X-ray transition energies can be used to measure the charge distributions. In the extreme case, the $1s$ level of the μ^- can disturb the nucleus by the action of the abnormal negative charge within it.

That is not quite the end of the story, for we should answer the question: what happens finally to the μ^-? It either suffers decay or is destroyed in a nuclear reaction with the nucleus. The decay is given in Table 9.6; the reaction is

$$\mu^- + p \rightarrow n + \nu_\mu$$

and is similar to electron capture (section 5.3) but sometimes more disruptive of the nucleus. This final process could be thought to affect the energy levels by the implication of a further perturbation apparently not present for electrons. That is true, but the perturbation is negligible as can be seen by considering the following argument. A μ^- in a $1s$ orbit in lead has a mean lifetime before nuclear absorption of about 10^{-8} s: in the semi-classical Bohr picture of well-defined orbits this time allows the traversal of about 10^{12} complete orbits. Thus the perturbation due to the nuclear absorbing interaction must be very small compared to the potential determining the orbit. We will find in Chapter 12 that this perturbation is also present for electrons, although undetectable in the spectroscopy of the atomic electrons, except in very unusual circumstances.

PROBLEMS

3.5 What is the mechanical potential energy of an electron at the centre of a gold nucleus ($Z=79$, $A=197$) given that the nucleus is a uniformly charged sphere of radius $R=1.2A^{1/3}$ fm.

3.6 Estimate the radius of the first Bohr orbit of a μ^--meson about

(a) a proton,

(b) a $^{12}_{6}C$ nucleus.

What is the principal quantum number of the Bohr orbit for a μ^- which lies just outside a lead nucleus ($Z=82$, $A=208$), if the nuclear radius is $1.2A^{1/3}$ fm?

Using Bohr theory, calculate the transition energy from the next higher orbit into this orbit.

3.7 Calculate two things for a μ^--meson in a $2s$ state about a $^{12}_{6}C$ nucleus:

(a) the binding energy, assuming a point-like nuclear charge distribution;

(b) the first-order perturbation of this energy using model I and assuming $a=1.2A^{1/3}$ fm.

3.8 Suggest how the calculation of Problem 3.7 can be improved.

Some information for these problems:

(1) 'Estimate' means get it right to one significant figure.

(2) The mass of the μ^--meson is 105.7 MeV/c^2.

(3) The wavefunction for the $2s$ state is

$$\psi(r) = \left(\frac{Z}{2r_0}\right)^{3/2}\left(2-\frac{r}{r_0}\right)e^{-Zr/2r_0},$$

where

$$r_0 = \frac{4\pi\varepsilon_0\hbar^2}{e^2 m}.$$

3.7 Nuclear scattering and nuclear size

As we indicated in Section 3.4 the study of electron scattering by a nucleus serves to determine the distributions of electric charge in that nucleus: the study of neutron scattering could serve to determine the distribution of the sources of the nuclear force. (For conciseness we have decided to call this the distribution of matter.) In fact these measurements need not be restricted to the use of neutrons: other particles can be employed, such as protons, α-particles ($^4_2$He), deuterons ($^2_1$H), tritons ($^3_1$H), helions ($^3_2$He), and even the long-lived elementary particles we shall meet later. However, the use of any but the simplest particles will add the uncertainty in the size of the incident particle to the meaning of the results obtained. These particles are charged so that scattering is caused by the combined effects of both the nuclear and the Coulomb forces. At the higher incident particle kinetic energies sensible for this work (> 10 MeV) the Coulomb effects are important for scattering at small angles: at other not-so-small angles its effect is a calculable correction to the dominant nuclear scattering.

In Fig. 3.6 we show the angular differential elastic scattering cross-section for 14 MeV neutrons by nickel. One immediately obvious feature is the diffraction peaks and valleys characteristic of scattering from an absorbing object with a moderately well-defined boundary.

The problem is now one of interpretation when our understanding of the nature of nuclear forces is slight in comparison to our understanding of the electromagnetic forces. In addition, even the lightest nuclear particles are known to be composite (see Chapter 10) and so have finite size. The simplest interpretation rests upon the **optical model** of the nucleus. This model describes the effect of the nucleus on the incident particle in terms of a potential well with an ability to absorb: given such a proposed potential the scattering can be predicted by solving the Schrödinger equation for unbound states of the incident particle. The optical analogy would be a semi-transparent sphere

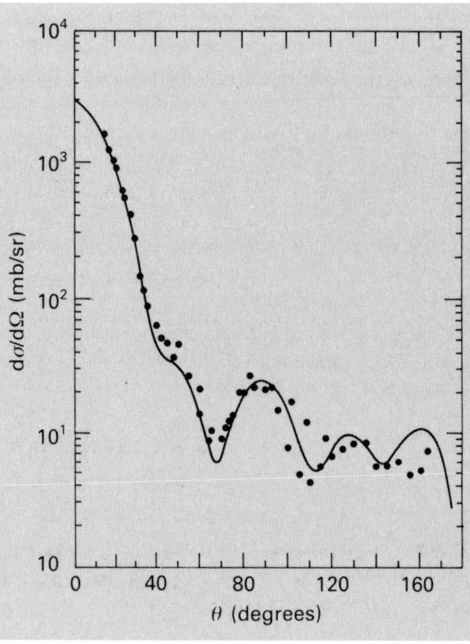

Fig. 3.6 The measured differential cross-section (solid points) against centre-of-mass angle for the elastic scattering of 14 MeV neutrons by nickel. The solid line is a prediction from a set of optical model parameters. The momentum transfer at the diffraction minima at 68° and 110° are about 179 MeV/c and 262 MeV/c ($q/\hbar = 0.91$ and $1.33\ \text{fm}^{-1}$) respectively. They do not coincide with the minima in the results for electron scattering (Fig. 3.4) because the absorption of the neutron wave by nuclear matter alters the effective diffracting aperture from that in electron scattering. (Figure adapted from Bauer *et al.*, 1963. The errors on the measured points have not been included. The relative errors vary from 5% at small angles to 15% at large angles. There is an overall uncertainty of 10% in the absolute cross-sections.)

scattering and absorbing light: this is why the model was sometimes called the *cloudy crystal ball* model.

In Table 3.3 we have given the terms of a typical potential which could apply to the scattering of neutrons and protons. The first detail we need to note is that it has an imaginary part which describes the absorption. The second is that we see that the Saxon–Woods form is used for the shape of the potential. Just as in the case of electron scattering the parameters which describe the potential can be varied to make the predicted scattering fit that observed. Those few words cover a multitude of efforts on the part of many people (see Hodgson (1971) for a chapter reviewing this work up to that date). The result is that the extraction of few encompassing measures is difficult, but roughly, for the Saxon-Woods parameters,

$$a = 1.2A^{1/3}\ \text{fm}$$

and

$$d = 0.75\ \text{fm}.$$

This value of a is the same as that obtained in electron scattering. The 'thickness' of the fall-off region is somewhat greater than found in the electron scattering ($d \simeq 0.55$ fm). The degree of agreement is surprising considering that the dimensions of the nuclear scattering potential depend not only on the matter distribution in the nucleus but also on the range of nuclear forces and on the size of the incident particle.

Table 3.3 An example of an optical model potential for neutron and proton scattering by nuclei.

$V(r) = V_c(r)$	The Coulomb potential for proton only.
$- Vf_1(r)$	The nuclear potential well.
$- iWf_2(r)$	The imaginary potential representing absorption of the incident nucleon.
$+ V_{LS}\left(\frac{\hbar}{m_\pi c}\right)^2 \frac{1}{r}\frac{df_{LS}(r)}{dr}\mathbf{L.S}$	The spin–orbit interaction.

Notes

(1) V, W and V_{LS} are expected to and do vary with energy.

(2) The factor $\left(\frac{\hbar}{m_\pi c}\right)^2$ has the dimension of area, is equal to $(2.00\ \text{fm})^2$, and cancels the two lengths in $\frac{1}{r}\frac{d}{dr}$. The mass m_π is the mass of the charged pion (see Section 9.10), so this factor maintains a sensible scale to the quantities involved in this term.

(3) The functions $f_1(r)$, $f_2(r)$, $f_{LS}(r)$ are usually taken to have the familiar Saxon–Woods form

$$f(r) = \frac{1}{\left(1 + \exp\left[\frac{r-a}{d}\right]\right)}$$

with $a = R_0A^{1/3}$. It is possible that R_0 and d will vary from term to term of the expression for the optical potential. The V_c will be formulated to allow for the distribution of the nuclear charge into a Saxon–Woods form.

DEFINITIONS AND KEYWORDS

The optical model This is a model of the nuclear charge and matter distribution and its effect on the scattering of light nuclear particles. The forces and the absorption are represented by a complex potential and problems are solved using the Schrödinger equation.

Cloudy crystal ball model A descriptive alternative name for the optical model.

Scale invariant If a cross-section or a rate has no dependence on a parameter that can have the dimension of length, it is said to be scale invariant. The Rutherford and Mott differential cross-sections are scale invariant.

3.8 Overview of size determinations

We have briefly described some of the more important methods of determining nuclear size and the reader should now have a picture of a spherical nucleus with approximately uniform charge and matter density except near a surface where both fall together to near zero in two to three fermis. A moment's reflection will show that this picture can only be described in this way for medium and heavy nuclei: for light nuclei ($A < 20$) the parameter a of the Saxon–Woods form will be comparable to or smaller than the thickness of the surface ($\sim 4d$) so that the nucleus is more surface than it is constant density nuclear matter.

We have not described any of the experimental methods employed to make the measurements which are the basis of the conclusions that we have discussed. Hofstadter (1956) has described the methods that he and his collaborators used in the first precision research performed using electron scattering. The principles have not changed since that time but there have been improvements in the technology. Recently the study of inelastic electron scattering over a wide range of energies and nuclei have been used to probe aspects of nuclear properties other than the straightforward charge distribution.

At this point there is an interesting point that we can make about the Rutherford and Mott formulae for the different angular scattering cross-section. The cross-section reflects the strengths of the charges on the colliding particles and the r^{-1} fall-off of the potential:

$$\frac{d\sigma}{d\Omega} = \frac{1}{4}\left(\frac{zZe^2}{4\pi\varepsilon_0 mv^2}\right)^2 \operatorname{cosec}^4(\theta/2).$$

There is nothing in this formula which depends on a length. The cross-sectional area comes, for its dimensions, from a factor ($\hbar c$/the incident kinetic energy)^2. Thus this formula describes the scattering of point-like objects. It is said to be **scale invariant**. When one or both of the colliding objects has size then, as we have found, the formula must be modified by a form factor, $F(q^2)$. This factor is itself dimensionless but is a function of a quantity, q^2, which has the dimensions ($\hbar$/length)^2; it must therefore contain a parameter which can be associated with a length. Therefore, including the form factor destroys the scale invariance.

This is our first brush with the concept of scale invariance: whenever it comes again we shall draw attention to it. Physicists in this business tend to contract the phrase and, instead of saying 'a cross-section is scale invariant', say 'a cross-section exhibits scaling' or 'a cross-section scales'. And whatever they say, it means that they are dealing with a process which shows the sign of involving interactions between point-like particles. Other examples are to be found in Fig. 10.10 and in Sections 9.5, 10.10, and 12.11.

3.9 The shape of nuclei

We hinted at the very beginning of this chapter that nuclei were not necessarily spherical, although we continued as if they were. In Fig. 3.7 the artist has taken liberties with the possibilities, and answered the shape question. There are many prolate nuclei. How does this show up? Firstly, since the electric charge distribution is no longer spherically symmetric the nucleus can have electric moments other than monopole. The first allowed is the electric quadrupole moment (Section 8.8). If this moment is non-zero, it is manifest by its effects on the optical spectroscopy of the atom. Higher even moments can exist but are very difficult to observe. Secondly, a non-spherical nucleus will have rotational states of motion which, by their regularity, are identifiable in the spectrum of excited states. We shall return briefly to that subject in Sections 4.5, 8.9 and 11.6.

The brevity of this section must not be taken to imply that non-spherical nuclei are unimportant. Understanding the why and how of their shape and the implications for nuclear physics has involved research, both theoretical and experimental, of great sophistication and difficulty.

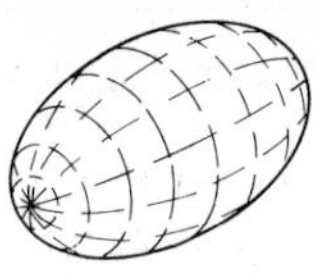

(a)

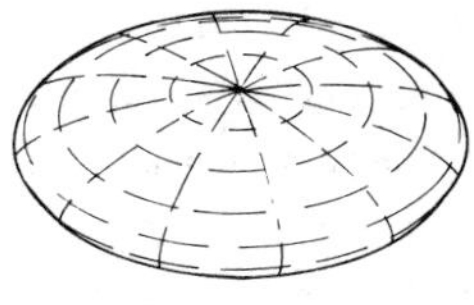

(b)

Fig. 3.7 The possible shapes of nuclei apart from spherical: (a) prolate like a rugby football; (b) oblate like a flattened sphere. Nuclei are normally spherical or prolate. A few may be slightly oblate.

References

Bauer, R. W., Anderson, J. D., and Christenson, L. J. (1963). *Nuclear Physics*, **48**, 152–8.

Dreher, B., Friedrich, J., Merle, K., Rothhaas, H., and Lührs, G. (1974). *Nuclear Physics*, **A235**, 219–48.

Hodgson, P. E. (1971). *Nuclear Reactions and Nuclear Structure*. Clarendon Press, Oxford.

Hofstadter, R. (1956). *Reviews of Modern Physics*, **28**, 214–54.

Hofstadter, R., and Collard, H. R. (1967). *Landolt Bornstein Tables*, New Series, Group I, Vol. 2, 21–54. Springer-Verlag, Berlin.

Sick, I, (1974). *Nuclear Physics*, **A218**, 509–541.

Sick, I., *et al* (1975). *Physical Review Letters*, **35**, 910–3.

Sick, I., McCarthy, J. S., and Witney, R. R. (1976). *Physics Letters*, **64B**, 33–5.

4

The Masses of Nuclei

4.1 The naturally occurring nuclei

Apart from three or four exceptions, the naturally occurring elements up to lead are stable and lie on or near a line (the **line of stability**) in the N, Z plane for which the neutron number N is equal to Z, $Z+1$, or $Z+2$ in nuclei up to $A=35$ (except for ^{1}H and ^{3}He, for which $N=Z-1$) but which thereafter increases faster than Z until in lead $N \simeq 1.5Z$. This is illustrated in Fig. 4.1 which shows N

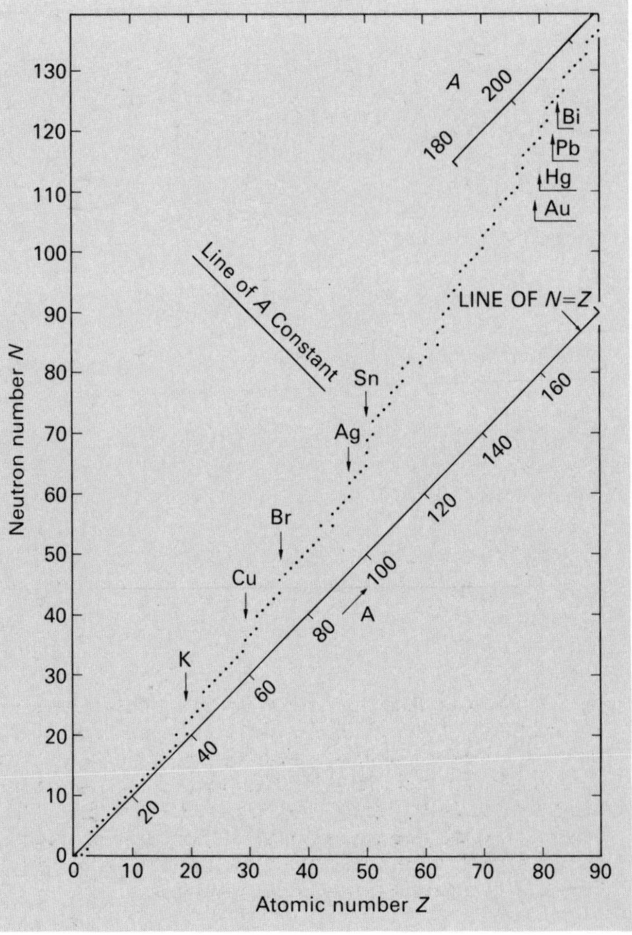

Fig. 4.1 The stable odd-A nuclei. Each dot represents a stable nucleus plotted with coordinates (Z,N). We shall show in Section 5.4 that there is only one stable isobar for each value of A for nuclei with odd A and therefore this plot is particularly simple. It demonstrates the increase in the number of neutrons relative to the number of protons as A increases. The same increase occurs for even-A nuclei but since many such A have two stable isobars (and three in a few cases) the figure would be very much more crowded with dots if these nuclei were to be included in the plot. These extra dots would crowd around the line of dots for the odd-A nuclei. The line which passes through the average position of the dots on the Z,N plane is called the **line of stability**.

against Z for the odd-A stable nuclei up to $A=209$. We would like to understand why the stable nuclei have this property and what happens if nuclei are produced in which N is greater or less than the stable optimum. We would also like to understand why it is that for $A>209$, there are no stable nuclei. In this chapter we take the first step in this programme; we consider nuclear mass and a nuclear model that gives a formula for mass with only a few parameters and which is surprisingly successful.

DEFINITIONS AND KEYWORDS

The line of stability This is the line on the Z,N plane which follows the average position for each A of the stable nuclei.

Nuclear binding energy This is the energy required to dismantle a nucleus into its constituent nucleons. It is the difference between the mass of all the separated nucleons and the mass of the assembled nucleus, multiplied by c^2.

The liquid drop model is a nuclear model which exploits the similarities between a nucleus and a drop of liquid.

The semi-empirical mass formula for nuclei is derived from the liquid drop model of the nucleus with some quantum-mechanical modifications.

4.2 The nuclear binding energy

We know that the atomic mass is approximately AM_H where M_H is the hydrogen atomic mass. Let us consider only the nuclear mass for the moment, which constitutes almost the entire atomic mass and which we represent by $M(Z,A)$. Such a nucleus, in principle, can be pulled apart into Z protons and $N=A-Z$ neutrons. If the nucleus is stable against break-up into these individual nucleons, its mass must be less than the mass of these constituents when separated by large distances, that is

$$M(Z,A)c^2 = ZM_pc^2 + NM_nc^2 - B.$$

The mass of proton and neutron are M_p and M_n respectively and B is the **nuclear**

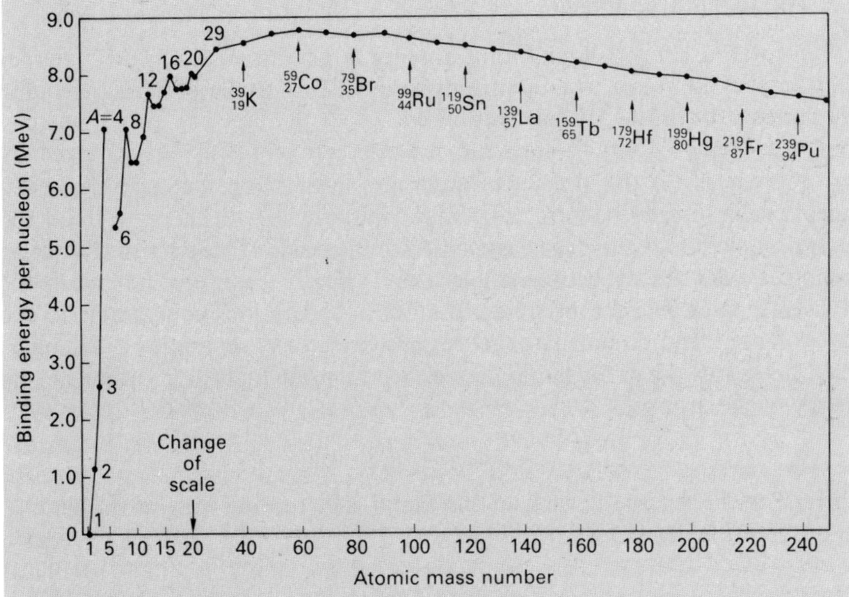

Fig. 4.2 The binding energy per nucleon as a function of A for all the stable nuclei up to $A=21$ and at $A=29$, 39, ... for the one stable nucleus at each of these mass numbers. Note the change of scale of the abscissa at $A=20$. The naturally radioactive nuclei are found at above $A=209$: for each of these heavy nuclei there is one that is stable against β decay or electron capture (Section 5.3) although it will decay by α-emission; it is the binding energy of this nucleus that is plotted. The line is drawn to guide the eye from point to point and must not be used for precise interpolation. (Source of the data: Wapstra and Audi (1985).)

binding energy: this is the energy required to dismantle the nucleus into Z protons and N neutrons. Note that we have used the Einstein equivalence of mass and energy and that, in future, we shall take mass (M) and rest mass energy (Mc^2) to be interchangeable without confusion. Figure 4.2 shows the nuclear binding energy per nucleon (B/A) as a function of A. We note some peaks at ${}^4_2$He, ${}^8_4$Be, ${}^{12}_6$C, ${}^{16}_8$O, and ${}^{20}_{10}$Ne but thereafter it is a moderately smooth function reaching a maximum of about 8.7 MeV at $A \simeq 60$ thereafter declining slowly to about 7.6 MeV at uranium. The detailed structure we will return to in a later chapter. The important fact is that over a large part of the periodic table the binding energy per nucleon is roughly constant.

The other fact which we uncovered in the last chapter is that the density of nuclear matter is approximately constant throughout most of the periodic table.

These two properties of nuclear matter are very similar to the properties of a drop of liquid, namely constant binding energy per molecule, apart from surface tension effects, and constant density for incompressible liquids. This analogy leads to a model of nuclei called the **liquid drop model** and to a formula for the mass based on it—namely the **semi-empirical mass formula**. We develop this theme in the next section.

We shall use the phrases 'more strongly bound' or 'less strongly bound' to mean greater or less binding energy in making comparisons. (There is usage of 'more stable' or 'less stable': since a system is either stable or not stable, this appears somewhat imprecise.)

4.3 The liquid drop model

A non-rotating drop of liquid in the absence of gravitational or other external fields adjusts its shape to minimize its energy. That shape is spherical and it minimizes the positive surface tension energy. If the liquid is incompressible, then the drop's density is constant, independent of radius R and given by $R \propto n^{1/3}$, where n is the number of molecules in the drop. Let each molecule, except one in or near the surface, be bound in the drop with energy a; this is the energy required to remove the molecule from the inside of the drop and is due to the forces that can exist between molecules. Typically these forces are negligible at large separations, can become attractive at separations comparable to the molecular size and become strongly respulsive at closer separations. Taking as zero the energy when all the molecules are separated by large distances, the energy of the drop is

$$-an + 4\pi R^2 T,$$

where T is the surface tension of the liquid. Or, turning this into a binding energy B of the drop as a function of n:

$$B = an - \beta n^{2/3},$$

where β contains all the constants of the surface term except the dependence on n. If the drop carries an electric charge Q, there is an extra term due to the mechanical potential energy of the charge distribution. If the charge is distributed uniformly in the surface, then that Coulomb energy is $Q^2/8\pi\varepsilon_0 R$. If it is distributed uniformly through the drop it is $3Q^2/20\pi\varepsilon_0 R$ (see Problem 4.1). This energy decreases the binding energy, which becomes

$$B = +an - \beta n^{2/3} - \gamma Q^2/n^{1/3},$$

where γ contains all the Coulomb effects except the dependence on Q and n.

We now examine the analogy with nuclei. We assume that

(1) the nucleus is spherical;
(2) the nucleons in the nucleus behave like the molecules in a drop—that is there is a short-range attractive force holding the nucleons together, and a shorter-range repulsive force which stops the nucleons collapsing into one another,
(3) the nuclear density is constant.

With these assumptions we can write down a formula for the nuclear binding energy $B(Z,A)$ by simple analogy, changing $n \rightarrow A$ and $Q \rightarrow Z$:

$$B(Z,A) = a_V A - a_S A^{2/3} - a_C Z^2/A^{1/3} ,$$

where a_V, a_S and a_C are now constants appropriate to the nuclear case; a_V for the **volume term**, a_S for the **surface term** and a_C for the **Coulomb term**.

This formula is inadequate: it predicts that, for fixed A, the nucleus with the $Z=0$ has the greatest binding energy. We shall find in studying β-decay that this process can change neutrons into protons or bound protons into neutrons, the change occurring in the direction which reduces the energy. If our formula were to be correct, a nucleus with $Z \neq 0$ would change so that $Z \rightarrow 0$. That has clearly not happened to the nuclei that are familiar to us. The effects we must add in extending the so-far too naïve model are quantum mechanical.

First we note an important assumption not so far stated: we have assumed that the nuclear binding of neutrons is identical to that for the protons once the effects of the Coulomb interaction have been moved to another term. Therefore imagine two potential wells each with an associated set of energy levels that are identical, one for the protons and one for the neutrons. These levels fill accordingly to the Pauli exclusion principle because both neutron and proton are fermions. If $Z=N$, then both wells are filled to the same level (the Fermi level). If we move one step away from that situation, say in the direction of $N>Z$, then one proton must be changed into a neutron (see Fig. 4.3). All other things being equal (including equal proton and neutron mass), this state has energy ΔE greater than the initial state, where ΔE is the level spacing at the Fermi level. A second step in the same direction causes the energy excess to become $2\Delta E$. A next step means moving a proton up three rungs as it changes from proton to neutron and the excess becomes $5\Delta E$. In fact, listing in units of ΔE, as each step (which changes proton to neutron) is made, we find the changes require energy in units of ΔE

$$1, 1, 3, 3, 5, 5, 7, \ldots$$

so that the cumulative effect is

$$\begin{array}{ll} & 1,\ 2,\ 5,\ 8,\ 13,\ 18,\ 25,\ 32, \ldots \quad \text{units of } \Delta E \\ \text{for} & N-Z=2,\ 4,\ 6,\ 8,\ 10,\ 12,\ 14,\ 16, \ldots \end{array}$$

Therefore to change from $N-Z=0$ to $N>Z$, with $A=N+Z$ held constant, requires an energy of $\sim(N-Z)^2\Delta E/8$. Of course, this is independent of whether it is N or Z that becomes larger and it means that, if all other things are equal, nuclei with $Z=N$ have less energy and are therefore more strongly bound than a

DEFINITIONS AND KEYWORDS

The five binding energy terms in the semi-empirical mass formula:

1 Volume term This term accounts for the binding energy of all the nucleons as if every one were entirely surrounded by other nucleons.

2 Surface term This term corrects the volume energy term for the fact that not all nucleons are surrounded by other nucleons but lie in or near the surface.

3 Coulomb term This term gives the contribution to the energy of the nucleus due to the mechanical potential energy of the nuclear charge.

4 Asymmetry term This term accounts for the fact that if all other factors were equal, the most strongly bound nucleus of a given A is that closest to having $Z=N$.

5 Pairing term This term accounts for the fact that a pair of like nucleons is more strongly bound than is a pair of unlike nucleons.

Separation energy The energy required to remove from a nucleus a named type of constituent, for example, neutron, proton, or α-particle.

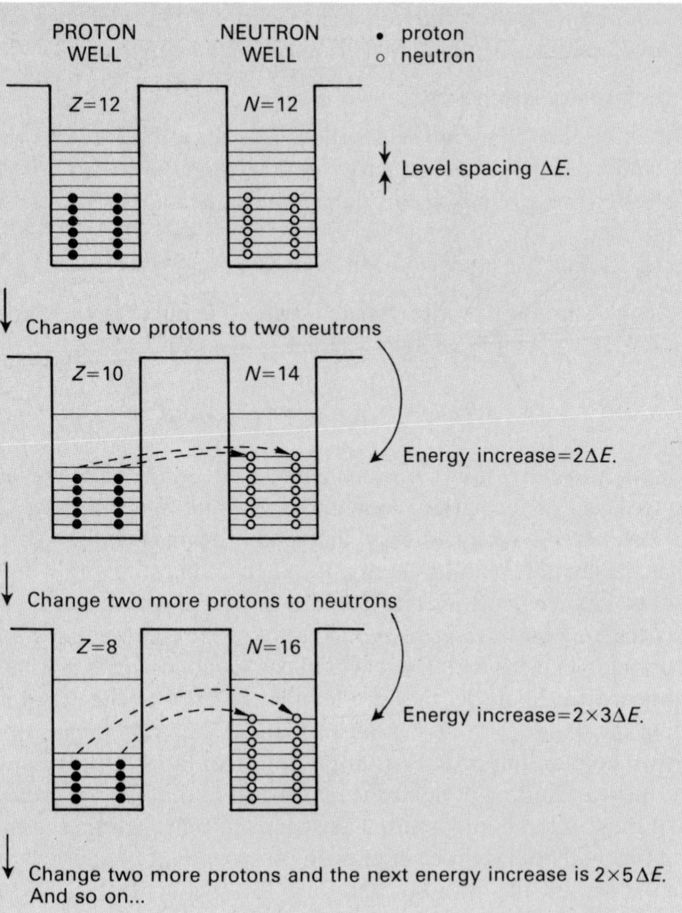

Fig. 4.3 The occupation of energy levels of a nucleus by protons (●) and by neutrons (○) according to the Pauli exclusion principle in a nucleus which is changing from $Z=N$ to $N>Z$, while $A=Z+N$ remains constant. The cost in energy of making the change of two protons into two neutrons and placing the latter in unoccupied neutron levels increases at every change. (The cost or gain in energy due to the neutron–proton mass difference is not included.)

nucleus with $Z\neq N$. Thus we must add a term which reduces the binding energy when $Z\neq N$. Since the energy levels of a particle in a potential well have a spacing inversely proportional to the well volume, we can put $\Delta E \propto A^{-1}$. Therefore we include a term which reduces the binding energy for nuclei for which $Z\neq N$. This is the **asymmetry term**:

$$-a_A(Z-N)^2/A$$

to be added to the binding energy formula.

One last term remains: it is called the **pairing term**. It reflects the fact that it is found experimentally that two protons or two neutrons are always more strongly bound than one proton and one neutron (see also Fig. 4.4). That is, like nucleons 'pair'. For odd A nuclei (Z even, N odd or Z odd, N even) this term is taken to be zero. For A even there are two cases;

(1) Z odd, N odd (oo),
(2) Z even, Z even (ee).

The binding energy will be greater for case 2 than for 1 so we add into the binding energy formula a quantity $\delta(Z, A)$ for case 2 and subtract it for case 1. Bohr and Mottleson (1969) show that the form

$$\delta(Z,A) = a_P/A^{1/2},\ a_P = 12\ \text{MeV},$$

NOTATION

Z even, *N* even nuclei are called even–even (ee)
Z odd, *N* odd nuclei are called odd–odd (oo)
Z even, *N* odd nuclei are called even–odd (eo) and, of course, vice-versa.
Even-*A* nuclei are ee or oo. Odd-*A* nuclei are eo or oe.

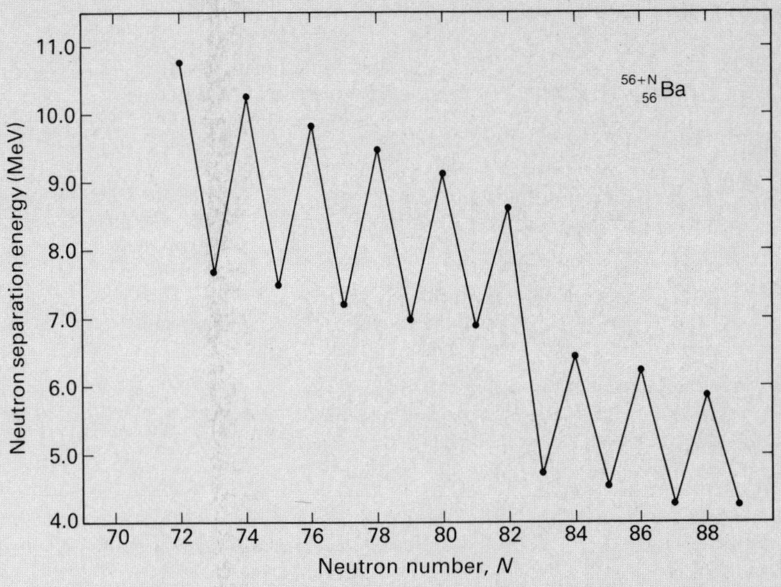

Fig. 4.4 The energy required to remove the last neutron (the **neutron separation energy**) from the isotopes of barium ($_{56}$Ba). This energy is about 2 MeV greater when N is even than when N is odd. This 2 MeV represents the energy required to break the pairing favoured for like nucleons. The irregularity at $N = 82$ is one of the effects that we leave for discussion in Chapter 8. (Source of data: Wapstra and Audi 1985.)

fits the rather scattered data with a precision better than 1 MeV for all $A > 20$.

Before putting the formula for binding energy together we note that there is one refinement that is sometimes made. The charge on the nucleus is carried in discrete units, one on each proton. The charge on the proton does not interact with itself (or if the proton constituents do interact, that energy is already included in the proton mass). It is therefore sensible to replace Z^2 appropriate to a continuous charge distribution by $Z(Z-1)$ which is appropriate to this discreteness of the nuclear charges. However, we do not do that: the reason is that the apparently best set of coefficients a_V, etc. has been determined using the formula with Z^2. In addition, the final precision of the formula is probably not sufficient to allow improvements at this level to be discerned.

So putting all our terms together, we have

$$B = a_V A - a_S A^{2/3} - a_C Z^2/A^{1/3} - a_A (A-2Z)^2/A + \delta(Z,A).$$

The values of the coefficients have to be found by fitting to the binding energy data for medium and heavy nuclei. The light nuclei ($A < 20$) are not included as there is no smooth curve of binding energy against A or Z due to the effects of shell closures (see Chapter 8). The fit is not perfect because these effects persist throughout the periodic table and because some nuclei are not spherical. Attempts have been made to improve the formula which we mention in Section 4.6.

We have written the whole formula for the nuclear mass in Table 4.1; but as rest mass energy—hence the c^2 attached to the real masses. Also shown is a favoured set of values for the coefficients.

Table 4.1 The nuclear semi-empirical mass formula summarized.

$M(Z,A)c^2$	Nuclear rest mass energy
$= ZM_pc^2 + (A-Z)M_nc^2 - B(Z,A)$	Rest of mass of constituents less the binding energy

where

$B(Z,A) =$	$+a_V A$	Volume binding term
	$-a_S A^{2/3}$	Surface energy term
	$-a_C Z^2/A^{1/3}$	Coulomb term
	$-a_A(A-2Z)^2/A$	Asymmetry term
	$\begin{cases} -a_P/A^{1/2} & \text{oo nuclei} \\ +0 & \text{eo and oe nuclei} \\ +a_P/A^{1/2} & \text{ee nuclei} \end{cases}$	Pairing term

M_pc^2 = rest mass energy of the proton = 938.280 MeV.
M_nc^2 = rest mass energy of the neutron = 939.573 MeV.
A favoured set of values for the coefficients:

$a_V = 15.56$ MeV,
$a_S = 17.23$ MeV,
$a_C = 0.697$ MeV,
$a_A = 23.285$ MeV,
$a_P = 12.0$ MeV.

To obtain the atomic rest mass energy, change M_P, the proton mass, to M_H, the mass of the hydrogen atom, thus:

$\mathcal{M}(Z,A)c^2$ = atomic rest mass energy
$\approx ZM_Hc^2 + (A-Z)M_nc^2 - B(Z,A)$.

(Note: $\approx$ because this formula neglects some atomic electron binding energy.)
M_Hc^2 = rest mass energy of the hydrogen atom = 938.791 MeV.

Figure 4.5 shows how the various contributions (except the pairing term) change with A throughout the periodic table. What is surprising is that this formula is good from $A \simeq 20$ to the end of the periodic table with a precision better than $1\frac{1}{2}\%$ on the binding energy. This is shown in the case of the odd A nuclei in Fig. 4.6.

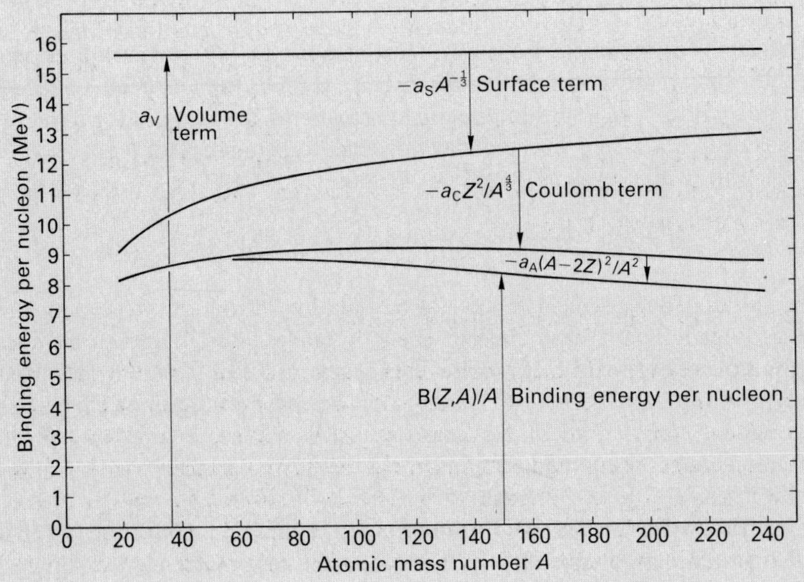

Fig. 4.5 This figure shows the various contributions, except the pairing term, of the semi-empirical mass formula to the binding energy per nucleon as a function of A.

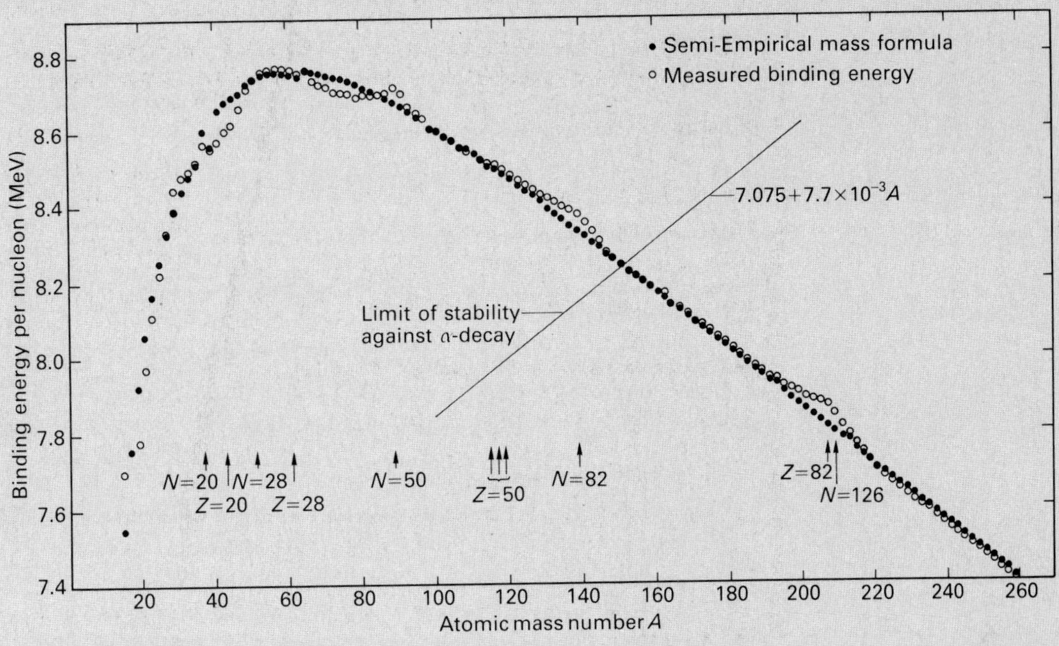

Fig. 4.6 The binding energy as a function of A for the odd-A nuclei from $A = 15$–259. The solid points are the prediction of the semi-empirical mass formula as given in Table 4.1. The open points are the measured values. The points for the formula do not lie on a smooth curve because Z for these nuclei is not a smooth function of A (see Fig. 4.1). Note that the zero of the ordinate is suppressed and its scale is much enlarged. Thus, in spite of the deviations from the formula, it is clear that the formula predicts the binding energy per nucleon for $A > 20$ with a precision which is, for the majority of cases, better than 0.1 MeV. The straight line crossing the curve at $A = 151$ gives the limit of stability of nuclei to α-decay (see Section 5.4).

We will discuss this success shortly.

Readers should beware of the possibility of one source of confusion. Many books and some examination questions use the atomic mass instead of the nuclear mass. We represent the former by $\mathscr{M}(Z,A)$ and continue to use $M(Z,A)$ for the nuclear mass. In Table 4.1 we show the relation between the two. We notice that using the mass of the hydrogen atom instead of the mass of the proton automatically includes the mass of Z electrons but not the atomic binding energy, apart for an incorrect and negligible $13.6Z$ eV. When we consider the energetics of β-decay we will see that the atomic mass has some advantages (see Section 5.4).

4.4 The Coulomb and asymmetry terms

Although we have said the a coefficients are determined from the experimental values of the binding energies, attempts have been made to determine these coefficients from a more fundamental approach to nuclear structure and forces. We shall not pursue these attempts here but there is one coefficient which can be estimated easily—the Coulomb coefficient a_C. We invite the reader to take up this challenge in Problems 4.1 and 4.2. There is a harder challenge in Problem 4.3: that of attempting to estimate the coefficient a_A of the asymmetry term. Do not be surprised if your result is too small.

PROBLEMS

4.1 A simple problem in electrostatics! Show that the potential energy due to electrostatic forces of a uniformly charged sphere of total charge Q and radius R is $3Q^2/(20\pi\varepsilon_0 R)$.

4.2 The Coulomb term in the semi-empirical mass formula is

$$a_C Z^2/A^{1/3}.$$

Using the result of Problem 4.1, calculate the value of a_C in MeV/c^2. Assume that the nuclear radius is given $R=1.24\times A^{1/3}$ fm.

Using the values of a_V, a_S, and a_A given in Table 4.2 and the fact that the binding energy of $^{181}_{73}$Ta is 1454 MeV, check your value of a_C. Comment on any discrepancy you may find.

The nucleus $^{235}_{92}$U can undergo spontaneous fission (see Chapter 6): one of the many fission channels is

$$^{235}_{92}\mathrm{U} \rightarrow {}^{87}_{35}\mathrm{Br} + {}^{145}_{57}\mathrm{La} + 3\mathrm{n}.$$

Estimate the energy released in this channel.

(Adapted from the 1979 examination in the Final Honours School in Natural Science, Physics, University of Oxford.)

4.3 The Fermi-gas model of the nucleus assumes that the nucleus is a sphere of volume $V=4\pi R^3/3$ (take $R=R_0\times A^{1/3}$ fm with $R_0=1.2$ fm) and solves the Schrödinger equation to find the number of states available up to a momentum P; that number is $4\pi P^3 V/3(2\pi\hbar)^3$. Show that for a nucleus with equal numbers of protons and neutrons the energy E_F of the Fermi level is given by

$$E_F = (\hbar^2/2MR_0)\ (9\pi/8)^{2/3},$$

where M is the mass of a nucleon. Estimate the total kinetic energy of the nucleons in a nucleus of $^{16}_{8}$O.

For a nucleus Z,N the asymmetry term is $a_A(N-Z)^2/A$. Assuming $|N-Z| \ll A$, use the model to estimate a_A.

(Adapted from the 1978 examination in the Final Honours School in Natural Science, Physics, University of Oxford.)

4.5 The implications of the semi-empirical mass formula

The fact that the mass formula works so well means that the assumptions built into it may possess some measure of truth. The most important fact is that provided by the volume binding term $a_V A$. If every nucleon interacted with every other in the nucleus that term would have to vary as $A(A-1)$, that is proportional to the number of nucleon pairings (analogously to the Coulomb term if put as proportional to $Z(Z-1)$). Putting the volume term as $a_V A$ means assuming that each nucleon interacts only with its nearest neighbours and that the constant density is equivalent to a separation from nearest neighbours that does not change with A. All this means that nuclear forces saturate. Forces that exhibit **saturation** have the property that, for a system that is held together by such a force, the total binding energy is proportional to the total mass or to the number of constituents, apart from surface effects. Familiar examples are solids and liquids: these have attractive, short-range (see Section 3.1) forces (Van der Waals') between the constituent molecules, but at small molecular separations the forces become strongly repulsive, preventing collapse of all the constituents into a small volume defined by the range of the attractive force. Examples of non-saturating forces are provided by the Coulomb and by the gravitational

DEFINITIONS AND KEYWORDS

Saturation A property of some forces such that a multibody system held together by such a force has a total binding energy (apart from surface effects) that is proportional to the mass or to the number of constituents.

Vibrational states Excited states of nuclei due to a collective motion of the constituents leading to an oscillating shape.

Rotational states Excited states of nuclei due to rotation of a deformed nucleus.

interactions: the Coulomb force has an effect that grows as $Z(Z-1)$, not as Z, in nuclei and the contribution of gravity to the binding energy of a massive body is proportional to the mass squared. In the absence of other forces, or if the mass becomes sufficient, it is believed that gravity will cause the collapse of a massive body into a black hole. Returning to the saturating force between nucleons we see that it must, as a consequence, be short range. The nuclear density of nucleons is approximately 1 in every 7 cubic fermis so that the average separation is about 1.9 fm. Thus the range must be 1–2 fm. Of course there must be something to prevent more and more nucleons crowding into this range, that is, to cause saturation: the effect of the Pauli exclusion principle is not sufficient. Although in the early attempts to understand nuclear forces it was suspected that there was a short-range repulsion, the required strength was considered excessive and more acceptable solutions were found. If the force between two nucleons is assumed to depend not only on their separation but also on their relative orbital angular momentum, on their total spin, and in the case of the unlike neutron and proton on the symmetry of their relative wave function, then it is possible to explain saturation and to quantitatively explain the binding energy of some light nuclei. However, when results became available from a study of nucleon–nucleon scattering at energies in the range 100–400 MeV, it was found that there is a strong repulsive force between nucleons at separations of less than about 0.5 fm, in addition to the properties proposed. Thus all the ingredients for saturation exist.

An example of the nuclear forces beginning to show properties relevant to saturation is provided by the $A=5$ system: it has no bound states; see Table 4.2. The helium nucleus with $Z=2$ and $A=4$ is strongly bound: a fifth nucleon will not stick since it has to go into a p-state of motion relative to the first four and this keeps it too much out of the range of the forces and it experiences a weaker attraction. A fifth and a sixth do bind if they are neutron and proton, to form ${}^6_3\text{Li}$ (or two neutrons to form ${}^6_2\text{He}$); however, it requires only 3.70 (0.97) MeV to remove both, whereas it requires 20.6 (19.8) MeV to remove one neutron

Table 4.2 The nuclear force in light nuclei.

Nucleus	A	Z,N	Binding energy (MeV)	Binding energy/A (MeV)
${}^2_1\text{H}$	2	1,1	2.22	1.11
${}^3_1\text{H}$	3	1,2	8.48	2.83
${}^4_1\text{H}$	4	1,3	unbound ${}^3\text{H}+\text{n}$	
${}^3_2\text{He}$	3	2,1	7.72	2.57
${}^4_2\text{He}$	4	2,2	28.30	7.07
${}^5_2\text{He}$	5	2,3	unbound ${}^4\text{He}+\text{n}$	
${}^6_2\text{He}$	6	2,4	29.27	4.88
${}^5_3\text{Li}$	5	3,2	unbound ${}^4\text{He}+\text{p}$	
${}^6_3\text{Li}$	6	3,3	32.00	5.33
${}^7_3\text{Li}$	7	3,4	39.25	5.61
${}^7_4\text{Be}$	7	4,3	37.60	5.37
${}^8_4\text{Be}$	8	4,4	56.50	7.06

Look at the jump in binding energy at ${}^4_2\text{He}$: the next nucleon can only be put in a state of motion relative to the first four in which it experiences a force too weak to bind it.

(proton) from a helium nucleus. Thus we observe the effect of forces changing with a change in the relative motion.

All our statements about saturation are dangerously qualitative and a proper assessment of the role of these possible causes requires a theoretical and quantitative approach beyond the scope of this book.

By its very nature, the liquid drop model cannot provide any information on some very important static nuclear properties except in rather vague terms. Many nuclei have spin (that is total angular momentum): the model cannot say much about this property except that the presence of the pairing and the asymmetry terms suggest that the nucleon spins pair up and that the net effect will be a small total spin. It follows that we expect the ground state spin of all even–even nuclei to be zero: this is found to be the case. The assumption of spherical nuclei means zero electric quadrupole moment, whereas we know many nuclei have such a moment (Section 8.8). This implies a permanent deformation (non-spherical nuclei) but there is nothing in the model which would cause such a deformation.

The model does accommodate the existence of excited states corresponding to the modes of vibration that a liquid drop may have. The conceptually simplest of such states correspond to standing waves on the surface of the nucleus and are called **vibrational states**. These are one kind of collective motion that nuclei may have, the collective implying that the motion of the constituent nucleons is correlated in some way to give the shape vibration. The non-spherical nuclei that exist can have **rotational states** (see Section 8.9) which are another form of collective motion. In addition, vibrational or rotational bands of levels based on the ground state of a nucleus can be repeated on intrinsically excited states. These are states that appear as a consequence of a change in the quantum-mechanical motion of the nucleons within the nucleus but which are far from being as obviously collective as are the vibrational and rotational levels. The liquid drop model cannot say much about the existence of such intrinsic excited states of nuclei.

The collective motions of nuclei must be very different from those of a drop of liquid. In the latter case the molecular motions keep each molecule localized to a relatively small part of the drop but in nuclei the uncertainty principle does not permit localization of a nucleon to the same extent relative to the size of a nucleus.

Only the asymmetry and pairing terms of this application of the liquid drop model acknowledge the existence of quantization. It is this absence of a proper quantum-mechanical approach that limits further useful developments of the model. However, these words should not detract from the value of the model and of the semi-empirical mass formula: this will be exemplified particularly in the next chapter where we examine in turn the role of β-decay in determining the line of stability and the role of α-decay and of spontaneous fission in the stability of heavier nuclei.

In Chapter 8 we shall examine the shell model of the nucleus. That model is so different from the liquid drop model that it is a great surprise that both should be attempting to describe the same system. We shall examine the similarities and differences in Chapter 8 and attempt to justify their simultaneous success.

4.6 Conclusions

The semi-empirical mass formula has been extended by the inclusion of many subtle nuclear effects. The motivation for this work is to be able to calculate the masses of nuclei far from the line of stability. For any given A there are two points beyond which the nucleus becomes unstable against the emission of a nucleon. The locus of these points on the high Z side of the line of stability is called the **proton drip line** and the locus of points on the high N side is called the **neutron drip line**. The latter is particularly important in the *r*-process of stellar nucleosynthesis (see Section 14.5) for which a full understanding requires a good knowledge of the masses of nuclei just inside the neutron drip line. These nuclei are not readily accessible experimentally and it is necessary to rely on the theoretical approach based on the semi-empirical mass formula.

DEFINITIONS AND KEYWORDS

Proton drip line The line on the Z, N plane where the proton separation energy is zero.

Neutron drip line The line on the Z, N plane where the neutron separation energy is zero.

References

Bohr, A. and Mottleson, B. R. (1969) *Nuclear Structure*, Vol. I, 169–71. W. A. Benjamin, New York.

Wapstra, A. H. and Audi, G. (1985). The 1985 determination of atomic masses. *Nuclear Physics*, **A432**, 1–54.

5

Nuclear Instability

DEFINITIONS AND KEYWORDS

α-decay The process in which a nucleus can spontaneously emit a nucleus of ^{4}He (α-particle).

β-decay The process in which a nucleus can spontaneously emit an electron or a position (β^- or β^+ respectively). An anti-neutrino or a neutrino is emitted simultaneously.

γ-decay The process in which a nucleus in an excited state spontaneously emits a photon (γ-ray).

5.1 Nuclear decay

In Sections 2.2–6 we described the radioactive decay law and some of its implications for nuclear and particle decay. In this chapter we return specifically to nuclear decay and look at those processes which were investigated early in the history of radioactivity, namely **α-decay** (the emission of the nuclei of helium), **β-decay** (the emission of electrons or positrons) or **γ-decay** (the emission of photons). The objective is to investigate the kinematics of these decays and, in particular, to elucidate the role of α- and of β-decay in determining which nuclei are stable and which are not. However, there are other decay processes which we will mention but which are not important in determining stability of the naturally occurring nuclei.

5.2 Energy-level diagrams

The reader will have a clear picture from atomic physics of energy-level diagrams and the possibility of radiative transitions between them. We have already drawn such a diagram in Fig. 2.2, where the vertical distance between two levels is the energy difference and the level X having the higher energy, is drawn above the one, Y, having the lower energy. A transition from X to Y normally involves the emission of a photon, the energy difference going into photon energy and energy of the recoil atom. This idea can be taken over immediately in nuclear physics: transitions involving γ-rays from nuclei (usually energies of about 10 keV up to 3 or 4 MeV) take place when the nucleus makes a transition from an excited state to a state of lower energy. We shall discover later examples of how a nucleus can be found in such an excited state. The integers Z and A are unchanged in photon emission and we normally represent this by placing all levels in the same vertical column. The vertical scale is usually marked in MeV and, of course, Einstein tells us that it is equivalent to a mass scale with the nuclear mass increasing with increasing energy of excitation. Figure 5.1 shows the example of the low-lying levels of $^{24}_{12}$Mg.

Let us now consider α-decay of a nucleus:

$$(Z,A) \rightarrow (Z-2, A-4) + {}^4_2\text{He}.$$

Normally an α-particle of unique energy is emitted. How to represent this? The complication is now that the daughter nucleus is different from the parent by

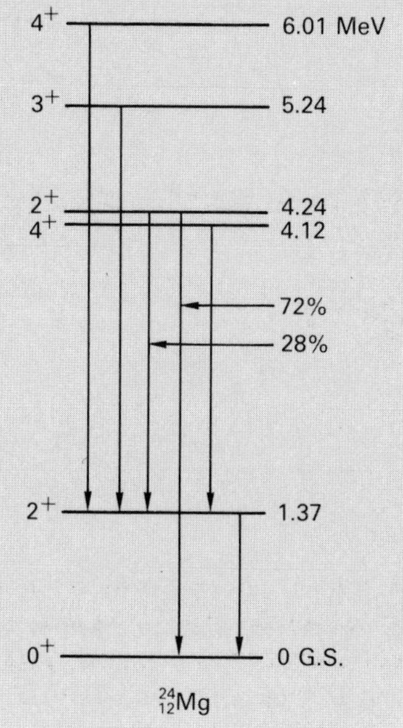

Fig. 5.1 The energy-level diagram which shows the ground and excited states up to 6.01 MeV of $^{24}_{12}$Mg. The levels are labelled by their spin-parity (e.g. $j^P = 2^+$) and energy (1.37 MeV). Directed arrows show the γ-ray emitting transitions. The 4.24 MeV level decays with branching fractions 28% to the 1.37 MeV level and 72% to the ground state.

$\Delta Z = -2$, $\Delta A = -4$. In Fig. 5.2(a) we show the basis of the style we shall use. Level X is the parent nucleus (Z,A), level Y is the daughter nucleus $(Z-2,A-4)$. The horizontal displacement emphasizes that Z and A have changed. The energy scale is such that the vertical position of X is $M(Z,A)c^2$ and that of Y is $M(Z-2,A-4)c^2 + M(2,4)c^2$, the quantity $M(2,4)c^2$ being the rest mass energy of an α-particle. The energy scale is normally marked in megaelectronvolts with zero at Y. The vertical height of X above Y is the energy difference and is called the Q_α-value:

$$Q_\alpha = [M(Z,A) - M(Z-2,A-4) - M(2,4)]c^2. \tag{5.1}$$

This available energy goes into the kinetic energies of the α-particle and of the recoil of the daughter nucleus (in this simple case). If $Q_\alpha > 0$, α-decay is energetically possible; however, it may not occur for other reasons but the transition passes that test of energy.

One of the facts of α-decay is that the process can involve the emission of α-particles of more than one energy; that is, in spectroscopic terminology, the α-particle energy spectrum is a line spectrum. This is illustrated by the case of $^{242}_{94}$Pu shown in Fig. 5.2(b). It is a particular example of an α-decay scheme and, by its nature, is one in which a daughter nucleus can be left in an excited state.

Energy-level diagrams involving β-decay are more complicated and to deal with them we require more facts about this process and these merit a dedicated section. We shall return to α-decay and its role in nuclear stability in Section 5.4.

5.3 More on β-decay

The process by which a nucleus emits an electron when we know nuclei do not contain electrons may appear at first sight difficult to understand. However, we

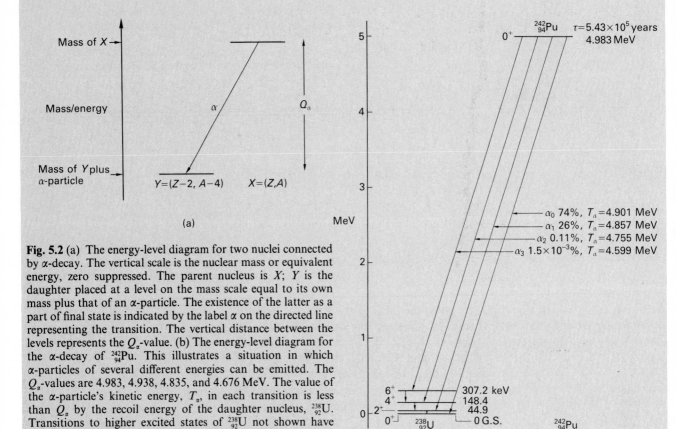

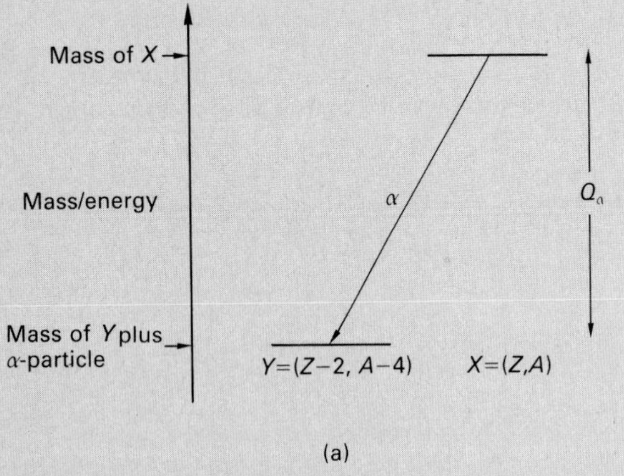

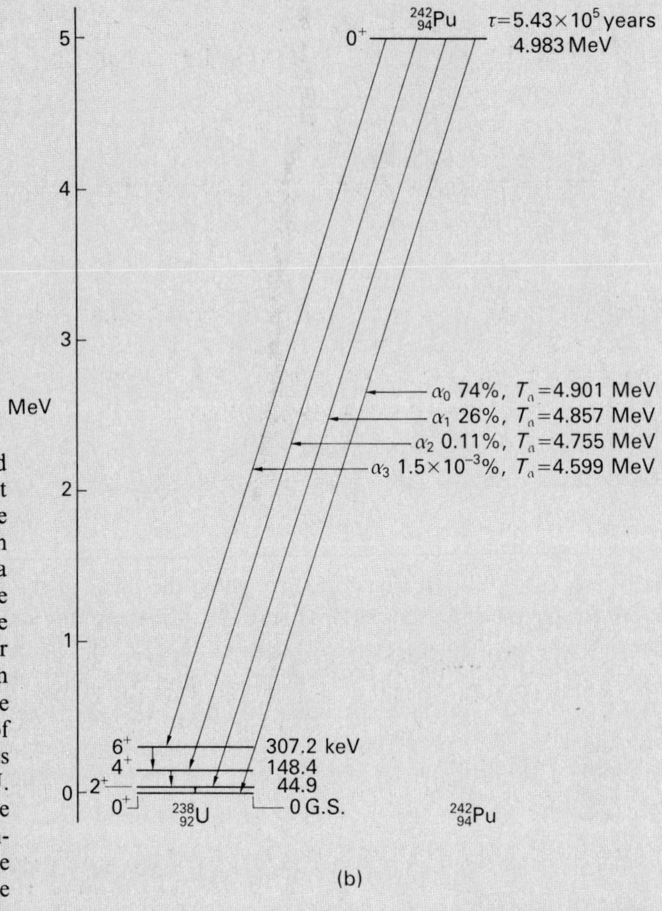

Fig. 5.2 (a) The energy-level diagram for two nuclei connected by α-decay. The vertical scale is the nuclear mass or equivalent energy, zero suppressed. The parent nucleus is X; Y is the daughter placed at a level on the mass scale equal to its own mass plus that of an α-particle. The existence of the latter as a part of final state is indicated by the label α on the directed line representing the transition. The vertical distance between the levels represents the Q_α-value. (b) The energy-level diagram for the α-decay of $^{242}_{94}$Pu. This illustrates a situation in which α-particles of several different energies can be emitted. The Q_α-values are 4.983, 4.938, 4.835, and 4.676 MeV. The value of the α-particle's kinetic energy, T_α, in each transition is less than Q_α by the recoil energy of the daughter nucleus, $^{238}_{92}$U. Transitions to higher excited states of $^{238}_{92}$U not shown have transition probabilities too small to be significant. The strongest line is α_0. If decay occurs by one of the less probable transitions, then the daughter nucleus is left in an excited state and decays by γ-ray emission. The vertical scale is in MeV.

are used to atoms emitting photons: if we think of this as the act of creating a photon, then there may be no reason why we cannot create a material (that is having rest mass) particle such as an electron. In fact the creation of massive particles from energy is possible given that certain conservation laws are satisfied. What those laws are will become increasingly clear. For the moment we will keep in mind that particle creation is possible.

The apparent process in β-decay is the conversion of nucleus (Z,A) into nucleus $(Z+1,A)$ and an electron (e^-):

$$(Z,A) \rightarrow (Z+1,A) + e^-.$$

In fact the better interpretation is the conversion of a bound neutron (n) into a bound proton (p):

$$n \rightarrow p + e^-.$$

(We shall find an even better interpretation in terms of quarks later.) However, what about proton-rich nuclei? It turns out that they can frequently undergo β-decay in which a positron (e^+, Section 1.6) is emitted:

$$(Z,A) \rightarrow (Z-1,A) + e^+,$$

which is a bound proton changing to a bound neutron:

$$p \rightarrow n + e^+.$$

So we now have two processes (**β^--decay** and **β^+-decay**) which, if energy allows, can respectively increase or decrease Z at constant A.

Now there is another process which can occur and has the same effect as β^+-decay. A proton-rich nucleus can capture an atomic electron and thereby change a proton into a neutron:

$$e^- + (Z,A) \rightarrow (Z-1,A), \qquad (5.2)$$

or

$$e^- + p \rightarrow n.$$

This is called **electron capture**: the electron is captured usually from the K-shell (X-ray nomenclature) but can be captured from the L, M, N, or an even higher shell. Here the reverse of creation occurs: the electron is annihilated!

Here the reader may suspect that something is missing: in electron capture (equation (5.2)) how is energy to be conserved? The parent and daughter nuclei (the latter might be in an excited state) have well-defined masses, as does the electron. Unless these add up to conserve energy the process cannot occur, yet it does. This is connected with an early mystery in β-decay: this was the continuous spectrum of emitted electron kinetic energies (see Fig. 5.3). If we consider a branching decay chain such as that of Fig. 5.4 then it is found that energy changes in the two branches would be the same if in both β-decay branches the electrons were always emitted with the maximum kinetic energy observed, which is not the case. A second mystery is best exemplified by the decay:

$$^{14}_{6}C \rightarrow {}^{14}_{7}N + e^-.$$

The nucleus $^{14}_{6}C$ is known to have spin 0 and $^{14}_{7}N$ to have spin 1 (see Problem 5.2). The electron has spin $\frac{1}{2}$. There is no way angular momentum can be conserved in the decay as written. And angular momentum conservation cannot be discarded lightly.

It was Pauli in 1930 who made the hypothesis that provided a solution to these difficulties and that has satisfied all experimental tests: he proposed that an electrically neutral particle of spin $\frac{1}{2}$ is created and emitted at the same time as the electron (or positron) in β-decay. The particle is called a neutrino (symbol, ν) and it can take a share of the energy because in β-decay there is now a three-

DEFINITIONS AND KEYWORDS

The neutrino hypothesis (Pauli, 1930)

There exists a neutral, spin $\frac{1}{2}$ particle which is emitted simultaneously with the electron in β-decay. This hypothesis resolved the following difficulties in understanding:

(1) the apparent non-conservation of energy in β-decay;

(2) the apparent non-conservation of angular momentum in certain β-decays.

This hypothesis has satisfied all tests, including the direct detection of neutrinos. The last was not achieved until 1956. We shall discover in Chapter 12 why it is difficult to detect the neutrinos emitted in β-decay.

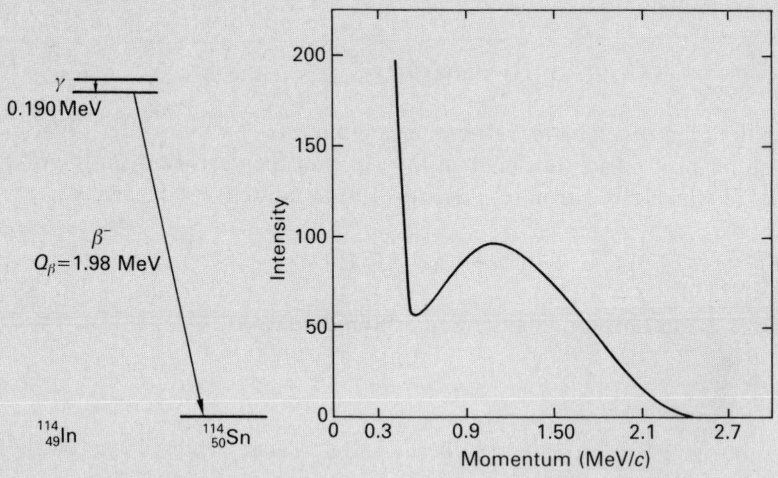

Fig. 5.3 The momentum spectrum of electrons emitted in the decay of an excited state of $^{114}_{49}$In. The absorption of slow neutrons by $^{113}_{49}$In leads to the production of an excited state of $^{114}_{49}$In which decays by γ-ray emission to the ground state with a mean life of 71 days. This ground state decays by β-decay with a mean life of 104 s. This figure shows the energy-level diagram and the momentum spectrum of electrons emitted by a sample of this excited state. From 0.50 MeV/c to the end point at 2.44 MeV/c the spectrum is continuous and due to the electrons emitted in the ground state decay. Below 0.50 MeV/c the continuous spectrum has superimposed on it a line spectrum of electrons emitted when the excited state of the $^{114}_{49}$In nucleus de-excites by directly ejecting an electron from the atom instead of emitting a γ-ray: this is the process of **internal conversion** (see Section 11.8). A single transition energy (190 keV in this case) will give rise to several lines corresponding to ejection from the K,L.M, . . . shells. In this case the measurements have not been continued into and below the most energetic internal conversion line near which is close to 190 keV. Such internal conversion lines made the early interpretation of β-decay spectra a confusing task; with better measurements it became clear that all β-decay spectra are continuous up to a maximum momentum, which was different in each case, but were similar in shape. (Data from Lawson and Cork, 1940.)

body configuration of the final state. Electron capture is now two body ($e^- + p$) to two-body ($\nu + n$) and energy and momentum can be conserved. The neutrino spin of $\frac{1}{2}$ allows angular momentum conservation where previously some appeared to be missing. There is, however, much more to the neutrino story: there are three kinds each with its antiparticle so that, until the more complete description can be given (Sections 12.2 and 12.3), we have to label the neutrinos in nuclear β-decay in what may appear to be a somewhat arbitrary fashion:

β^--decay: $$(Z,A) \rightarrow (Z+1,A) + e^- + \bar{\nu}_e, \tag{5.3}$$

where $\bar{\nu}_e$ is an electron-type **antineutrino**.

β^+-decay: $$(Z,A) \rightarrow (Z-1,A) + e^+ + \nu_e, \tag{5.4}$$

where ν_e is an electron-type **neutrino**.

Electron capture: $$e^- + (Z,A) \rightarrow (Z-1,A) + \nu_e. \tag{5.5}$$

DEFINITIONS AND KEYWORDS

Electron capture The process in which a nucleus captures an electron from an atomic shell and emits a neutrino.

Neutrino An electrically neutral particle with spin $\frac{1}{2}$ emitted simultaneously with the positron in β^+-decay.

Antineutrino An electrically neutral particle with spin $\frac{1}{2}$ emitted simultaneously with the electron in β^--decay.

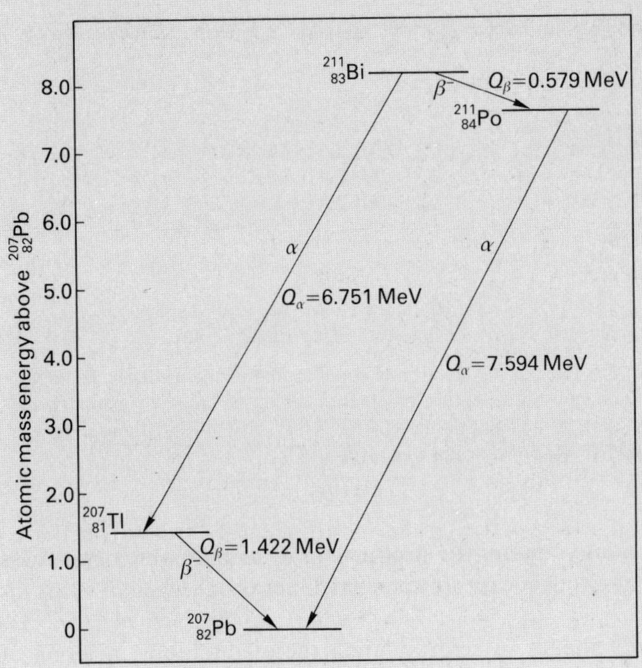

Fig. 5.4 A bimodal decay of $^{211}_{83}$Bi in which the daughters decay to the same final nucleus $^{207}_{82}$Pb. Consistency with energy conservation requires that the energy release in the β-decays is the maximum electron kinetic energy observed. However, most of the electrons emitted do not have that energy.

In order to consider the energetics of β-decay, we need to know the mass of the neutrino emitted in β-decay. It is known to be less than 18 eV and since this is small compared to the total energy released in most β-decays we shall assume that the mass is zero. In fact the neutrino mass is important and a non-zero but small mass would be of considerable significance in cosmology and in theories of elementary particles (see Sections 14.12 and 13.8).

We wish to note that the free neutron undergoes β-decay:

$$n \to p + e^- + \bar{\nu}_e + 0.782\ \text{MeV}, \qquad \tau = 898\ \text{s}. \tag{5.6}$$

The Q_β-value, 0.782 MeV, is the energy available to distribute as kinetic energy among the decay products. What about free proton β-decay which might be $p \to n + e^+ + \nu_e$? This cannot occur because neutron decay tells us that the proton is lighter than the neutron. This is fortunate for us as the stability of protons (on the time scale of $\gg 10^{14}$ years) is essential to the existence of the universe and of ourselves. However, we must check up on another potentially universe-extinguishing reaction: electron capture in a hydrogen atom:

$$e^- + p \to n + \nu_e.$$

Again we are safe as the products are heavier than the hydrogen atom by 782 keV, but it is a very small margin on which our existence depends. Again, course, the safety of the proton against electron capture follows immediately from the existence of free neutron decay, equation (5.6). However, it is interesting to point out these facts explicitly.

In Table 5.1 we have written down the nuclear mass–energy relations which must be satisfied if β-decay or electron capture is to occur. If a condition is satisfied, then the appropriate decay is possible and the excess energy available

Table 5.1 Energy conditions in β-decay and electron capture in terms of nuclear masses, $M(Z,A)$:

Decay

β^-: $(Z,A) \Rightarrow (Z+1,A) + e^- + \bar{\nu}_e$,	$Q_\beta = (M(Z,A) - M(Z+1,A) - m_e)c^2 \quad > 0.$
β^+: $(Z,A) \Rightarrow (Z-1,A) + e^+ + \nu_e$,	$Q_\beta = (M(Z,A) - M(Z-1,A) - m_e)c^2 \quad > 0.$
EC: $(Z,A) + e^- \Rightarrow (Z-1,A) + \nu_e$,	$Q_{EC} = (M(Z,A) + m_e - M(Z-1,A))c^2 \quad > 0.$

In terms of atomic masses ($\mathcal{M}(Z,A)$) these conditions become:

β^-:	$Q_\beta = (\mathcal{M}(Z,A) - \mathcal{M}(Z+1,A))c^2 \quad > 0.$
β^+:	$Q_\beta = (\mathcal{M}(Z,A) - \mathcal{M}(Z-1,A) - 2m_e)c^2 > 0.$
EC:	$Q_{EC} = (\mathcal{M}(Z,A) - \mathcal{M}(Z-1,A))c^2 \quad > 0.$

[We have assumed that the mass of the electron is equal to that of the positron. There is a fundamental theorem in relativistic quantum mechanics which is based on sound basic principles that requires particle and antiparticle to have the same mass. There is neither experimental nor theoretical reason to doubt this result.]

is shared as kinetic energy among the products in a manner which conserves momentum. Note that electron capture can sometimes occur when β^+-decay is impossible.

There are some advantages to reformulating these conditions in terms of atomic masses $\mathcal{M}(Z,A)$ instead of nuclear masses. Substituting

$$\mathcal{M}(Z,A) = M(Z,A) + Zm_e,$$

and likewise for $Z-1$ or $Z+1$, the conditions may be found: they are also given in Table 5.1. The first thing to note is that these conditions neglect the differences in atomic electron binding energies. The second thing to note is the particular simplicity of the conditions for β^--decay and electron capture expressed in atomic masses. We represent these conditions in the energy-level diagrams of Fig. 5.5. Thus in Fig. 5.5(a) we show the basic energy-level diagram

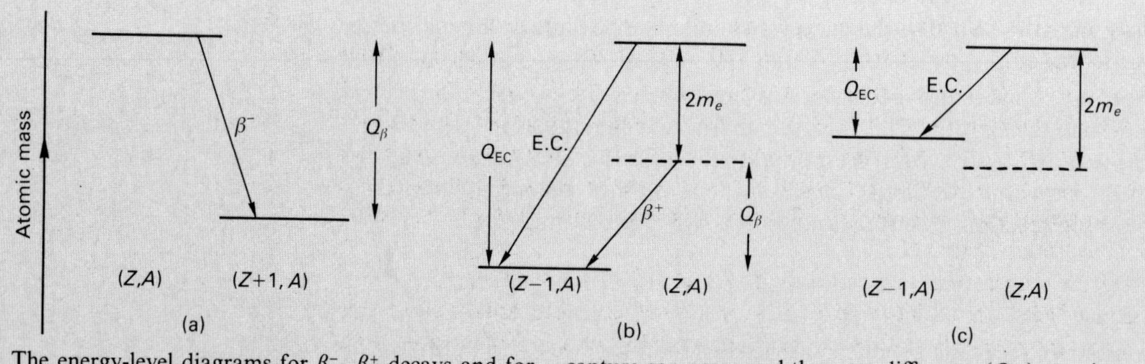

Fig. 5.5 The energy-level diagrams for β^-- β^+-decays and for electron capture (E.C.). Here it is most convenient to use a vertical scale which gives the atomic masses of the levels involved. In (a) the parent level has to be above the level of the daughter for β-decay to be possible, the level difference, Q_β, being the energy available to share among the products as kinetic energy, which, neglecting nuclear recoil, will be the maximum kinetic energy the electron can have. In (b) electron capture can occur and the mass difference (Q_{EC}) goes into total energy of the neutrino and recoil of the daughter atom (branch labelled E.C.). For β^+-decay to occur the mass difference must be greater than $2m_e$; what is left is available for kinetic energy (Q_β). This situation is represented by the right-hand branch of (b), labelled β^+. For an atomic mass difference less than $2m_e$, β^+-decay is impossible and only electron capture can occur, as shown in (c).

for the atomic masses involved in β^--decay. If the level representing the daughter atom is at a point on the vertical scale giving the atomic mass which is below that for the atomic mass of the parent atom, then the decay is energetically allowed. In Fig. 5(b) we show the basic energy-level diagram for the atomic masses involved in β^+-decay. The atomic masses have to differ by more than twice the electron mass and this can be represented by putting a virtual level $2m_e$ below the parent. If this virtual level is above the mass of the daughter atom, then β^+-decay is energetically possible; in this circumstance electron capture (EC) is also possible and competes with β^+-decay to decrease Z by one. In Fig. 5.5(c) we show a circumstance in which only electron capture is possible: the parent atom (Z,A) has a mass greater than the daughter atomic mass but by less than $2m_e$.

We shall use the conditions we have described in the next section. However, a warning to the reader: the interpretation of β-decay energy-level diagrams depends on whether the masses are those of the atoms or of the nuclei.

5.4 The stability of nuclei

We want in this section to consider the role of β-decay and of α-decay in determining the stability, or otherwise, of nuclei in the wider context of the whole range, and beyond, of values of Z and of A of the naturally occurring nuclei.

The stable nuclei lie on or near a curve of N against Z as shown in Fig. 4.1. What happens to a nucleus not in this region of stability? The answer is, of course, β-decay, which, keeping A constant, can step Z to bring the nucleus onto a stable position. We must therefore be interested in the nuclear mass of isobars (fixed A) as a function of Z. In Fig. 5.6 we show the nuclear masses of the isobars of $A=101$: these masses have been calculated using the semi-empirical mass formula and a smooth curve of no physical significance has been drawn through the points. Since $N>Z$ for most nuclei, an increase in Z decreases the asymmetry contribution to the mass. However, the increase in Z increases the Coulomb energy contribution. On the left of the minimum in the mass the decrease wins over the increase and the atomic mass decreases with Z. On the right the increase in Coulomb energy wins and the mass increases.

In the particular case of Fig. 5.6, on the left limb as Z changes from $41\rightarrow42\rightarrow43\rightarrow44$ the atomic mass decreases at every step so that β^--decay can occur at each. The mass changes on the right as Z changes from $47\rightarrow46\rightarrow45$ are also greater than $2m_ec^2$, so β^+-decay or electron capture can occur at each step. The steps $44\rightarrow45$ or $45\rightarrow44$ for the real nuclei cannot occur by β-decay but $45\rightarrow44$ can go by electron capture. The result is that $Z=44$ (ruthenium) is the only stable isobar of $A=101$.

This result is quite general for any odd-A nucleus: if, in the change $Z\rightarrow Z+1$, there is insufficient mass decrease for β^--decay, then electron capture can occur in $Z+1$ and Z is the only stable nucleus. Thus the prediction: odd-A nuclei have only one stable isobar. This prediction is borne out by fact with one qualification: many nuclei at the heavy end of the periodic table are unstable against α-decay so the stability implied here is not absolute but is stability against β-decay or electron capture.

The even-A nuclei are different. Those with Z even have a binding energy advantage arising from the pairing term, whereas the odd-Z nuclei have a lower binding energy due to the opposite contribution from this term. Thus there are

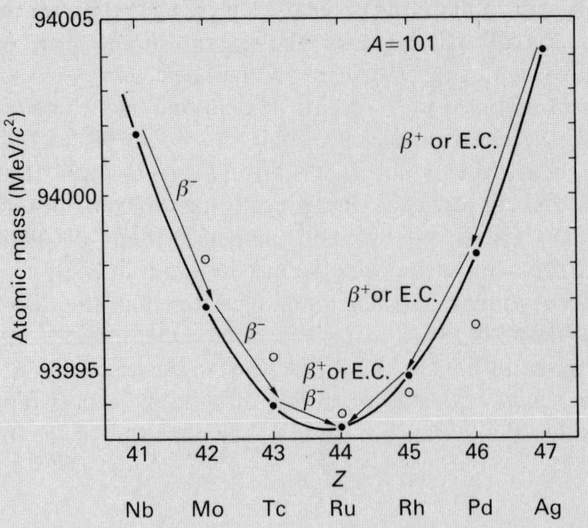

Fig. 5.6 The atomic mass of the isobars of $A = 101$ as a function of Z in the region of the line of stability. The solid points are calculated using the semi-empirical mass formula (Table 4.2); the line drawn through the points has no physical significance. The energy changes in β-decay given in Table 5.1 permit the transitions indicated so that the lowest atomic mass is thereby expected to be the only stable isobar, ruthenium in this case. This is the situation in all odd-A nuclei and the conclusion is that there is only one stable isobar for odd-A nuclei. The actual atomic masses are given by open points: the conclusion is the same. However, it is clear that even the relatively small errors in the result of the semi-empirical mass formula may not permit, in all cases, a prediction of which Z has the lowest atomic mass at the bottom of a shallow curve. For the real nuclei the transition from $A = 45$ to $A = 44$ can occur only by electron capture.

two curves of isobar atomic mass against Z and alternate Z lie on different curves. See Fig. 5.7 for $A = 100$. On the left and right extremities of the curve, β^-- or β^+-decay take the nuclei towards the bottom of the two curves. There the stability problem is more complex: nucleus $Z = 43$ can decay by electron capture to $Z = 42$ or by β^--decay to $Z = 44$. The prediction is that there are now two stable isobars for $A = 100$, namely $Z = 42$ and 44, which is true.

This situation in even-A nuclei sometimes leaves an odd–odd isobar energetically able to decay by all modes, β^-- or β^+-decay or electron capture; Fig. 5.8 shows an example, that of $^{40}_{19}K$.

An examination of this even-A situation suggests two conclusions:

1 there are no stable odd–odd nuclei;

2 many even–even nuclei can have more than one stable isobar.

Conclusion 2 is true. For example: of the isobars of $A = 96$, the nuclei $^{96}_{44}Ru$. $^{96}_{42}Mo$, $^{96}_{40}Zr$ are all stable. Conclusion 1 is almost true; there are four real exceptions among the light nuclei, namely $^{2}_{1}H$, $^{6}_{3}Li$, $^{10}_{5}B$ and $^{14}_{7}N$. In the case of $^{2}_{1}H$, it is the only bound system with $A = 2$. At the other A, the masses are changing rapidly with Z and the two parabolae of $A = 100$ shown in Fig. 5.7 become, at these low Z, narrower, and have steeper sides: they are also distorted by shell model effects (Chapter 8). The result is that each of these values of A, the nuclei one place away in Z from each of these odd–odd nuclei are heavier and there

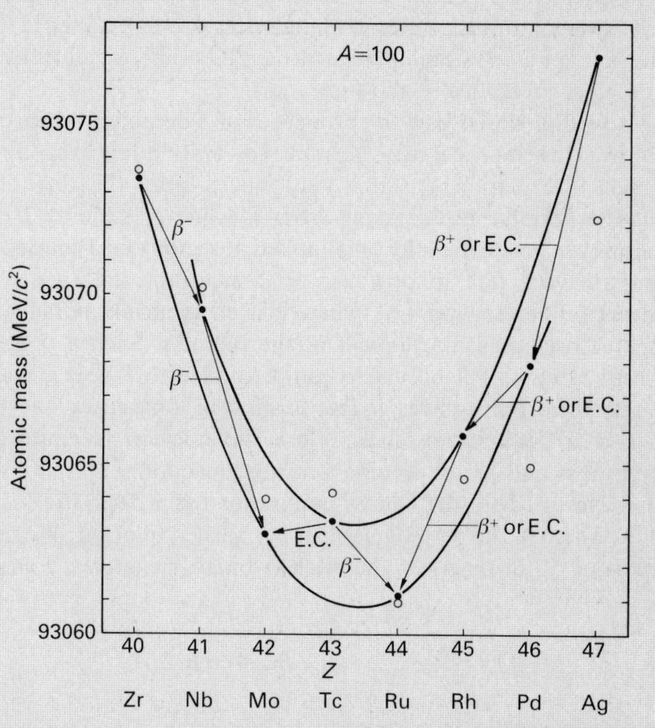

Fig. 5.7 The atomic mass of the isobars of $A = 100$ as a function of Z in the region of the line of stability. The solid points are calculated using the semi-empirical mass formula (Table 4.2). The pairing term contributes an opposite amount to the even- and odd-Z masses with the result that alternate mass points lie on different parabolae. The energy changes in β-decay given in Table 5.1 predict that the transitions indicated will occur and that molybdenum and ruthenium will be stable. As in Fig. 5.6 the actual masses are indicated by the open points. The conclusions are not changed in this case. The general conclusion is that even-A nuclei can have two or more stable isobars.

Fig. 5.8 The energy levels involved in the decay of the oo nucleus $^{40}_{19}$K (potassium). This is an example of a nuclide with a trimodal decay: β^+, β^-, and electron capture.

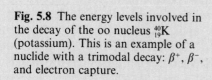

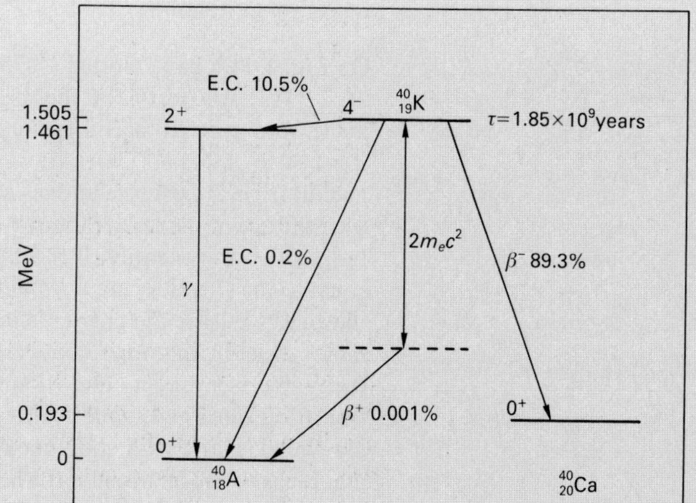

can be no decay to these neighbours. There are two heavier examples, $^{50}_{23}V$ and $^{180}_{73}Ta$ which occur naturally (isotopic abundances 0.25% and 0.0123% respectively), but their apparent stability is due to unobservably small transition rates although decay is energetically allowed.

We have now discovered how the properties of β-decay and electron capture taken with the semi-empirical mass formula can (with a few exceptions among the light nuclei) explain the stability of nuclei near the line of stability throughout the periodic table. Away from the line of stability β-decay is a certainty: however, we have really only looked at regions fairly close to the line. Nuclei that are very rich in protons, or in neutrons, may be beyond the appropriate drip line (Section 4.6), where it is energetically possible to emit a proton (or neutron) as a direct relief of the richness. Such a process occurs relatively very rapidly, if it occurs in competition with β-decay, and so these nuclei have very short mean lives, in fact so short in some cases that the nucleus may not have a distinct existence before the nucleon is emitted and the producing process and decay become a nuclear reaction.

We now have to apply the energy conditions for α-decay to occur to real nuclei to find where in the periodic table it is expected to occur. We first rewrite the definition of Q_α in terms of the nuclear binding energies. Equation (5.1) becomes

$$Q_\alpha = B(Z-2, A-4) + B(2,4) - B(Z,A). \tag{5.7}$$

Thus α-decay is energetically allowed if

$$\begin{aligned} B(2,4) &> B(Z,A) - B(Z-2,A-4) \\ &\approx 4\frac{\mathrm{d}B}{\mathrm{d}A} = 4\left\{A\frac{\mathrm{d}(B/A)}{\mathrm{d}A} + \frac{B}{A}\right\}. \end{aligned} \tag{5.8}$$

It is B/A which is plotted in Fig. 4.5. Above $A \simeq 120$, $\mathrm{d}(B/A)/\mathrm{d}A$ is about -7.7×10^{-3} MeV. Now $B(2,4)$, the helium nuclear binding energy, is 28.3 MeV, so we are looking for the critical A at which

$$28.3 = 4\,\{B/A - 7.7 \times 10^{-3}A\}. \tag{5.9}$$

On Fig. 4.6 we have plotted $7.075 + 7.7 \times 10^{-3}A$ and it crosses the curve of B/A at $A = 151$. Above this A the inequality of equation (5.8) is satisfied by most nuclei and α-decay becomes, in principle, energetically possible. In fact from $A = 144$ to $A = 206$, 7 α-emitters are known amongst the naturally occurring nuclides. Their existence implies mean lifetimes comparable to or greater than at least the age of the earth (about 4×10^9 years). When α-emitters are found in this range of A, the energies of the emitted α-particle (which in each case is almost equal to the Q_α value) are normally less than 3 MeV. We shall find that the lower the energy release the greater is the lifetime: it is therefore certain that although most nuclei in this range on the line of stability may be energetically able to decay by α-emission, they do not do so at a detectable level because the transition rate is too small. Above $Z = 82$ many naturally occurring α-emitters are found, many with short mean lives. Why are they to be found when their lifetime is so short? Most of the heavy nuclei to be found on earth were probably produced in one or more supernova explosions of early massive stars (see Section 14.5). Such explosions can produce very heavy nuclei including **trans-**

ranic elements and their subsequent decay by α-emission will take them down he periodic table in steps of $\Delta A = -4$. Each α-decay increases the ratio N/Z ıntil a β^--decay intervenes to restore the nucleus closer to the line of stability against β-decay alone in this region). This descent is delayed significantly if any ıucleus encountered has a mean life comparable to the present age of the naterial. Such nuclei are $^{235}_{92}$U (mean life $= 1.0 \times 10^9$ years), $^{238}_{92}$U (6.5×10^9 y) and $^{232}_{90}$Th (2.0×10^{10} y). The delays for these nuclei mean that significant quantities of each have survived on Earth, sufficient to permit discovery by man. Their slow decay feeds a **secular equilibrium** of fast-decaying daughter products down o a stable lead isotope. Note that three bottle-necks occur for A (modulo 4) equal to 3,2, and 0. The three naturally occurring radioactive series with these nodulo values end on $^{207}_{82}$Pb, $^{206}_{82}$Pb, and $^{208}_{82}$Pb respectively. The members of the series with A (modulo 4) equal to 1 have been produced artificially; there are no bottle-necks of sufficient long mean life to ensure that these nuclei will have survived naturally, except as the end of the series, $^{209}_{83}$Bi. Returning to equations (5.7)–(5.9) and looking at the straight line on Fig. 4.6, we see that as A increases, the inequality of equation (5.8) increases. This means that we expect that the Q_α-

DEFINITIONS AND KEYWORDS

Spontaneous fission The spontaneous splitting of a nucleus into two equal, or near equal parts.

Transuranic elements Elements having atomic numbers greater than that of uranium ($Z = 92$).

Secular equilibrium Consider a decay chain $A \to B \to C \to D \to \cdots$ where A has a mean life very large compared with those of B, C, D, ... In an undisturbed sample of A there are present amounts of B, C, D, etc. determined by the equilibrium between the production by the slow decay of A and the loss by fast decay of these products. This equilibrium is long term (secular) but not permanent since the amount of A present is slowly declining. If N_A, N_B, N_C, ... are the numbers of A, B, C, ... present and ω_A, ω_B, ω_C, ... are the decay transition rates, then the secular equilibrium is given by $\omega_A N_A = \omega_B N_B = \omega_C N_C = \cdots$.

Table 5.2 The decay steps for the naturally occurring radioactive series with A(modulo 4)=2. Of particular interest are the α-decay Q_α-values and the mean lives. There is clearly a strong correlation between increasing Q_α and decreasing mean life.

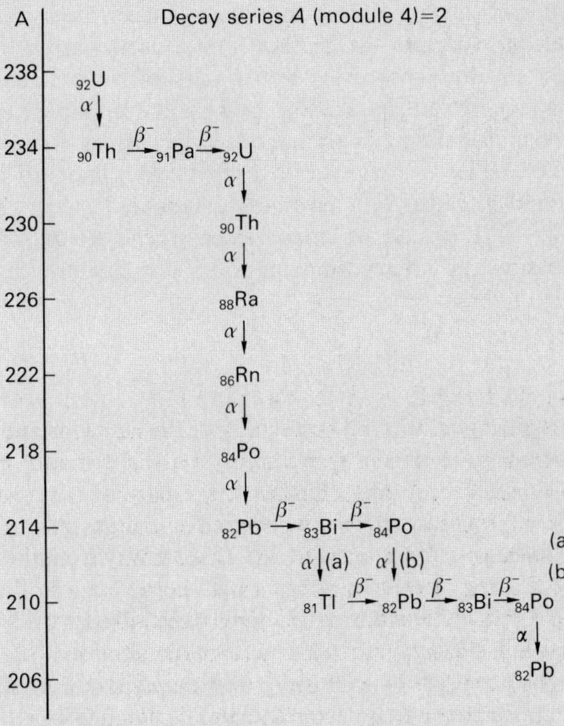

Decay	Q_α (MeV)	Mean life†
$^{238}_{92}$U $\xrightarrow{\alpha}$ $^{234}_{90}$Th	4.27	6.45×10^9y
$^{234}_{92}$U $\xrightarrow{\alpha}$ $^{230}_{90}$Th	4.86	3.53×10^5y
$^{230}_{90}$Th $\xrightarrow{\alpha}$ $^{226}_{88}$Ra	4.77	1.12×10^5y
$^{226}_{88}$Ra $\xrightarrow{\alpha}$ $^{222}_{86}$Rn	4.87	2.31×10^3y
$^{222}_{86}$Rn $\xrightarrow{\alpha}$ $^{218}_{84}$Po	5.59	5.51 d
$^{218}_{84}$Po $\xrightarrow{\alpha}$ $^{214}_{82}$Pb	6.11	4.40 m
(a) $^{214}_{83}$Bi $\xrightarrow{\alpha}$ $^{210}_{81}$Tl	(a) 5.62	94 d
(b) $^{214}_{84}$Po $\xrightarrow{\alpha}$ $^{210}_{82}$Pb	(b) 7.83	2.37×10^{-4}s
$^{210}_{84}$Po $\xrightarrow{\alpha}$ $^{206}_{82}$Pb	5.41	200 d

The mean life given for $^{214}_{83}$Bi$\Rightarrow ^{210}_{81}$Tl$+\alpha$ is the reciprocal of the transition rate for this particular decay. The decay of this nuclide is bimodal so that its actual mean life is less than this value due to the effect of the competition from the β^--decay mode to $^{214}_{84}$Po.

† y=years, d=days, m=minutes, s=seconds.

value of decays increase also with A. However, binding energies, and hence Q_α-values, do not vary smoothly with A because of shell model effects (Chapter 8). These fluctuations are of the order of 1 MeV and represent large fractional fluctuations in the values of Q_α for α-decays. We shall discover in Chapter 6 that the α-decay rate is very sensitive to the value of Q_α with the result that the lifetimes can fluctuate with changing A by many orders of magnitude. This is illustrated in Table 5.2 where the Q_α-values and lifetimes are given for the principal α-decays in the naturally occurring series from $^{238}_{92}U$ to $^{206}_{82}Pb$. The mean lives span 2.4×10^{-4} seconds to 6×10^9 years.

It is a shell-model effect that stops these series abruptly. Lead (Pb) has $Z = 82$, $^{209}_{93}Bi$ has $N = 126$: both these numbers are 'magic' and, as we shall find, that means a greater binding energy than expected from the semi-empirical mass formula. That in turn means that, although the Q_α-value for the next α-decay is positive, it is small, and the transition rate is so small that no significant decay has been detected or has occurred in the lifetime of the earth.

5.5 Spontaneous fission

In α-decay four nuclear constituents are lost from the parent nucleus. It is the large binding energy per nucleon of the α-particle which makes this possible, 7.08 MeV. However, the nucleons in $^{12}_{6}C$ are even more strongly bound (7.6 MeV per nucleon) than are those in $^{4}_{2}He$, and as a result decay by $^{12}_{6}C$ nucleus emission is energetically possible in some heavy nuclei. Decay by $^{14}_{6}C$ emission has been found but it is very rare. Other similar decays are actively being sought. However, all these processes are a part of a spectrum of possibilities in which a heavy nucleus breaks into two (or more) parts; at one end of the spectrum the result is one small part and one large part, at the other it is two almost equal parts. The last process is called **spontaneous fission**. Decay by α-particle emission could be called very asymmetric fission!

An examination of Fig. 4.2, the curve of binding energy per nucleon, shows that it is energetically possible for a nucleus having $A > 100$ to fission into two equal parts. For example the fission

$$^{238}_{92}U \rightarrow {}^{145}_{57}La + {}^{90}_{35}Br + 3n$$

releases about 156 MeV, which goes into kinetic energy of the neutrons and the fission products. Heavy nuclei are rich in neutrons so that their fission must produce very neutron-rich fission products. In fact the supply of neutrons is great enough that three or four are normally emitted simultaneously with fission. And, of course, the fission products are normally some way from the line of stability and reach that line by several steps of β^--decay. In the above example $^{145}_{57}La$ reaches $^{145}_{60}Nd$ in three steps with summed Q_β of 8.5 MeV. The other product has a special property to be described in Section 7.13, of sometimes emitting a neutron after one β^--decay: some details are given in Fig. 7.15. The result is that $^{90}_{35}Br$ becomes $^{89}_{39}Y$ with a sum of Q_β equal to 14.4 MeV or $^{90}_{40}Br$ with the sum of Q_β of 23.6 MeV. Some of all of this β-decay energy is carried away by neutrinos, leaving about 15 MeV available as kinetic energy of electrons. Another property of fission is that the production of equal or near equal mass fission products is unlikely and somewhat asymmetric fission is the usual outcome. The most probable mass number difference is about 45.

Although spontaneous fission is expected to become more probable as A increases, it is still a rare process even in uranium. For example the α-decay of $^{238}_{92}U$ has a transition rate of about 5×10^{-18} s^{-1}; the transition rate for spontaneous fission is about 3×10^{-24} s^{-1}, a branching fraction of approximately 6×10^{-7}. For heavier elements the spontaneous fission transition rate and branching fraction increase with increasing A and it may become the dominant decay mode at A greater than about 260.

It is clear that spontaneous fission does not play a significant role in determining the stability or abundance of the naturally occurring elements. It will have played a role in depleting some of the heavier transuranic elements produced at the time of heavy element formation (see Section 14.5).

5.6 Tricks with transition rates

Throughout this chapter we have been meeting nuclear decay processes with transition rates which vary considerably. We have promised the reader that we shall investigate the factors that control these rates and in the next chapter we shall make good that promise for the case of α-decay; the case of β-decay must wait until Chapter 12. However, the reader will have noticed that α-decay transition rates increase with the energy released and that the same is true for β-decay. Let us now see how this last fact, together with the mass properties of nuclei, can be used.

Radioactive sources are widely used for a variety of industrial, medical, and research purposes and it is clearly of importance to have sources of a useful life. Most γ-ray transitions are very fast ($>10^{12}$ s^{-1}) and such active nuclei cannot be

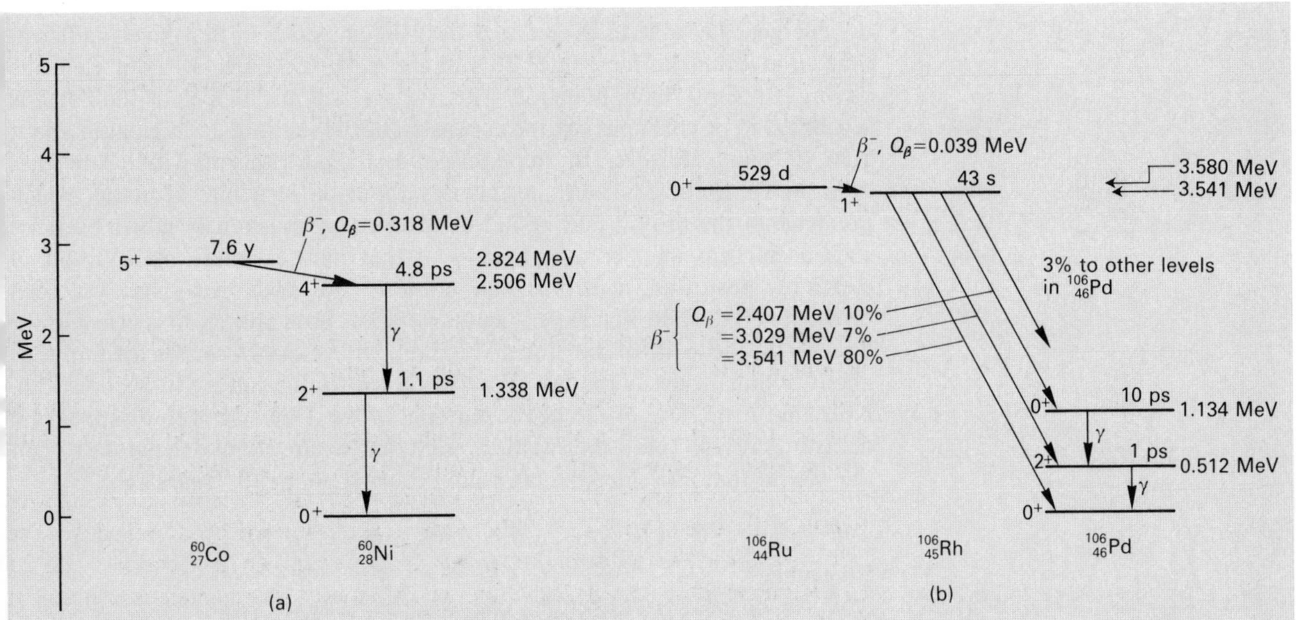

Fig. 5.9 (a) The energy-level diagram for the atomic masses of $^{60}_{27}Co$ and $^{60}_{28}Ni$ showing the transitions that make the former a source of γ-rays near 1 MeV with a useful life. (b) The energy levels of $^{106}_{44}Ru$, $^{106}_{45}Rh$ and $^{106}_{46}Pd$ showing the arrangement of levels that makes the first a source of β-particles with kinetic energies up to 3.5 MeV and a useful life. These diagrams show the atomic masses in MeV above the lowest mass in each case. The symbol 'y' stands for years and 'd' for days: ps is the SI system picosecond $=10^{-12}$ seconds.

kept on a shelf! However, if γ-ray emission follows a low-energy β-decay of a reasonably long life then a useful source of γ-rays is possible. The well-known example is $^{60}_{27}Co$: this is readily produced by neutron capture in naturally available and stable $^{59}_{27}Co$. Its decay scheme is shown in Fig. 5.9(a): it has a mean life of 7.6 years and its β^--decay to an excited state of $^{60}_{28}Ni$ is followed by a cascade of two γ-rays, the levels involved having mean lives of less than 10 ps, and transition energies of 1.17 and 1.33 MeV. Thus $^{60}_{27}Co$ is a useful γ-ray source.

The provision of useful β sources of high energy but long life makes use of the arrangement of alternate nuclei on the curves of the atomic mass for even–even and odd–odd nuclei. Look at Fig. 5.7: if $^{100}_{42}Mo$, $^{100}_{43}Tc$ and $^{100}_{44}Ru$ were to be slightly more to the left on the two parabolae then the first could have been slightly heavier than the second, resulting in a slow β^--decay from the first to the second followed by a fast energetic β^--decay from second to last. Fig. 5.9(b) shows just such a circumstance in $A = 106$: $^{106}_{44}Ru$ decays with a mean life of 529 days to $^{106}_{45}Rh$, which in turn decays to $^{106}_{46}Pd$ with a mean life of 43 s. The values of Q_β are 39.4 keV and 3.54 MeV respectively: thus $^{106}_{44}Ru$ is a source of energetic β-particles with a reasonable life. This isotope is a product of the fission of uranium and other fissile elements (see Section 7.12).

5.7 Conclusion

One of the main objectives of this chapter has been to discuss the properties of nuclear mass and of β-decay in order to show how these properties explain the existence of the line of stability of the naturally occurring nuclei up to lead. In addition we have shown that above lead α-decay becomes important and that, with β-decay, it degrades all these heavier elements to lead or bismuth, over time. But what of the distant future? We know that above $A \simeq 150$ many nuclei are energetically able to decay by α-particle emission, so that given sufficient time, that could be their fate. This might leave the Earth without elements having A greater than about 150. This, in fact, will not happen as the Earth is scheduled to be consumed by the expansion of the sun into a red giant in about 1.5×10^9 years. However, for material not destined for this particular fate what does the future hold? There are theories about the structure of matter which predict that the proton and bound neutron decay into much lighter particles such as mesons and leptons. Decays of that nature would finally result in electrons, positrons, neutrinos and photons. No such decay has yet been detected and the best that experimental work has been able to do is give a lower limit on the mean life of the protons of about 10^{32} years (for certain assumptions about dominant decay modes). We will return to this subject in Section 13.6. Note that if the fate of the heavy material of the Universe is to disappear by decay in, say, 10^{34} years, then at the present time we are about one part in 10^{23} on the way to that fate since the big bang occurred about 10^{11} years ago!

Reference

Lawson, J. L. and Cork, J. M. (1940). *Physical Review* **57**, 982–94.

PROBLEMS

5.1 Explain the terms in the semi-empirical atomic mass formula:

$$\mathcal{M}(Z,A) = ZM_{\rm H} + NM_{\rm n} - a_{\rm V}A + a_{\rm S}A^{2/3} + a_{\rm C}Z^2/A^{1/3} + a_{\rm A}(A-2Z)^2/A \pm a_{\rm P}/A^{1/2}.$$

Show for large A and Z that the energy released when a nucleus (Z,A) emits an α-particle is given by

$$Q_\alpha = -4a_{\rm V} + 8a_{\rm S}/3A^{1/3} + 4a_{\rm C}Z(1-Z/3A)/A^{1/3} - 4a_{\rm A}(N-Z)^2/A^2 + B(2,4).$$

where $B(2,4)$ is the binding energy of the α-particle, 28.30 MeV.

The only naturally occurring isotopes of silver and gold are $^{107}_{47}$Ag and $^{197}_{79}$Au. Discuss the stability of these nuclei in the light of the expression for Q_α. Use the values for the coefficients ***a*** given in Table 4.1.

(Adapted from the 1980 examination in the Honour School in Natural Science, Physics, University of Oxford.)

5.2 How can the spin of the nucleus $^{14}_{7}$Nitrogen be measured by molecular spectroscopy?

5.3 Discuss the experimental evidence for the existence of the neutrino.

The nuclide $^{21}_{11}$Na decays to $^{21}_{10}$Ne with the emission of a positron. The radius of the nucleus with the mass number 21 is 3.6 fm. Estimate the maximum kinetic energy the positrons can have.

(Adapted from the 1976 examination in the Honour School in Natural Science, Physics, University of Oxford.)

5.4 In an undisturbed ore containing 0.1% by weight of $^{238}_{92}$U there will be some $^{226}_{88}$Ra. Calculate the weight of this isotope of radium to be found in one metric ton of ore. What is the rate of generation of helium gas in kg per year in this amount of ore? Use the data given in Table 5.2.

5.5 Calculate values for $a_{\rm C}$ and $a_{\rm A}$ of Problem 5.1 making use of the following facts: $^{35}_{18}$Ar emits positrons with a maximum kinetic energy of 4.95 MeV and $^{135}_{56}$Ba is the stable isobar of mass number 135. Express your answers in MeV/c^2.

(Adapted from the 1967 examination in the Honour School of Natural Sciences, Physics, University of Oxford.)

5.6 Find some nuclear tables and trace the most likely route from $A = 245$, $Z = 96$ to $^{209}_{83}$Bi by a combination of successive decays, each either α- or β-decay. Which is the longest lived active nucleus on this route?

In principle there are four decay series (A(modulo 4)=0, 1, 2, 3). If the progenitors were produced equally 10^{10} years ago, estimate the ratio of the natural abundances on Earth of the longest lived active nuclide of each series. Do you see a problem in answering this question for the even A series?

6

Alpha Decay

6.1 Introduction

We have described the process of α-decay in Sections 1.1 and 5.2. The theory of α-decay is an application of simple quantum mechanics and its presentation to the reader will not, in the simple approach that we shall adopt, add much to our knowledge of nuclear structure. However, we have stressed the need for a quantitative approach and in this case we show how to answer the question: why do the mean lives of α-emitting nuclei vary so dramatically from 2.03×10^{10} years for ${}^{232}_{90}\text{Th} \to {}^{228}_{88}\text{Ra} + \alpha$ to 4.3×10^{-7} seconds for ${}^{212}_{84}\text{Po} \to {}^{208}_{82}\text{Pb} + \alpha$? So we have 24 orders of magnitude variation in transition rates to be explained for what is fundamentally the same process.

6.2 Other properties of α-decay

Early observations on α-decay established that, for a unique source, the majority of the emitted α-particles had the same kinetic energy. For each α-emitter, this kinetic energy, T_α, is a fraction $M_D/(M_D + M_\alpha)$ of Q_α where M_D and M_α are the masses of the daughter nucleus and of an α-particle respectively; this factor allows for the energy of the recoiling daughter. The quantity Q_α was defined in Section 5.2 when we were discussing the energetics of α-decay. In Table 5.2 we have given the values of Q_α and the mean life of the principal α-emitters in one of the naturally occurring radioactive series: those few examples make clear that the transition rates (ω) are a strong function of the kinetic energy. The empirical rule connecting the two is known as the Geiger–Nuttal rule (1911). It states, in a slightly modernized form, that

$$\log_{10} \omega = B \log_{10} R_\alpha - C,$$

where R_α is the range (see Section 11.2) in air at 15°C and 1 atmosphere pressure of the α-particles emitted in a decay with transition rate ω. The constant B is 57.5 (for R_α in centimetres and ω in seconds^{-1}) and C depends on the series (e.g. about 41 for the members of the series that includes ${}^{238}_{92}\text{U}$.). This formula is described more fully in the caption to Fig. 6.1 which shows that the rule is not a very good fit to the data; however, it does indicate a connection between decay rates differing by factors as great as 10^{15}. The explanation in terms of physical principles was given independently by Gamow and by Condon and Gurney in 1929.

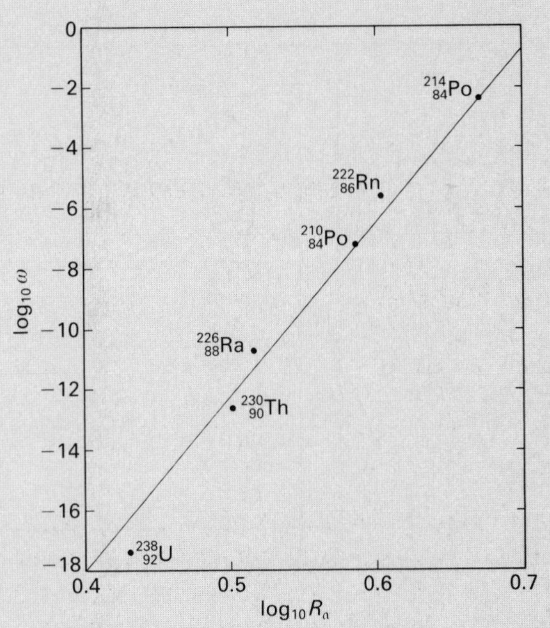

Fig. 6.1 The Geiger–Nuttal rule. This figure shows a plot of values of $\log_{10}\omega$ against $\log_{10} R_\alpha$ for some of the members of the naturally occurring radioactive series with $A=4n+2$, where ω is the transition rate (seconds^{-1}) for the ground state to ground state transition and R_α is the range (centimetres) of the α-particle in dry air at 15°C and one atmosphere pressure. From such data Geiger and Nuttal derived the relation

$$\log_{10} t_{1/2} = -57.5 \log_{10} R_\alpha + C,$$

where $t_{1/2}$ is the half-life ($=0.693/\omega$ if transitions other than to the daughter ground state are negligible in their contribution to the total transition rate) and C is a constant that depends on which of the three series the relation is required to describe. The straight line on the figure has a slope of 57.5.

The range was an important measure of the kinetic energy of α-particles: it is the distance moved in matter before coming to rest. For monoenergetic α-particles from a radioactive source this distance is always the same apart from small fluctuations of the order of a few percent. The range increases smoothly with initial kinetic energy and for the naturally occurring active materials is 2.5–6 cm in air. Look at Fig. 6.6 for an example of range and at Section 11.2 for more on the subject.

Let us start by suggesting a model of what is happening inside the nucleus: Figures 6.2(a) and (b) show two potential wells, one for the neutrons inside the nucleus, the other for the protons. For heavy nuclei, the nucleon separation energy is about 6 MeV, so the nucleons fill energy levels up to about 6 MeV below zero total (kinetic plus potential) energy. If two protons and two neutrons from the top of the filled levels amalgamate into an α-particle, the binding energy of 28.3 MeV is sufficient to provide the four separation energies and leave the α-particle with positive energy but leaving it in a potential well (Fig. 6.2c) similar to that experienced by the protons. This picture represents the energy situation: a proper treatment will give a quantum-mechanical probability amplitude for finding an α-particle at a given energy. Thus the effective mechanical potential for an α-particle as a function of distance between the centre of the α-particle and the centre of the system which is the parent nucleus less the α-particle, is proposed to be as shown in Fig. 6.2(c). It has three regions.

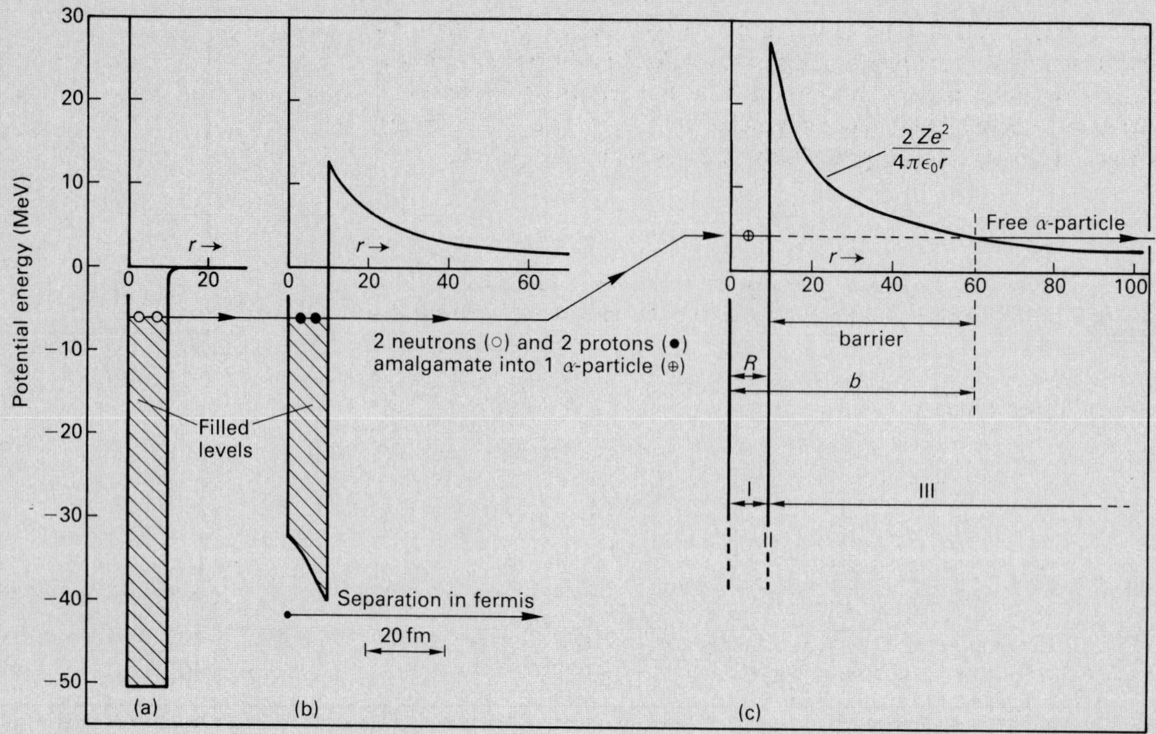

Fig. 6.2 Diagrams of the form of the potential energy versus distance r from the nuclear centre for (a) neutrons and (b) protons in and near a nucleus of $Z=90$, $A\approx 236$. The changing of two protons and two neutrons as they amalgamate to form an α-particle is shown. If an α-particle is formed, then it is presumed to have a potential energy as a function of distance, as shown in (c). The regions I, II, and III are described in the text.

I At distances less than R, approximately the nuclear radius, the α-particle is in a potential well of unspecified depth but representing the effect of the nuclear binding force on the α-particle.

II At a distance R this potential becomes positive and reaches a maximum value of $U(R)=zZe^2/4\pi\varepsilon_0 R$, where $z=2$ and Z is the atomic number of the remaining nucleus.

III At distances greater than R the potential is Coulomb, $U(r)=zZe^2/4\pi\varepsilon_0 r$.

If the parent nucleus $Z+2$, is energetically capable of emitting an α-particle of kinetic energy T_α, then there are two possibilities:

(1) $T_\alpha > U(R)$: the α-particle, if inside the nucleus, is free to leave and will do so almost instantaneously. (That means in a time comparable to the time taken for the α-particle to cross the nucleus, which is less than 10^{-21}.)

(2) $T_\alpha < U(R)$: classically the α-particle is confined to the nucleus. Quantum mechanically it is free to tunnel through the **potential barrier**, emerging with zero kinetic energy at radius b (where $b = zZe^2/4\pi\varepsilon_0 T_\alpha$, $z = 2$) and to move to large r, where it will have the full kinetic energy T_α.

When the barrier is caused by the charges on the particles involved, as in this case, it is called the **Coulomb barrier**. The problem is to find the barrier penetration probability.

6.3 The simple theory of Coulomb barrier penetration

The details of this calculation are based on simple wave mechanics which must be familiar material. The reader should go back to look at the wave mechanics of barrier transmission and check the result we have given in Table 6.1. A particle of kinetic energy T and mass M is incident on a square potential barrier of thickness t and height U: the probability of transmission is approximately

$$e^{-2Kt} \tag{6.1}$$

where $K=\surd[2M(U-T)/\hbar^2]$. If now we have an arbitrarily shaped barrier, the overall transmission is approximately the product of factors like this, where K is

Table 6.1 Barrier penetration

Consider the potential barrier:

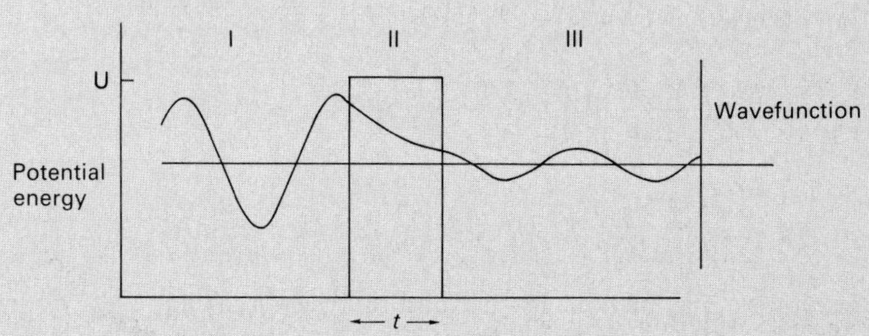

Let there be in region I a wave $\exp(ikx)$, incident from the left; there will be a reflected wave $B\exp(-ikx)$, where k is the wave number in that region.
Then

$$\hbar k = P = \surd(2MT)$$

where P is the momentum of the particle represented by the wave, T is its kinetic energy and M is its mass. In region II the particle has a wavefunction

$$\alpha e^{Kx} + \beta e^{-Kx} \qquad \text{where} \qquad \hbar K = \surd[2M(U-T)].$$

In region III, there is a transmitted wave Ce^{ikx}. The probability for transmission through the barrier is $|C|^2$. Using continuity at the boundaries, we find that

$$e^{-ikt} = \frac{C}{4}\left[\left(1+\frac{ik}{K}\right)\left(1+\frac{K}{ik}\right)e^{-Kt} + \left(1-\frac{ik}{K}\right)\left(1-\frac{K}{ik}\right)e^{+Kt}\right].$$

We shall be considering barriers and kinetic energies for which K is mostly greater than 3 fm^{-1} so that even for a thin barrier of 1 fm $Kt \gg 1$, the second term dominates and we have

$$|C|^2 = 4e^{-2Kt}\left(1+\frac{k^2}{K^2}\right)^{-1}\left(1+\frac{K^2}{k^2}\right)^{-1}.$$

We want to use such an expression through a barrier in which U and, therefore, K are varying. The exponential will be the most important term and since T and $U-T$ are of the same order of magnitude we approximate $k \simeq K$ and obtain

$$|C|^2 \simeq e^{-2Kt}$$

given by the potential at a point, $U(t_i)$, and where we consider a thickness Δt_i; that is, this transmission probability is the product

$$e^{-2K_1\Delta t_1}e^{-2K_2\Delta t_2}\ldots e^{-2K_i\Delta t_i}\ldots$$

$$=e^{-2\int K\mathrm{d}t},$$

where $K=\sqrt{[2M(U(t)-T)/\hbar^2]}$ and the integration is over the thickness for which $U(t)>T$.

In this approximation we use the simple picture of the effective potential for the α-particle shown in Fig. 6.2(c). If R is the nuclear radius and the α-particle emerges at radius b, the transmission probability is

$$\exp\left\{\frac{-2}{\hbar}\int_R^b \sqrt{[2M_\alpha(U(r)-T_\alpha)]}\mathrm{d}r\right\} \equiv \exp(-G), \tag{6.2}$$

where M_α, T_α are the α-particle mass and kinetic energy respectively and this equation defines G. Now $U(r)=zZe^2/4\pi\varepsilon_0 r$, $z=2$, so that

$$G=2\sqrt{\left(\frac{M_\alpha Ze^2}{\pi\varepsilon_0\hbar^2}\right)}\int_R^b\sqrt{\left(\frac{1}{r}-\frac{1}{b}\right)}\,\mathrm{d}r$$

$$=2\sqrt{\left(\frac{M_\alpha Ze^2 b}{\pi\varepsilon_0\hbar^2}\right)}\left\{\cos^{-1}\sqrt{\left(\frac{R}{b}\right)}-\sqrt{\left(\frac{R}{b}-\frac{R^2}{b^2}\right)}\right\}. \tag{6.3}$$

[The integration may be done by the substitution $r=b\cos^2\theta$.]

This result applies to an α-particle which approaches the barrier. To obtain the transition rate we must estimate the number of trials made by the α-particle to penetrate the barrier in one second. We do not know sufficient details of the α-particle existence in the nucleus to do more than make a crude estimate. We assume that it has velocity V_0 inside the nucleus: if it travels diagonally across the nucleus it makes $V_0/2R$ trials per second at the barrier. Therefore the transition probability is given by

$$\omega=\frac{V_0}{2R}e^{-G}. \tag{6.4}$$

We have made so many approximations that we have lost all fear of making more. If $U(R)$ is large compared to T_α (true for all the α-emitters of interest to us) we shall have $b\gg R$ and then the curly bracket in equation (6.3) approximates to $\pi/2-2\sqrt{(R/B)}$ and we have

$$G=2\sqrt{\left(\frac{M_\alpha Ze^2 b}{\pi\varepsilon_0\hbar^2}\right)}\left(\frac{\pi}{2}-2\sqrt{\frac{R}{b}}\right)=\frac{4\pi\alpha Zc}{V}-8\sqrt{\left(\frac{M_\alpha c^2 RZ\alpha}{\hbar c}\right)}, \tag{6.5}$$

where $V=\sqrt{(2T_\alpha/M_\alpha)}$, the α-particle velocity at a large distance after emission, and we have used $b=Ze^2/2\pi\varepsilon_0 T_\alpha$. Summarizing we have

$$\omega=\frac{V_0}{2R}\exp\left[-\frac{4\pi\alpha Zc}{V}+8\sqrt{\left(\frac{M_\alpha c^2 RZ\alpha}{\hbar c}\right)}\right], \tag{6.6}$$

where α is the fine structure constant (do not confuse this α with subscript α as in M_α and T_α indicating quantities belonging to the α-particle). This equation can be written

$$\ln \omega = f - gZ/\surd T_\alpha, \tag{6.7}$$

where $g = 4\pi\alpha\surd(M_\alpha c^2/2) = 3.97\ \text{MeV}^{1/2}$ and $f = \ln(V_0/2R) + 8\surd(RZ\alpha M_\alpha c^2/\hbar c)$. The quantity g is a constant but f will vary through V_0 and R. We do not know what to do about V_0 but $\ln R$ and $\surd(RZ)$ will vary only slightly over the range of natural α-emitters. Thus for simplicity we assume f is a constant. That leaves us with equation (6.7) as the first attempt to describe the connection between ω and T_α.

Equation (6.7) can be related to the Geiger–Nuttal rule. The latter has $\ln \omega \propto \ln R_\alpha$. Since for the α-particle energies of interest the range $R_\alpha \propto T_\alpha^{3/2}$, the rule gives $\ln \omega \propto \ln T_\alpha$. The barrier penetration theory has $\ln \omega \propto -1/\surd T_\alpha$. In the energy range $4.0 < T_\alpha < 7.0$ MeV, which covers most of the α-emitters, a quantity that is linear in $\ln T_\alpha$ is within 3% of being linear in $-1/\surd T_\alpha$. The scatter of the data about the predictions of the theory or of the rule are much greater than this 3% and the Geiger–Nuttal rule is vindicated.

At this point we want to make a small correction to allow for an implicit assumption that has been made: that assumption is that the daughter nucleus is very massive compared to the mass of the α-particle. This is not the case, so that the wave function describing the outgoing α-particle and the recoiling daughter has to contain the reduced mass $M\ (= M_\alpha M_D/(M_\alpha + M_D))$ instead of M_α and the total kinetic energy, which is Q_α, instead of T_α. In equations (6.5) and (6.6) the velocity V is the relative velocity of the two particles instead of that of the α-particle: this velocity is $\surd(2Q_\alpha/M)$ and thus in equation (6.7) $\surd(2T_\alpha/M_\alpha)$ must be replaced by $\surd(2Q_\alpha/M)$ and we have

$$\ln \omega = f' - gZ(M/M_\alpha Q_\alpha)^{1/2}, \tag{6.8}$$

where g is the same as in equation (6.7). The quantity f' is slightly different from f.

In Fig. 6.3 we have plotted $\ln \omega$ against $Z/\surd(Q_\alpha M_\alpha/M)$ for the ground state to ground state transitions of many of the naturally occurring and man-made α-active elements. There are deviations from a single straight line but there is a general tendency for the points to cluster near a linear relation between $\ln \omega$ and $Z/\surd(Q_\alpha M_\alpha/M)$ with a slope somewhat less steep than $-3.97\ \text{MeV}^{1/2}$. The deviation is due to the neglect of the term f' of equation (6.8). The quantity $8\surd(RZM_\alpha c^2\alpha/\hbar c)$ changes by about 7 in the range plotted. An examination of the plot shows that this would correct the points to bring them to a line of steeper slope. The extremities of the data for the ground state to ground state transitions of naturally occurring radioactive nuclides are ${}^{232}_{90}\text{Th}$ with $\omega = 1.2 \times 10^{-18}\ \text{s}^{-1}$ and $Q_\alpha = 4.08$ MeV to ${}^{212}_{84}\text{Po}$ with $\omega = 2.3 \times 10^6\ \text{s}^{-1}$ and $Q_\alpha = 8.95$ MeV. Thus we have a theory which goes some way towards adequately explaining the range of these mean lives. In addition, the theory shows that, for example, at $Q_\alpha < 3$ MeV for $Z = 80$ or at $Q_\alpha < 2$ MeV for $Z = 60$ the mean life is expected to exceed the age of the Universe ($\sim 2 \times 10^{10}$ years) and although α-decay may be energetically possible, it becomes so improbable as to be impossible to detect. Thus our statement in Section 5.4 about the mean life of active elements in the range $160 < A < 208$ is understood.

However, it is clear that the theory is oversimplified and several points come to mind:

1. We could have used the full expression of equation (6.3) instead of assuming that $R/b \ll 1.0$.

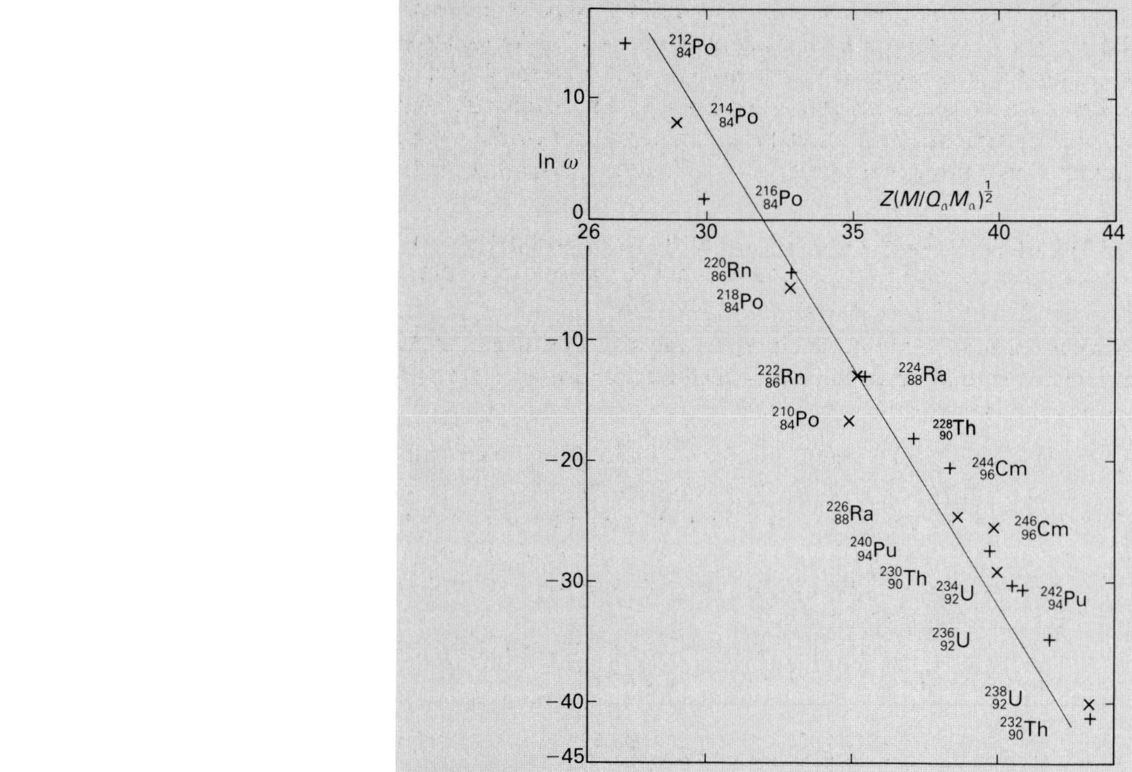

Fig. 6.3 Values of $\ln \omega$ plotted against the values of $Z(M/M_\alpha Q_\alpha)^{1/2}$ for the principal α-emitters of the two radioactive series having A(modulo 4) equal 0 (points +) and equal 2 (×). Each point is for the ground state to ground state transition only. As usual Z, M_α are the atomic number of the daughter and the mass of the α-particle respectively; M is the reduced mass of daughter and α-particle. The quantity $M_\alpha Q_\alpha/M$ is nearly equal to the kinetic energy of the emitted α-particle, T_α; the use of the former is equivalent to using the final particles' relative velocity, as required equation (6.8), rather than the velocity of the α-particle, as in equation (6.5). A line of slope $-3.97\ \text{MeV}^{1/2}$ has been drawn to show the variation expected from the simple theory: it is clear that there are considerable departures from this elementary theory but that it does roughly explain the gross variation of transition rate over 24 orders of magnitude.

2. The treatment of the α-particle inside the nucleus is totally unsatisfactory: it may not exist continually in the nucleus. The correct thing to do is to find the complete wavefunction for all the nucleons and to project out the probability amplitude for finding the nucleus as α-particle plus potential daughter nucleus. That amplitude is a wavefunction that must be matched in three dimensions to the falling wavefunction for the α-particle in the barrier which in turn must be matched to the outgoing wave outside the Coulomb barrier. This is clearly impossible to do rigorously.

3. We could expect that the probability an α-particle is to be found in the nucleus will depend on whether the nucleus is even–even, odd–odd, or odd-A. In fact the decay rates are a couple of order of magnitudes greater for the ee nuclei than for other nuclei, if all other factors are equal or allowed for.

4. An extension of the last idea will come with the shell model of the nucleus (Chapter 8). We shall find that the nucleons have shell closures, as in atomic physics, and there is an important one for α-decay when $N=126$. For example, the value of Q_α for the α-decay of a parent nucleus (Z,N) when $N=128$ is greater by several megaelectron volts than when $N=126$, for the same Z. Thus the details of nuclear structure will affect the decay rate through the energy release and less obviously through the probability of finding an α-particle in the nucleus.

5. The shape of the potential of Fig. 6.2(c) is idealized and will probably be softer in its change from the nuclear to Coulomb potential. The sensitivity of the decay rate to T_α suggests this will have an important effect.

6. The theory has assumed that nuclei are spherical. However, many of the nuclei heavier than lead are known to be deformed (see Section 3.9); this will probably have a significant effect on the transition rates.

Summarizing, we can see that our simple approach is indeed very oversimplified. Attempts have been made to improve the theory of α-decay and to understand better the systematics of the connection between the decay energy and the transition rate, but these are beyond the scope of this book. However, there is one physical effect that we have not mentioned and which does affect α-decay. That is described in the next section.

In spite of all the difficulties mentioned, when we wish to make an estimate of a transition rate we shall use the full version of equation (6.8):

$$\ln \omega = \ln\left(\frac{V_0}{2R}\right) + 8\sqrt{\left(\frac{RZ\alpha Mc^2}{\hbar c}\right)} - 4\pi Z\alpha\sqrt{\left(\frac{Mc^2}{2Q_\alpha}\right)}, \tag{6.9}$$

where $V=\sqrt{(2Q_\alpha/M)}$ and we take $V_0=V$ and $R=1.53\times A^{1/3}$ fm. This last number comes from a study of the systematics of α-decay and is the radius effective in decay and presumably a convolution of nuclear and α-particle radius.

6.4 The angular momentum barrier

We want to use α-particle decay as a context in which to introduce another idea which is significant for α-decay but also is very important in all processes involving nuclear and particle reactions and decays. Consider the α-decay

$$(Z+2, A+4) \rightarrow (Z, A) + {}^4_2\text{He}.$$

Suppose the parent and daughter nuclei have spin (quantum numbers j_P and j_D. In each case that spin is made of the vector sum of all the orbital and spin angular momenta present among the constituent nucleons. The α-particle has spin 0. Total angular momentum must be conserved so that if $j_\text{P} \neq j_\text{D}$, the α-particle must emerge with relative orbital angular momentum (quantum number l) with respect to the recoiling daughter nucleus. The conservation of the vector of angular momentum requires that

$$|j_\text{D} - j_\text{P}| \leqslant l \leqslant j_\text{D} + j_\text{P}.$$

The quantum number l must be zero or a positive integer: this is no problem as both j_P and j_D are integer for even A nuclei and both are half-odd integer for odd-A nuclei.

Let us now write down the Schrödinger equation for an α-particle ($z=2$) leaving a recoiling nucleus (Z,A). For $r>R$,

$$-\frac{\hbar^2}{2M}\nabla^2\psi(\mathbf{r})+\frac{zZe^2}{4\pi\varepsilon_0 r}\psi(\mathbf{r})=Q_\alpha\psi(\mathbf{r}). \tag{6.10}$$

Here M is the reduced mass of the system. We do the usual separation of variables: let

$$\psi(\mathbf{r})=R(r)Y(\cos\theta,\varphi).$$

The angular part Y will be the spherical harmonic $Y_l^m(\cos\theta,\varphi)$ if the α-particle has orbital angular momentum l. The quantum number m, as is usual, gives the z-component of l. Putting $R(r)=U(r)/r$ and substituting into equation (6.10) gives, for the radial function $U(r)$,

$$-\frac{\hbar^2}{2M}\frac{\mathrm{d}^2U(r)}{\mathrm{d}r^2}+\left\{\frac{zZe^2}{4\pi\varepsilon_0 r}+\frac{l(l+1)\hbar^2}{2Mr^2}\right\}U(r)=Q_\alpha U(r). \tag{6.11}$$

We see immediately that the barrier at r is higher than pure Coulomb by an amount $l(l+1)\hbar^2/2Mr^2$. For example, it is higher by about 0.14 MeV for $l=2$ at $r=15$ fm where the Coulomb term is 17.3 MeV for $Z=90$ and $z=2$. This is the **angular momentum barrier**.

It is clear what will happen: the total barrier is harder to penetrate and the transition rate will be lower (and the mean life longer) than with the Coulomb barrier alone. It is also clear that even at small l and large Z the effect, although far from a major contributor to the barrier, could be an important consideration. Blatt and Weisskopf (1952) have given some figures for the suppression factors in α-decay transition rates due to the angular momentum barrier: we show these in Table 6.2. Thus, for example, for a hypothetical nucleus, $Z+2=88$, and spin 2 and $Q_\alpha=4.88$ MeV, without considering angular momentum we expect a mean life on our simple theory of approximately 6×10^8 s. If the daughter nucleus has spin 0 and we include the angular momentum barrier, this mean life will increase to approximately 1.7×10^9 s.

Moving on we see that even in the absence of a Coulomb potential we can still have an angular momentum barrier. The qualitative rule that we have discovered applies equally: the probability of a particle penetrating the angular momentum barrier is relatively small at low particle energies and increases rapidly with increasing energy. The barrier is more impenetrable the greater is l.

DEFINITIONS AND KEYWORDS

Potential barrier There can exist regions classically inaccessible to a given particle because in such regions its potential energy would be greater than its total (potential plus kinetic) energy. Such a region that is preventing the passage of the particle from one accessible region to another is called a potential barrier. Classically such a barrier is impenetrable, but in circumstances in which quantum mechanics is applicable the barrier becomes penetrable.

Coulomb barrier This is the effective barrier that exists between two like charged particles and reduces the probability of close approach. It also suppresses the probability of separation from a bound state which can decay to two such particles.

Angular momentum barrier This is the effective quantum-mechanical barrier that is present when two particles have non-zero relative orbital angular momentum and which reduces the probability of close approach or of separation from a bound state which can decay to two such particles.

Table 6.2 Values of the suppression factor due to the angular momentum barrier in an α-decay for which Z=86, T_α=4.88 MeV, R=9.87 fm

l	0	1	2	3	4	5	6
ω_l/ω_0	1.0	0.7	0.37	0.137	0.037	0.0071	0.0011

From Blatt and Weisskopf (1952).

These matters have been put on a proper quantitative basis but that cannot be our business now (see Blatt and Weisskopf).

Note that we have assumed that the particle is emerging from the nucleus. However, both the Coulomb and angular momentum barrier effects can apply also to particles entering the nucleus, and the reader will see that this will be relevant to the rates of nuclear reactions where the first step is the penetration into the nucleus by an incident particle.

There is a classical picture of the action of the angular momentum barrier: consider the decay of a system A into two final particles B and C with, perforce, equal and opposite momentum, P (Fig. 6.4). If the conservation of angular momentum requires these particles to have relative angular momentum $l\hbar$, then the momentum vectors do not lie in a line but are separated by a certain perpendicular distance called the impact parameter, b (see Section 1.2), such that $Pb = l\hbar$. Thus l and P define b. However, if the range of the interaction between these two particles is less than b, it is less likely that it will be able to place the momentum vectors at the required b than if b could be zero, and therefore the transition rate will be less than in the absence of l.

The quantum-mechanical picture is as follows. Consider the case of neutron emission by a nucleus: this case is simpler to discuss than α-particle emission as there is no Coulomb barrier. But, of course, now we will assume that there is an angular momentum barrier due to the presence of relative orbital angular momentum $l\hbar$ between the outgoing neutron and the recoiling nucleus. The radial part of the wave function that describes this outgoing system is like $j_l(kr)$, the spherical Bessel function of order l, where k is the reduced wave number of the momentum of the emitted neutron ($P = \hbar k$). At small r these functions behave like $(kr)^l$, so that for small kr (< 1) the overlap between the outgoing wave function and the wave function of the initial nucleus is smaller the greater is l. This overlap determines the value of the matrix element connecting the initial and final states in the transition. As l increases at fixed k (fixed outgoing momentum), that matrix element decreases and so does the transition rate, all other things being equal. Conversely, at fixed l, the matrix element and the transition rate increase with increasing momentum.

We should note, however, that the effect of the angular momentum barrier relative to that of the Coulomb barrier in α-decay must not be exaggerated. The differences in nuclear structure associated with the shell model or in nuclear radii between different α-emitters may have a much larger effect than the differences in the changes of the angular momentum between different decays.

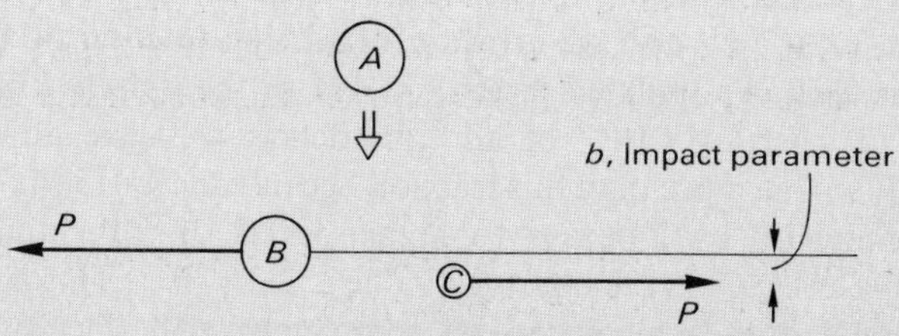

Fig. 6.4 The classical picture of relative orbital angular momentum: A disintegrates into two particles, B and C, separating with equal and opposite momentum P, and impact parameter b, which will have angular momentum Pb. Usually P is fixed by the kinematics and the angular momentum by conservation: then it follows that b is determined. In quantum mechanics, the particles' perpendicular separation is uncertain but the outgoing wave function concentrates it around b.

6.5 Decay schemes involving α-particle emission

Fig. 5.2(b) shows the decay scheme from the ground state of the parent ${}^{242}_{94}Pu$ to the daughter ${}^{238}_{92}U$ either in its ground state or in one of three of its many excited states. The branching fractions are shown and as expected the dominant decay mode is to the spin 0 ground state of ${}^{238}_{92}U$ since this gives the most energetic α-particle. Decays to higher excited states are decreasingly probable as the energy of the α-particle decreases and the l-value increases. Our simple formula based on the energy alone leads us to expect, for example, that the ratio of the branching fraction of the $0 \to 6$ transition to that of the $0 \to 0$ should be

$$\frac{\omega_6}{\omega_0} = \frac{\exp\left(\dfrac{-4\pi Z\alpha c}{V_6}\right)}{\exp\left(\dfrac{-4\pi Z\alpha c}{V_0}\right)} \tag{6.12}$$

where V_6 and V_0 are here the relative velocities of the α-particle and recoiling daughter nucleus at large separations for the two transitions. The result is 5.4×10^{-3}. The observed ratio is 2.0×10^{-5} so the angular momentum barrier is suppressing the $0 \to 6$ transition by a factor of $(265)^{-1} = 3.8 \times 10^{-3}$. We see from Table 6.2 that this is close to what would be expected.

What about the actual rate for the ground state to ground state transition? From Fig. 5.2(b) we find that it is observed to be 4.3×10^{-14} s^{-1}. Using the formula of equation (6.9) we obtain 3.4×10^{-12} s^{-1}, too great by a factor of 79! Although better agreement would be pleasing, this result is hardly surprising. The scatter of points in Fig. 6.2 reminds us that we do not have a complete theory.

Fig 6.5 shows a more complicated scheme where the α-emitting nucleus ${}^{214}_{84}Po$ is fed by β-decay from the nucleus ${}^{214}_{83}Bi$ (99% of the latter's decays). Those decays leave the ${}^{214}_{84}Po$ in one of its excited states in 82% of all ${}^{214}_{83}Bi$ decays. Such excited states are expected to decay by γ-ray emission to the ${}^{214}_{84}Po$ ground state, but in this case the alternative α-decay to the ground state of ${}^{210}_{82}Pb$ is so energetic that it has a transition rate which can begin to compete with γ-decay. For example, the expected transition rate for the $2^+ \to 0^+$ decay releasing an α-particle with $Q_\alpha = 9.249$ MeV ($T_\alpha = 9.076$ MeV allowing for recoil) from the 1.415 MeV excited state of ${}^{214}_{84}Po$ can be found by comparison with the known rate of the ground state to ground state transition, using equation (6.12); neglecting the small effect of the angular momentum barrier the result is 4.1×10^7 s^{-1}. The mean life of this excited state is about 0.3 ns, so its γ-ray decay rate is about 3×10^9 s^{-1} and we expect a branching fraction of 1.4×10^{-2} for α-decay. This level is populated directly in 1% of the β-decays of ${}^{214}_{83}Bi$, so we expect that at least 1.4×10^{-4} of all ${}^{214}_{83}Bi$ delays will give an α-particle of 9.076 MeV. It will be more than this fraction because the 1.415 MeV level is also populated indirectly by β-decays to higher levels of ${}^{214}_{84}Po$ and subsequent γ-decay to this level. Thus the observed value is 2.2×10^{-2}. This level in ${}^{214}_{84}Po$ is one example where α-decay is not lost completely in the face of competition from γ-decay. (There is a level in ${}^{214}_{84}Po$ for which $Q_\alpha = 10.76$ MeV and for which α-decay competes with γ-decay on almost equal terms.) Several factors decide the overall probability of the emission of one of these unusually energetic α-particles. The probability β-decay goes to a certain excited state (of ${}^{214}_{84}Po$ in this example) is less the greater is the excitation energy of that state: the reason for this will become clear when we look at transition rates in β-decay in Section

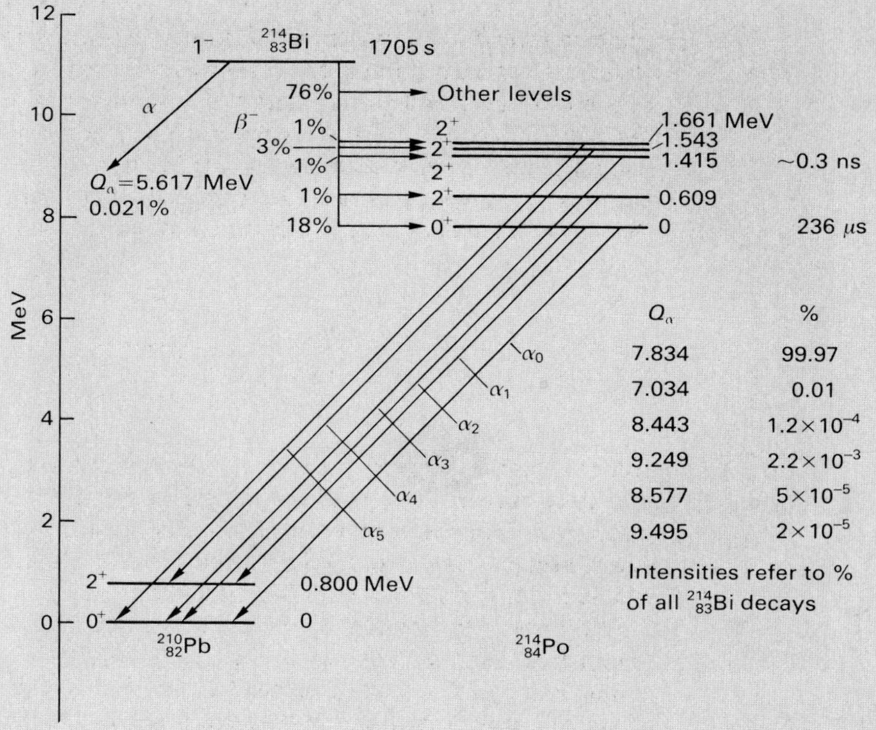

Fig. 6.5 A simplified energy-level diagram for the sequence $^{214}_{83}\text{Bi} \xrightarrow{\beta\text{-decay}} {}^{214}_{84}\text{Po} \xrightarrow{\alpha\text{-decay}} {}^{210}_{82}\text{Pb}$. The β-decay sometimes goes to an excited state of $^{212}_{84}$Po which would normally be expected to reach the ground state of this nucleus by single or multiple γ-ray emission or internal conversion. However, the Q_α-values for α-emission from these excited states are so large that in some cases their transition rates can compete with the γ-ray and internal conversion transition rates. Thus occasionally these states emit an α-particle instead of heading for their ground state and a small fraction of α-particles are observed to have a kinetic energy greater than the 7.69 MeV observed from the ground state to ground state transition (α_0). This figure shows four of twelve extra-energy transitions that have been observed. An example of one of these *long range α-particles* is shown in Fig. 6.6. (Many of the excited states of $^{214}_{84}$Po and $^{210}_{82}$Pb have been omitted. In addition no γ-ray transitions are shown).

12.5. The emission of an α-particle is increasing in probability as that excitation energy increases, and the competing process of γ-ray emission depends on the disposition of lower-lying levels. These factors only come together to give extra-energy α-particles in a few nuclei. Thus α-decay from excited states is rare. This example of $^{214}_{83}\text{Bi} \rightarrow {}^{214}_{84}\text{Po} \rightarrow {}^{210}_{82}\text{Pb}$, has a total of about 1 in 420 decay sequences giving a final α-particle more energetic than the $T_\alpha = 7.688$ MeV of the $^{214}_{84}$Po ground to $^{210}_{82}$Pb ground state transition.

These unexpectedly energetic α-particles were discovered early in the investigations of the properties of α-decay. Because range in air of the α-particles was a measure of their energy, they were called **long range α-particles**. Figure 6.6 is a cloud chamber photograph of one such α-particle among many hundred others.

6.6 Barriers in other decays

Spontaneous fission was described in Section 5.6: we shall now briefly look at a model of this process and find that it involves barrier penetration. However,

Fig. 6.6 A cloud chamber photograph of α-particles from the decay of $^{214}_{83}Bi$. The almost constant range of particles of 7.69 MeV from the decay of the ground state of $^{214}_{84}Po$ is clear (see Section 11.3 and Fig. 11.4). The outstanding particle is the product of the rare decay from an excited state of $^{214}_{84}Po$, as described in the text and in Fig. 6.5.

[What is a cloud chamber? A volume of gas saturated with a suitable vapour (e.g. ethyl alcohol) is enclosed in a cylinder. The gas is expanded adiabatically so as to become supersaturated and condensation of the vapour onto drops takes place: the first stage is condensation on charged ions so that small drops of liquid form along the trajectories of ionizing particles which have traversed the gas just previously to or immediately after the expansion. The tracks are illuminated through side windows and photographed through a window in the cylinder end. By careful timing of a flash photograph with respect to the expansion, the tracks may be recorded at their most visible and before general condensation occurs. In this example the radioactive source was fixed on an inside wall of the chamber and collimated to give a fan of visible tracks. The cloud chamber was invented by C. T. R. Wilson in 1911. Gentner *et al.* (1954) have published an atlas of cloud chamber photographs illustrating many nuclear particle processes. The above photo is in this atlas and was published originally by Philipp (1926).]

first we shall look at the status of spontaneous fission relative to α-decay. The isotope $^{238}_{92}U$ has one of the smallest detected spontaneous fission rates: it is $2.7 \times 10^{-24}\ s^{-1}$, which corresponds to a branching fraction of 5.5×10^{-7}, where the principal decay mode is α-emission. Spontaneous fission is not significant as a decay mode for uranium or for any light nuclei, but it does become increasingly important among the transuranic elements as Z increases. At $Z = 100$, fermium has an isotope ($A = 256$) stable against both β-decay or electron capture, which has a mean life of 3.8 hours and which decays 8% by α-emission or 92% by spontaneous fission; that is a transition rate for the latter mode of $6.7 \times 10^{-5}\ s^{-1}$.

We must now investigate the mechanism of the decay to understand how the rates vary to this extent. The liquid drop model from which we derived the semi-empirical mass formula provides the basis of a model for fission. The mass of a nucleus will increase if it is deformed from spherical into prolate because the surface term increases, and although the Coulomb term decreases, it changes less. (We note that many nuclei are already slightly deformed in their ground state and so here we mean a deformation increasing from that situation.) However, as the deformation further increases the increase in energy stops, and as the nucleus breaks into two parts the energy decreases, and continues to decrease until the parts are two spherical, or near spherical, nuclei far apart. This is represented by the potential energy curve of Fig. 6.7. Thus the system which is the initial nucleus is confined into a potential well which classically it cannot leave. As in α-decay, the system can, however, tunnel through this **fission barrier**. Fission becomes another barrier penetration problem. This one is not easy to solve as the actual shape of the potential is even less well-known than it is for α-decay. The height of the top of the barrier above the well bottom is called the **activation energy**. This energy is about 6–8 MeV for $A=238$ but decreases in height as Z^2/A increases. It is believed that, when Z^2/A reaches about 49 ($Z=115$, $A=270$?), the Coulomb term decreases faster than the surface term increases with increasing deformation and there is no barrier. Such a nucleus, if produced, would suffer fission within about 10^{-22} s. For nuclei just below this limit the barrier would be small and transition rates for fission would be much greater than the largest known (about $3\times10^3\ \mathrm{s}^{-1}$ for $Z=102$, $A=250$, nobelium).

In Section 5.5 we also mentioned other possible decays between the extremes of fission and of α-decay regarded as very asymmetric fission. First thoughts suggest that these are most likely to involve the emission of light nuclei with particularly large binding energy such as $^{12}_{6}C$, $^{16}_{8}O$, and $^{20}_{10}Ne$. However, there is a

DEFINITIONS AND KEYWORDS

Fission barrier This is the barrier which prevents almost instantaneous spontaneous fission in those nuclei energetically capable of underoing fission.

Activation energy The height of the top of the fission barrier above the ground state energy of a nucleus.

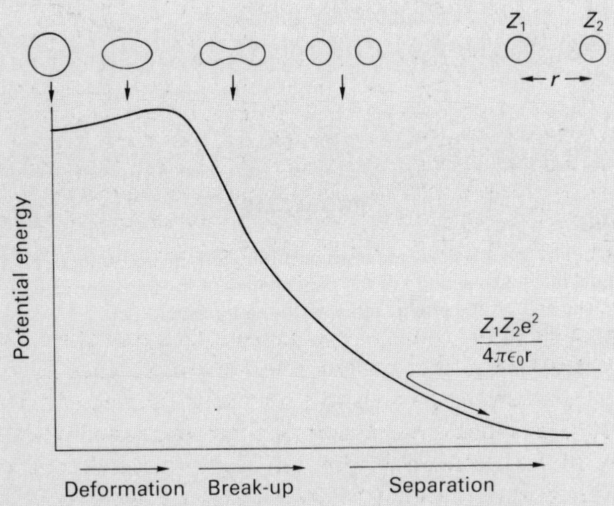

Fig. 6.7 The fission barrier. The expected form of the potential energy curve as a nucleus deforms, breaks up, and the products separate during the process of fission. The abscissa is not well defined until the fission products are well separated, when we can expect it to vary Coulomb-wise. At zero abscissa the nucleus has an energy greater than that at infinite separation of the products at rest, by the energy released in fission. There is a barrier to fission which is probably about 6–8 MeV higher than the undeformed energy in nuclei near uranium. This height is the activation energy and it decreases as Z^2/A increases and may reach zero near $Z=115$. (The heaviest nucleus so far observed has $Z=106$.) This barrier determines the spontaneous fission transition rate.

Q-value advantage for the emission of heavier isotopes of these nuclei. Decay by the emission of $^{14}_{6}C$, $^{24}_{10}Ne$, $^{25}_{10}Ne$ has been established. Thus, for example,

$$^{223}_{88}Ra \to ^{209}_{82}Pb + ^{14}_{6}C + 31.87\ \text{MeV},$$

has a branching fraction of about 8×10^{-10}. Such decay modes are obviously similar to α-decay and will be controlled by an appropriately scaled up Coulomb barrier.

6.7 Some conclusions

The modest success of the simple theory of α-decay reinforces the simple model of the nucleus which we have been building in the last few chapters. The nucleus for protons and for α-particles, appears to act like a simple potential well for separations less than the nuclear radius and to have a normal Coulomb potential at larger separations.

One of the important things we have included in this chapter is a section on the angular momentum barrier. The reader is urged to pay particular attention to this subject as it will appear again in later chapters.

References

Blatt, J. M. and Weisskopf, V. F. (1952). *Theoretical Nuclear Physics.* John Wiley & Sons.

Gentner, W., Maier-Leibnitz, H., and Bothe, H. (1954). *An Atlas of Typical Expansion Chamber Photographs.* Pergamon Press, London.

Philipp, K. (1926). *Natürwissenschaft*, **14**, 1203–4.

PROBLEMS

6.1 Describe briefly the physical processes occurring in α-decay. Without detailed calculation, give a qualitative explanation of the dependence of the transition rate on the Z of the daughter nucleus and on the energy released in the transition, Q.

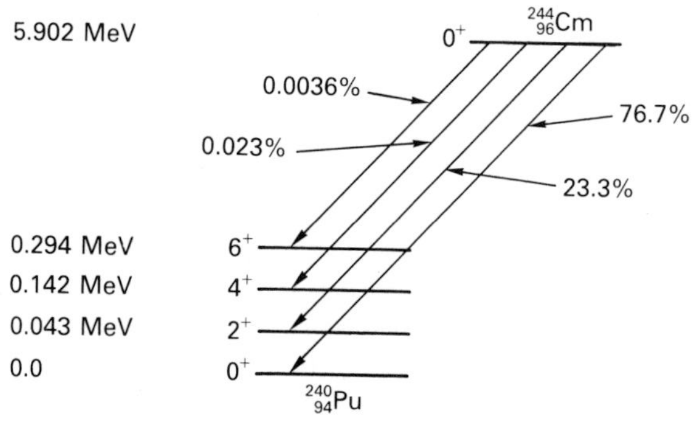

The figure shows the α-decay scheme of $^{244}_{96}$Curium to $^{240}_{94}$Plutonium. The transitions are marked with the branching fractions in per cent. According to a simple formula the transition rate, λ, for the ground state to ground state transition is given by $\ln \lambda = C - DZ/Q^{1/2}$, where $C = 132.8$ and $D = 3.97$ (MeV)$^{1/2}$ when λ is in s^{-1}. Calculate the mean life of $^{244}_{96}$Cm.

If the same C and D are assumed to apply to the transition from the ground state of $^{244}_{96}Cm$ to the 6^+ level of $^{240}_{94}Pu$, show that this formula for α-decay transition rate overestimates the rate for that transition and suggest a reason for the discrepancy.

(Adapted from the 1988 examination in the Honour School of Natural Science, Physics, University of Oxford.)

6.2 Explain why naturally occurring nuclei of mass number $A \approx 60$ are not α-unstable.

With neglect of nuclear spins, and for Coulomb potential barriers large compared with the α-particle energy, the dependence of the mean life τ of an α-unstable nucleus on the charge Z and α-particle velocity v is represented by

$$\tau \propto \exp\left(\frac{4\pi Z e^2}{4\pi\varepsilon_0 \hbar v}\right).$$

Discuss the basic assumptions made in the derivation of this expression.

In the decay $^{238}_{92}U \rightarrow {}^{234}_{90}Th + \alpha$, the emission of α-particles of energy 4.195, 4.147, 4.038 MeV is observed. Use the above formula to estimate the relative strengths of these three lines. What are the most probable spin-parity assignments for the states of ^{234}Th produced in this reaction? State briefly why consideration of the spins of the states of ^{234}Th would change your estimates of the relative line strengths.

(Adapted from the 1975 examination in the Honour School of Natural Science, Physics, University of Oxford.)

7

Nuclear Collisions and Reactions

7.1 Historical introduction

It was clear to Rutherford and to other workers in nuclear physics that it would be essential not only to study the spontaneous changes in nuclei but also to investigate the possibility of changing nuclei by external means. The obvious way to do this was by causing collisions between nuclei. The first observation of a nuclear reaction was by Rutherford in 1919 when he detected protons produced in the bombardment of nitrogen by α-particles. The nuclear reaction is

$$^{4}_{2}\mathrm{He} + ^{14}_{7}\mathrm{N} \rightarrow ^{17}_{8}\mathrm{O} + ^{1}_{1}\mathrm{H}.$$

In 1925 Blackett succeeded in photographing examples of this reaction in a cloud chamber: Fig. 7.1 is a reproduction of a photograph containing an event that is interpreted as an example of the above reaction.

Progress in the study of nuclear reactions required two things:

1. The development of controlled sources of high-speed projectiles which could be used to bombard targets of chosen nuclei.
2. The development of an understanding of the behaviour of fast particles and of the methods that could be used to detect them.

This book does not cover these subjects except sometimes to make limited remarks about the technique used when we illustrate some physics with a diagram showing experimental results or a photograph of a nuclear or particle process.

However, we need to define the terms used in the requirement (1) above.

(a) *Projectile:* any nucleus, subatomic particle or subnuclear particle can be used to cause nuclear changes or to probe nuclei, if given an appropriate kinetic energy. Early experiments used protons, α-particles; recent developments have included the use of heavy ions (i.e. $Z > 3$) as projectiles. We have already met the use of electrons, protons and neutrons (Chapter 3) and we will meet later the use of subnuclear particles such as π-mesons, K-mesons as projectiles.

(b) *High speed:* this is a slightly old-fashioned term used to imply a sufficient or an appropriate kinetic energy. What is that? From our discussion or α-particle scattering we can say that for charged particles it is probably sufficient

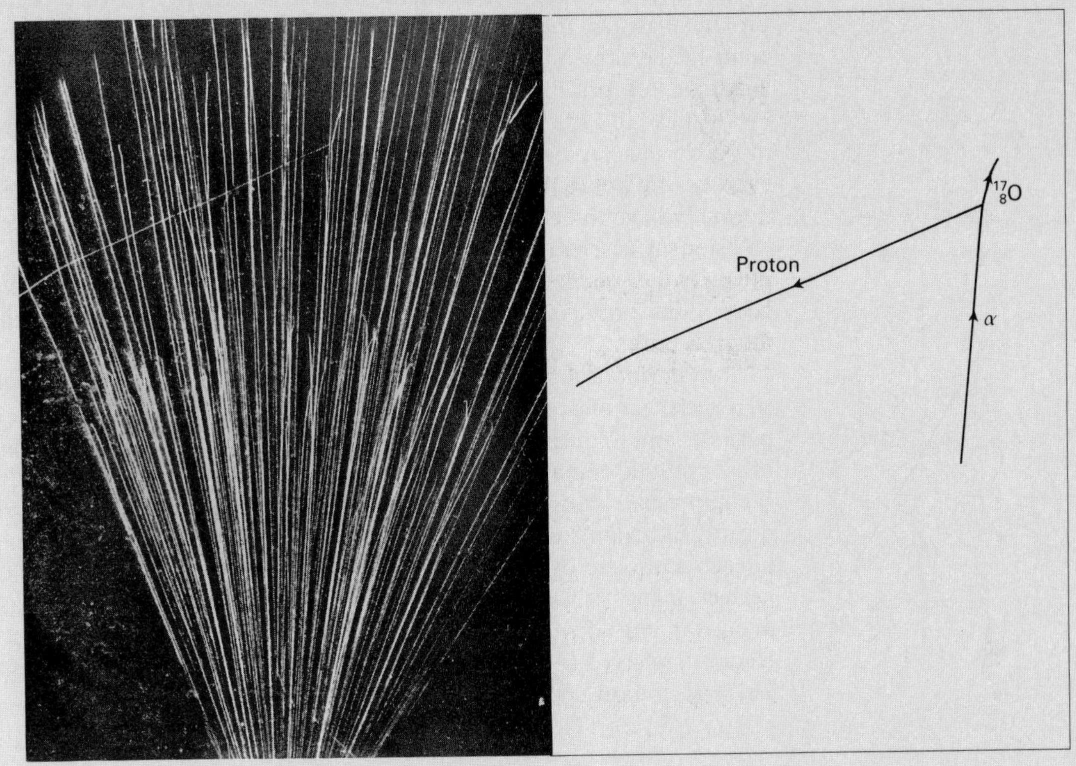

Fig. 7.1 An example of one of the cloud chamber photographs of the first nuclear reactions observed by a visual technique (Blackett and Lees, 1932). The gas in the chamber is exposed to α-particles from a source of $^{212}_{82}Pb$ (ThB in the early notation). The daughter $^{212}_{83}Bi$ (ThC) gives α-particles of 6.09 MeV. The alternative β-decay mode of ThC gives $^{212}_{84}Po$ which in turn gives α-particles of 8.78 MeV. The two groups can be separated by their different range. One longer-range α-particle has caused a nuclear reaction interpreted to be

$$^{14}_{7}N + ^{4}_{2}He \rightarrow ^{1}_{1}H + ^{17}_{8}O.$$

The α-particle energy just before the collision was about 3.6 MeV. The proton is emitted at 112° with the respect to the incident α-particle and with a kinetic energy of 1.1 MeV. The recoil $^{17}_{8}O$ is projected at 15° with an energy of about 1.1 MeV. Blackett and Lees determined these energies from the range (see Section 11.2) of these particles and deduced that the Q-value for this event was -1.41 ± 0.23 MeV. The modern value is -1.19 MeV.

The cloud chamber method is described in the caption to Fig. 6.6. It is ironic that Blackett and Lees found eight of these reactions in a cloud chamber filled with nitrogen and some oxygen (to keep the tracks fine). This picture, which is one of the clearest and most complete, was obtained in an argon/oxygen/hydrogen mixture; the only consistent explanation offered by the authors is that there was impurity nitrogen present. Blackett and Lees do not appear to have analysed their gas and state that the argon was 99% pure.

energy so that the particle can surmount the Coulomb barrier. However, from considering α-decay we can assert that incident particles not surmounting the barrier might still tunnel into the nucleus quantum mechanically and cause nuclear changes. We expect therefore that incident charged particles need to have energies or order 1 MeV and up, depending on the Z of the target nucleus. This restriction does not apply to uncharged particles, neutrons for example: there is no barrier, and even neutrons of thermal energy can cause nuclear reactions.

(c) *Controlled:* this means that the incident particles have an energy and direction under the control of the experimentalist who wishes to use them.

It is the technology of accelerators and of charged particle beams which provide these facilities. Present technologies can only provide projectiles of well-controlled energy if the projectile is charged. The maximum laboratory energies available for protons have grown exponentially with time from 7×10^5 eV (= 700 keV) in 1932 to near 10^{12} eV (= 1 TeV) in 1986. The choice of projectiles available has expanded from proton or α-particle to the ion of almost any element (but not in the entire energy range possible for protons). Neutral beams normally have to be engineered from the secondary or tertiary products of the collision of accelerated particles with targets. Such techniques imply limited intensity in a usable range of energy and direction (phase space of the required beam), and problems of unwanted (background) radiation accompanying the tailored beam.

The detection of individual fast particles is a technology which has grown in step with that of accelerators: ideally one would like to identify each of many particles and to measure their vector momenta with precision in a large area or in a large solid angle. However, in most circumstances the design of a detector is a compromise between conflicting requirements. In this book we shall frequently illustrate the physics by showing photographs that were obtained using, in each case, one of the visual techniques that are available. We will have to neglect many other techniques and thereby deprive the reader of a full understanding of what can and what cannot be measured. To correct this the reader is referred to the books by Kleinknecht (1986) and by England (1974). There also many specialized books on particular techniques.

7.2 Matters of definition

For the present we discuss nuclear collisions in the context of the most common arrangement of investigation: Fig. 7.2 represents a well-collimated beam of incident particles striking a fixed target. This target is normally thick enough to give a sufficient rate of useful collisions but not thick enough to degrade the

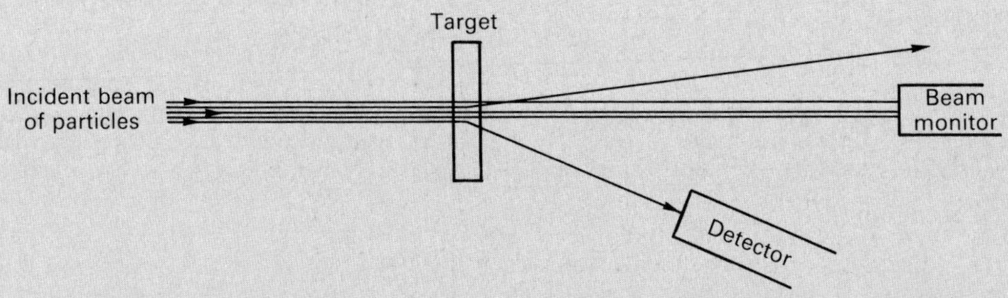

Fig. 7.2 The basic arrangement for the investigation of nuclear collisions. A beam of particles is incident on a suitable target. If the target is 'thin' most of these particles will traverse the target and suffer only a small deflection. However, a small fraction will be scattered through large angles in single nuclear collisions. The energy of the scattered particle will be that characteristic of elastic scattering, or less, thereby indicating inelastic scattering. In the case of inelastic scattering and nuclear reactions, products other than the incident particle will, in general, be observable. Frequently two, or more, detectors will be operated to detect simultaneously two, or more, products of a collision. Ideally each detector will identify the product detected and measure its vector momentum. Practically it will fall short of this: for example, it may only be capable of determining the kinetic energy of any charged particles arriving at its entrance aperture.

It is also necessary to measure the integrated intensity of the incident beam. This is sometimes done by measuring the incident flux by a method which does not significantly affect the beam or, as here, by measuring the transmitted beam.

energy and collimation of the incident beam or to impede significantly the emergence of collision products. These requirements are not always compatible. Detectors will be positioned to detect and perform required measurements on these collision products.

We shall try to keep to the following usage. **Scattering** of an incident projectile by collision with a target particle occurs when the incident particle is found among the products. **Elastic scattering** occurs if the projectile and target particles are intrinsically unchanged in the collision and no other particles are produced. **Inelastic scattering** occurs if the target particle becomes excited, breaks up, or if other particles are produced. A **nuclear reaction** occurs if there is a rearrangement of nuclear constituents between the colliding particles. This vocabulary will be extended and modified when we consider reactions between subnuclear particles.

There is a shorthand notation for simple collisions which is useful. The reaction:

$$a + A \rightarrow b + B$$

represents a nuclear reaction where a is an incident particle, A is a target nucleus. The symbols b and B represent the products and b is, by convention, the lighter of b and B. In this case we have a two-body final state. The shorthand representation is:

$$A(a,b)B$$

For many light nuclei there exists a shorthand symbol for the nucleus: for example, α for the helium nucleus, p for the proton. We list these symbols in Table 7.1. Thus the reaction

$$^{4}_{2}\mathrm{He} + ^{14}_{7}\mathrm{N} \rightarrow ^{1}_{1}\mathrm{H} + ^{17}_{8}\mathrm{O}$$

can be written

$$^{14}_{7}\mathrm{N}(\alpha,\mathrm{p})^{17}_{8}\mathrm{O}.$$

Table 7.1 Symbols for some light nuclei and other particles used in nuclear reactions.

Nucleus	Name of nucleus or of particle	Symbol
$^{1}_{0}$n	neutron	n
$^{1}_{1}$H	proton	p
$^{2}_{1}$H	deuteron	d
$^{3}_{1}$H	triton	t
$^{3}_{2}$He	helion	h (uncommon name and notation)
$^{4}_{2}$He	α-particle	α
—	photon	γ
—	electron	e^-
—	positron	e^+

Do not confuse d for deuteron with d for down quark to be used in Chapter 10 and onwards.

DEFINITIONS AND KEYWORDS

Scattering A collision of an incident particle with a target particle in which the former appears among the products of the collision.

Elastic scattering A collision in which the incident and target particle remain unchanged, no other particles are created and no energy is lost by any means from the collision.

Inelastic scattering A collision in which the target particle becomes excited or breaks up. This term can also be applied to a collision in which it is the incident particle alone or both which undergo such changes.

Nuclear reaction A collision between nuclei in which the nuclear constituents are rearranged.
Note that there is not a clear distinction between some nuclear reactions and inelastic scattering.

If a reaction has more than one light product, the extra products are grouped after the comma inside the bracket: thus

$$n + {}^{12}_{6}C \rightarrow {}^{11}_{5}B + n + p$$

would be written

$${}^{12}_{6}C(n,\ np){}^{11}_{5}B.$$

Photons and electrons can also cause nuclear reactions in the right circumstances so we include them in our list for which shorthand symbols exist.

7.3 Kinematics of nuclear collisions

There is no evidence contrary to the principles of the conservation of energy and of linear momentum and therefore we accept them as true. However, it is observed that, just as many chemical reactions produce energy, so do many nuclear reactions; by this we mean that the kinetic energy of the products is greater than that of the colliding particles. In nuclear and particle physics the change in the kinetic energy is called the ***Q*-value** of that reaction. For reactions in which there is an increase in the kinetic energy Q is positive. For other nuclear reactions there is a decrease in the kinetic energy and Q is negative. The positive Q reactions are said to be **exothermic**. Negative Q reactions are **endothermic**. If energy is conserved what is the origin of change in the kinetic energy? Of course, in both chemistry and physics, it is in the difference between the total rest mass of the colliding particles and that of the products so that the Q-value is just that mass difference multiplied by c^2. In Table 7.2 we give some examples of nuclear reactions and their Q-values.

In this chapter we shall draw several energy-level (term) diagrams; the first is shown in Fig. 7.3. The vertical scale is energy so that for the reaction $A(a,b)B$ there are two important levels; one is the the rest mass energy of A plus the rest mass energy of a; the other is the same for B plus b. If $Q>0$, then the first is above the second by an energy Q. On these diagrams the zero of total energy is suppressed but this is not significant for nuclear reactions.

Having invoked $E=Mc^2$, which comes from special relativity, we become inconsistent and do not bother about the other effects of relativity. In this chapter we shall deal only with low-energy collisions where velocities are much

Table 7.2 Examples of nuclear reactions.

We follow the usual practice of adding the Q-value as an arithmetic quantity to the reaction equation.

Exothermic reactions	$p+{}^{7}_{3}Li \rightarrow \alpha+\alpha+17.34$ MeV.
	$n+{}^{6}_{3}Li \rightarrow \alpha+t+4.79$ MeV.
	$d+{}^{10}_{5}B \rightarrow n+{}^{11}_{6}C+6.47$ MeV.
Endothermic reactions	$p+{}^{14}_{7}N \rightarrow {}^{3}_{2}He+{}^{12}_{6}C-4.78$ MeV.
	$\alpha+{}^{7}_{3}Li \rightarrow n+{}^{10}_{5}B-2.79$ MeV.
	$\gamma+d \rightarrow n+p-2.23$ MeV.

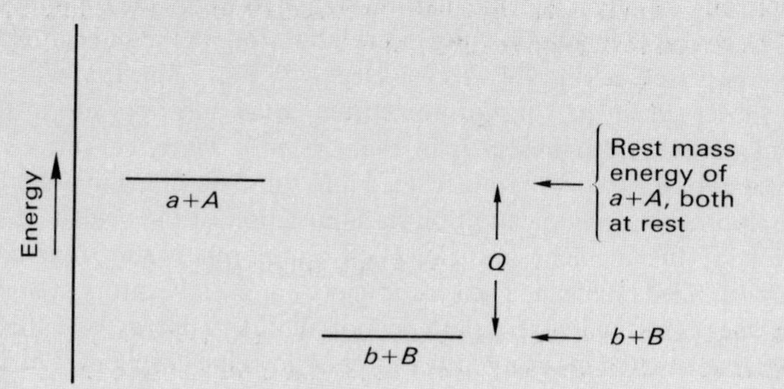

Fig. 7.3 Energy-level diagram for the reaction $a+A \rightarrow b+B$ for the case of $Q>0$ (exothermic reaction). Endothermic reactions will have the rest masses of $b+B$ greater than that of $a+A$.

less than c, and although Q must be included in the energy conservation equations, other relativistic effects may be neglected. So, beware, the kinematics which follows is non-relativistic! Note, however, that relativistic kinematics is straightforward, but at this point is an unnecessary luxury which would add nothing to the nuclear physics.

Many tables of atomic masses give the **mass defect**: it is defined as $M-A$, where M is usually in atomic mass units. Since total A is unchanged in low-energy nuclear reactions, Q is equal to the amount by which the total mass defect of the initial state particles exceeds that of the final state particles, multiplied by 931.5 MeV (see Section 2.2).

Consider now the collision of a moving particle label 1 with a stationary particle 2. This is shown in diagrammatic form in Fig. 7.4. This is a typical experimental situation and we call the coordinate system in which 2 is stationary the **laboratory system** (abbreviation: lab.). The coordinate system in which the two particles have equal and opposite moments is called the **centre-of-mass**

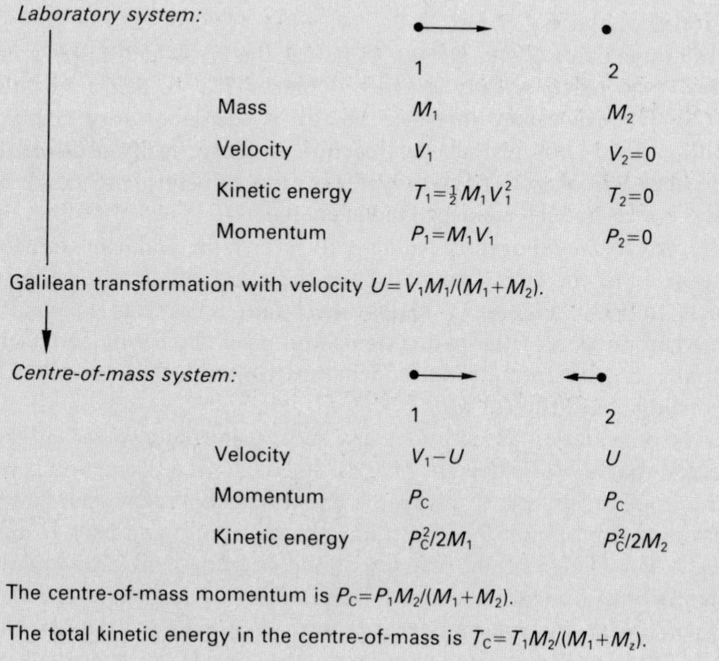

Fig. 7.4 In the laboratory a moving particle 1 approaches a collision with a stationary particle 2. This figure shows the workings of the Galilean (non-relativistic) transformation to the centre-of-mass of the collision.

system (c.m.): strictly it is defined as the system in which the total momentum is zero. The centre of mass is moving in the laboratory in the same direction as the incident particle (see Fig. 7.4). Its velocity (U in Fig. 7.4) is that which gives the target (observed in the c.m.) a momentum equal and opposite to that of the incident particle (also observed in that system). Thus the various velocities observed in one system are transformed into those observed in the other system by the appropriate vector addition or subtraction of the velocity U along the direction of the incident particle. Such a non-relativistic transformation is sometimes called Galilean. The kinetic energy in the laboratory system is T_1, the kinetic energy of the incident particle; not all of this energy is available for the collision because the motion of the centre of mass locks up a part of T_1. What is left is available and is the total kinetic energy in the centre of mass (T_C). The formula for T_C is given in Fig. 7.4. The formula for the centre-of-mass momentum (P_C) is also given and you are asked to verify these formulae in Problem 7.1.

The next step is to consider a simple two-body reaction:

$$1+2 \rightarrow 3+4$$

The Q-value is given by

$$Q=[(M_1+M_2)-(M_3+M_4)]c^2$$

where the M_i are the masses. The energy and momentum tied up in the motion of the centre of mass is unchanged by the reaction, so that the change in kinetic energy from $1+2$ to $3+4$, which is Q, is the same in both the laboratory and in the centre of mass. Thus the kinetic energy available to 3 and 4 in the centre of mass is $T'_C=T_C+Q$. In the c.m. system the two products 3 and 4 must separate with equal and opposite momentum (for the magnitude of these quantities we use the symbol P'_C, the prime distinguishing this final state c.m. momentum from that of the initial state) sharing the available energy T'_C so as to satisfy this condition. Look at Fig. 7.5

Of the three cases $Q=0$, $Q>0$, $Q<0$, it is the last which raises a new idea. Obviously the reaction cannot proceed unless $T'_C>0$, hence we must have $T_C>|Q|$. The **laboratory threshold energy** is the laboratory energy of the incident particle above which the reaction is energetically possible and it is greater than $|Q|$. Reactions with $Q>0$ can, in principle, proceed even if the collision occurs at zero incident kinetic energy.

To return to the laboratory we have to reverse the Galilean transformation that took us to the c.m. This will have the effect of throwing the two final products forward. Figure 7.5 shows the relation between the final state as observed from each of the two systems and gives the formulae which can be used make the transformation of kinematic quantities. You are asked to prove these formulae in Problem 7.2.

All the kinematics are most easily done using energy for all kinematic quantities: that is Mc^2 instead of M, Pc instead of P. The exception is velocity v, which is replaced by v/c to make the equations consistent and simple. Some details are given in Table 7.3 The transformations into and back from the c.m. require a knowledge of the masses of the reacting particles: it is normally sufficiently precise to use the atomic mass number, A, neglecting the mass defect (using, of course, the rest mass energy instead of mass (so that $Mc^2=938.5\times A$ MeV). If the precision implied in using A is considered inadequate, then it is

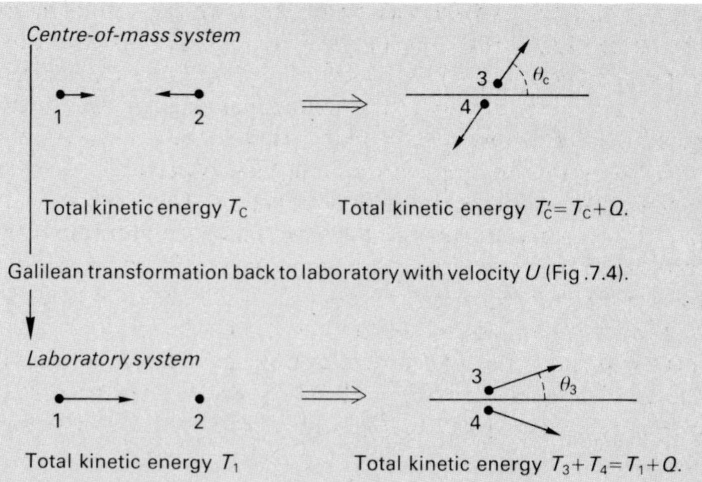

Fig. 7.5 The reaction $1+2 \rightarrow 3+4+Q$ before and after as seen in the centre-of-mass system and in the laboratory system. Note that in the centre of mass this two-body final state must have the products 3 and 4 parting company with equal and opposite momentum, P'_C. This is given by

$$\frac{P'^2_C}{2}\left(\frac{1}{M_3} + \frac{1}{M_4}\right) = T'_C.$$

The transformation back to the laboratory throws these products forward so as to regain the incident momentum in the direction of the incident particle. To see how the transformation is done, consider particle i ($i=3,4$) at angle θ_C. It has momentum P'_C which resolves into a transverse component $P'_C \sin\theta_C$ and a longitudinal component $P'_C \cos\theta_C$, where transverse and longitudinal refer to the direction of the incident particle. Only the longitudinal component changes in the transformation and in the laboratory particle i is at an angle θ_i and has momentum P_i given by

$$P_i \sin\theta_i = P'_C \sin\theta_C,$$

$$P_i \cos\theta_i = P'_C \cos\theta_C + M_i U.$$

probably also necessary to use relativistic kinematics; we do not take that position in the low energy (< 10 MeV incident energy) reactions.

The reader should attempt Problem 7.3: the second part involves finding the maximum energy of neutrons to be observed in the laboratory produced in each case of two well-defined two-body nuclear reactions with fixed incident particle energy. The incident energy and the Q-value determine the kinetic energy of the neutrons in the c.m. It is then a matter of deciding the direction of neutrons observed in the lab. That done, the c.m. energy and direction must be transformed to the lab. The alternative is to set up and solve the conservation of energy and momentum equations in the laboratory frame.

The discovery of the neutron by Chadwick (1932) is an example of the use of kinematics to make a significant discovery. At that time it was known that the bombardment of many light nuclei by α-particles from natural sources produced a penetrating radiation. This radiation was observed to eject protons from hydrogen-bearing materials and the initial interpretation was that the α-particles caused (α,γ) reactions and that the γ-rays were scattered from the protons in the proton equivalent of Thompson scattering (now called Compton scattering, Section 11.4) of photons by electrons. However, the energy of the

Table 7.3 Hints for doing kinematic calculations.

Instead of	(units)	use	(units)
Mass M	(kg)	Rest mass energy $= Mc^2$	(MeV)
Momentum P	(kg m s^{-1})	Momentum $\times$ $c = Pc$	(MeV)
Velocity V	(m s^{-1})	Velocity $\div$ $c = \beta$	(dimensionless)

Then, suppressing the c everywhere we have

$$\beta = \surd(2T/M),$$
$$P = \surd(2TM) = M\beta,$$

and all the formulae given in Figs. 7.4 and 7.5 become simpler to use. The reader should check the simplicity and consistency of the equations. Of course mass is often given in MeV/c^2 or GeV/c^2: the number is the rest mass energy in MeV or GeV respectively. Thus the velocity of a 1 MeV proton is given by

$$\beta = \surd(2/938.3) = 0.046.$$

But bear in mind that these are non-relativistic relations.

ejected protons was known approximately, and a simple calculation (Problem 7.4) showed that the γ-rays would have to have energies of order 50 MeV, unexpectedly high for γ-rays emitted by nuclei; in addition, if the producing reaction was $^{9}_{4}\text{Be}(\alpha,\gamma)^{13}_{6}\text{C}$ for example, the known binding energies, allowing for errors, prohibited γ-ray energies in excess of 14 MeV. Chadwick measured the range and hence estimated the velocities of protons and of $^{14}_{7}\text{N}$ nuclei recoiling from impacts by this radiation. The ratio of these velocities gave the mass of the particles of this radiation in terms of the proton mass M_p (Problem 7.5). Chadwick found $1.16 \pm 10\% \times M_p$. He correctly interpreted this as evidence for the existence of the neutron. Measurements made shortly after improved the precision on the value of the neutron to proton mass ratio.

DEFINITIONS AND KEYWORDS

Q-value This is the amount by which the sum of the rest mass energies of the initial participants of a nuclear reaction exceeds the sum of the rest mass energies of all the products of the reaction.

Exothermic reaction A reaction for which $Q > 0$.

Endothermic reaction A reaction for which $Q < 0$.

Mass defect The mass of an atom in atomic mass units less the atomic mass number. Sometimes multiplied by 931.5 MeV to give it the dimensions of energy.

Laboratory system The inertial frame in which a collision is observed.

Centre-of-mass system The inertial frame in which two colliding particles have equal and opposite momentum. More generally for a system of two or more particles it is the frame in which the vector sum of their momenta is zero.

Threshold An endothermic nuclear reaction cannot proceed unless the incident kinetic energy is sufficient to make the centre-of-mass kinetic energy $T_c \geq |Q|$. The equality signals the threshold.

Laboratory threshold energy The kinetic energy of the incident particle in the laboratory at the threshold for an endothermic reaction.

PROBLEMS

7.1 Prove the formulae for U, P_c, and T_c given in Fig. 7.4.

7.2 Prove the formulae given in the caption to Fig. 7.5.

7.3 Compute the Q-values for the reactions

$$d+d \rightarrow n+{}^3_2He,$$
$$d+t \rightarrow n+\alpha,$$

given:

mass defect of the neutron	=0.008 665 u,
mass defect of deuterium atom	=0.014 102 u,
mass defect of tritium atom	=0.016 050 u,
mass defect of helium 3 atom	=0.016 030 u,
mass defect of helium 4 atom	=0.002 603 u.

What are the maximum energies (in the laboratory) of neutrons that can be produced using 4 MeV deuterons incident on stationary targets of deuterium and of tritium? Use non-relativistic kinematics and assume nuclear masses are $A\times931.5$ MeV/c^2 in the transformations, but use the Q-values already calculated.

7.4 Show that, if the proton recoils forward with a kinetic energy of 2 MeV from photon scattering by an initially stationary proton, the incident photon energy is 32 MeV. (This problem requires relativistic treatment.)

7.5 Show that if a particle of mass m and velocity v elastically scatters from a nucleus of mass M, the greatest velocity the recoiling nucleus can have is $2mv/(m+M)$.

Hence the ratio of the greatest recoil velocities of protons (v_p) and of nitrogen nuclei (v_N) bombarded by neutrons of the same energy is given by

$$\frac{v_p}{v_N}=\frac{M_n+M_N}{M_n+M_p},$$

where M_n, M_p, and M_N are the masses of the neutron, proton, and nitrogen nucleus respectively. Chadwick's values were $v_p=3.3\times10^7$ m s^{-1} and $v_N=4.7\times10^6$ m s^{-1}. Find his value for M_n/M_p.

7.4 Conservation laws in nuclear collisions and reactions

We have accepted conservation of energy and momentum. In all the examples given we have made one other assumption and that is that the number of protons and the number of neutrons is separately conserved. We shall find circumstances in which this is no longer true but for this chapter, where we are considering non-relativistic induced nuclear reactions which are observable terrestrially, it is essentially true. Implicit in that assumption is the conservation of charge.

Angular momentum is also conserved. This influences the angular distribution of the products of a collision. However, we cannot discuss this subject any further in this book except in some simple cases.

Parity is another quantity which is conserved in nuclear reactions; this leads to some selection rules which can sometimes forbid reactions which would otherwise be possible, see Problem 7.6.

These conservation laws are summarized in Table 7.4. However, when we come to consider collisions at relativistic energies or those involving the weak interactions we shall find that these rules must be qualified and extended.

Table 7.4 Conservation laws in non-relativistic nuclear collisions

The conserved quantities are:

(1) energy and linear momentum;
(2) angular momentum;
(3) parity;
(4) number of neutrons and protons separately.

Warning! This is a list specific to low-energy nuclear scattering and reactions.

7.5 What can we learn from studying nuclear reactions?

Nuclear reactions are a physical phenomenon and as such deserve study and understanding. However, they help us to increase our knowledge in other areas and we have given a four-point list, surely incomplete, of such areas alongside. Let us look at it and see what each item means.

> *Comment*
>
> The areas of knowledge and of application that are increased by a study of nuclear reactions:
>
> (1) nuclear masses and energy levels;
>
> (2) nuclear size and structure;
>
> (3) nucleosynthesis in astrophysics and cosmology;
>
> (4) power generation and artificial isotope production.

1. *Nuclear masses and energy levels.* Nuclear reactions provide an excellent means for measuring nuclear mass differences: this happens because the loss or gain of kinetic energy is this mass difference and the conversion is

$$1\ \text{keV} \simeq 10^{-6}\ \text{atomic mass units}$$

A precision measurement of reaction Q-values gives very precise mass differences. In addition, as we shall find shortly, excited energy levels may be produced and studied in nuclear reactions and the excitation energies measured with considerable precision. This is the subject of **nuclear spectroscopy**.

2. *Nuclear size and structure.* Rutherford scattering is elastic scattering and deviations from the expected cross-section provided information on the size of nuclei; in Sections 3.2–3.5 we showed how elastic scattering of neutrons and electrons, in particular, provided information on the matter and charge distribution in nuclei. Many nuclear reactions have total and partial cross-sections which depend strongly on the state of internal motion of the constituents of the colliding particles and of the products. Thus understanding the mechanism of nuclear reactions allows us to probe the details of nuclear structure. This subject is almost completely beyond the scope of this volume. However, bear it in mind.

3. *Nucleosynthesis in astrophysics and cosmology.* The abundances of the naturally occurring isotopes on earth, in the stars, and throughout the universe are witnesses to their history. The fractional amount of helium created in the 'big bang' is a measured quantity and any model of that event must correctly account for that quantity. The conversion of hydrogen into heavy elements in the stars and supernovae is an obvious feature of the present development of the universe. A quantitative knowledge of nuclear reactions is an essential tool in understanding these subjects and we shall discuss some aspects in Chapter 14.

4. *Power generation and artificial isotope production* Nuclear power generation is now a matter of social and industrial significance. Many artificial isotopes have industrial and medical uses. We shall discuss briefly the physics of power generation by nuclear fission in Section 7.12 and the reactions of fusion, which are a potential future energy source, in Section 7.14.

7.6 Nuclear spectroscopy

Consider the reaction, $A(a,b)B$:

$$a + A \rightarrow b + B + Q.$$

The symbol B means the nucleus B in its ground state: however, it could be produced in one of a series of increasingly excited states, B_1^*, B_2^*, . . . , etc.

$$\begin{aligned} a + A &\rightarrow b + B_1^* + Q_1, \\ &\rightarrow b + B_2^* + Q_2, \\ &\rightarrow \text{etc.} \end{aligned}$$

The Q-values satisfy $Q > Q_1 > Q_2 > Q_3 > \ldots$. The energy-level diagram is shown in Fig. 7.6: clearly only those excited states, B_i^* can be produced for which $T_C + Q_i > 0$. For a fixed T_C, the kinetic energy of b decreases as Q_i decreases or goes more negative. Experimentally, if the beam of incident particles a is maintained at a fixed energy and the particles b are observed emerging from a target at a fixed angle, then the kinetic energy spectrum of this product will reflect the spectrum of excited states. This is shown in Fig. 7.7, where this situation is idealized:

1. The height of the peaks may vary from level to level under the influence of dynamic effects. Such influences or the conservation laws may even suppress the production of certain levels.
2. There may well be an experimental background on which the peaks will sit. Weak peaks may become impossible to see.
3. Imperfect energy resolution in the spectrometer, incident beam energy spread and thick target effects can all contribute to widening the observed peaks far beyond their natural line width and spoil the precision measurement of their energy.

Of course, we can have $b = a$ and then $B = A$, or A^*, or A^*, or A^*, or ..., and we have elastic or inelastic scattering of a by A. The Q-value is zero for the former and negative for the latter. Given adequate T_C the energy levels of A may be investigated.

What happens to the states B^* or A^*? We know already that if the energy of

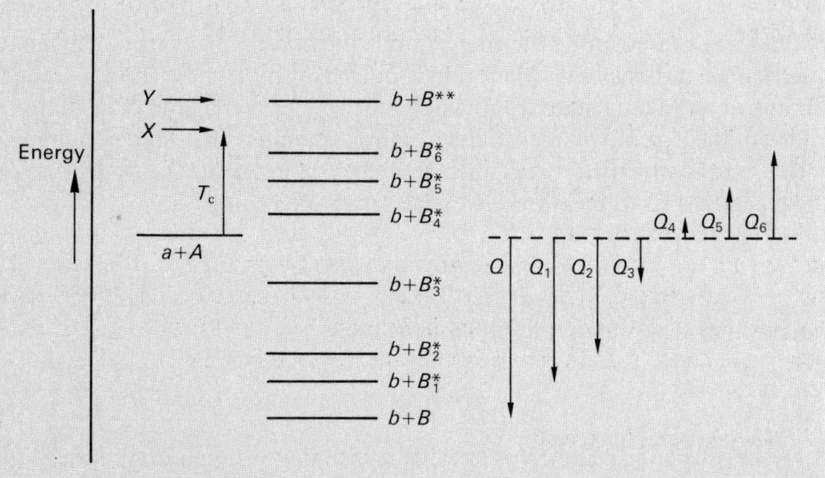

Fig. 7.6 This is an extension of Fig. 7.3 for the reaction $a + A \rightarrow b + B$ and is a diagram of the energy levels of the final state where nucleus B can be in its ground state or in one of its excited states B_i^*, $i = 1, 2, 3, \ldots$ In this example Q_1, Q_2, Q_3 are > 0, but Q_4, Q_5, and Q_6 are < 0. If T_C represents the centre-of-mass energy in the collision of $a + A$, then all the levels of $b + B^*$ can be reached which are below the energy X. Given that the level Y, labelled $b + B^{**}$, is the lowest energy at which it is possible for B to disintegrate, then the height of Y above $a + A$ gives the threshold for a three-body final state. The least expensive disintegration is usually the emission of a nucleon or an α-particle; then the height of $b + B^{**}$ above $b + B$ is the relevant separation energy (Section 4.3).

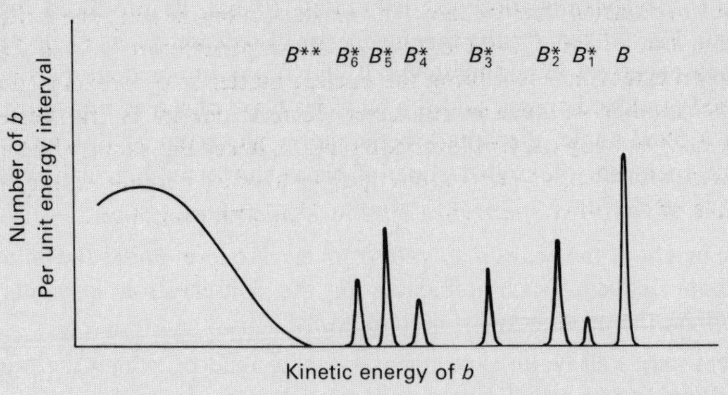

Fig. 7.7 The kinetic energy spectrum of particles b produced at a fixed angle with respect to the incident beam in the reaction $a + A \rightarrow b + B$ or B^*, corresponding to the energy levels of Fig. 7.6 and with an incident energy above the threshold (at Y in Fig. 7.6) for producing disintegration of B. Looked at from the view of the spectrum of b, decreasing kinetic energy for b corresponds to increasing mass for B^*. Below a certain energy for b, B can disintegrate and the total final state is at least three-body, and the spectrum of the kinetic energy of b is therefore continuous down to zero.

excitation is not too high (< the separation energy of a nucleon or of an α-particle), the most likely mode of de-excitation is by the emission of γ-rays. Devices exist which can measure such γ-ray energies in the range of a few kilovolts up to a few megavolts with a precision of the order of 1 keV (lithium-drifted germanium detectors, see Table 11.3). These devices are also usefully efficient. Thus a measurement of the spectrum of γ-rays from a target is a feasible way of measuring the energy levels. We have already met an analogous situation in radioactivity where an α or β transition may leave a daughter nucleus in an excited state from which it decays by γ emission (Fig. 5.2b).

One of the experimental difficulties in this technique is that of determining the nucleus from which the γ-rays are emitted: for example, the hopeful experimenter may wish to measure the excited state of B using

$$a + A \rightarrow b + B^*,$$

by detecting the χ-ray in the subsequent decay

$$B^* \rightarrow B + \gamma.$$

However, detecting γ alone will not distinguish that reaction from, for example,

$$a + A \rightarrow a + A^*$$

followed by $$A^* \rightarrow A + \gamma,$$

or any other reactions producing excited nuclei. The experimenter may have to do subsidiary experiments to check that his interpretation, or his method, is yielding the data required.

A very useful technique for exciting nuclear levels in preparation for measurement of the de-excitation γ-rays is **Coulomb excitation**. It is most useful

DEFINITIONS AND KEYWORDS

Nuclear spectroscopy The science of measuring and understanding nuclear energy levels and the transitions between them.

Coulomb excitation The inelastic scattering of one large-Z nucleus by another in which the nuclei do not come sufficiently close together to allow the nuclear force to act but in which the long range and strength of the Coulomb interaction can cause the excitation.

in heavy nuclei at low excitations. The nucleus to be studied is bombarded by heavy ions of sufficient energy to permit excitation by the long-range Coulomb interaction, but of insufficient energy to allow penetration of the Coulomb barrier to an extent that will bring the nuclear matter into contact and cause a significant number of nuclear reactions. Kinetic energy is transferred into energy of excitation by the Coulomb interaction, hence the name. The high Z of the nuclei involved helps to make for a high yield of excited nuclei. Figure 7.8 is an example of the γ-ray spectrum found in Coulomb excitation.

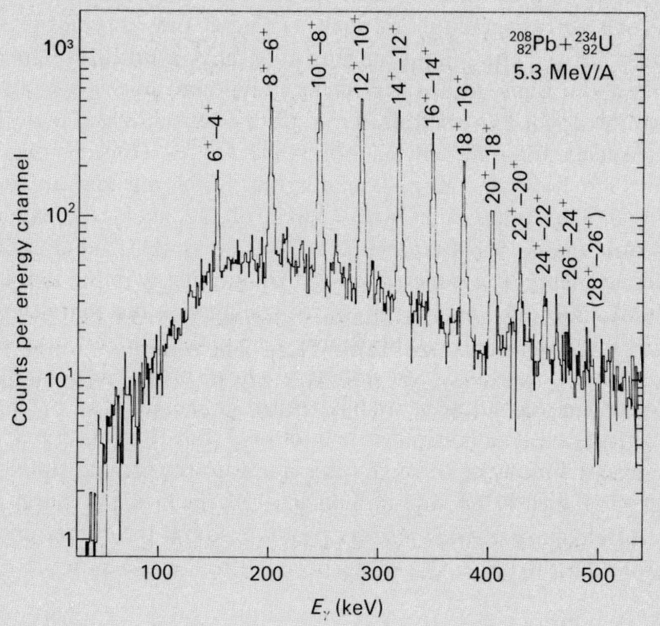

Fig. 7.8 The spectrum of γ-rays seen in a Ge(Li) detector from a target of $^{234}_{92}U$ bombarded by ions of $^{208}_{82}Pb$ with a kinetic energy of 5.3 MeV per nucleon (total 1.1 GeV) (Ower *et al.*, 1982). The peaks are due to γ-rays from de-excitation of a series of levels in the target; these levels are populated by Coulomb excitation of the target nuclei in not-too-close encounters by the incident particles. The peaks sit on a continuum of events due to γ-rays for which the energy is not totally contained in the detector and to background radiation. There are no γ-rays from $^{208}_{82}Pb$ in the range shown since the first excited state of this nucleus is at 2.6 MeV. Each line is labelled with the spin and parity of the initial and final nuclear state involved in the transition. An examination of the lines shows that a series of levels are involved having spin parity 4^+, 6^+, 8^+, 10^+, 12^+, ... These are associated with a rotational band in $^{234}_{92}U$. We shall discuss such bands in Section 8.10. The de-excitation by photon emission is discussed in Section 11.6.

In this particular example there is one complication not evident in the figure: the excited $^{234}_{92}U$ nuclei are recoiling (velocity about $0.1c$) when the γ-rays are emitted. It is necessary to detect both nuclei after the collision and de-excitation in order to correct for the Doppler shift of the observed photon energies. The spectrum shown is corrected for this effect.

The operation of Ge(Li) detectors is described in Table 11.3.

7.7 The compound nucleus model

In this section we discuss the **compound nucleus model** of nuclear reactions, introduced by Bohr in 1936. Consider an incident particle a and a target nucleus A. The model assumes that a enters the nucleus, where it suffers collisions with constituent nucleons of A until it has lost its incident energy and becomes an indistinguishable part of the nuclear constituents. This system is called a **compound nucleus**, compound because the nucleus contains both the incident particle a and the target nucleus A, amalgamated so that neither retains its identity. The compound nucleus is in an excited state, excited by both the kinetic energy of the incident particle and by the binding energy released when it is absorbed into the target nucleus. The compound nucleus must be unstable since, by definition, it can disintegrate into $a+A$, if not into other final states. The compound nucleus may execute many cycles of its natural period before it disintegrates or emits a γ-ray. Thus we expect that a compound nucleus is a distinct object if its mean life is much longer than the time of transit of the incident particle across the target nucleus. The latter time is typically $<10^{-22}$ s, which means that the compound nucleus must have a mean life greater than about 10^{-21}s and a line width, $\Gamma < 1$ MeV. Compound nuclei are observed to have line widths down to a small fraction of an electron volt.

How are such compound nuclei observed? Let us consider the reaction $A(a,b)B$: Fig. 7.9 shows an energy-level diagram which includes, in addition to the initial and final states, the energy levels of the nucleus C which is the compound of a and A. For example, C is $^{20}_{10}\mathrm{Ne}$ if a is $^{4}_{2}\mathrm{He}$ and A is $^{16}_{8}\mathrm{O}$. If the incident kinetic energy of a is increased, the kinetic energy in the centre of mass raises the total centre-of-mass energy (which includes the rest mass energies of a and A) of the initial state: if that total energy is equal to the rest mass energy of an excited state in C, that is C^*, then there is a large probability that nucleus C will be produced in that state. As we have noted, that state must be unstable: in this hypothetical case it can disintegrate into $a+A$, but also into $b+B$, or emit a γ-ray to de-excite directly or through other states to the ground state of C. The excited state C^* is clearly a compound nucleus. As the incident energy is varied, so that the total centre-of-mass energy passes through the compound nucleus rest mass, we expect to see a peak in the partial cross-sections for scattering and

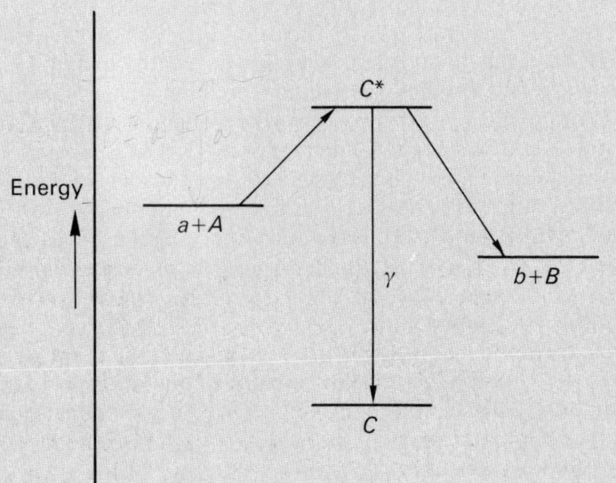

Fig. 7.9 An energy-level diagram showing the levels involved in two reactions proceeding through the same compound nucleus:

$$a+A \to C^* \to b+B$$

and

$$a+A \to C^* \to \gamma + C.$$

The compound state C^* will be formed when the incident kinetic energy raises the total energy (including rest mass energies) in the centre-of-mass to the rest mass energy of the state C^*.

for reactions that can proceed through the compound nucleus C^*, namely reactions such as

$$a+A \rightarrow C^* \rightarrow a+A \quad \text{(compound elastic scattering)},$$
$$\rightarrow b+B \quad \text{(compound nuclear reaction)},$$
$$\rightarrow \gamma+C \quad \text{(compound radiative capture)}.$$

If there is an excited state of A^*, then energy conservation may permit

$$a+A \rightarrow C^* \rightarrow a+A^* \quad \text{(compound inelastic scattering)}.$$

If there are peaks in partial cross-sections, then there must be peaks in the total cross-section: such total cross-sections for slow neutrons are rich in compound nucleus effects. Figure 7.10 shows the total cross-section for neutrons incident on indium as a function of kinetic energy in the range 5 meV to 10 eV. There are very pronounced peaks, each manifesting a compound nucleus: these peaks and the associated compound nucleus are frequently called **resonances**.

This process of preparation of a compound nucleus using a two-body collision in which the two bodies combine to give the compound nucleus is called **formation**. Note that a compound nucleus is defined by this mode of preparation. It is one of a subset of all nuclear states, ground and excited, whose members are prepared in a certain way. Thus a given compound state could

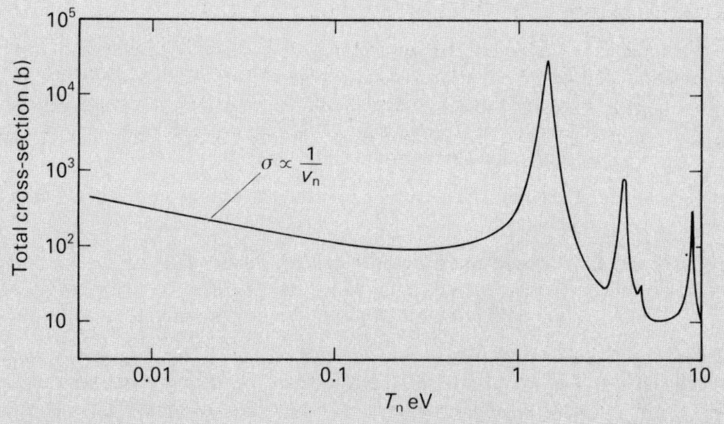

Fig. 7.10 The total cross-section for neutrons in natural indium (96% $^{115}_{49}$In) as a function of neutron kinetic energy T_n, in the range 5 meV to 10 eV. Note that both scales are logarithmic so that, in particular, the prominence of the peaks above the continuum is attenuated in the figure. These peaks are called resonances and each corresponds to the formation of a state of the compound nucleus ($^{116}_{49}$In). Between $T_n = 0$ and 1 eV there are no resonances, but the behaviour of the total cross-section is dominated by the effect of nearby resonances, including any excited states of $^{116}_{49}$In below $T_n = 0$. These effects give a total cross-section varying as v_n^{-1}, where v_n is the neutron velocity. At 4.7 eV there is a small peak due to a resonance for neutrons in $^{113}_{49}$In (4%).

alternatively be prepared in a different way: for example, in the hypothetical reaction

$$a + A \rightarrow c + C^*,$$

the nucleus C^* which we previously thought of as a compound nucleus is produced but it is not a compound of the colliding reactants; this is called **production**. It is important to understand that the *production* of such a nucleus does not lead to resonance peaks in the production cross-section.

The disintegration of a compound nucleus we shall refer to as **decay**.

The concept of the compound nucleus model can be extended to elementary particle physics, where some collision cross-sections exhibit resonances. In this area the formation process does not lead to a nucleus but to an excited state of quarks. Therefore we shall replace the term compound nucleus by the **compound state**.

In fact the compound nucleus idea has a wider implication than the excitation of a single, well-defined state. At incident energies in the range typically 1–20 MeV, the cross-section may not show peaks but none the less a compound nucleus may be formed. At these energies, the number and width of states is such that they overlap and at low resolution (for example, a large incident beam energy spread) the cross-section observed will be an average over a certain range. If the resolution is increased the cross-section may be observed to fluctuate with energy as the effects of many individual states come and go. We return very briefly to this point in Section 7.10. In this range, the number of different ways the compound nucleus can decay may be very large. This regime is called the **continuum** (or **statistical**) **region**.

DEFINITIONS AND KEYWORDS

Compound nucleus model A model of nuclear reactions in which an intermediate state is formed from the combination of incident particle and target nucleus.

Compound nucleus The intermediate state formed in a compound nucleus reaction. It is normally one of the excited states of the nucleus formed by the combination of the incident particle and target nucleus.

Resonance A peak in a cross-section that can be associated with a compound nucleus or state. Sometimes refers to the compound nucleus or state itself.

Formation The preparation of a compound nucleus by the combination of the incident and target particles at the appropriate energy.

Production In contrast to formation, a compound state in question is prepared by a mechanism in which it is not a compound of the incident and target particles.

Comment The appearance of a peak (resonance) in the cross section that has a total energy which identifies it as the compound state in question occurs only in formation. In production the cross section will be a smooth function of incident energy unless a compound state of the colliding particles is formed.

Decay The term used to mean the distintegration of the compound nucleus.

Compound state A more general term than *compound nucleus* as it includes the possibility, for example, of particles undergoing formation not into nuclei but into other particle states.

Continuum (or **statistical**) **region** Compound state formation in a region where resonances are becoming wide and dense, and where they overlap to an extent as to be unresolvable individually.

7.8 Compound state properties

First another matter of terminology: the identity of two particles which, in a collision, can combine to form a compound state defines what is called a **channel**. A state reached by decay is also a channel. There must be at least one channel and there can be more. In the hypothetical example of Section 7.8, (aA) (bB), (γC), and (aA^*) would all be channels. Channels of more than two bodies are also possible. All channels will occur on decay. Only some two-body channels are usable as entrance channels due to the difficulties of arranging three-body collisions or collisions of a particle with an excited nuclear state. Since, by definition, a channel (for example aA) communicates with its compound state (C^*), the compound state mass must be greater than the **channel mass**, that is in this example, $M_{C^*} > M_a + M_A$.

The decay of a compound state is identical in physical principles to the decay of a radioactive nucleus. Multichannel compound state decay equals multimodal radioactive decay. Therefore all the statements about transition rates, branching fractions, line shape and line width which were made in Sections 2.3 and 2.8 apply equally to compound states. The difference is that most compound states have mean lives too short to be directly observable and that most radioactive decays have a line width too small to be directly observable.

Implicit in the model is the idea that the compound nucleus very soon loses memory of how it was prepared. This is Bohr's **hypothesis of independence**: preparation and decay of a compound state are independent. This means that the transition rates into different channels are independent of how the state was prepared and we are confirmed in using the ideas of radioactivity. In fact, one of the ideas implicit but never stated in the usual treatment of radioactivity is that the intrinsic properties of a radioactive material are independent of how or when it was generated. Figure 7.11 shows the situation used in an early test of Bohr's hypothesis.

There is an exception to the compound state's amnesia. If it has non-zero

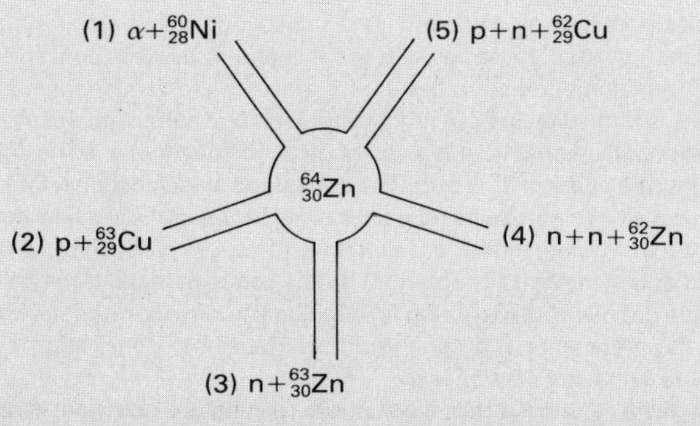

Fig. 7.11 The compound nucleus in ${}^{64}_{30}Zn$ with five of its channels. The test of the hypothesis of independence was made by Ghoshal (1950): he measured the branching fractions as a function of excitation energy into channels 3, 4, and 5 and showed that, at each energy, they were independent of whether the state was formed by entry through channel 1 or 2. The experiments that Ghoshal did were designed to cover the same energy range of excitations of ${}^{64}_{30}Zn$ and this involved incident proton energies from 5 to 33 MeV in channel 2 and α-particle energies from 12 to 40 MeV in channel 1. These energies put the regime into that described at the end of Section 7.8: many overlapping states in the compound nucleus. Ghoshal detected the reactions by the β-activity of the final nucleus in each of the channels 3–5, so that he was summing over all subchannels which could lead to the same final nucleus; an example of one such channel would be that in which the compound nucleus decayed to a nucleus in one of its excited states rather than in its ground state. His results were consistent with Bohr's hypothesis of independence.

total angular momentum (spin, quantum number j) and if this spin does not interact with its surroundings to an appreciable extent during its lifetime, then conservation of angular momentum can often ensure that it remembers the axis along the direction of the incident particle in the case of preparation by formation. How does this memory work? We label the incident direction the z-axis: the magnetic substates of the compound state, j_z, of j are populated in the formation process. (Of course, we mean many examples of the compound state, all identically prepared.) If the population is uniform then, on decay, the intensity of any decay product is the same in all directions (isotropic) in the centre of mass. If non-uniform, then the distribution can be non-isotropic but has to be symmetric about 90° to the z-axis, and the information about the initial direction is available in the final state (see the Comment). The important fact is that the decay products are distributed uniformly or, if non-uniformly, symmetrically forward–backward. This memory is made more complex if either or both of the colliding particles is polarized or if other reaction mechanisms are present and have amplitudes that interfere with the compound state amplitude or if the state is not sufficiently isolated from neighbouring compound states. We are not in a position to explore further the consequences of this situation in this volume, but we can note that it permits the extraction of information about the spins of the states involved.

The Lorentzian line shape of Fig. 2.6 applies to a single, isolated compound state: however, its manifestation as a peak in a formation cross-section is influenced by other factors which are to do with, among others, the probability of formation and decay varying with energy, with the effect of nearby states and

Comment

Why are compound state decays symmetric about 90°? Let us consider the collision of two spinless particles leading to a compound state and set up a z-axis along the direction of the momentum of the lightest of these particles in the centre of mass. By conservation of angular momentum, if the compound state has spin j, then the incident channel from which the compound state is formed must have orbital angular momentum with quantum number $l=j$. But that orbital angular momentum is a vector perpendicular to the z-axis and therefore the eigenvalue (l_z) of the operator L_z is zero. The compound state must therefore have $j_z=0$ and its wave function allows no observable distinction between the $+z$ and the $-z$ directions. The decay products will therefore be emitted with equal probability in the $+z$ and $-z$ hemispheres, that is symmetrically about 90° in the centre of mass. If the colliding particles have spin but are unpolarized, they bring no additional information about a direction into the collision and the arguments and conclusion are the same. Polarization complicates the situation and the symmetry may be lost.

However, if there is some other mechanism causing the reaction in addition to the compound state mechanism, then other scattering amplitudes of different l and j may be involved and interference effects can destroy the symmetry.

(This argument about the symmetry in the decay of a compound state depends on the assumption that parity is conserved in the decay process. In Chapter 12 we shall show that parity is not conserved in the weak interactions. The formation and decay of the spin 1 Z^0 bosons in electron–positron annihilation near 91 GeV

$$e^+ + e^- \rightarrow Z^0 \rightarrow e^+ + e^-$$

is an electroweak process in which parity is not conserved (Sections 12.4 and 12.9). The result is that the intensity of final state positrons is not symmetric about 90° in the centre of mass with respect to the collision axis. An asymmetry is also expected in other decay channels.)

with the possibility of interference with alternative reaction mechanisms. This is another subject beyond our present aims.

We now use a few examples to look at some features of compound state properties.

1. *Formation by neutrons.* An example of a low-energy neutron total cross-section that exhibits peaks at certain energies was used in the last section to illustrate the effects of compound nucleus formation. It is the absence of any Coulomb barrier that leads to such effects with low-energy neutrons. Technically it is now possible to produce controlled beams of low-energy neutrons and here low means down to thermal energies. (Thermal is the word used because such beams are sometimes produced by cooling neutrons by diffusion through cold matter. For orientation, 273K = 35 meV kinetic energy). Many medium and large A nuclei can **capture** such neutrons very readily. The neutron separation energy for the nucleus $A+1$ is normally in the range 5–8 MeV; thus the capture of a slow neutron by nucleus A leads to a compound state with an excitation energy above the ground state by this separation energy. Such excitation often occurs in a region of high density of narrow states, which in turn shows up as a rich structure of resonances in the neutron total cross-section. Note that these circumstances permit an examination of energy levels at an excitation of about 8 MeV with a resolution which can be of the order of 1 meV.

Now look again at Fig. 7.10, which shows the total cross-section for neutrons on natural indium (96% $^{115}_{49}$In, 4% $^{113}_{49}$In) in the energy range 50 meV to 10 eV. This shows three resonances. Look at that at 1.46 eV: it reaches a total cross-section of 28 000 barns (2.8×10^6 fm^2) when the projected nuclear area is about 1 barn (100 fm^2). Therefore, to neutrons of this energy the cross-section to capture the neutron to form this corresponding compound state, an excited state of $^{116}_{49}$In, is nearly thirty thousand times the nuclear cross-section. How does this come about?

Of course, it is a quantum-mechanical effect. The collision cross-section at low energies is dominated, not by the target size, but by the reduced de Broglie wavelength of the incident particle ($\lambda\!\!\!^{-} = \hbar/P$). The relevant momentum is that of the centre of mass, P_{C} (see Fig. 7.4). The associated area is $\pi\lambda\!\!\!^{-2}$. For slow neutrons on medium and heavy nuclei the laboratory and centre-of-mass momentum are almost the same, so we can give a simple formula for this area in terms of the neutron's laboratory kinetic energy, T_{n}:

$$\pi\lambda\!\!\!^{-2} = 6.5 \times 10^7/T_{\mathrm{n}}(\mathrm{eV})\ \mathrm{fm}^2$$
$$= 6.5 \times 10^5/T_{\mathrm{n}}(\mathrm{eV})\ \mathrm{barns}.$$

Thus at $T_{\mathrm{n}} = 1.46$ eV (Fig. 7.9) the cross-section could reach 4.5×10^5 barns. So the observed 28 000 barns is really quite small compared to what could be. In fact, for neutron total cross-sections the maximum cross-section is $4\pi g\lambda\!\!\!^{-2}\Gamma_{\mathrm{n}}/\Gamma$, where g is a statistical factor depending on the spins of the neutron, target and compound nucleus, $\Gamma/\hbar$ is the total decay transition rate of the compound state to all channels and $\Gamma_{\mathrm{n}}/\hbar$ is the transition rate for the compound state to the neutron plus target channel. Since $\frac{1}{4} \leqslant g \leqslant \frac{3}{4}$ (depending on the target and the compound state spins), we see that for this case $\Gamma_{\mathrm{n}}/\Gamma < 0.06$. Thus at the most 6% of this compound nucleus returns to the incident channel. This branching fraction has to be less than 1 but the fact that it is usually very much less than one for slow neutron resonances is what we shall now discuss.

2. *Decay by photon and neutron emission.* We put these two together since the relative probabilities have important consequences. For compound nuclei where the emission of either a high energy (>1 MeV) neutron or γ-ray are possible, the former is much more likely. However, for compound states formed by slow neutron capture the situation is different. If a neutron is emitted it can only be of the same low kinetic energy as that of the captured neutron. The compound state is usually several megaelectronvolts above the ground state so that photon emission is relatively very energetic and as such can become much more likely than slow neutron emission (normally there are no other open channels). This energy situation is exemplified in $^{115}_{49}\text{In}+\text{n}\rightarrow^{116}_{49}\text{In}^*$ at the 1.46 eV resonance shown in Fig. 7.10. This compound nucleus is 6784 keV above the ground state of $^{116}_{49}\text{In}$; there are many more levels below this particular one and many routes for de-excitation by emitting γ-rays, most involving energies in the range 100 keV to several MeV. One of these is shown in Fig. 7.12. Once formed this compound nucleus has a 4% probability of emitting a neutron (of 1.46 eV kinetic energy, neglecting recoil) and a 96% probability of taking one of these routes to the ground state of $^{116}_{49}\text{In}$ ($\Gamma_n/\Gamma_\gamma=0.04$, where $\Gamma_\gamma/\hbar$ is the total transition rate for the radiative decay of the compound state). This is typical and the consequence is that **radiative capture** of slow neutrons is the dominant mode in slow neutron absorption by medium and heavy nuclei.

There is an interesting but rather mechanical argument as to why neutron emission is so unlikely. On capture, the slow neutron shares its 6–8 MeV of binding energy among A nucleons. The probability that the consequent motions reconcentrate that energy on to one neutron in a certain time must be small and become smaller as A increases. The conclusion is that Γ_n will decrease as A increases. The reader will have no difficulty in recognizing this as unsatisfactory because it is not an argument based on quantum mechanics. (Can you think of a better one? See Problem 7.11.) However, the conclusion is correct. For example: at $A\simeq70$, Γ_n is typically 0.3 eV for a resonance at a neutron kinetic energy of

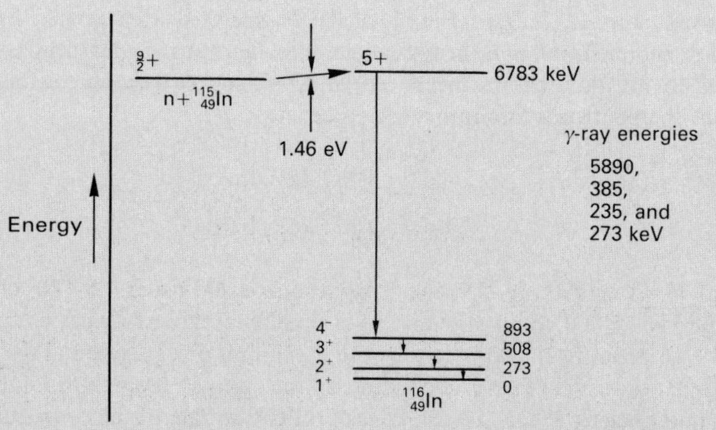

Fig. 7.12 Some of the energy levels involved in compound radiative capture of 1.46 eV neutrons by $^{115}_{49}\text{In}$. The compound nucleus is an excited state at 6783 keV above the ground state of $^{116}_{49}\text{In}$. A very large number of other excited states are not included in the figure. The dominant decay of the compound nucleus is by γ-ray emission and the figure shows one of many routes to its ground state.

1 eV but has fallen to about 0.002 eV at $A \simeq 230$. Of course, we really need a proper quantitative comparison of neutron and photon emission transition rates to have a convincing and full understanding of the importance of radiative neutron capture.

3. *Decay of a compound state by charged particle emission.* In this case the Coulomb barrier inhibits the emission of charged particles. As the energy increases the barrier is more easily penetrated. We expect therefore that at low energies the appropriate partial cross-section will be small and any resonances will not be strong: as the energy increases resonances should increase in strength. Figure 7.13 shows the cross-section for the reaction $^{23}_{11}Na(n,p)^{23}_{10}Ne$ as a function of the kinetic energy of the incident neutron. There is the expected increase in the visibility of the effects of resonances as that energy increases.

4. *Decay of a compound state by fission.* This is a very important process, particularly following formation by the absorption of a slow neutron, and is discussed in Section 7.12.

5. *Compound state formation by charged particles.* This is the reverse of the decay by charged particle emission and we expect that the formation of resonances or compound states will be suppressed at low incident kinetic energies and will become more evident as that energy increases. In this case, and in a previous one of decay by charged particle emission, further increase in energy may take a reaction beyond the resonance and continuum regions.

DEFINITIONS AND KEYWORDS

Channel A system which is produced in the decay of a compound state is called a channel to that state. Some of the two-body channels can be used for the formation of the compound state.

Channel mass The sum of the rest masses of the nuclei or particles in that channel.

Hypothesis of independence That the preparation and decay of a compound state are independent processes.

Capture The process in which a target nucleus absorbs an incident particle to form a compound state.

Radiative capture Capture in which the compound nucleus formed subsequently decays by emitting one or more γ-rays.

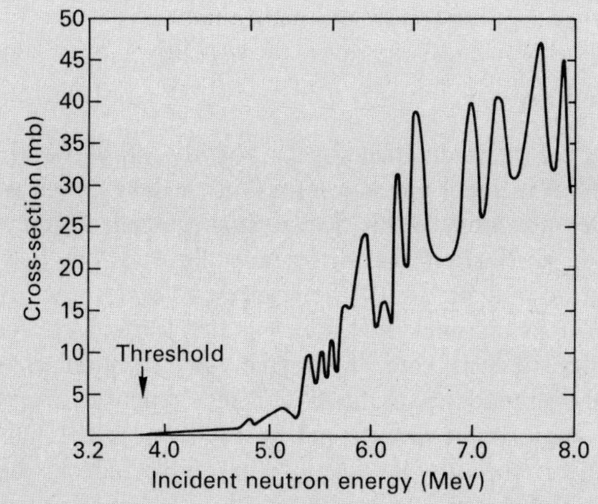

Fig. 7.13 The cross-section for the reaction $^{23}_{11}Na(n,p)^{23}_{10}Ne$ ($Q = -3.59$ MeV) as a function of incident neutron energy from 4 to 8 MeV. The threshold is at 3.75 MeV. The Coulomb barrier to the outgoing proton has a maximum height of about 4 MeV. Throughout the energy range the cross-section shows considerable structure due to compound nucleus formation: however, such states do not show up between threshold and 5 MeV because the Coulomb barrier suppresses the probability of proton emission. Above neutron energies of 5 MeV the barrier becomes increasingly transparent to the outgoing proton. This graph is based on data obtained by Williamson (1961). A smooth curve has been drawn through experimental points which have errors of about 15% and it therefore does not give an accurate representation of the actual cross-section. The structure in this curve is probably due to the fluctuations caused by the sum of many overlapping resonances.

6. *Compound states in elementary particle physics.* Many compound states have been found in the cross-sections for the scattering of π- and K^--mesons by nucleons (see, for example, Fig. 10.2). These were called resonances and are known to be excited states of three-quark systems. Resonances of other three-quark and of quark–antiquark systems that are not accessible by formation were discovered in production. The transit time is now 10^{-24} s, so these resonances have mean lives greater than this: the great majority of well-established resonances have total widths less than 200 MeV ($\tau > 3 \times 10^{-23}$ s). The role of quarks in particle physics is the subject of Chapter 10.

Bohr's original formulation of the compound nucleus model of nuclear reactions envisaged the incident particle sharing its kinetic energy among the constituents of the nuclear target and this compound having a mean life long relative to the time the incident particle would take to traverse this nucleus if it suffered no collision. The concept has been modified by stressing the compound state rather than the compound nucleus. In 1989, the high energy collisions of electrons and positrons have been observed to produce the Z^0 particle (Fig. 12.8 and Table 12.5):

$$e^+ + e^- \to Z^0$$

This reaction gives a resonant peak in the annihilation cross-section at a centre of mass energy of 91 GeV, with width about 2.6 GeV. This is clearly compound state formation although there is no sharing of incident energy among constituents since all the particles involved appear to be point-like (in so far as there is no evidence that any has internal structure). Thus we have moved on from the idea that the colliding particles survive as entities in the compound state.

7.9 Direct reactions

The compound state is characterized by its long life compared with the expected time of transit of an incident particle across the nucleus. Nuclear reactions that occur in a time comparable to this transit time are called **direct reactions**. The cross-sections for such reactions vary smoothly and slowly with energy in contrast to the resonant nature of the compound state reactions observed as peaks in the incident channel. In addition, direct reaction cross-sections are comparable to the nuclear area. They also have another property which is important. The energetic products of a direct reaction are not distributed isotropically in angle but concentrated at angles near the incident direction (peaked forward, in the jargon of the trade). This reflects the fact that the incident particle makes only one, or very few, collisons with nucleons in the target nucleus and its forward momentum is not transferred to an entire compound state. This property serves to distinguish direct from compound state reactions where, as we have seen, the distributions are isotropic or symmetric forward–backward.

Direct reactions are an important source of information about nuclear structure. The **stripping reactions** (d,p) and (d,n) occur more readily if there is an obvious vacancy in the nucleus for the nucleon stripped from the passing deuteron: then the cross-section and angular distribution of the ongoing nucleon (proton or neutron respectively) give information about the wave function and angular momentum of the captured nucleon in the vacancy filled. Conversely, the **pick-up reactions** such as (p,d) and (n,d) give information about

DEFINITIONS AND KEYWORDS

Direct reaction A nuclear reaction which occurs in a time comparable with the time of transit of the incident particle across the nucleus. This time is typically 10^{-24} to 10^{-22} s.

Stripping reaction A direct reaction in which one or more nucleons are transferred to a target nucleus from a passing, necessarily complex, incident fast nucleus.

Pick-up reaction A direct reaction in which one or more nucleons are transferred from a target nucleus to a passing incident fast particle.

the wave function of the nucleon (neutron or proton respectively) prior to pick-up by the incident nucleon.

PROBLEMS

7.6 Define the terms **atomic mass unit, mass defect,** and **average binding energy per nucleon** as applied to the neutral atom. Draw a diagram showing how the average binding energy per nucleon varies with atomic mass number and discuss briefly its significance for the stability of nuclei of large mass number.

When the ${}^{12}_{6}C(\alpha,\gamma)$ cross-section is measured, a peak is found at an α-particle energy of 7045 keV. Explain this observation and find the energy of the γ-ray emitted. Discuss whether it would be possible to confirm your explanation by measurement of the ${}^{15}_{7}N(p,\alpha)$ cross-section as a function of proton energy.
(The mass defects of the proton, α-particle, ${}^{15}N$, ${}^{16}O$ are 7289, 2425, 100, and −4737 keV respectively.)

(Adapted from the 1977 examination of the Final Honours School of Natural Science, Physics, University of Oxford.)

7.7 The reaction ${}^{13}_{6}C(d,p){}^{14}_{6}C$ has a Q-value of 5.952 MeV and a resonance when the incident deuteron has a laboratory energy of 2.45 MeV. Do you expect the reaction ${}^{11}_{5}B(\alpha,n){}^{14}_{7}N$, Q=0.158 MeV, to have a resonance? If so, estimate the laboratory kinetic energy of the α-particle at which it would occur. The following pieces of information are needed to answer:

1. The β-decay ${}^{14}_{6}C \rightarrow {}^{14}_{7}N + e^- + \bar{\nu}_e$ has a Q-value of 0.156 MeV.
2. The neutron-hydrogen atom mass difference is 0.782 MeV/c^2.

7.8 Define the **Q-value** and the **threshold** of a nuclear reaction.
Outline the compound nucleus hypothesis in nuclear reactions and give one piece of experimental evidence for the hypothesis. Describe conditions under which it will fail.

The table shows the energies of the gamma rays and alpha particles emitted when resonances in the ${}^{19}_{9}F$+p reaction are studied.

Proton resonance energy (laboratory) keV	Gamma ray energies MeV	Alpha particle energies (centre of mass) MeV
668	6.13	1.30
	6.92	1.47
	7.12	2.10
843	no gamma rays	7.14
874	6.13	1.46
	6.92	1.62
	7.12	2.25

Draw a diagram to show the levels in ${}^{20}_{10}Ne$ and ${}^{16}_{8}O$ involved in these reactions. Suggest why there is no gamma radiation accompanying the (p,α) reaction at 843 keV. What experiments might be carried out to verify your interpretation of the levels in ${}^{16}_{8}O$?

(Adapted from the 1980 examination of the Final Honours School of Natural Science, Physics, University of Oxford.)

7.9 The radius of a nucleus of mass number A, determined by the measurement of the elastic scattering of high-energy electrons, can be approximated by the expression $R=1.2\times A^{1/3}$ fm. The cross-sections for the radiative captue of 0.025 eV neutrons by many nuclei are of the order 10^{-24} to 10^{-23} m^2. Explain why these two statements are not inconsistent.

The radiative capture cross-section, as a function of neutron energy, has the following general form for many nuclei: for energies <1 eV the cross-section is proportional to v^{-1}, where v is the neutron velocity; for energies >1 eV, the cross-section varies rapidly with energy, having narrow peaks superimposed on a relatively small background. Give a qualitative explanation of this energy dependence.

Describe an experimental arrangement suitable for the measurement of a radiative capture cross-section.

(Adapted from the 1977 examination of the Final Honours School of Natural Science, Physics, University of Oxford.)

7.10 Describe briefly and explain the principle of operation of detectors for (a) thermal neutrons and (b) neutrons of energy 100 MeV.

How may a suitable neutron beam and a target with atomic number and mass number (Z,A) be used to investigate

(1) the excited states of the nuclide $(Z,A+1)$,
(2) the excited states of the nuclide $(Z-1,A)$?

An anomaly is observed in the scattering of 10 MeV α-particles by hydrogen. At what energy does this anomaly appear in the scattering of protons by helium?

(Adapted from the 1975 examination of the Final Honours School of Natural Science, Physics, University of Oxford.)

7.11 Can you find an explanation for the fact that radiative de-excitation is more likely than re-emission of the neutron after slow neutron capture, other than that given in the text, Section 7.8?

7.10 Compound state to direct

By their nature compound nucleus reactions occur at excitations above the ground state of the compound nucleus up to about 8–10 MeV. At these excitations the energy levels are very dense and are becoming wider so that they begin to overlap. Well before this situation arises the analysis of the energy levels in terms of nuclear structure and states of motion of the constituents is an impossible task for the many-body system that is a nucleus. However, the average of the density of the states, of the widths and of the strengths can be predicted satisfactorily. As the densities increase the states can no longer be resolved and at any one energy a reaction cross-section is the sum of the effects of several resonances. As energy increases that sum (cross-section) will fluctuate but the fluctuations decrease as more and wider resonances sum until the fluctuations disappear and the energy variation of the cross-section becomes smooth. In this way the reaction mechanism changes from a regime of compound state formation to one of essentially a direct reaction mechanism. In the region of overlapping resonances, the mean and root mean square of the fluctuations can be explained.

However, the picture painted is certainly over-simplified. In compound nucleus formation we envisage the incoming particle interacting many times with nucleons in the target nucleus until its energy is shared between all the constituents. In contrast we see a direct reaction involving only perhaps one collision between the incident particle and one constituent nucleon. However, there must be a whole range of possibilities from a single collision through two, three, four, or more, to the many of compound state formation.

7.11 Elastic scattering

Not only nuclear reactions but also elastic scattering is an integral part of the properties of nuclear collisions. The reader will be aware that an absorbing optical target diffracts light; that diffraction is the elastic scattering of light. The situation is identical in nuclear and particle physics: if the target acts in any way as an absorbing target then there must also be elastic scattering. For example, consider neutron scattering by an absorbing nucleus which captures every neutron with classical impact parameter (Section 1.2) less than some r (which might be the nuclear radius R). Then the absorption cross-section would be expected to be about πr^2. It can be shown that, in this case, the elastic cross-section is also about πr^2, giving a total cross-section of about $2\pi r^2$. However, although it is true that if there is absorption there must be elastic scattering, there can be elastic scattering without absorption—for example Rutherford scattering by a nucleus.

Compound state formation always contains some elastic scattering because if the state can be reached by a certain channel, it can also decay by that channel. As energies increase in nuclear reactions and the compound states become more dense and wider, there will be, as mentioned already, fluctuations in the elastic scattering cross-section with energy which finally smooth out, leaving total elastic sections varying smoothly with energy. This brings us back to one of the subjects of Chapter 3, the optical model, which is used to describe this situation. The reader who wishes to be reminded is directed back to Section 3.7.

7.12 Induced fission and the fission reactor

In Chapter 5 we briefly discussed spontaneous fission, a process occurring with increasing probability as Z increases, significantly so above $Z=92$. In the large-Z nuclei, fission can be induced by energy transfer to the nucleus. A very important way of doing this is by neutron capture. The compound nucleus formed is excited above its ground state by the neutron separation energy of the ground state if the compound nucleus has been formed by slow neutron capture (see Section 7.8). This separation energy is greater in even–even nuclei than it is in Z-even, N-odd nuclei due to the effect of the pairing term. Thus the excitation energy is greatest for capture by Z-even, A-odd nuclei, and can be close to or greater than the activation energy. Thus in these nuclei, slow neutron capture has a large probability of causing fission. Such nuclei are said to be **fissile**: examples are $^{235}_{92}U$ and $^{239}_{94}Pu$. The products of fission are energy, two fission products and several neutrons (between 2 and 3, on the average, depending on the nucleus and on the energy of excitation). These neutrons can cause further fissions and the cycle can lead to a self-sustaining or explosive chain reaction in a macroscopic volume containing fissile material under the right conditions. The parameter which describes the regeneration of neutrons is called the **reproduction factor** and it is usually represented by the symbol k. It is essentially the ratio of the number of fissions in one generation to the number in the previous generation. Clearly k must be equal to 1 for a steady, self-sustaining chain reaction and greater than 1 for an explosive chain reaction. Let us briefly consider the conditions that can influence the value of k.

Figure 7.14 shows the probabilities of various reactions in the two major isotopes of uranium $^{235}_{92}U$, $^{238}_{92}U$ as they occur in natural uranium (0.7% and 99.3% respectively). The probabilities are expressed as probability per 10^{-6} metre of path in the metal. The important features are:

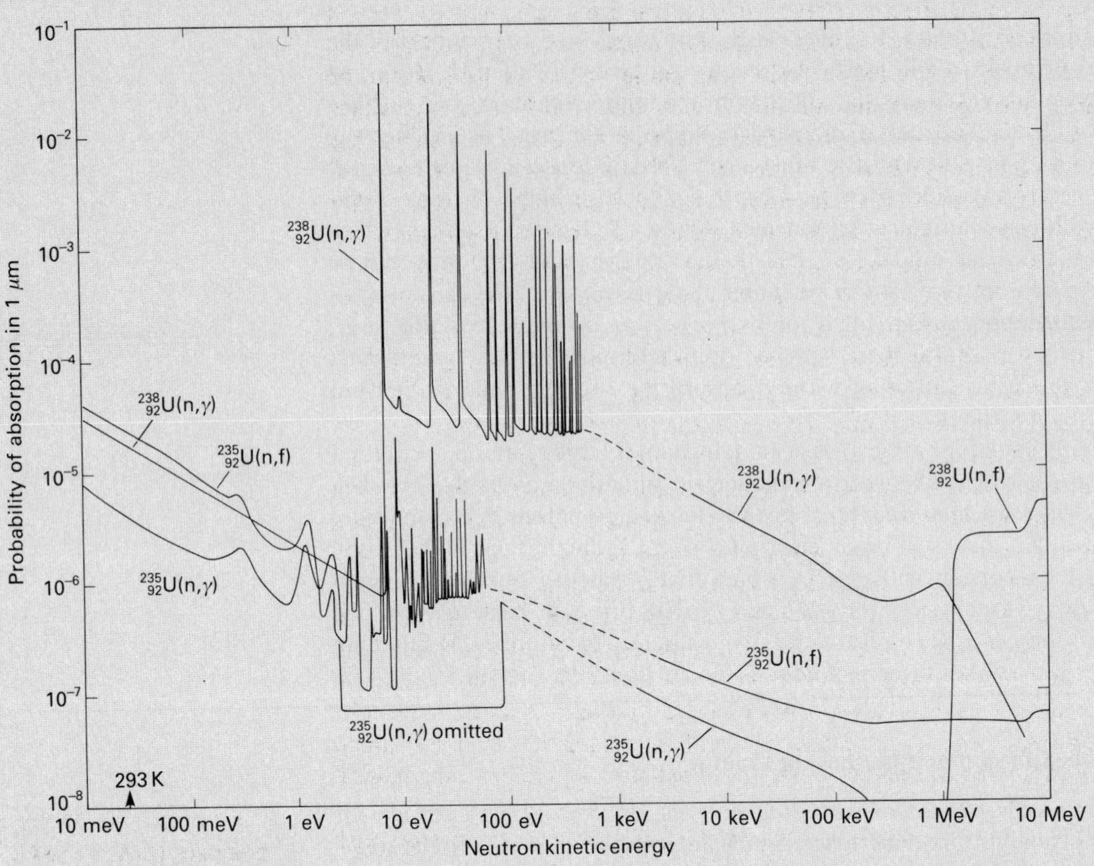

Fig. 7.14 The partial probabilities for various reactions for neutrons in natural uranium metal of normal density. The ordinate is the probability that in 10^{-6} metre of path in uranium a neutron is absorbed to cause the reaction labelling the curves. The abscissa is the neutron kinetic energy.

(n,γ) = radiative capture,

(n,f) = capture followed by fission.

The average separation of resonances in $^{238}_{92}U$ is about 18 eV and in $^{235}_{92}U$ is about 0.6 eV. The abscissa, which is logarithmic in incident neutron energy, therefore shows an increasing apparent density of resonances in $^{238}_{92}U$ at 10 eV and up; the onset with $^{235}_{92}U$ is at 0.3 eV. We run out of ability to show the resonant structure at about 50 eV and 6 eV for $^{238}_{92}U$ and $^{235}_{92}U$ respectively. Above these energies we indicate by a dotted line the average effect of many close resonances. The ability to observe these resonances depends on the resolution of the measurements: as the energy increases the resonances increasingly overlap and the net effect is fluctuations which in turn are smoothed away at still higher energies.

This figure is intended to give only an overall picture of the structure of the cross-sections in each of the two major isotopes of uranium. It is not a faithful representation of the actual cross-sections.

1. Apart from elastic scattering the most likely fate of a neutron below 1 eV is to cause fission in $^{235}_{92}U$ although that isotope is only 0.7% of the total.
2. The large radiative capture cross-section in $^{238}_{92}U$ particularly in the range 10–1000 eV.
3. The threshold for induced fission is $^{238}_{92}U$ is just below 1 MeV.

Another fact is that neutrons emitted in fission have an energy spread with a mean below 1 MeV. The majority cannot cause fission in $^{238}_{92}U$, $k < 1$ and a self-sustaining chain reaction in a volume of natural uranium is impossible. This is not the case in pure $^{235}_{92}U$. There is no threshold and fission is about three times more likely than γ-ray emission after neutron capture at all neutron energies below 1 MeV: thus, unless lost by radiative capture or leakage from the material, all neutrons emitted in fission can cause further fissions, k can be greater than one and a chain reaction is possible. The chance of leakage decreases as the ratio of volume to surface area increases: therefore a sphere has the most economic geometry. As a sphere of fissile material is increased in radius at constant density, k will increase. The sphere which has $k = 1$ has what is called the critical mass. An increase of mass gives $k > 1$ and the growth of the neutron flux in the material and the fission rate can be very rapid and the energy release explosive. The principle of an atomic or fission bomb is to assemble from a non-critical arrangement of fissile material a super-critical arrangement, injecting a carefully timed burst of neutrons to initiate the chain reaction. The assembly must be done sufficiently rapidly that the k is large enough for long enough to give many generations of neutron multiplication before the energy released causes disruptive forces that blow the assembly apart.

The controlled release of fission energy in natural uranium requires a careful use of the data shown in Fig. 7.14: consider a spontaneous fission in uranium giving rise to several neutrons with a kinetic energy in range 0.1–1 MeV. Suppose the neutrons can be allowed to escape from the uranium and cooled to thermal energies (about 0.03 eV) then passed back into the uranium. They will escape probable capture in $^{238}_{92}U$ in that dangerous energy range 10–1000 eV and be made available for absorption at low energy where fission-causing capture in $^{235}_{92}U$ is more likely than radiative capture in $^{238}_{92}U$. If this state of affairs can be tuned so that more fissions are caused per generation of neutrons than required to produce them, then a self-sustaining chain reaction is possible using natural uranium.

How is this done? Consider how neutrons may be reduced in energy (cooled): if the fast neutrons (from fission) are allowed to enter a material of atomic mass number A they will undergo elastic scattering, and if the bulk of the material is sufficient, will continue to suffer such elastic scatters losing, on the average, a fraction $2A/(1+A)^2$ of their kinetic energy per collision (see Problem 7.9). Therefore small A is to be preferred, and in carbon, for example, about 110 collisions are required to reduce the kinetic energy of a 1 MeV neutron to the temperature of the carbon at 273 K. This process is called **moderation** and the material in which it takes place is called a **moderator**. Obviously, if the neutron is to survive many collisions without capture, the moderator must have a very low capture cross-section relative to its elastic scattering cross-section. Carbon (as graphite) and heavy water (D_2O) satisfy this criterion and are the most common moderators for reactors fuelled with natural uranium. We can now see the reason behind the arrangement of materials in such a reactor. In a graphite moderated reactor, uranium rods about 1 cm in diameter are placed in channels in the graphite on a grid of about 20 cm. Most fission neutrons escape from the uranium rods, cool in the graphite moderator, and diffuse until chance takes them back into a uranium rod, where the probability of fission-causing capture is high. Fission produces several neutrons which continue the process.

This explanation hides many problems of neutron economy: obviously the reactor will not work if too many of the neutrons produced by fission are radiatively captured or diffuse out of the bulk of the moderator plus uranium. To sustain the chain reaction so that the fission rate is constant requires an average of just one neutron from those produced in each fission to cause another fission. Thus if $k = 1$ the reactor is running at constant fission rate: if

DEFINITIONS AND KEYWORDS

Fissile material This is a material of which the nuclei have a dominating probability of suffering fission after slow neutron capture.

Reproduction factor, *k* This is the number of neutrons available to cause fission at one generation divided by that number in the previous generation in a nuclear fission reactor.

Moderation The process of reducing the energy of neutrons.

Moderator A material in which fast neutrons lose energy by elastic collisions with nuclei and, if not captured or escaping from the material, finally reach the same temperature as the material.

$k<1$ the reactor will stop and if $k>1$ the fission rate increases. The existence of many types of reactor indicate that the problems of neutron economy can be solved and k kept equal to 1. The problem of keeping $k=1$ is discussed in Section 7.14.

If we look again at Fig. 7.14 we see that the radiative capture in $^{238}_{92}U$ must be the fate of a fraction of the neutrons in a nuclear reactor fuelled with natural uranium. This capture leads to the following changes:

$$^{238}_{92}U + n \rightarrow {}^{239}_{92}U + \gamma \text{ (radiative capture)},$$

$$^{239}_{92}U \rightarrow {}^{239}_{93}Np + e^- + \bar{\nu}_e,$$

$$^{239}_{93}Np \rightarrow {}^{239}_{94}Pu + e^- + \bar{\nu}_e.$$

The mean lives of $^{239}_{92}U$ and $^{239}_{93}Np$ (neptunium) are 34 minutes and 81 days respectively. The relatively long-lived $^{239}_{94}Pu$ (plutonium) (35 000 years) is an even-Z, odd-A heavy nucleus and, as we have already noted, is fissile. Thus the reactor produces fissile material as it consumes it. In principle, the neutron economy in a natural uranium reactor might be arranged so that of an average of N neutrons produced in the fission of $^{239}_{92}U$, one is used to maintain the chain reaction, one or more to produce $^{239}_{94}Pu$ and $<N$-2 to cover neutron losses. Thus the reactor can produce more fissile material than it consumes and all the uranium becomes potentially available for power generation, if plutonium-powered reactors are available. This is the principle of breeder reactors. The difficulties of realizing this system are very considerable.

PROBLEMS

7.12 Neutrons of kinetic energy below 1 MeV are scattered isotropically in the centre of mass by nuclei, mass number A. Show that, averaged over many collisions, the energy lost per collision is a fraction $2A/(1+A)^2$ of the incident neutron kinetic energy.

7.13 The resonance at 1.46 eV for neutron absorption by $^{115}_{49}In$ (Fig. 7.10) is almost completely isolated from any other resonance and the cross-section may be described by the Breit–Wigner formula. In this case it describes the variation with neutron kinetic energy T_n of the neutron total cross-section $\sigma(T_n)$ by the formula

$$\sigma(T_n) = \pi \lambda\!\!\!^{-2} g \frac{\Gamma_n \Gamma}{(T_n - T_0)^2 + \Gamma^2/4},$$

where $\lambda\!\!\!^{-} = \hbar/P_c$, P_c is the neutron centre-of-mass momentum, T_0 is the value of T_n at resonance (1.46 eV). The quantities $\Gamma_n/\hbar$ and $\Gamma/\hbar$ are the transition rates for the resonance to decay into the channel n+$^{115}_{49}In$ and into all channels respectively (Γ_n is called the neutron partial width and Γ is the total width of the resonance.) The statistical factor g is given by

$$g = \frac{(2J+1)}{(2s+1)(2j+1)},$$

where $s=\frac{1}{2}$ is the neutron spin, j is the spin of $^{115}_{49}In$, J is the spin of the compound nucleus formed.

Investigate the meaning of this formula (start at Section 2.9) and calculate the value it gives for the cross-section maximum and compare that with the value in Fig. 7.10.

$[j=\frac{9}{2},\ J=5,\ \Gamma_n=3.0\text{ meV},\ \Gamma=75\text{ meV}.]$

The fission reactor provides a means of generating radioactive materials which are β^--active or γ-ray emitters. The fission products themselves are neutron-rich and therefore β-active, and some are convenient sources: for example $^{106}_{44}Ru$ for which the relevant level scheme is shown in Fig. 5.9(b). Other sources may be made by neutron absorption by stable nuclei exposed to neutrons in an operating fission reactor. Thus natural cobalt exposed produces $^{60}_{27}Co$, which is a useful source of γ-rays: see Fig. 5.9(a) for the energy-level scheme. Positron sources cannot be produced in this way and nuclear reactions caused by protons or other heavy charged particles have to be used to produce the proton-rich nuclei required.

7.13 Reactor control and delayed neutron emission

How are nuclear reactors controlled? Most are built so that the temperature coefficient of k is negative and any power excursion leading to reactor heating takes k below 1 and therefore such excursions are self-quenching. In addition, it is possible to intervene in the neutron economy by deliberately increasing or decreasing the loss of neutrons by absorption. This is done by the insertion or withdrawal of control rods containing material with a large capture cross-section for thermal neutrons (e.g. boron, cadmium). The fission rate of a reactor

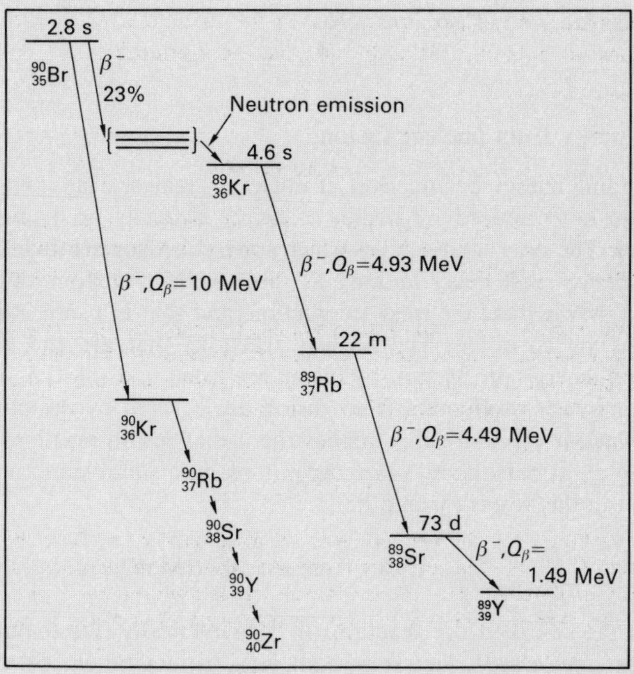

Fig. 7.15 An example of the decay chain of a fission product $^{90}_{35}Br$ that is a delayed neutron emitter. The delay is due to the mean life, 2.8 s, of this nucleus; the neutron is emitted by some excited states of $^{90}_{36}Kr$ which are populated by a fraction of the $^{90}_{35}Br$ β-decays. The mean life of each state is given in seconds (s), minutes (m) or days (d). The β-decays are marked as if going to the ground states in the steps from $^{89}_{36}Kr$ to $^{89}_{39}Y$ but, in fact, a fraction in all cases go to excited states. To prevent overcrowding no details have been included of the $^{90}_{36}Kr$ to $^{90}_{40}Zr$ decays.

can be regulated in this way. The problem concerns the time available to make such changes in the face of a malfunction or in order to prevent a power excursion due to some fluctuation. The average neutron lifetime between production and fission capture could be about 1 ms. A 1% increase in k will allow a doubling of the fission rate in 30 ms. Thus the time available for adequate control is apparently very short for the mechanical adjustment of control-rod position. However, there is another property of nuclei which makes reactors much easier to control. This is the existence of **delayed neutron emitters**. All fission products are neutron rich and are therefore β^--active, normally requiring several successive decays to reach stability. There are, however, some in which a β^--decay leaves the daughter nucleus unstable to neutron emission. The latter process will have a transition rate which is large compared with the rates for β-decay. Therefore the delay from production of the fission fragment to neutron emission is determined by the transition rate of the intervening β-decays. These delays can amount to several seconds. These neutrons are sufficiently numerous to provide a contribution to the neutron economy that takes k from below to 1. Thus the fission rate multiplication is determined by the delayed neutrons and can be very much more leisurely than anticipated above. One decay chain containing a delayed neutron source is shown in Fig. 7.15.

Although not relevant to the control of nuclear reactors it is worth noting that there are also some delayed proton emitters. Proton-rich nuclei normally move to the line of stability by successive β^+-decays; however, a few undergo decay to a state for which proton emission can compete with the next β-decay. Examples are ${}^{77}_{37}Rb$, ${}^{108}_{51}Sb$, and ${}^{118}_{54}Xe$.

DEFINITIONS AND KEYWORDS

Delayed neutron emitter This is a fission product that emits a neutron after one or more steps of β-decay have occurred.

Nuclear fusion Nuclear reactions leading to the production of helium from lighter nuclei.

Controlled thermonuclear reactions Nuclear fusion reactions taking place under controlled conditions involving the reactants in intimate contact at high temperatures rather than as incident particles colliding with stationary target particles.

7.14 Energy from nuclear fusion

The binding energy per nucleon of nuclei is greatest near $A = 60$. Energy can therefore be produced by fission of heavy elements or by **fusion** of lighter elements. The stars are machines which convert hydrogen into heavier elements: however, we shall delay looking at these stellar reactions until Chapter 14 because they involve the weak interaction, and that is a subject of Chapter 12, which is yet to come. The nuclear reactions that are the possibilities for terrestrial energy production by fusion are listed in Table 7.5. The hopes that exist for power production from fusion are covered by the phrase **controlled thermonuclear reactions**: this implies the use of fusion reactions which can be made to go at reasonably low temperatures in a suitable mixture. The easiest reaction in this respect would be

$$d + t \rightarrow {}^{4}_{2}He + n + 17.6\ \text{MeV}.$$

This is the so-called d–t reaction. It is energetically uneconomic to consider firing an accelerated deuteron beam at a tritium target, or vice versa. The method seen, at the present, most likely to be successful is to compress and heat sufficiently a mixture of tritium and deuterium gases or liquids; sufficiently means so that enough d–t reaction occurs to provide real energy gain. However, although there is plenty of deuterium in ordinary water, tritium is radioactive (${}^{3}_{1}H \rightarrow {}^{3}_{2}He + e^- + \bar{\nu}_e$, $\tau = 17.7$ years) and must be made, most readily from the reaction

$${}^{6}_{3}Li + n \rightarrow {}^{4}_{2}He + t + 4.79\ \text{MeV}.$$

Therefore the neutron from the d–t reaction must be used to replenish the tritium supply. This cannot be done with 100% efficiency and, assuming the absence of other neutron sources (eg fission), the aim would be to use the reactions

$$d + d \rightarrow {}^{3}_{2}He + n + 3.27\ MeV,$$

$$d + d \rightarrow t + p + 4.03\ MeV.$$

These produce less energy per reaction than the d–t reaction and require about six times the temperature at the same density to attain the same rate of energy production. Of course the tritium and neutrons produced could be the part of a more complex process involving all the reactions. Whatever schemes form the basis of energy production, the development problems are not nuclear but are those concerned with reaching the necessary reactant densities and temperatures for a time sufficient to give energy gain. Present developments are based on either heating of millimetre size spherical targets containing the mixture by multiple laser beam compression, or by joule heating of a magnetically confined plasma of the mixture. These methods are not the subjects of this book. Interested readers are referred to Motz (1979) and to Chen (1984).

In the so-called hydrogen bomb it is likely that a mixture of the heavy hydrides of the $A = 6$ isotope of lithium, ${}^{6}_{3}LiD$ (D for deuterium) and ${}^{6}_{3}LiT$ (T for tritium), is compressed and heated to a sufficient extent to cause explosive burning through the reactions mentioned. The compression and heating is probably effected by exposing a suitably packaged mixture to the X-rays from an exploding fission bomb (Section 7.12).

Table 7.5 The possible controlled thermonuclear reactions:

(1) $d+d \rightarrow {}^{3}_{2}He+n+3.27$ MeV,

(2) $d+d \rightarrow t+p+4.03$ MeV,

(3) $d+t \rightarrow n+\alpha+17.59$ MeV,

and for the production of tritium

(4) ${}^{6}_{3}Li+n \rightarrow t+\alpha+4.79$ MeV.

This table is based on the assumption of the natural availability of deuterium and lithium (natural lithium is 7 to 9% ${}^{6}_{3}Li$).

7.15 Conclusion

Although this is one of the longer chapters of this volume, it can only be a brief survey of a subject which on its own has many weighty books devoted to analysing and synthesizing its many aspects. The problem is that there is no unified theory of nuclear reactions that is of practical use. This is not surprising in view of the difficulty of dealing theoretically with many-body collision problems, particularly when the forces between the bodies are not completely understood. In addition, the huge range of phenomena engendered by the possibilities inherent in the variables of energy, of incident particle, of target, and of possible outcome of a collision, conspire to make a unified view impossible. Theories of reactions tend to deal with a limited class of reactions; we have mentioned two, compound and direct, but there are subdivisions and we have indicated that there are reactions that are intermediate.

There are many kinds of reactions that we have not mentioned, let alone discussed. The reader will appreciate that although some things are forbidden, for example reactions violating electric charge conservation, anything not forbidden can and may happen. Thus nuclear reactions can be caused by the absorption of energetic photons, by the scattering of charged leptons (electrons, muons), and by the interaction of neutrinos. In these cases the electromagnetic interaction or the weak interaction or both become partners with the strong interaction in the forces that must exist between particles if reactions are to occur.

This chapter also fails in one special respect: this author wanted to keep readers close to being able to do calculations in nuclear and particle physics. Yet to do that with nuclear reactions (apart from kinematics) requires introducing subjects that are not appropriate to a volume that must be kept down to a reasonable size. They are straightforward and in Problems 7.11 and 7.13 reader is invited to undertake a project (or two) which will go someway to opening new vistas and to stimulating curiosity.

References

Blackett, P. M. S. and Lees, D. S. (1932). *Proceedings of the Royal Society,* London, **A136**, 325–338.

Chadwick, J. (1932). *Proceedings of the Royal Society,* London, **A136**, 692–708.

Chen, F. F. (1984). *Introduction to Plasma Physics and Controlled Fusion,* Vols 1 and 2 (2nd edn). Plenum Press, New York.

England, J. B. A. (1974). *Techniques in Nuclear Structure Physics,* Parts 1 and 2. Macmillan, London.

Ghoshal, S. N. (1950). *Physical Review,* **80**, 939–942.

Kleinknecht, K. (1986). *Detectors for Particle Radiation.* Cambridge University Press.

Motz, H. (1979). *The Physics of Laser Fusion.* Academic Press, New York.

Ower, H. *et al.* (1982). *Nuclear Physics,* **A388**, 421–444.

Williamson, C. (1961). *Physical Review,* **122**, 1877–82.

8

Nuclear Models

8.1 Introduction

We have already in Chapter 4 discussed one model—the liquid drop model of the nucleus—which led to the successful semi-empirical mass formula. This formula gives a good description of the masses of the stable nuclei and of the nuclei on or near the line of stability. However, it does not have much to say about many important nuclear properties. In Table 8.1 we give a list of some of these properties of nuclei and in this chapter we want to look at two models which go some way to providing a description of a few of them.

Table 8.1 Nuclear properties not described by the liquid drop model.

(1) Ground state spins and parities
(2) Excited state spins and parities
(3) Existence of magic numbers
(4) Magnetic moments
(5) Density
(6) Values of coefficients in semi-empirical mass formula (apart from Coulomb)

8.2 The magic numbers 1

The reader will be familiar with the magic numbers of atomic physics. These are evident in a plot of the first ionization energy against atomic number Z, which is shown in Fig. 8.1. There are peaks at the atomic Z numbers of

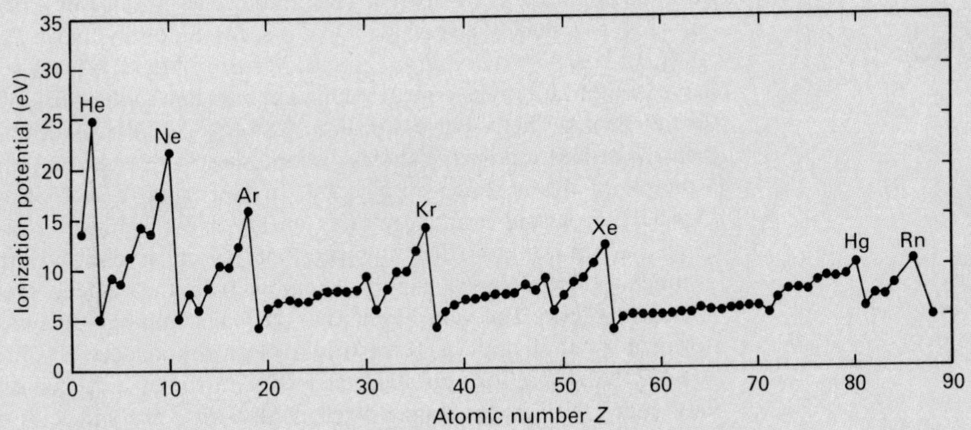

Fig. 8.1 The first ionization potential (electron separation energy) against atomic number for the elements. The well-known 'magic numbers' are $Z=2$, 10, 18, 36, 54, 80, 86. Helium is the element for which two electrons fill the 1*s* level: mercury at $Z=80$ is the element for which the last two electrons fill the 6*s* level. The remainder are the other noble gases for which electrons fill the levels so that the configuration of the last eight electrons is $(ns)^2(np)^6$, $n=2, \ldots\ldots 6$. The occurrence of magic numbers in neutron and proton separation energies for nuclei is one of the pieces of evidence which pointed the way to the nuclear shell model.

$$2, 10, 18, 36, 54, 80, \text{ and } 86,$$

corresponding to the elements

$$\text{He, Ne, Ar, Kr, Xe, Hg, and Rn.}$$

> **DEFINITIONS AND KEYWORDS**
>
> **Nuclear magic numbers** These numbers are
>
> 2, 8, 20, 28, 50, 82, and 126.
>
> When the number of protons (Z), or of neutrons (N), or both, in a nucleus has one of these values the nuclear binding energy is greater than expected and other nuclear properties behave as if the magic number of nucleons formed a closed shell of states analogous to the closed shells of atomic structure.
>
> **Shell model** The nuclear model which has energy levels for nucleons in a suitable potential well. A group of levels closely spaced in energy is called a shell, as in atomic physics: hence the name.

Apart from mercury (Hg), these are all noble gases and we know that their electronic structure is characterized either by a closed shell of 2 electrons in helium or by one or more closed shells plus eight electrons filling the *s* and *p* levels of a principal quantum number. There are also less obvious peaks when there are closed shells plus two electrons in an *s*-level: they can be seen at $Z=4$, 12, 30, 48, and 80. Mercury is the last of these. The spacing between peaks is determined by the existence of the transition elements where inner shells of *d* and *f* electrons fill up as Z increases. The full complement of eight electrons in *s* plus *p* levels is one that is hard to break by removing one electron, leading to a large value of the ionization potential (the electron separation energy). The **nuclear magic numbers** are

$$2, 8, 20, 28, 50, 82, \text{ and } 126,$$

and they apply to proton and neutron numbers separately. What is the evidence that these numbers matter? The semi-empirical mass formula provides a smooth prediction for the binding energy, smooth if we neglect the pairing term and allowing for the integer nature of Z and N. Figure 4.6 shows the binding energy per nucleon for the odd-A nuclei: the measured values deviate from the prediction of the semi-empirical mass formula in the direction of increased binding energy in several regions, marked on this figure by the associated magic number. Most evident is that in the region where both Z and N are close to or equal to magic, at $Z=82$, $N=126$. A more careful study shows that if either Z or N is kept fixed and the other is varied then total binding energy is a maximum when the latter is one of the magic numbers. Of course the semi-empirical mass formula applies only for $A>20$. The binding energy at lower A has peaks at ${}^{4}_{2}\text{He}$ ($Z=2$, $N=2$, **doubly magic**) at ${}^{16}_{8}\text{O}$ ($Z=8$, $N=8$ doubly magic again) as well as at ${}^{12}_{6}\text{C}$ ($Z=6$, $N=6$) and ${}^{20}_{10}\text{Ne}$ ($Z=10$, $N=10$), ${}^{24}_{12}\text{Mg}$ ($Z=12$, $N=12$). The latter have greater binding energies than neighbours due to the effect of the pairing of like nucleons. There is also a peak at ${}^{40}_{20}\text{Ca}$ ($Z=20$, $N=20$, doubly magic). The neutron separation energy shows discontinuities at points where N is magic: an example of this is shown in Fig. 4.4. In this example as N increases towards $N=82$ the pairing term keeps this energy saw-toothing up and down until $N=82$, when it takes a dive by several megaelectronvolts and the 83rd neutron is much less expensive in energy to remove from the nucleus, even allowing for the pairing term. The same happens at the magic numbers 50 and 126 for N, and for protons at 50 and 82. Terrestrial nuclear abundances for Z or N magic are greater than those for non-magic elements nearby in A. Elements with Z magic have many more isotopes than do elements with Z non-magic. (For example, Sn with $Z=50$ has 10 stable isotopes.) Elements with N magic have more isotones than elements with N non-magic. Elements with N magic have neutron capture cross-sections very much less than neighbours with N non-magic.

The even–even nuclei have ground states with spin-parity $j^{\mathrm{P}}=0^{+}$ and, with two exceptions, first excited states with spin-parity $=2^{+}$. The two exceptions are ${}^{40}_{20}\text{Ca}$ and ${}^{208}_{82}\text{Pb}$, both doubly magic, with 0^{+} and 3^{-} respectively. The excitation energy of the first 2^{+} state in the nuclei in the region of $A=208$ is shown in Fig.

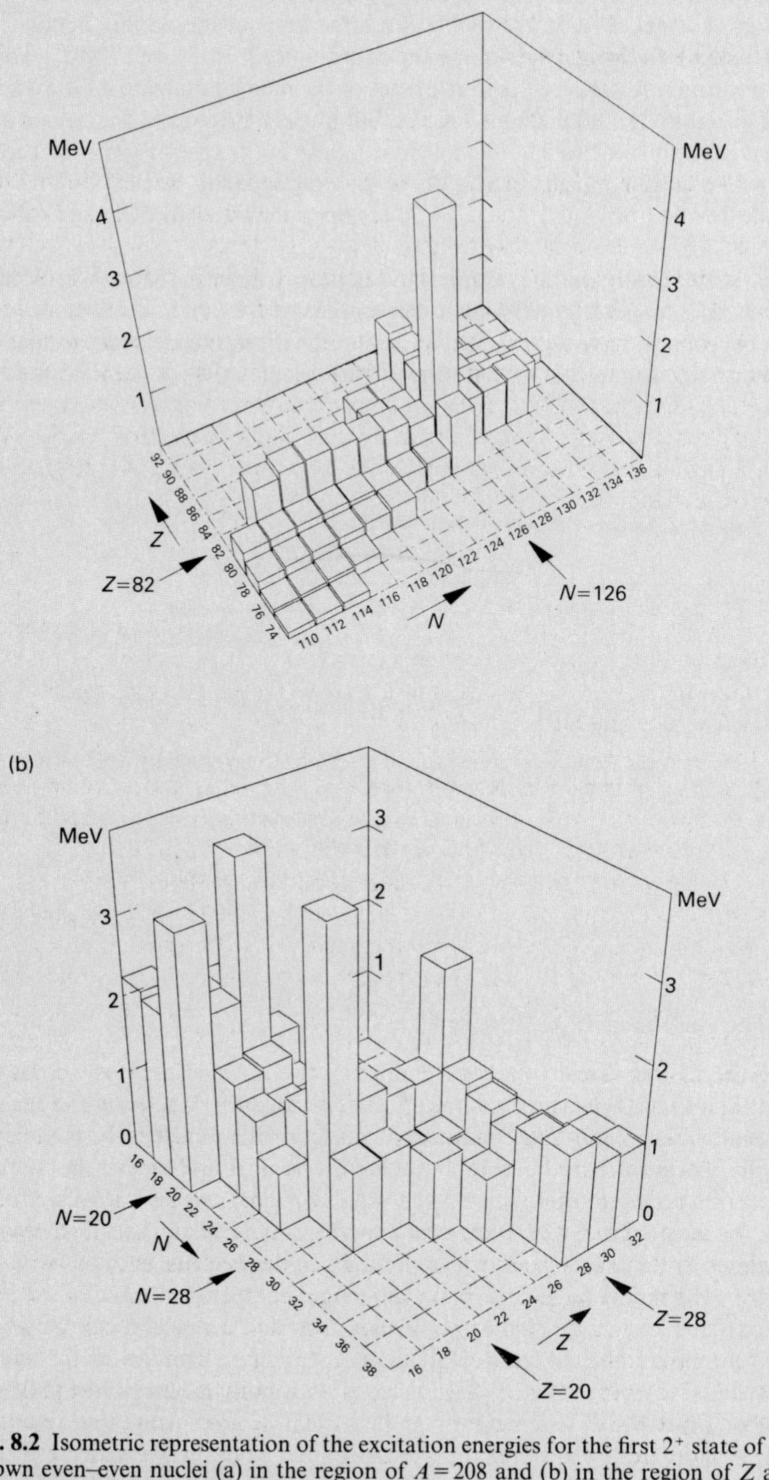

Fig. 8.2 Isometric representation of the excitation energies for the first 2^+ state of the known even–even nuclei (a) in the region of $A=208$ and (b) in the region of Z and $N=20$–28. In both regions there are peaks at the nuclei with both Z and N magic: $Z=82$, $N=126$ ($^{208}_{82}$Pb); $Z=20$, $N=20$ ($^{40}_{20}$Ca); $Z=20$, $N=28$ ($^{48}_{20}$Ca); $Z=28$, $N=28$ ($^{56}_{28}$Ni). There are ridges of large excitation along lines of constant $N=126$, 28, or 20, and along lines of constant $Z=82$, 28, or 20. In both cases some nuclei with low excitation energies are hidden by the high land.

8.2(a), which is a two-dimensional histogram ('lego' plot) of this excitation as a function of Z and N. Clearly visible is a large peak at the doubly magic $^{208}_{82}\mathrm{Pb}$ ($N=126$) with flanking ridges along the coordinates $N=126$ and $Z=82$. This is a dramatic manifestation of the properties of the magic numbers. At the doubly magic $^{40}_{20}\mathrm{Ca}$ and $^{16}_{8}\mathrm{O}$ there are also peaks, but nuclei do not exist to provide such extensive foothills with such long magic ridges as in the case of those near $^{200}_{82}\mathrm{Pb}$. At $N=28$ a ridge connects two peaks at the doubly magic nuclei $^{56}_{28}\mathrm{Ni}$ and $^{48}_{20}\mathrm{Ca}$ and another, less obvious, along $Z=20$ connects the latter to $^{40}_{20}\mathrm{Ca}$, all as shown in Fig. 8.2(b).

This evidence for the magic numbers is summarized in Table 8.2. What is required is a nuclear model which can explain these magic numbers and we might expect it to have a great deal in common with the models that we have of the electronic structure of atoms and of the periodic table. It is called the **shell model**.

Table 8.2 Evidence for the nuclear magic numbers
2, 8, 20, 28, 50, 82, 126.

(1) Deviations in nuclear binding energies near magic numbers.
(2) Neutron (proton) separation energies have peaks when $N(Z)$ is magic.
(3) Elements with $Z(N)$ magic have more isotopes (isotones) than usual.
(4) Elements with Z magic have natural abundances greater than those of nearby elements.
(5) N magic nuclei have slow neutron absorption cross-sections very much lower than those for N non-magic nuclei.
(6) The first 2^+ excited state of even–even nuclei have exceptionally large excitation energies if Z or N or both are magic.
(7) The existence of islands of isomerism (see Section 11.9).

8.3 The shell model: preliminaries

The basic assumption of the shell model is that a single nucleon moves in a potential which is the average of the effect of all the other nucleons and that this potential varies smoothly. Since every nucleon of a nucleus is bound, this potential is expected to be a potential well. The next assumption is that each nucleon moves in an orbit which is that of a single particle in that well. Given a shape for the potential well, these orbits can be calculated and specified: they are analogous to the orbits of electrons in atoms. Our objective is to construct the details of the model so as to explain the magic numbers.

A question comes up at once: how can it be that a nucleon can occupy an orbit for a time sufficient to allow that concept to make sense when the nucleon is in an environment crowded with nucleons? Normally the mean free path of an energetic ($\gtrsim 10$ MeV) nucleon moving in nuclear matter during the course of a reaction is about 2 fm. If this applies to bound nucleons, then even a once-around-the-nucleus orbit without a collision is almost impossible. It is the Pauli exclusion principle which rescues us from this difficulty: collisions that change orbits cannot occur since that normally implies an exchange of energy with one nucleon losing energy and that, in turn, implies a transition for that nucleon to an orbit of lower energy: but all those orbits are occupied and orbit-destroying collisions cannot occur. This qualitative explanation has been put on a sound footing in a many-body treatment of nuclear matter (see Section 8.11).

We shall now proceed to build the model in steps.

1. Propose some simple potentials.
2. Look at the nature of solutions of Schrödinger's equation in these potentials.
3. Fill the energy levels with neutrons and protons according to Pauli's exclusion principle.
4. Look for the magic numbers. If that does not succeed we will have to modify the potential, and go back to 2.

Three potentials are shown in Fig. 8.3. The first two are the finite square well and the infinite harmonic oscillator potential; both are physically unreasonable for the nucleus but we will look at them as they can be solved. The third, the Saxon–Woods potential, has the same shape as has been found to fit the nuclear charge distribution, and is likely to be physically reasonable. These potentials are for the neutrons. The potentials for protons will have an additional contribution from the Coulomb interaction. However, we will not include this addition but we must not forget that it raises the energy of all the proton levels relative to those for the neutrons.

The solutions of the Schrödinger equation are found in the usual manner. These solutions are conveniently functions of the polar coordinates r, θ, and φ and we assume that the stationary state wavefunctions can be separated to give

$$\psi(\mathbf{r}) = R(r)\,Y_l^m(\theta,\varphi). \tag{8.1}$$

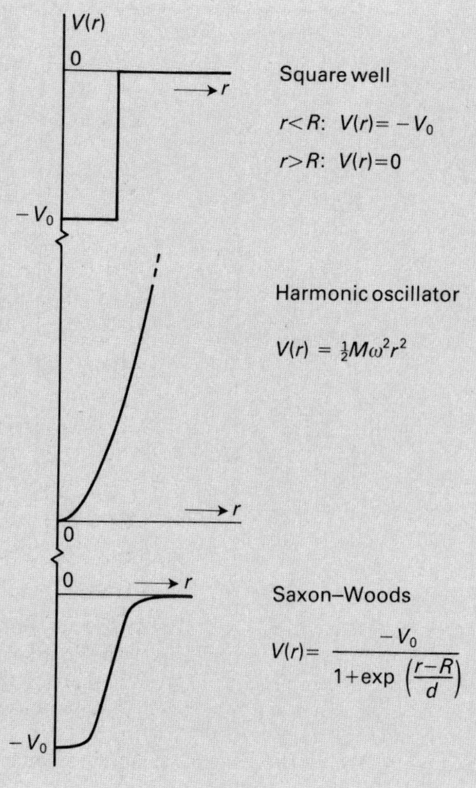

Fig. 8.3 Three spherical potential wells. The harmonic oscillator can be solved analytically and the finite square well has well-understood solutions. However, the Saxon–Woods shape (Section 3.7) provides a more reasonable model for the nuclear potential. By examining the solutions for the harmonic oscillator and for the square well we can obtain a good idea of the level ordering for a realistic nuclear potential.

The function $Y_l^m(\theta,\varphi)$ is a spherical harmonic of order l,m. Therefore this sta has orbital angular momentum quantum number $l\,(=0, 1, 2, \ldots)$ and magnet quantum number $l_z=m(-l\leqslant m\leqslant l)$. The eigenvalue of the orbital angul momentum squared ($\boldsymbol{L}^2$) is $l(l+1)\hbar^2$ and that of the z-component (L_z) is l_z Consider now the radial part, $R(r)$: if we put $R(r)=U(r)/r$, then $U(r)$ satisfies

$$-\frac{\hbar^2}{2M}\frac{\mathrm{d}^2U(r)}{\mathrm{d}r^2}+\left\{V(r)+\frac{l(l+1)\hbar^2}{2Mr^2}\right\}U(r)=EU(r), \qquad (8.$$

where M is the nucleon mass, $V(r)$ is the potential. An eigenfunction $U_{nl}(r)$ specified by l, and by n, which we call the **principal quantum number**. I eigenvalue E_{nl} is the energy (kinetic plus potential) of the state specified by n an l. The third term on the left is a contribution we have met before in Chapter (equation (6.11)) when, in the context of α-decay, we called it the angul momentum barrier. For stationary states it acts like a contribution to $V(r$ Figure 8.4 shows schematically the form of the effective potential and we can se that the addition of this angular momentum term to a square well will have th effect of producing a potential which will confine the nucleon increasingly clos to the surface of the nucleus as l increases. Figure 8.4 also shows schematicall

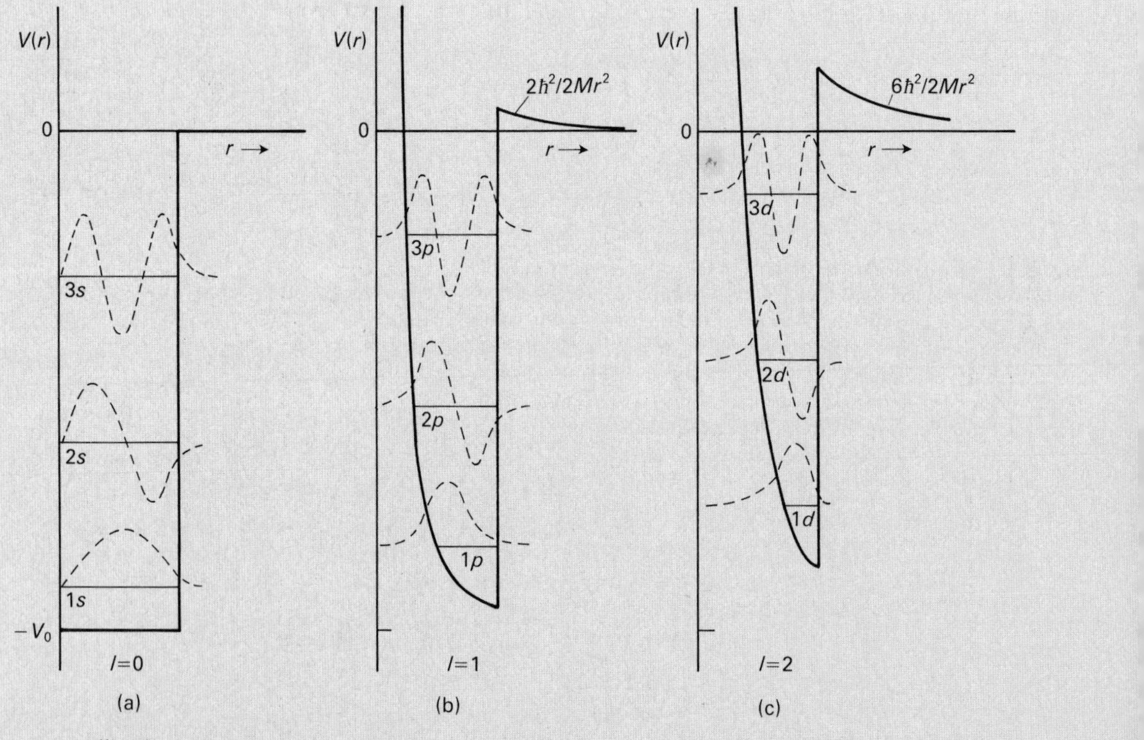

Fig. 8.4 A square-well spherical potential as a function of radius r is shown in (a). It is modified by the angular momentum barrier and the new shapes are given in (b) for $l=1$ and in (c) for $l=2$. The single particle energy levels are shown by horizontal lines for the states nl, $n=1, 2, 3$, and $l=0, 1, 2$ (s, p, d respectively), where n is the principal quantum number and l is the orbital angular momentum quantum number. The broken line shows the form of $rR(r)$ for the appropriate wavefunction, plotted about its energy level line as zero. Remember that the probability density distribution as a function of radius for the particle is proportional to $(rR(r))^2$.

he radial function $rR(r)$ of $l=0$, 1, and 2 for the principal quantum numbers 1, , and 3. We can see the surface confining effect of increasing l and that the onfinement increases the curvature of the wave function for fixed n and this ncreases the energy of the states as l increases.

Now a matter of notation. We use the usual spectroscopic notation (nl) for tates having principal quantum number n and angular momentum $l=s, p, d, f$, .. (quantum number $l=0, 1, 2, 3, \ldots$). Note that the lowest-energy p-state is alled $1p$, the next $2p$, the lowest f level is $1f$, and so on. This is in contrast to the otation used for states in atomic physics where the lowest p-state is called $2p$, he lowest d is $3d$, the lowest f is $4f$, and so on. The reason for this is that the pure Coulomb potential is unique in having the following degeneracies (counting pwards from lowest energy):

(1) the second s-level and the first p-level,

(2) the third s-level, the second p-level and the first d-level,

(3) and so on.

This is shown in Fig. 8.5(a) and historically it led to the labelling of atomic

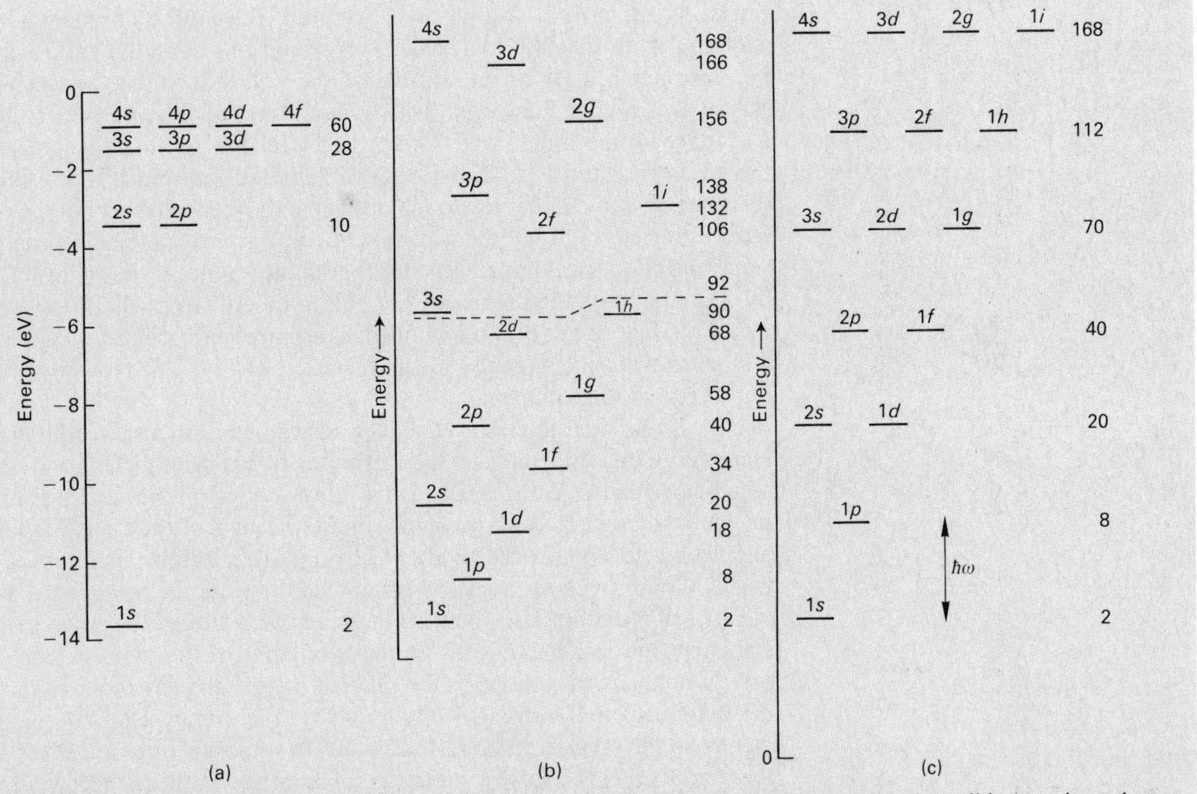

Fig. 8.5 The single-particle energy levels in three spherical potential wells: (a) the Coulomb, (b) the infinite square well, and (c) the harmonic oscillator. In (a) the ordinate gives the actual energies. In (b) and (c) the energy scale is arbitrary but the relative energies above zero at the well bottom in each case are shown correctly. On the right of each we show the accumulated occupancy starting with the 1s level and working upwards in energy. The magic numbers 2, 8, 20 occur for (b) and (c).

states which will be familiar to the reader. In the nuclear case we continue from Fig. 8.4 the usual but different from Coulomb convention, in which the principal quantum number indicates the ordinal appearance—see Fig. 8.5(b).

The harmonic oscillator potential $V(r)=\frac{1}{2}M\omega^2r^2$ has its own degeneracies—see Fig. 8.5(c). A few moments examination shows the following properties. A gross structure of energy levels separated by an energy $\hbar\omega$ with the lowest, the 1s level, $3\hbar\omega/2$ above the bottom of the well. States of a fixed l are separated by $2\hbar\omega$ with the first occurrence at an energy $l\hbar\omega$ above 1s. Thus 1f is $3\hbar\omega$ above 1s and 2f is $5\hbar\omega$ above 1s. All this is summarized by the formula

$$E_{nl}=(2n+l-\tfrac{1}{2})\hbar\omega$$

for the energy of the level nl above well bottom.

Figure 8.5 also shows the accumulated occupancy of the levels filling from bottom up for the cases of the infinite square well and the pure harmonic oscillator. Filling with neutrons or protons, each level can be occupied by two (two for the two spin-$\frac{1}{2}$ orientations) multiplied by $2l+1$ for each l state. Apart from 2, 8, and 20, the remaining magic numbers are not immediately visible. (Here we are assuming that states of different l_z or s_z are degenerate since there is no external field.)

Figure 8.5 shows the levels in two unreasonable wells. If the wells are truncated so as to approach what could be the real situation in a nucleus, then the levels are moved. The physical situation which allows a qualitative understanding is as follows. Increasing the orbital angular momentum confines a nucleon increasingly to the outside of the well and thus to places where the potential is higher than usual for the harmonic oscillator or near the infinite hard edge for the square well. This gives a level of a given l a higher energy than the levels $l-1$ and $l-2$. Truncating the potential so that it has a finite depth allows these wavefunctions to spill out beyond the effective end of the potential, reduces the curvature of the wavefunction and so lowers the level energy. Thus stage one of our well-improvement program now reduces the energy of all states by amount which increases with l. Thus the square well states are moved although they do not become shuffled. The harmonic oscillator degeneracies are lost as shown schematically in the first step of Fig. 8.6. However, the magic numbers still do not appear.

How is this situation improved? The answer is to add a **spin–orbit interaction**. Here spin and orbit refer to the attributes of a single nucleon moving in the assumed potential. Such a term is found in atomic physics and it is due to the magnetic interaction of the magnetic dipole moment of the electron spin and the magnetic field experienced by the electron in its rest frame as it moves through the Coulomb field. In nuclei it cannot be magnetic in origin as it would be insufficiently strong. The potential between two nucleons is known to contain a spin–orbit term: here the spin is the total spin of the two nucleons and the orbital angular momentum is the relative orbital angular momentum between the two nucleons. Translated into a nucleus the average effect on one nucleon due to all the others will be zero deep inside a nucleus since the effect of one of these will be cancelled by the effect of another at the same distance in the opposite direction. Near the surface, however, that cancellation does not occur and a single nucleon experiences a spin–orbit interaction due to the other nucleons. The cancellation effect causes the strength of the interaction to be greatest where the density of nucleons is changing most rapidly, so that frequently the radial variation of the spin–orbit term is put proportional to

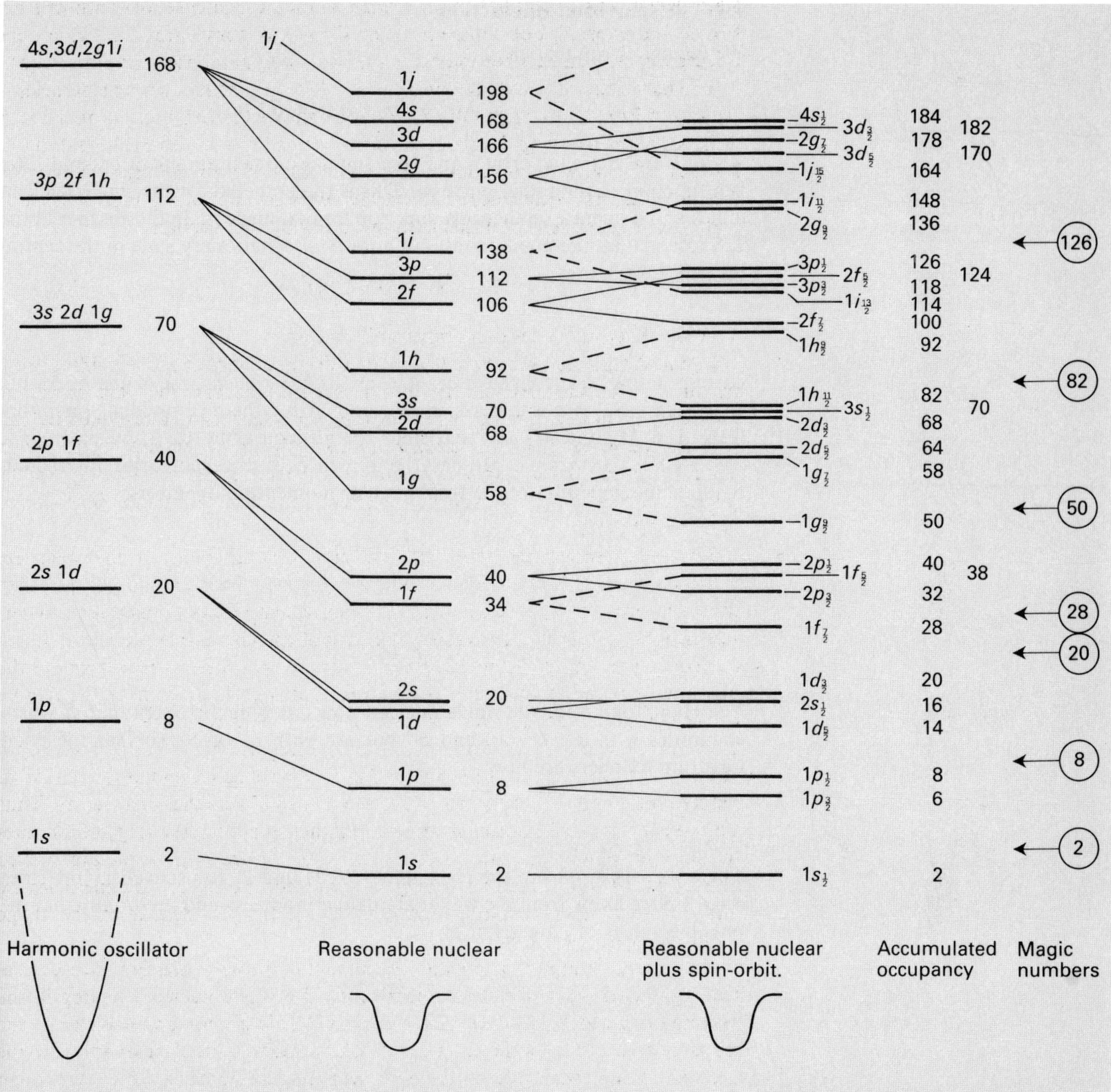

Fig. 8.6 A schematic representation of the change in single-particle level ordering in moving from (a) the harmonic oscillator to (b) a reasonable nuclear potential to (c) the same with an inverted spin-orbit interaction added. On the right of each are shown the accumulated occupancy numbers. An ordering is given in (c) which gives the magic numbers. Between these numbers the residual interactions between nucleons will alter the ordering as the levels are filled. The splittings caused by the spin–orbit interaction which are vital to the explanation of the observed magic numbers are emphasized by heavy broken lines: they are $1f \rightarrow 1f_{7/2} + 1f_{5/2}$, $1g \rightarrow 1g_{9/2} + 1g_{7/2}$, $1h \rightarrow 1h_{11/2} + 1h_{9/2}$, and $1i \rightarrow 1i_{13/2} + 1i_{11/2}$. These are the splittings mainly responsible for the gaps defining the magic numbers.

$$\frac{1}{r}\frac{\mathrm{d}V(r)}{\mathrm{d}r}$$

(see Table 3.3) where $V(r)$ is the central component of the potential, as in equation (8.2).

8.4 The spin–orbit interaction

We change the potential

$$V(r) \to V(r) + W(r)\,\boldsymbol{L}.\boldsymbol{S},$$

where $\boldsymbol{L}$ and $\boldsymbol{S}$ are the orbital and spin angular momentum operators and $W(r)$ is a function of the radial position. This is the potential for the single nucleon and $\boldsymbol{L}$ and $\boldsymbol{S}$ operate on the wave function for that nucleon. In the absence of the $\boldsymbol{L}.\boldsymbol{S}$ interaction, the good quantum numbers of a stationary state in the central potential, $V(r)$, are

$$l,\ l_z,\ s,\ s_z$$

corresponding to the eigenvalues $l(l+1)\hbar^2$, $l_z\hbar$, $s(s+1)\hbar^2$, $s_z\hbar$ of the operators $\boldsymbol{L}^2$, L_z, $\boldsymbol{S}^2$, S_z. (Note that we use L_z and S_z for the z-component operators instead of M_L and M_s and that since the nucleon spins are $\frac{1}{2}$, we have $s=\frac{1}{2}$, $s_z=\pm\frac{1}{2}$.) As in atomic physics the presence of the spin–orbit interaction requires the definition of the total angular momentum operators:

$$\boldsymbol{J}=\boldsymbol{L}+\boldsymbol{S},$$

and

$$J_z=L_z+S_z.$$

The Hamiltonian for the single nucleon now contains the operator $\boldsymbol{L}.\boldsymbol{S}$ which commutes with $\boldsymbol{J}^2$, $\boldsymbol{L}^2$, J_z, and $\boldsymbol{S}^2$ but not with L_z or S_z, so that the good quantum numbers are now

$$j,\ j_z,\ l, \text{ and } s,$$

where $j(j+1)\hbar^2$, $j_z\hbar$ are the eigenvalues of $\boldsymbol{J}^2$ and J_z respectively. Now since $\boldsymbol{J}=\boldsymbol{L}+\boldsymbol{S}$ we know from the way that angular momenta add vectorially that the possible values of j are given by

$$j=l+\tfrac{1}{2}$$

or

$$j=l-\tfrac{1}{2}.$$

unless $l=0$, for which $j=\frac{1}{2}$ is the only possibility. The $\boldsymbol{L}.\boldsymbol{S}$ term means that for $l>0$ the potential is different for these two levels. The eigenvalue of $\boldsymbol{L}.\boldsymbol{S}$ for a stationary state with quantum numbers l, j and s $(=\frac{1}{2})$ is

$$\tfrac{1}{2}[j(j+1)-l(l+1)-s(s+1)]\hbar^2$$

(see Problem 8.1) which becomes

$$\tfrac{1}{2}l\hbar^2 \text{ and } -\tfrac{1}{2}(l+1)\hbar^2 \quad \text{for } j=l+\tfrac{1}{2} \text{ and } l-\tfrac{1}{2} \text{ respectively, } l\neq 0.$$

Thus the potential for $j=l+\frac{1}{2}$ states is

$$V(r)+\tfrac{1}{2}l\hbar^2 W(r)$$

and for $j=l-\frac{1}{2}$ states is

$$V(r)-\tfrac{1}{2}(l+1)\hbar^2 W(r).$$

Thus states of a given l are split into two states of different j unless $l=0$, when there is no change in energy. Once $\boldsymbol{L}$ and $\boldsymbol{S}$ have been combined, the maximum occupancy allowed by Pauli of a level of a given j is $2j+1$.

To obtain agreement with observation it is necessary to make $W(r)$ negative (inverted spin–orbit interaction) which puts a level with $j=l+\frac{1}{2}$ below that with $j=l-\frac{1}{2}$. The result is that the grouping of levels, as in the case of the harmonic oscillator, for example, breaks up and new gaps appear as some states of high j are pushed down by amounts proportional to $j-\frac{1}{2}$ $(=l)$ and others up by amounts proportional to $j+\frac{3}{2}(=l+1)$. We see the results in Fig. 8.6, to be described in the next section.

8.5 The magic numbers 2

Figure 8.6 shows the rearrangement of levels from those of the pure harmonic potential through to those of a realistic nuclear potential with a spin–orbit term. The first step is to truncate the harmonic oscillator potential as described in the last section and that is the first stage of our well-improvement program: the levels no longer have the simple degeneracies and the previous accumulated occupation numbers 2, 8, 20, 40, 70, 112 have broken up without leading to the appearance of the magic numbers. The second step in the well-improvement program is to add the inverted spin–orbit term. We expect high l states to split considerably. The $1f$ $(l=3)$ level splits to $1f_{7/2}$ $(j=l+\frac{1}{2})$ and $1f_{5/2}$ $(j=l-\frac{1}{2})$ with the former becoming lower in energy; it does not come down enough to tangle with the $2s$ and $1d$ levels but is left sufficiently far from its origin in the $2p$ $1f$ group to form an individual shell of $(2j+1=8)$ levels. Thus the next step in accumulation from the $2s$ $1d$ group with 20 takes us to 28, the next magic number. The initially unsplit group $3s$ $2d$ $1g$ loses by downward movement its highest angular momentum state, $1g_{9/2}$, to the $2p$ $1f$ (less $1f_{7/2}$) group to make that a group with 22 levels, thereby completing a shell at 50 levels. At the same time the $3s$ $2d$ $1g$ (less $1g_{7/2}$) group gains by a descent from above of the $1h_{11/2}$ level, making a new shell with 32 levels taking 50 to 82, the sixth magic number. The next shell with 44 levels is constituted from the $2p$ $2f$ $1h$ group less $1h_{11/2}$ but plus $1i_{13/2}$ from above; that shell takes us to a total of 126 levels to give the last magic number that is accessible in the known nuclei.

The vital changes that occur when the spin–orbit force is turned on is the splitting of the $1f$, $1g$, $1h$, and $1i$ levels; these splittings are picked out in Fig. 8.6. It is easy to see that they break the harmonic oscillator groupings by moving the highest angular momentum state down to form a shell of its own ($1f_{7/2}$) or to join the group below to constitute a shell ($1g_{9/2}$, $1h_{11/2}$, $1i_{13/2}$).

We would make a point about filling these levels. The procedure is the same as in atomic physics: a level with total angular momentum quantum number j has $2j+1$ substates corresponding to the possible values of the z-component of the total angular momentum: $-j\leqslant j_z\leqslant j$. In the absence of an external field, these states are degenerate and we shall call this a single level and bear in mind that it can accommodate $2j+1$ identical nucleons.

The final ordering of levels given in Fig. 8.6 is not fixed. As the well grows in radius with increasing A, the ordering inside shells changes as the levels are filled, due to the effect of residual interactions between the nucleons, but not to the extent of spoiling the magic numbers. We do not have space to investigate these matters here and will restrict our discussion to applications which do not depend on a detailed knowledge of the ordering or energies. We shall also remember that there are two wells, one for the neutrons and another with a similar nuclear part but with the addition of a Coulomb potential for the protons. The two sets of levels are filled with neutrons and protons respectively, but as is clear from our discussion of the asymmetry term plus Coulomb term in the semi-empirical mass formula (Section 4.3), we expect that the top level containing a neutron will be at nearly the same energy as the top level containing a proton, for nuclei stable against β-decay.

Some terminology: a nucleus (Z,A) has Z protons and $N=A-Z$ neutrons and if either Z or N is equal to a magic number the nucleus is said to be a **singly-closed-shell** nucleus. One in which both Z and N are magic is said to be a **doubly-closed-shell** or **doubly-magic** nucleus. We can make some straightforward predictions about the latter nuclei. Another type of nucleus for which simple predictions can be made are those with doubly-closed shell plus one nucleon or less one nucleon. In the latter case the **hole** in the doubly-closed shell acts like a single nucleon of opposite charge and reversed j_z, added to a doubly-closed-shell nucleus.

DEFINITIONS AND KEYWORDS

Principal quantum number The quantum number indexing the radial wave function of a stationary state in a spherical potential.

Spin–orbit interaction The nuclear interaction between the spin and the orbital angular momentum of a single nucleon in a nucleus.

Singly-closed shell Either Z or N, but not both, is equal to a magic number.

Doubly-closed shell (or doubly magic) Both N and Z are equal to a magic number, not necessarily the same.

Hole The absence of a particle from an otherwise full shell.

Configuration A presumed or possible occupancy of nuclear shell levels by neutrons and protons.

Later we shall want to describe the **configuration** of nucleons in some nuclei: we shall use the usual notation $(nl_j)^k$, which means a state with principal quantum number n, orbital angular momentum quantum number l, total angular momentum quantum number j, occupied by k neutrons (or k protons). Normally, if we are considering an odd-A nucleus we shall specify only the level occupied by the unpaired nucleon. If we are looking at the effect of a **hole** in a closed shell we shall use $k=-1$. Thus $^{17}_{8}O$ has a configuration for the protons

$$(1s_{1/2})^2(1p_{3/2})^4(1p_{1/2})^2$$

and the configuration

$$(1s_{1/2})^2(1p_{3/2})^4(1p_{1/2})^2(1d_{5/2})^1$$

for the neutrons. When many levels are involved we may list only the levels filled above a closed shell and the holes, if any, in the closed shell. Thus $^{209}_{82}Pb$ is probably neutron configured $(2g_{9/2})^1$ but $^{207}_{82}Pb$ probably has a neutron configuration $(3p_{1/2})^{-1}$. These abbreviations should not confuse.

8.6 The spins and parities of nuclear ground states

We have considered the motion of a single nucleon in a potential and called the vector sum of it spin and its orbital angular momentum its total angular momentum. In a nucleus, the total angular momenta of the constituent nucleons add vectorially to make the **nuclear spin.** Any level (specified by n,l,j) which is fully occupied ($2j+1$ occupants) contributes nothing to a nuclear spin since the angular momenta of the occupants sum to zero. Since $2j+1$ is even for all single-particle levels, we expect that nuclei in which protons and neutrons separately fully occupy a set of single-particle levels will have nuclear spin zero. Such nuclei are even–even and we know already that these have no nuclear spin: in particular, this zero spin prediction applies to, and is of course correct for, doubly closed-shell nuclei. It follows that any doubly-closed plus one or less one nucleon has a nuclear spin equal to the total angular momentum of the extra nucleon or of the hole.

We can make further progress by a simple hypothesis: it is that if two neutrons or two protons are occupying a level with the same j and l, then their total angular momenta couple to give zero contribution to the nuclear spin. Two predictions follow from this:

(1) the spin of the ground state of *all* even–even nuclei is zero;

(2) the spin of the ground state of any odd-A nuclei is that of the unpaired nucleon.

Result (1) is true in all cases. Result (2) is true if the correct ordering of the levels is known (remember, that ordering can vary).

The **nuclear parity** (see comment below) is the product of the parity of all the single nucleon wavefunctions, $\prod_A(-1)^l$. It follows that in the shell model:

(3) the parity of the ground state of all even–even nuclei is even;

Comment

A reminder about parity. The parity of a mathematical function is even (+) or odd (−) as $f(-x)=\pm f(x)$. Of course some functions have mixed parity $f(x)=x^2+x$, for example. Similarly the wavefunction of a particle in a stationary state can have even or odd parity:

$$\psi(\boldsymbol{r})=\pm\psi(-\boldsymbol{r}).$$

It is convenient to think of the parity operator P which has the effect

$$P\psi(\boldsymbol{r})=\psi(-\boldsymbol{r}).$$

If ψ has a unique parity this means that

$$P\psi(\boldsymbol{r})=\pm\psi(\boldsymbol{r}),$$

and the parity expressed as ±1 is the eigenvalue of the parity operator. Nuclear states are eigenstates of P and therefore have parity ±1. The parity of the wavefunction of a single particle moving in a fixed potential is $(-1)^{-l}$, where l is the orbital angular momentum quantum number. For more than one particle in the same potential the overall parity is the product of the parities of all the individual particle wavefunctions.

In Chapter 10 we shall find that the hadrons have an intrinsic parity which must be included if the parity of a state is to be accounted for. In the case of nucleons their parity is arbitrary (Section 10.3) but conventionally taken to be +. Therefore we do not have to worry about any intrinsic parity for nucleons bound in a nucleus.

The strong nuclear forces conserve parity. Without proof we state that this implies that each nuclear state must have a unique parity. Electromagnetism also conserves parity and therefore the Coulomb interaction between protons in the nucleus does not change this situation. However, we shall find in Section 12.9 that the weak interaction which is responsible for, among other things, β-decay, does not conserve parity: the consequence is that there is a very weak additional force between nucleons which has the effect of mixing into the wavefunction of any nuclear state a very small amount of wavefunction of opposite parity. The effects of this are very difficult to detect and we can safely neglect them.

(4) the parity of the ground state of all odd-A nuclei is that of the wavefunction of the unpaired nucleon.

These results are also correct, given correct identification of the levels involved in prediction 2.

The odd–odd nuclei will have one proton and one neutron in levels of the same or different quantum numbers. The two total angular momenta do not have to couple to zero, even if equal, and the model can make no prediction except that the nuclear spin will be in the range allowed by vector addition:

$$|j_1 - j_2| \text{ to } j_1 + j_2.$$

An example is $^{14}_{7}N$ the last neutron and the last proton occupy a $1p_{1/2}$ level so that the nuclear spin is expected to be 0 or 1. The parity is predicted to be even: in fact the spin-parity is 1^+.

8.7 Electromagnetic moments: magnetic dipole

Nuclei with spin $\geqslant \frac{1}{2}$ will have magnetic dipole moment and, if the spin is $\geqslant 1$, can have an electric quadrupole moment (see Section 8.8). Of all the ground state nuclei the even–even nuclei are trivial with zero spin and no moments. The uncertainty in how the angular moments couple in odd–odd nuclei means the simple model will have difficulty calculating their moments. So we concentrate on the odd-A nuclei and we will start by attempting to calculate their magnetic dipole moments.

A magnetic moment is a vector and for a quantum system with angular momentum $\hbar\sqrt{j(j+1)}$, the value discussed or tabulated is the z-component of the magnetic moment when the z-component of the angular momentum $j_z\hbar$ has the maximum value possible, namely $j\hbar$. If we define the **gyromagnetic ratio** γ by the equation

$$\mu_z = \gamma j_z \hbar,$$

then the magnetic moment μ is given by

$$\mu = \gamma j \hbar.$$

For orbital motion ($j = l$) of a singly charged particle of mass m

$$\gamma = e/2m.$$

For motion containing both spin and orbital angular momentum γ is not so simple and we put

$$\gamma = ge/2m$$

where g is called the **Landé g-factor**. Then in atomic physics, μ is usually given in units of **Bohr magnetons** ($e\hbar/2m_e$) so that

$$\mu = gj.$$

For nuclear and particle physics the unit is the **nuclear magneton** $e\hbar/2M_p$ where e

and M_p are the charge and mass of the proton respectively. Then the equation $\mu = gj$ again applies with μ the nuclear magnetic moment in nuclear magnetons, j the nuclear spin quantum number and g the **nuclear *g*-factor**. These magnetic moment units are quantified in Table 8.3. The problem is now to predict the g-factors for nuclei.

Now consider the moments of the odd-A nuclei. In each case there will be no contributions except from the unpaired nucleon and that gives, in principle, two: the first due to its intrinsic magnetic moment and the second due to the orbital motion of its charge; of course, the second contribution is zero in the case of the neutron. In the absence of spin magnetic moment, the orbital motion (angular momentum quantum number l) of a proton would have a magnetic moment of $e\hbar l/2M_p$, that is l nuclear magnetons. For the uncharged neutron the state would have no magnetic moment. If we express this as

$$\mu_l = g_l l \text{ magnetons},$$

then formally the orbital g-factor $g_l = 1$ for a proton and 0 for a neutron (Table 8.4.). For the magnetic moment due to spin (spin angular momentum quantum number s) of a nucleon alone, we write, by analogy,

$$\mu_s = g_s s \text{ magnetons}.$$

DEFINITIONS AND KEYWORDS

Nuclear spin The total angular momentum of a nucleus.

Nuclear parity The parity of the wavefunction that describes the motion of all the nucleons in the nucleus.

Gyromagnetic ratio The ratio of magnetic moment to angular momentum.

Landé *g*-factor The ratio of the magnetic moment in Bohr magnetons to the total angular momentum quantum number for atomic states.

Bohr magneton The unit of atomic magnetic moment. See Table 8.3.

Nuclear magneton The unit of nuclear magnetic moment. See Table 8.3.

Nuclear *g*-factor The ratio of the magnetic moment in nuclear magnetons to the total angular momentum quantum number for nuclear and particle states.

Table 8.3 The units of magnetic dipole moment.

The energy E of a magnetic dipole $\boldsymbol{\mu}$ in a magnetic field $\boldsymbol{B}$ is given by

$$E = -\boldsymbol{\mu}.\boldsymbol{B},$$

so that the units of μ are joules per tesla. The magnetic moment of a classical electron (no spin) in an orbit with angular momentum $\hbar$ is $-e\hbar/2m_e$, e being the magnitude of the electronic chage and m_e the mass of the electron. The magnitude of this quantity is one **Bohr magneton.** Therefore

$$1 \text{ Bohr magneton} = 9.274 \times 10^{-24} \text{ J T}^{-1}, \\ = 5.788 \times 10^{-11} \text{ MeV T}^{-1}.$$

In nuclear physics the same quantity for a proton is $e\hbar/2M_p$, which is one **nuclear magneton**,, M_p being the mass of the proton.

$$1 \text{ nuclear magneton} = 5.051 \times 10^{-27} \text{ J T}^{-1} \\ = 3.152 \times 10^{-14} \text{ MeV T}^{-1}.$$

Table 8.4 Values of the nucleon intrinsic magnetic moments and g-factors.

	μ_s nuclear magnetons	g_s	g_l
Proton	2.7928	5.5856	1
Neutron	−1.9130	−3.8261	0

For the nucleons $s=\frac{1}{2}$, and the measured values of μ_s in nuclear magnetons are given in Table 8.4. These values fix the spin g-factors g_s for neutron and proton separately (Table 8.4 again). This situation is more complicated than for an electron: its magnetic moment is different from -1 Bohr magnetons by approximately 0.1%, so for all but precision work it has $g_s=-2$. This magnetic moment has been calculated with a precision of about 1 in 10^{10} using quantum electrodynamics (see Section 9.6), giving a result in agreement with measurements of comparable precision. Such a happy situation does not exist for the nucleons: their moments are not close to integer multiples of the nuclear magneton and it has not proved possible to predict theoretically the nucleon magnetic moments, except in an approximate way by using the quark–parton model (Chapter 10) or by attempts to solve the problem of bound quarks using quantum chromodynamics (Sections 9.10 and 9.11).

The calculation of the odd nucleon's magnetic moment when both spin and orbital motions are taking place is, in principle, analogous to the calculation of the magnetic moment of an electron orbit in atomic physics. A key phrase which will remind the reader about that calculation is the Landé g-factor.

The unpaired nucleon's g-factor is defined by

$$\mu=g_j j. \tag{8.3}$$

where μ is the magnetic moment in nuclear magnetons and j is the nucleon's total angular momentum. (Of course, μ becomes the nuclear magnetic moment and j the nuclear spin.) Then g_j is given by

$$g_j=g_s\frac{j(j+1)-l(l+1)+s(s+1)}{2j(j+1)}+g_l\frac{j(j+1)+l(l+1)-s(s+1)}{2j(j+1)}. \tag{8.4}$$

This looks very complicated but a few moments' study shows that it has a simple repeated structure and two negative signs for which the reader can find a mnemonic that will place them in their correct position. You are invited to prove this formula (Problem 8.2).

There is one case in which the formula simplifies considerably. If $j=l+s$, $s=\frac{1}{2}$, then

$$\mu=g_j j=g_l l+g_s\tfrac{1}{2}\text{ magnetons.}$$

Physically this corresponds to the vectors of l and s adding to give the maximum possible j. The result is just the arithmetic sum of the spin and orbital magnetic moments contributing.

How well does this formula (equations (8.3) and (8.4)) work for the odd-A nuclei? The measured values of the magnetic moment for all such nuclei up to $^{19}_{9}$F are shown in Problem 8.3. There are two empty columns to fill, one of which is for the values of the magnetic moment predicted by equations (8.3) and (8.4). Once you have done that you will find the predictions are not good. Further up the periodic table the results are generally bad with the prediction usually lying between the two values calculated from the assignments $l=j\pm\frac{1}{2}$ for a given j. That is inside the range -1.913 to $1.913j/(j+1)$ nuclear magnetons for an unpaired neutron and between $j(j-1.293)/(j+1)$ and $j+2.293$ nuclear magnetons for an unpaired proton. This is an unexpected result because once the j and parity of an odd-A nucleus is given, then l for the single unpaired nuclear is

PROBLEMS

8.1 A single nucleon state is an eigenstate of the operators $\boldsymbol{J}^2$, $\boldsymbol{L}^2$, $\boldsymbol{S}^2$ and J_z with quantum numbers j, l, s, and j_z respectively. Show that it is an eigenstate of the operator $\boldsymbol{L}.\boldsymbol{S}$ with the eigenvalue

$$\tfrac{1}{2}[\,j(j+1)-l(l+1)-s(s+1)\,]\hbar^2.$$

8.2 Prove the formula of equation (8.4).

8.3 Now put the shell model to work for yourself. Below is a table of the odd-A nuclei up to $^{19}_{9}$F. Determine the type, neutron or proton, of the odd nucleon and use the level ordering for the 'nuclear potential with spin–orbit interaction' of Fig. 8.6 to decide its configuration. Pencil your results lightly into column 2. Then put the expected spin-parity into column 3 and work out your prediction for the nuclear magnetic moment and pencil that into column 4. Column 5 contains the measured values of the magnetic moment. If other readers have been here before you, erase their pencil marks or check them!

Nucleus	Odd nucleon type and configuration	Nuclear spin-parity	Magnetic dipole moment nuclear magnetons Calculated	Measured
$^{3}_{1}$H				2.9788
$^{3}_{2}$He				−2.1276
$^{7}_{3}$Li				3.2564
$^{9}_{4}$Be				−1.1776
$^{11}_{5}$B				2.6885
$^{11}_{6}$C				−1.0300
$^{13}_{6}$C				0.7024
$^{13}_{7}$N				0.3221
$^{15}_{7}$N				−0.2831
$^{15}_{8}$O				0.7189
$^{17}_{8}$O				−1.8937
$^{17}_{9}$F				4.7224
$^{19}_{9}$F				2.6288

8.4 Explain briefly why a model of the nucleus in which the nucleons are considered to be moving in a central potential predicts 'shell closure' effects in nuclear properties for neutron or proton numbers 2, 8, 20, 34, 40, 58, 92, and 138. The observed neutron or proton numbers at which these closure effects actually occur are 2, 8, 20, 28, 50, 82, and 126 (126 for neutrons only). Explain how a modification to the single-particle potential accounts for this observation.

Assign spins and parities to the ground states of the nuclei:

$$^{20}_{10}\text{Ne},\ ^{27}_{13}\text{Al},\ ^{41}_{21}\text{Sc}.$$

State any assumptions that you have made.

(Adapted from the 1982 examination of the Final Honours School of Natural Science, Physics, University of Oxford.)

determined. Parity is a good quantum number for nucleus so that the state of the other l cannot mix in and it is not obvious why the observed magnetic moment should lie between the expected value and the value it cannot have in this extreme single-particle picture. However, the effects we are about to mention can cause the observed values to deviate in this way from the predictions of the extreme single particle model. In view of the very simple model of what are truly complex nuclei that we are using it is not surprising that such discrepancies exist. It is useful to list possible reasons.

1. We have assumed the bound nucleons have the same intrinsic magnetic moment as do the unbound nucleons. This is probably incorrect.

2. The single particle shell model which predicts correctly the nuclear spin and parity of odd-A nuclei assumes that the nuclear spin is due to the single, unpaired nucleon. However, the same spin and parity can be constructed from other more complicated motions in which the remaining, normally paired nucleons are not so paired and contribute to the angular momentum. Thus the actual nuclear wavefunction is a superposition of all states of the same total angular momentum and parity. This is called **configuration mixing** (see Section 8.9). Clearly this situation makes the single particle calculation of magnetic moment of doubtful validity.

3. The existence of nuclear forces means that there are currents in the nucleus due to the exchange of charged π-mesons (Chapter 9). These currents contribute to the magnetic moment.

4. There is an effect in light nuclei which should be taken into account. The assumption is made that the single nucleon moves in a fixed potential but in all nuclei this cannot be the case: the single nucleon and the remaining nucleons move about their mutual centre of gravity so that the effective g-factor, g_l, for the orbital motion is no longer 0 or 1 for neutron or proton respectively.

DEFINITIONS AND KEYWORDS

Configuration mixing The situation in which the true state is a superposition of two or more possible configurations, each of which on its own can give the observed nuclear spin and parity.

These are complications we are not in a position to deal with. However, we can note that whatever improvements are introduced there remain no completely successful methods of calculating nuclear magnetic moments.

8.8 Electromagnetic moments: electric quadrupole

The contribution to the electromagnetic energy of a system which lies in a non-uniform electric field is given by a series of terms. The first depends on the electrostatic potential and the monopole moment (= charge of the system). The second depends on the gradient of the potential and the electric dipole moment. The third depends on the second space derivatives of the potential and the electric quadrupole moment.

The monopole moment is the volume integral of the charge density over the system. The dipole moment has three components p_x, p_y, p_z with respect to arbitrary cartesian coordinates and

$$p_x = \int x\rho(\boldsymbol{r})\mathrm{d}V$$

where $\rho(\boldsymbol{r})$ is the charge density at the point $\boldsymbol{r}$ and $\mathrm{d}V$ is a volume element; similar expressions give the other components. More precisely the moments are found by calculating the moments over the charge distribution of an appropriate power of r times a spherical harmonic. Thus

$$p_z = \left(\frac{4\pi}{3}\right)^{\frac{1}{2}} \int Y_1^0(\theta,\varphi) r\rho(\boldsymbol{r}) \mathrm{d}V,$$

$$p_x \pm \mathrm{i}p_y = \left(\frac{8\pi}{3}\right)^{\frac{1}{2}} \int Y_1^{\pm 1}(\theta,\varphi) r\rho(\boldsymbol{r}) \mathrm{d}V.$$

(The spherical harmonics $Y_l^m(\theta,\varphi)$ will be familiar to the reader as the eigenfunctions of the operators of orbital angular momentum and as the angular part of the wavefunctions of the hydrogen atom.) The moments found using $r^2 Y_2^m(\theta,\varphi)$, $-2 \leqslant m \leqslant 2$ are the components of the quadrupole moment.

Nuclei can only have electric dipole moments that are so small as to be undetectable by existing techniques; a possible exception is the neutron, which certainly has a very small moment but it may be within reach of the latest measurement techniques (see Problem 13.2 for a clue to the explanation of this situation). However, nuclei can have a prolate- or oblate-like shape which will give the nucleus an electric quadrupole moment. What this means in physical terms is shown in Fig. 8.7. If the axis of symmetry is along the z-axis, then the quadrupole moment is defined by

$$\mathcal{Q} = \int \rho(\boldsymbol{r})\{3z^2 - r^2\}\mathrm{d}V \qquad (8.5)$$

where $\rho(\boldsymbol{r})$ in the charge density and $\mathrm{d}V$ is a volume element. The integral is, in our case, over the nuclear volume. $\mathcal{Q}$ has the dimension of charge times length squared, but for nuclei it is normally given in barns (see Section 2.9) meaning, of course $\mathcal{Q}/e$, where e is the magnitude of the electron charge.

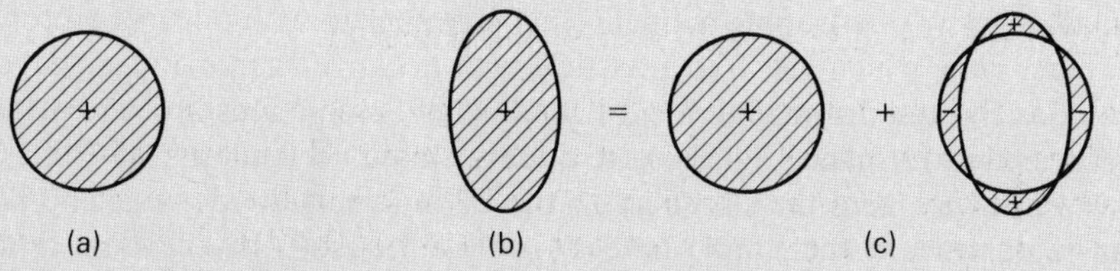

Fig. 8.7 A spherically symmetric positive charge distribution (a) has electric monopole moment but no higher electric moments. A prolate charge distribution (b) is equivalent to a spherically symmetric charge distribution plus some positive charge shared between the two polar regions and a band of equal negative charge added around the equator. This total addition (c) on its own has neither monopole moment nor dipole moment but does have electric quadrupole moment. The quadrupole moment is defined by

$$\mathcal{Q} = \int \rho(\boldsymbol{r})\,(3z^2 - r^2)\,\mathrm{d}V$$

where $\rho(\boldsymbol{r})$ is the charge density at the point $\boldsymbol{r}$. Thus the distribution in (c) has positive $\mathcal{Q}$. Similarly an oblate shape of positive charge would have negative $\mathcal{Q}$. The quadrupole moment for nuclei given as an area (usually barns) means $\mathcal{Q}/e$.

Table 8.5 Some electric quadrupole moments

Odd-A nuclei, doubly-closed shell ± one proton or neutron

	Nucleon single particle state	$\mathcal{Q}$(barns)
$^{17}_{8}O$	Neutron $(1d_{5/2})^1$	−0.026
$^{39}_{19}K$	Proton $(1d_{3/2})^{-1}$	0.11
$^{209}_{83}Bi$	Proton $(1h_{9/2})^1$	−0.35

Deformed nuclei

	$\mathcal{Q}$ (barns)
$^{115}_{49}In$	1.16
$^{165}_{67}Ho$	2.82
$^{175}_{71}Lu$	5.68
$^{181}_{73}Ta$	4.2
$^{197}_{79}Au$	0.59

This definition requires the moment to be oriented so that the axis of symmetry is along the z-axis and the nuclear spin z-component $j_z = j$. For an odd-A nucleus these quantum numbers refer to those of the unpaired nucleon and $j_z = j = l \pm \frac{1}{2}$. This implies that the orbital quantum number $l = j - \frac{1}{2}$ or $j + \frac{1}{2}$ and l_z can take on its maximum value l, or $l - 1$. The wavefunction of a nucleon with these quantum numbers is concentrated equatorially and we expect the nucleus to have an oblate shape and a small negative quadrupole moment.

A doubly-closed-shell nucleus (zero nuclear spin) must have zero quadrupole moment, but the argument above says that the next proton will give the nucleus a small negative quadrupole moment if $j \geqslant \frac{3}{2}$. A nucleus with one less proton has an empty state which acts like a particle with negative charge and we expect a small positive quadrupole moment if $j \geqslant \frac{3}{2}$. These predictions are supported by measurements for nuclei right next to doubly-closed-shell nuclei (see Table 8.5). However, away from the closed shells the E2 moments are positive and larger than is possible in the simple model (Table 8.5 again). In addition, doubly-closed-shell nuclei plus or minus one neutron (for example, $^{17}_{8}O$) have small negative quadrupole moments where our (extremely) simple model would predict zero. These observations point to nuclei being strongly deformed into prolate shapes away from closed shells and to being slightly deformed elsewhere. These conclusions lead to the collective model to be discussed in Section 8.9.

PROBLEM

8.5 Find out how nuclear magnetic dipole and electric quadrupole moments may be measured.

8.9 Excited states in the shell model

Consider a nucleus, doubly-closed shell plus one nucleon, for example ${}^{17}_{8}O$ as illustrated in Fig. 8.8. The shell model predicts that eight neutrons are in the configuration $(1s_{1/2})^2(1p_{3/2})^4(1p_{1/2})^2$ and the last unpaired neutron is in the state $1d_{5/2}$. The next single-particle levels available, working up in energy, are $2s_{1/2}$, $1d_{3/2}$, $1f_{7/2}$, $2p_{3/2}$, $1f_{5/2}$, and so on. It is natural to expect that the excited states of ${}^{17}_{8}O$ would correspond to the excitation of the unpaired nucleon into these states; thus we expect a $\frac{5}{2}^+$ ground state with excited states of $\frac{1}{2}^+$, $\frac{3}{2}^+$, $\frac{7}{2}^-$, $\frac{3}{2}^-$, ... However, the actual spectrum, as shown in Fig. 8.8, is more complicated, and this indicates that motions involving nucleons other than the unpaired neutron are very important. This is the usual pattern and it is clear that the simple model is not going to predict these levels.

What is happening? One possibility is that it is cheaper to create a hole in a normally filled level than to promote an unpaired nucleon. Thus in ${}^{17}_{8}O$ a neutron hole can be created in the $1p$ level, the promoted neutron pairing with the previously unpaired neutron, both then occupying states in the levels $2s_{1/2}$ or $1d_{5/2}$. The hole, $(1p_{1/2})^{-1}$ then gives a $\frac{1}{2}^-$ excited state: there is such a state at 3.055 MeV. The doubly magic plus one neutron nucleus ${}^{41}_{20}Ca$ has a ground state

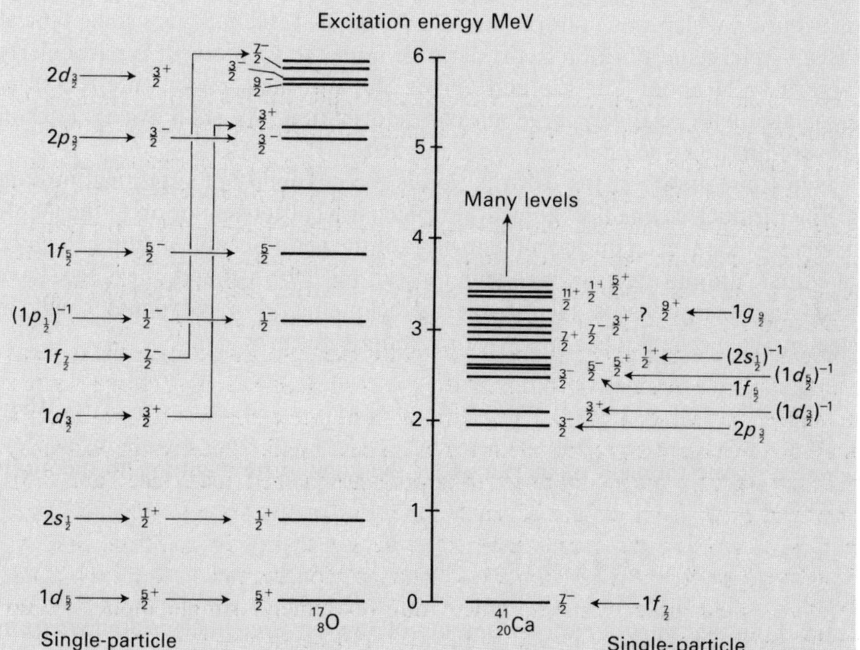

Fig. 8.8 The energy levels for the nuclei (${}^{17}_{8}O$ and ${}^{41}_{20}Ca$) of the type doubly-closed shell plus one nucleon. Each level is labelled with its spin-parity, if known. On each side some possible identification of levels as single particle states of the odd nucleon are given. There are also identifications as single holes due to the promotion of a nucleon out of the closed shell, the nucleon then pairing with the previous odd nucleon. It is obvious that although assignments may be made they do not occur in an expected order and that the spectrum of levels is more complicated than the single-particle states allow, particularly in ${}^{41}_{20}Ca$.

of $\frac{7}{2}^-$ expected for the configuration $(1f_{7/2})^1$ of the last neutron and the first excited state has $\frac{3}{2}^-$ consistent with a configuration $(2p_{3/2})^1$. In both these examples there are many states above the first excited states not readily or sensibly identified with single-particle states. The reason is that excitations can involve many more nucleons than the unpaired neutron in these two particular cases, with the result that there is a much richer spectrum of excited states than that first expected. In addition, each state will be a superposition of all those configurations such that each can combine all its own single-particle total angular momenta and parities to make that nuclear state's spin-parity. Thus $\frac{7}{2}^+$ in $^{41}_{20}Ca$ will be a superposition of, among many others, the configurations

$$(1g_{7/2})^1, \qquad (1d_{3/2})^{-1}(1f_{5/2})^1(2p_{3/2})^1, \qquad (1d_{3/2})^{-1}(1f_{5/2})^1(2p_{1/2})^1, \ldots .$$

This is an example of the configuration mixing defined in Section 8.7. To predict the degree of mixing requires a very sophisticated theoretical approach to describing the nucleus.

8.10 The collective model and other developments

In Section 8.8 we described the electric quadrupole ($E2$) moment and showed how positive values were characteristic of a prolate shape for a nuclear charge distribution which was uniform inside that shape. Table 8.5 gives some typical values: in terms of position in the periodic table the $E2$ moment is particularly large from A about 150–190 and above 220, and these regions are therefore associated with relatively large nuclear deformation. We shall discuss later in this section how such deformation can occur.

One consequence of the deformation is the possibility of rotational motion giving the nucleus angular momentum which is a collective motion of the whole assembly. Even–even nuclei in the regions of the periodic table mentioned have, of course, ground states of zero spin but can be deformed and, if excited, have low-lying energy levels identifiable as belonging to a **rotational band.** In elementary mechanics the energy of rotation is given by

$$E = I^2/2\mathscr{I} \tag{8.6}$$

where $\mathscr{I}$ is the moment of inertia and I the angular momentum. In quantum mechanics

$$I^2 = j(j+1)\hbar^2, \qquad j = 0, 1, 2, 3, \ldots .$$

It turns out that the symmetry properties of an even–even nucleus do not permit odd values of j. Thus we obtain for the energy levels

$$E_j = j(j+1)\hbar^2/2\mathscr{I}, \qquad j = 0, 2, 4, 6, \ldots . \tag{8.7}$$

In Fig. 8.9 we show some of the excited states of $^{172}_{72}Hf$ alongside the pattern of values of E_j for an empirical value of $\mathscr{I}$. In Problem 8.7 you are invited to calculate an expected value of $\mathscr{I}$ and compare it with that derived from a pattern of excited states. The latter will be much less than expected from a rotating rigid sphere, or, more correctly, a rotating rigid spheroid: this suggests that the motion is not entirely that of a rigid prolate nucleus but that it is partly

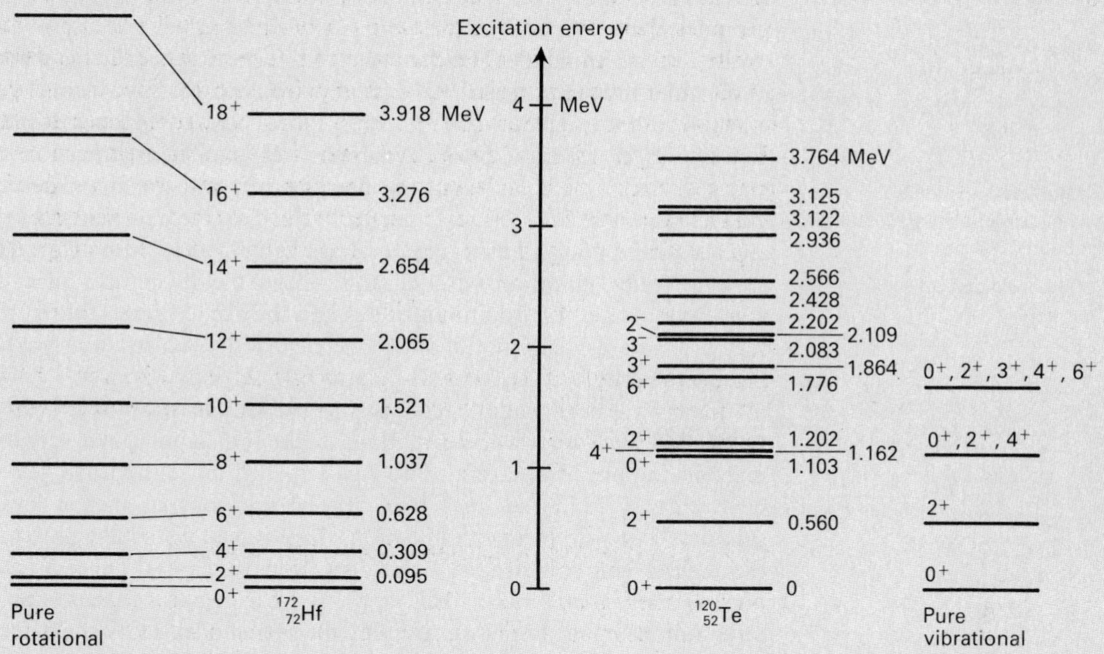

Fig. 8.9 Some of the energy levels of the nuclei $^{172}_{72}$Hf and $^{120}_{52}$Te. The former is a good example of a rotational band: alongside is the pattern of levels expected from a pure rotator normalized at the ground state and at the hafnium 10^+ state: there are deviations of the real case from the ideal spacing of levels. The nucleus $^{120}_{52}$Te has a pattern of excited states that begins as expected for vibrational but, as is usually the case, fails beyond the two phonon states, in this case at about 1.15 MeV.

that of a prolate-shaped wave moving around the surface of the nucleus—collective motion! In addition it is found that the levels diverge from the simple equation (8.7) as if $\mathscr{I}$ were changing as j increases. And why not? As j increases centrifugal forces will cause a change in the deformation and therefore in the moment of inertia. In addition it is possible for a nucleus to have excited states which are not rotational and then to have a band of rotational states built on each of these states. Rotational bands are not confined to even–even nuclei: however, in these other nuclei the pattern of energy levels in the bands is not as simple as that of equation (8.6).

The de-excitation of rotational levels is discussed in Section 11.6. An example of the line spectrum of photons emitted after Coulomb excitation of levels in a rotational band is shown in Fig. 7.8.

The observation that the rotation is not of a rigid body but the effect of a collective motion suggests that we could go back to the ideas of the liquid drop model. A classical drop of liquid can have surface waves which give the motion the same appearance as the rotation of a permanently deformed drop: this is the analogy of the rotational state of a nucleus which does not involve rigid rotation of the whole nucleus. In addition there can be vibrations about an equilibrium shape. In the case of nuclei there is the possibility of such **shape oscillations**. The simplest is that in which the radius oscillates about its mean value: this is the 'breathing mode' but, since the nuclear matter is very incompressible, such a mode would have a very high excitation energy and none has been identified with certainty. The next allowed vibration of shape is one in which the nucleus

DEFINITIONS AND KEYWORDS

Rotational band A set of levels with energies of excitation increasing in step with a regularly increasing quantized angular momentum wholly attributable to a collective rotation of the nucleus.

Shape oscillations Periodic changes in the shape or size of a nucleus by the collective motion of its constituents.

Phonon One quantum of collective vibration.

oscillates between prolate and oblate: this is called a quadrupole mode because the radius deviates from spherical by an amout time-varying with an amplitude proportional to $P_2(\cos\theta)=\frac{1}{2}(3\cos^2\theta-1)$, where θ is the polar angle with respect to the axis of symmetry. The quantum of this motion is called a **phonon** and it has angular momentum-parity 2^+. Ideally two phonons have twice the energy of one and couple to give $J^{\mathrm{P}}=0^+$, 2^+, or 4^+; three have three times the energy and give $J^{\mathrm{P}}=0^+$, 2^+, 3^+, 4^+, or 6^+. Approximately similar organization of excited states is seen in some nuclei but the ideal is far from the reality. The example of $^{120}_{52}\mathrm{Te}$ is shown in Fig. 8.9: as is normally the case the degeneracies are broken and the three phonon states are not recognizable. Nuclei with deformed ground states also have vibrational excited states. The difficulty in this subject is that the vibrational states, the rotational states and the excited states due to the one or several single-particle motions are not clearly decoupled so as to be always distinguishable.

Efforts have been made to reconcile the shell model with the collective model. Let us first consider how deformations might come about, in a very qualitative manner. Suppose we have a doubly-closed-shell nucleus with $Z=N=20$. The next vacancy is in the level $1f_{7/2}$. Several nucleons in such a level have a spherically non-uniform wavefunction which acts to distort the potential well of the nucleus from spherical. If such a distortion decreases the energy of the whole nucleus then it will occur: this appears to be what happens, the favoured distortion being prolate. By examining the shell model states in a deformed well it is possible to show that the energy of the whole nucleus will be minimized, for a given assembly of nucleons, at a certain deformation. Such approaches have been successful in describing the many properties of the ground and excited states of nuclei in the regions of the periodic table marked by deformed nuclei.

For the reader, this section has done no more than introduce the collective and unified models. Clearly there is a great deal more to these subjects than can be even mentioned here.

PROBLEM

8.6 The spin-parity j^{P} and excitation energy E of the ground state and a sequence of excited states of the nucleus $^{170}_{72}\mathrm{Hf}$ are given in the table.

j^{P}	0^+	2^+	4^+	6^+	8^+
E(keV)	0	100	321	641	1041

Account for this series of states, and calculate the moment of inertia of the nucleus in each of the excited states. Comment on your results and compare them with the moment of inertia of the nucleus considered as a rigidly rotating sphere. The moment of inertia of a sphere is $\frac{2}{5}MR^2$. Take $R=1.3\times10^{-15}A^{1/3}$ m.

(Adapted from the 1983 examination of the Final Honours School of Natural Science, Physics, University of Oxford.)

8.11 Reconciliation

The very simple shell model that we have described is sometimes known as the extreme single-particle-shell model, for evident reasons. There are improvements which take into account residual interactions that will occur as shells fill, with the result that the level ordering changes. Configuration mixing is another complication. One improvement we have indicated is that of allowing the possibility of deformed nuclei and thus deformed wells, the deformation coming about because the energy levels in such wells permit lower energies than for the undeformed case. However, the whole idea of energy levels of independently or nearly independently moving particles should be reconciled to the successful liquid drop model, which pays scant attention to the idea of energy levels, except in the asymmetry term.

The shell model and the liquid drop model are so unlike that it is astonishing that they are models of the same system. Of course, each model has limited objectives and limited successes. Neither model contains any assumptions that will allow, firstly, the prediction that for $A > 20$ the central nuclear density is almost constant, or secondly, the value of that density. An alternative approach, the **theory of nuclear matter,** starts with information about the inter-nucleon potentials and attempts to predict these properties of nuclei and the value of the average potential in which an individual nucleon can be thought of moving for the purposes of developing the shell model. The inter-nucleon potentials have some simple properties: there is a hard repulsive core at separations less than about 0.5 fm, an attractive short-range force at separations greater than this distance, falling perhaps exponentially with an e-folding distance of about 1.4 fm. There are other features which we will discuss in Sections 9.7–9.9, but these will suffice for our immediate purposes.

DEFINITIONS AND KEYWORDS

Theory of nuclear matter The theory which attempts to understand and to predict the properties of nuclei from the properties of inter-nucleon forces.

We can get a feel for the consequences of inter-particle forces of this nature for a many-body system by considering the wave-mechanical treatment of a gas of free hard sphere molecules of radius $C/2$. They interact only when the separation is C, where they experience an infinite repulsive force. The gas can be taken to be infinite in extent but we define a number density $\rho = (4\pi R_0^3/3)^{-1}$. (Thus A molecules occupy a volume $4\pi R_0^3 A/3$ and if that volume is a sphere it has a radius $R_0 A^{1/3}$: this links us to the nuclear density.) As the density of molecules increases, the average kinetic energy increases and approaches infinity as $R_0 \to C$ (curve (a) Fig. 8.10). If we now add a sufficient attractive potential for separations $> C$ but decreasing with further separation, then the energy (kinetic plus potential) of the average molecule could vary with R_0 as shown in curve (b) of Fig. 8.10. This curve has a minimum, with a negative energy at a certain R_0. The strength and range of the attractive part of the potential together with C are the quantities which determine the position and value of this minimum in the average energy, which, in turn, fix the equilibrium density and the binding energy. Although no quantitative argument has been given, this is a physically reasonable result.

A programme of this kind for nuclear matter has to take into account many factors: the system has to have a wavefunction which is totally antisymmetric under the exchange of any two identical nucleons and the inter-nucleon force depends not only on the distance of separation but also on the spin of the two nucleons and on their state of relative orbital angular momentum. There may also be velocity-dependent terms and forces which depend on the presence nearby of a third, or a fourth, ... nucleon. A great deal of attention has been given to developing and refining this theory of nuclear matter and it is now possible to go some way towards making reasonable predictions about the properties of nuclei from our existing knowledge of the inter-nucleon forces. However, it has not yet proved possible to obtain simultaneously the correct values of the binding energy and of the density. In addition, in Chapter 10 we shall learn that nucleons are not point-like but are extended objects containing

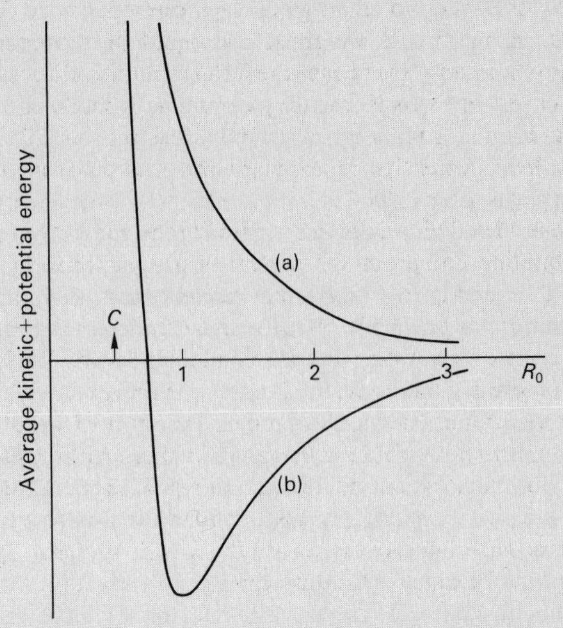

Fig. 8.10 The average energy (kinetic plus potential) per molecule for a gas of hard-sphere molecules as a function of R_0, where the number density is $(4\pi R_0^3/3)^{-1}$. Curve (a) is for the case of no inter-molecule interaction except for the effect of the infinite repulsive force at separations C. Curve (b) is for the case of the same hard core but with the addition of a square potential well extending from $R_0 = C$ to $R_0 = 2$. In this case the gas would have an equilibrium density at $R_0 = 1$. The actual equilibrium density and the average energy would depend on the detailed form of the potential energy curve. The potential well has to be sufficiently deep and wide to give a minimum for the average energy. That minimum will be at a negative energy and the system will be bound.

quarks: attempts are being made to understand the role that these constituents have in determining nuclear forces, but at present this structure has not had a great influence on nuclear matter theory.

Returning to our very simplified approach we note that the nucleon hard core prevents separations less than $C \simeq 0.5$ fm, so we assume each nucleon has an impenetrable core of radius $C/2$. In nuclear matter of normal density, for which $R_0 \simeq 1.2$ fm, the fraction of the nuclear volume that is hard core is therefore

$$(C/2R_0)^3 = 0.009.$$

Thus hard core collisions occur very infrequently and the motion of an individual nucleon can be described by assuming it moves independently in a smoothly varying potential well provided by the remaining nucleons.

In fact the attractive part of inter-nucleon potential is not very strong: if it were stronger the equilibrium density would increase until it approached the maximum possible at close packing of hard spheres. This would occur at $R_0 = C/1.81$; for existing stable nuclei $R_0 = 2.4C$. Thus nuclear matter behaviour is closer to that of a gas than it is to that of a crystal where densities are near those of close packing.

The qualitative arguments that we have made provide a reasonable physical picture of how the shell-model potential is a consequence of the known inter-nucleon force. How now do we get to the liquid drop model and the semi-empirical mass formula? It is in fact the latter we must discuss since the liquid part of the idea served to justify the simple approach that we made in Chapter 4, whereas we are looking for a more realistic basis for formula. In fact we are almost there: the minimization of the energy per molecule for a given density of our imaginary gas of hard-sphere molecules was independent of the total number of molecules. This implies a system with the property of saturation, that is the total binding energy of the system is proportional to the number of constituents. By analogy, we expect that the leading term in the total binding energy of a nucleus is proportional to A, as in the mass formula. The surface term can be thought of as the first correction and is justified by its success in the semi-empirical formula and in understanding the energy changes in the nuclear deformations which can lead to the fission of heavy nuclei (Section 6.6).

We must stress that the discussion in this section has overplayed the role of the hard core in causing saturation. It is known that the tensor force, the dependence on spin and orbital angular momentum of the nuclear force (see Sections 9.8 and 9.9) and, of course, the Pauli exclusion principle, all have an important role in the causes of saturation.

8.12 *Au revoir* to nuclei

The theories of nuclear structure and behaviour do not form a unified model in the sense that all properties can be predicted or explained from a set of assumptions of a fundamental nature. Nuclei are complicated many-body objects held together by poorly understood forces: different areas of structure or of behaviour have to be considered theoretically under a set of assumptions which may restrict that theory's applicability to a given range of A or to a given type of reaction, for example. Such limitations make the whole subject of theoretical nuclear physics appear to be a pastiche of slightly related ideas and techniques. Of course, it is this variability that provides the challenge to find a way to unification.

On that somewhat inconclusive note we leave the subject of complex nuclei and in the next sections we examine the knowledge and understanding that we now have of the nature of the strong nuclear forces and of the forces responsible for β-decay, and the relation of both to the more familiar electromagnetism. Only in Chapter 14 will we meet complex nuclei again, in connection with nucleosynthesis in the universe.

9

Forces and Interactions

9.1 Introduction and some nomenclature

A thread that has run through earlier chapters has been the idea of the existenc of a strong force (relative to the Coulomb force) which acts between nucleon and is capable of binding these particles into nuclei. It is also obvious that thi force must be the major force operating in the course of nuclear reactions o scattering. However, except at distances of closest approach below the range o nuclear forces, the α–particle scattering studied by Rutherford and his collabor ators was caused by the Coulomb force. That force is a part of electromagne tism, and that is familiar ground which we will discuss in the next section in way that will begin to acclimatize us to modern ways of discussing forces. Bu first let us dispose of some nomenclature.

The first item is the classification of particles into **fermions** and **bosons.** Thes are particles which obey the Fermi–Dirac and Bose–Einstein statistics respec tively. Therefore particles with half-odd-integer spin are fermions and thos with integer spin are bosons. The next term we want to introduce is the idea o an **interaction**. Electrically charged particles interact with the electromagneti field; this is a manifestation of the **electromagnetic interaction**. We shall soor meet some other interactions.

We wish also to extend the idea of real and virtual photons (Section 1.6) tc

DEFINITIONS AND KEYWORDS

Fermion A particle which obeys the Fermi–Dirac statistics. Particles with half-odd-integrer spin are in this class.

Boson A particle which obeys the Bose–Einstein statistics. Particles with integer spin are in this class.

Interaction This is a general term indicating the possibility of the exchange of energy and momentum between particles or of the possibility of the creation or annihilation of particles.

Electromagnetic interaction The interaction that exists between charged particles, or particles with electric structure, and the electromagnetic field, and that causes the electric and magnetic forces between such particles and the emission or absorption of photons. It can also cause the creation or annihilation of a pair consisting of a charged particle and its antiparticle.

Real particle A particle having energy total E and momentum P satisfying

$$E = +\sqrt{(M^2c^4 + P^2c^2)},$$

where M is the rest mass of the particle.

Virtual particle A particle having a transitory existence and not satisfying

$$E = +\sqrt{(M^2c^4 + P^2c^2)}.$$

where M is the rest mass of the particle.

We have met the same definitions of real and virtual applied specifically to the photon: see Section 1.6.

articles. The Heisenberg uncertainty principle allows a system to have an nergy uncertainty ΔE for a time Δt given by $\Delta E \, \Delta t \simeq \hbar/2$. A particle which is ee and stable (not interacting with external fields or with other particles) has $t=\infty$, so that $\Delta E=0$. Translating the idea to the inertial frame in which the article is at rest, this means that the particle must have its rest mass energy Mc^2), no more and no less. In any other inertial frame the total energy (E) and iomentum (P) must satisfy

$$E^2 - P^2c^2 = M^2c^4$$

uch a particle is said to be **real**. However, a particle may have only a brief xistence and not become truly free of sources; then the Heisenberg uncertainty rinciple allows that the values of total energy, momentum and its rest mass do ot satisfy this equation. Such a particle is said to be **virtual**.

Armed with real or virtual interacting fermions and bosons we can proceed.

).2 Electromagnetism

Classical electromagnetism is one of the great achievements of nineteenth-century physics. That structure is based upon experimental observations which are embodied in the laws of Coulomb, Ampère, and Faraday. Essentially there are two facts.

1. There are forces (Coulomb) between stationary charges and there are additional forces (magnetic) between moving charges.
2. Electromagnetic radiation is emitted by accelerating charges and can be absorbed or scattered by bodies with electrical structure.

Classical electromagnetism defines the **fields** $\boldsymbol{E}$ and $\boldsymbol{B}$ which obey Maxwell's equations (Table 9.1). It is convenient to define the vector potential $\boldsymbol{A}$: then

$$\boldsymbol{B} = \text{curl}\, \boldsymbol{A}. \qquad (9.1)$$

and

$$\boldsymbol{E} = -\text{grad}\, \varphi - \frac{\partial \boldsymbol{A}}{\partial t}, \qquad (9.2)$$

where φ is the (scalar) electrostatic potential. The properties of $\boldsymbol{A}$ are such that the four-vector $\boldsymbol{A}, \varphi/c$ (c = speed of light) transforms relativistically exactly as does any four-vector: familiar examples are the space–time coordinates $\boldsymbol{r}, ct$ or the momentum-total energy $\boldsymbol{P}$, E/c. The electromagnetic field has a very important property, that of **gauge invariance**. The observable fields $\boldsymbol{E}$ and $\boldsymbol{B}$ are unchanged even if the field quantities **A** and φ are simultaneously changed in the following way:

$$\begin{aligned} \boldsymbol{A} \rightarrow \boldsymbol{A}' &= \boldsymbol{A} + \text{grad}\, \Lambda, \\ \varphi \rightarrow \varphi' &= \varphi - \frac{\partial \Lambda}{\partial t} \end{aligned} \qquad (9.3)$$

where Λ is a scalar function of space and time. It is straightforward to show that

Table 9.1 Maxwell's equations

$$\text{Div } \boldsymbol{E} = \rho/\varepsilon_0,$$

$$\text{Div } \boldsymbol{B} = 0,$$

$$\text{Curl } \boldsymbol{E} = -\frac{\partial \boldsymbol{B}}{\partial t},$$

$$\text{Curl } \boldsymbol{B} = \mu_0 \left(\boldsymbol{j} + \varepsilon_0 \frac{\partial \boldsymbol{E}}{\partial t} \right),$$

where ρ is the charge density and $\boldsymbol{j}$ is the current density. These are the microscopic equations. The macroscopic equations are those that apply in the presence of large amounts of material and in which the fields are defined as the average over a limited volume and over a limited time such that atomic and thermal fluctuation effects are not visible. The microscopic equations apply to non-averaged quantities: therefore they do not include the $\boldsymbol{D}$ and $\boldsymbol{H}$ fields.

DEFINITIONS AND KEYWORDS

Field A single or multicomponent variable that is a function of the space–time coordinates and which represents the behaviour and properties of particles and of the forces acting between them.

Gauge invariance Applied to the electromagnetic field, it is a property by which the observable consequences of the existence of the field are unaltered by certain changes in the scalar and vector potentials that describe the field.

Gauge transformation The change in field variables which are associated with a gauge invariance.

Coupling constant A number which is a measure of the strength of an interaction.

the changes in $\boldsymbol{A}$ and φ of equation (9.3) cause no change in $\boldsymbol{E}$ and $\boldsymbol{B}$ (Probler 9.1). The change of equation (9.3) is called a **gauge transformation**.

The work of Planck on the spectrum of black-body radiation, of Einstein o the photoelectric effect and of Compton on the scattering of X-rays by electron showed that the electromagnetic field is quantized; of course we call th quantum of the field a photon. We shall not discuss the theoretical technicalitie of setting up a quantum-mechanical version of the electromagnetic field However, we know that the field cannot change unless it can interact wit electrically charged particles (or neutral particles with electrically charge constituents). In general the strength of an interaction is measured by a **couplin constant** which, in the simple case that we have here, is the value of the charge o the particle. This theory is quantitatively correct and predicts that electricall charged particles can emit or absorb photons, corresponding to the classica emission or absorption of radiation. The vector nature of the field means tha photons have intrinsic angular momentum, that is spin, of $1\hbar$, and the fact tha photons have no rest mass means that the three possible pure states o polarization associated with spin 1 became two for real photons; these are ofte most conveniently chosen to be states of left and right circular polarization Virtual photons have one additional degree of freedom, which may be though of as a state of polarization that can include a longitudinal component o electric field. Longitudinal means a component of electric field along th direction of the momentum of the virtual photon. A classical electromagneti wave at large distances from its source may only have transverse electric an magnetic fields: real photons must have the same property.

9.3 The Dirac equation

In 1928 Dirac discovered a first-order linear differential equation which describes the quantum mechanics of point-like, spin-$\frac{1}{2}$ particles in a manner completely consistent with special relativity. The solutions to this equation are four-component wavefunctions which may be thought of as representing a Dirac field in the same way as solutions of Maxwell's equations represent an electromagnetic field. Applied to electrons, the equation predicted a magnetic ıoment of 1 Bohr magneton (wrong by about 1 part in 10^3, for reasons we shall ıeet). It also predicted the existence of an electrically positive version of the ımiliar negative electron, which is called the positron (see Section 1.6). ?ositrons were discovered in cosmic radiation in 1931.) The theory we now call **uantum electrodynamics** (QED) describes the interaction of the charged Dirac eld with the quantized electromagnetic field. As a theory it was essentially omplete by 1950 by the hands of many people but particularly the Nobel ıureates Feynman, Schwinger, and Tomonaga. It is one of the most successful ıeories in physics judged by the astonishing quantitative accuracy of its redictions and the huge energy range in which it is applicable. In fact, the range ıight have been thought it be unlimited except that we now know that at high nergies it must be combined with the theory of the weak interaction if correct redictions are to be obtained: more on this subject will come in Section 9.13.

The calculation of quantities from quantum electrodynamics means using erturbation theory. The accuracy of measurements was such that the results of rst-order perturbation calculations were good but were insufficiently precise. Iowever, higher-order calculations gave results that were infinite. The cure of uch difficulties is called **renormalization** and it involves a redefinition of the arameters of the theory (mass and charge of the electron) in such a way that alculations made to any order give finite results. It is a fact that many theories, ther than QED, of particle processes have been used which are not renormalizble. However, renormalizability is now considered to be an essential property f any theory of elementary particles and their interactions. We shall not have ny more to say about this subject except that we shall give the status with espect to renormalizability of the theories that we shall meet. That QED is enormalizable has been proved.

DEFINITIONS AND KEYWORDS

Quantum electrodynamics (QED) This is the fully quantized, relativistic theory of electrons and positrons in interaction with the electromagnetic field.

Renormalization This is a procedure applied to quantized field theories whereby the appearance of infinite results for physical quantities that have to be finite is properly avoided by re-definition of the masses and coupling constants.

.4 Feynman diagrams

'hese diagrams provide a picture of the mechanisms operating in particle rocesses and we will describe them in the context of QED. The basic building lock is a vertex as shown in Fig. 9.1. The solid lines represent electrons or ositrons, the wavy line is a photon. We can define a time axis and orient the hree lines in any way we wish as, for example, in Figs. 9.2(a)–(d). The vertex ules are that energy and momentum are conserved and that electric charge is onserved (Table 9.2). A rule of interpretation is that lines which have one end

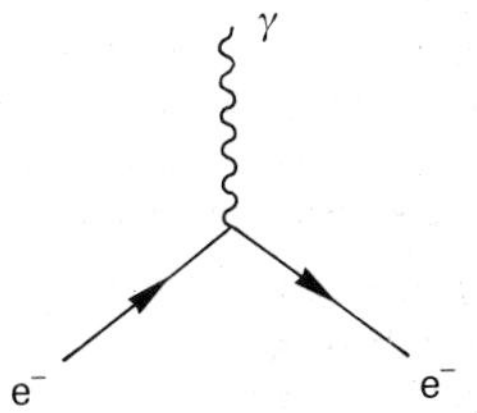

Fig. 9.1 The basic vertex of quantum electrodynamics. The arrowed line represents an electron, the wavy line a photon.

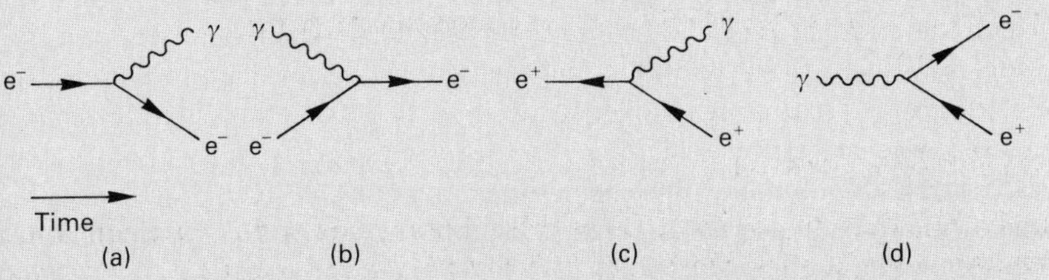

Fig. 9.2 Four examples of time-oriented vertices showing some of the basic processes of QED: (a) photon emission by an electron; (b) photon absorption by an electron; (c) photon emission by a positron; (d) photon materialization into an electron–positron pair.

Table 9.2 The rules for the construction and interpretation of Feynman diagrams.

1. Energy and momentum are conserved at a vertex.
2. Electric charge is conserved.

There are other conservation rules that apply in other circumstances and that we shall meet later. These two are universal.

3. Solid straight lines with arrow heads pointing in the direction of increasing time are used to represent fermions propagating forward in time. Arrow heads pointing in the reverse direction represent antifermions propagating forward in time.
4. Broken, wavy, or curly lines are used to represent bosons.
5. Lines having one end at the boundary of the diagram represent free (that is real) particles approaching or leaving a reaction. (We shall not actually draw boundaries to our diagrams as it is normally clear which are the in-going and which are the out-going particles.)
6. Lines that join two vertices (internal lines) normally represent virtual particles. (There are exceptions to this when an internal line represents a real but unstable particle which is a compound state of the initial particles; for example, the diagram for $e^+e^- \rightarrow Z^0 \rightarrow \nu\bar{\nu}$ in Fig. 12.7.)
7. The time ordering of the vertices connected by an internal line is not determined, so that two diagrams having an internal line apparently oriented differently with respect to time, but otherwise the same, are not different diagrams.
8. Every particle at the boundary should be labelled with a momentum. If this is done two diagrams which might otherwise appear to be the same become different diagrams (see Problem 9.3). However, we do not include momentum labels unless necessary. Time increases from left to right in our diagrams.

at the boundary of a diagram represent particles which are real or bound (e.g. electron bound in an atom) and which are either approaching an interaction or leaving it. Particles (for example, an electron) have arrows attached to the lines which point in the direction of time increasing. Antiparticles (for example, a positron) have arrows which point in the direction of time decreasing, and therefore appear in the diagrams as particles moving backwards in time (Table 9.2). Thus in Fig. 9.2(c) at the boundary and at a time after an interaction, two free particles appear, one an electron (e^-), the other a positron (e^+). See Table 9.2.

Now look at Problem 9.2: its result is that all the processes of Fig. 9.2 are impossible if all three particles in each are real. This is not a problem because, as we shall find shortly, the vertices of QED are combined to give a representation of the mechanism of a physical process and only lines ending on the boundary are constrained to be free.

These diagrams have one feature: the QED vertex is the conjunction of two fermion lines and one boson line. Translated into a physical process we can see that if interactions are represented by such three-arm vertices then conservation of angular momentum requires that the number of fermions (half-odd-integer spin) involved must be zero or two. In addition to the three-arm QED vertex, we shall be meeting examples of other kinds of vertex.

Now let us use the QED vertex as a building block for bigger diagrams. Any physical process involving the interaction of electrons, positrons and photons can be represented and the first step is to find the diagram with the least possible number of vertices. In Fig. 9.3(a) we show the only (two) two-vertex diagrams which describe positron–electron elastic scattering (Bhabha scattering). In Problem 9.3 we ask the reader to find the two two-vertex diagram for electron–electron (e^-e^-) scattering (Møller scattering). These diagrams are the 'simplest' combination of vertices which give the required process and we shall refer to such simplest diagrams as the **leading-order diagrams**. A given process may have one or more leading-order diagrams. If more than one, then, by definition, all have the same nmber of vertices.

Let us return to the existence of at least one virtual arm at a vertex: in the diagrams for e^+e^- and e^-e^- scattering, the in and out particles are real which requires the exchanged photons and the annihilation photon (in Fig. 9.3 and Problem 9.2) to be virtual. An example of a virtual electron is given in the two

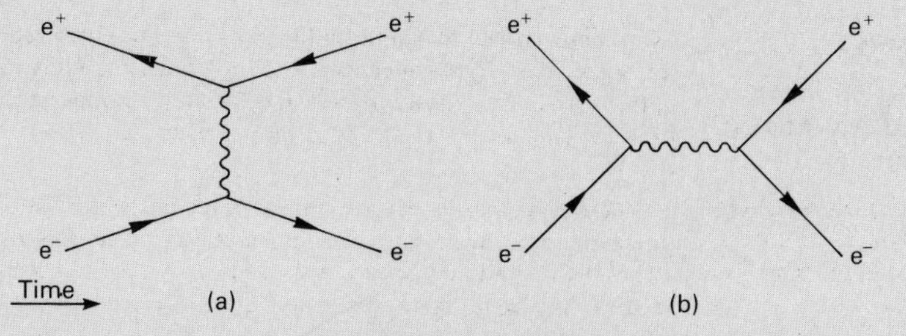

Fig. 9.3 Leading-order Feynman diagrams for positron–electron elastic scattering (Bhabha scattering); (a) represents the straightforward mechanism of single photon exchange; (b) represents the annihilation diagram. Special relativity does not allow the time order of the vertices in these diagram to be defined; thus any diagram which appears to change only the time ordering of the two vertices is not a new diagram. This is true generally; see Table 9.2.

Generally, all lines should be labelled with a four-momentum (momentum and total energy) and spin, particularly lines representing free particles. This is to ensure that all diagrams involving identical particles with different momenta are taken into account: two diagrams which have two final-state fermions, identical apart from labelling, will each contribute to the scattering amplitude. That amplitude must be antisymmetric under the exchange of labels. Two identical bosons will require a symmetric amplitude.

Such labelling is not required in this figure and normally we shall omit it. See, however, Problem 9.3.

leading-order diagrams for the radiation of a photon by an electron scattering at an atomic nucleus, Fig. 9.4.

The Feynman diagrams are important in the QED calculation of cross-sections and of transition rates. Every diagram represents a contribution to the total probability amplitude for the process. The techniques for calculating such amplitudes are not our business but we can mention one useful thing. The amplitude of a given diagram contains several factors, including an e (the charge on the electron) for every vertex; thus the amplitude for e^+e^- elastic scattering is proportional to e^2, or in dimensionless terms, to $\alpha(=e^2/4\pi\varepsilon_0\hbar c)$. It follows that the cross-section is proportional to α^2. The bremsstrahlung process (Fig. 9.4) has an amplitude which contains e for each vertex and Ze for the virtual photon

DEFINITIONS AND KEYWORDS

Feynman diagrams Diagrams introduced by Feynman to represent the mechanisms of elementary particle interactions.

Leading-order diagram This is a diagram which represents a given physical process and which has the minimum possible number of vertices. There may be more than one such diagram.

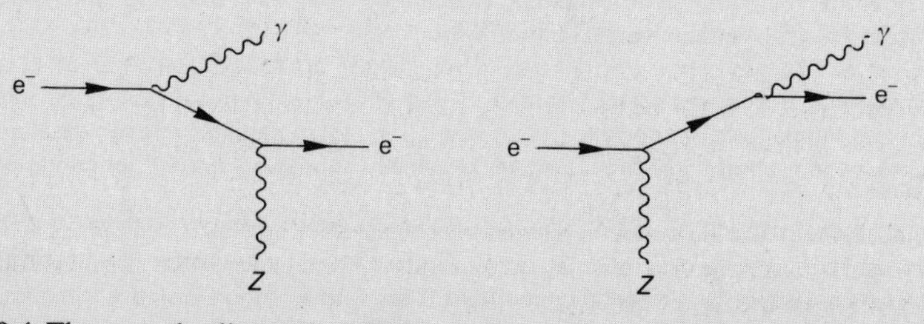

Fig. 9.4 The two leading-order diagrams for photon emission by an electron scattering at a nucleus (bremsstrahlung process). The Z represents the source of virtual photons, namely the electric charge of the nucleus, $Z|e|$. The electron propagating between the two vertices is virtual.

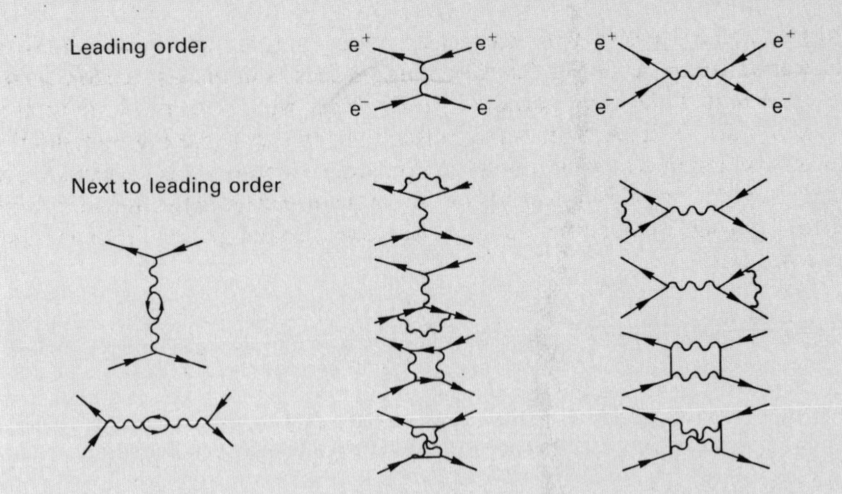

Fig. 9.5 Again the leading-order diagrams for electron–positron elastic scattering but, added to them, all the next-to-leading-order diagrams.

attaching to the nuclear charge: the cross-section is therefore proportional to $Z^2\alpha^3$.

As we mentioned in the last section, the calculations of cross-sections and of transition rates depend on the use of perturbation theory. The Feynman diagrams are a geometrical way of representing the terms in the perturbation calculation. The leading-order diagrams correspond to the lowest-order perturbation calculation. Of course, in addition to the leading orders there are higher-order diagrams, each of which will contribute extra amplitudes to the total. Since they can only be drawn by adding internal lines they must always involve two more vertices for each step in increasing order. In the cross-section the next to leading-order amplitudes appear most importantly in cross terms with the leading-order amplitudes, so these contributions are, in magnitude, about α times the leading order. It is the smallness of α (1/137) relative to 1 which normally makes all but the next-to-leading-order contributions negligible and the calculations manageable. In Fig. 9.5 we show all the two- and four-vertex diagrams for positron–electron elastic scattering.

PROBLEMS

9.1 Show that the gauge transformation of equations (9.3) leave unchanged the magnetic and electric fields of equations (9.1) and (9.2).

9.2 Show that the processes represented by Fig. 9.2(a)–(d) cannot satisfy the rule that energy and momentum are conserved at a vertex if all three particles in each case are free. (This means that all of these processes are physically impossible under that condition.)

9.3 Find the two leading-order diagrams for electron–electron elastic scattering. (The two to be found are both two-vertex diagrams.) Find one and label all the in-going and out-going particles with type and four-momentum. Then find a second which is different; note that there can be no annihilation diagram like that of Fig. 9.3(b).

9.4 Find the two leading-order diagrams for the production of an electron–positron pair by a real photon in the electric field of a nucleus.

9.5 More fun with Feynman diagrams

'More fun' here means some semi-quantitative work. Let us consider electron-nuclear scattering as in Fig. 9.6. Counting vertices as in the last section we see that the amplitude is proportional to $Z\alpha$. Now we can throw in one other property: the virtual photon has to transfer momentum q (see Fig. 9.6) and this makes the amplitude proportional to $1/q^2$, and the cross-section is expected to behave as $Z^2\alpha^2/q^4$. This q depends on the angle of scatter (θ) so we are talking about a partial cross-section. Putting things together and getting the dimensions right suggests

$$\frac{d\sigma}{dq^2} \propto \frac{Z^2\alpha^2(\hbar c)^2}{q^4c^2}.$$

In Table 3.1 we quoted the Mott scattering cross-section for a fixed (no recoil) target and relativistic electrons. For very relativistic electrons ($v \to c$) that formula can be rewritten as

$$\frac{d\sigma}{dq^2} = \frac{4\pi Z^2\alpha^2(\hbar c)^2}{q^4c^2}\cos^2\frac{\theta}{2}.$$

Thus our semi-quantitative guesswork is not far from correct. To get those other factors correct needs the full apparatus of QED. Let us try one other: consider the reaction

$$e^+ + e^- \to \mu^+ + \mu^-,$$

in collisions at high energies. (Reminder: in Section 3.6 we introduced the μ-mesons or muons. The μ^+ and μ^- behave, as far as their electromagnetic interactions are concerned, just like heavy versions of the positron and electron, so we can include them in an appropriate manner into QED.) Figure 9.7 shows the only leading-order diagram: clearly the total cross-section is proportional to α^2. All the particles are point-like, so there is no real size available to set a cross-sectional area (the cross-section is scale invariant—see section 3.8) At energies

Fig. 9.6 Diagram for electron scattering by a nucleus. If $\boldsymbol{P}_1$ and $\boldsymbol{P}_2$ are the incident and scattered electron momenta, the momentum carried by the photon is $\boldsymbol{q}=\boldsymbol{P}_1-\boldsymbol{P}_2$. However, unless the nucleus is very heavy, the photon may also transfer a significant amount of energy $\nu=E_1-E_2$. The relativistic four-momentum transfer, q, is $(\boldsymbol{q},\nu/c)$. The square of this four-momentum transfer is defined by $q^2=(\nu/c)^2-\boldsymbol{q}.\boldsymbol{q}$.

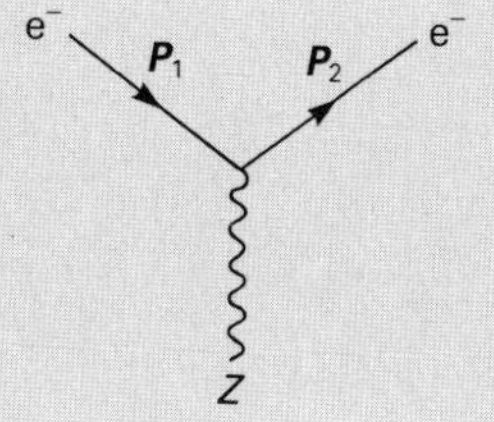

Fig. 9.7 The sole leading-order diagram for e^+e^- annihilation to $\mu^+\mu^-$. This process can occur only if the total centre-of-mass energy exceeds $2m_\mu c^2$, where m_μ is the mass of the muon.

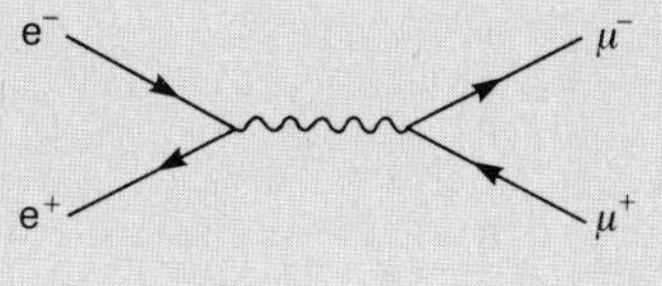

far above threshold and at which the centre-of-mass energy W is large compared to the rest masses of both μ-meson and electron, the only quantity with the dimension of area is $(\hbar c/W)^2$, so we guess

$$\sigma \sim \alpha^2\left(\frac{\hbar c}{W}\right)^2.$$

In fact the full calculation yields $4\pi/3$ times this. The scattering of photons by free electrons (Compton scattering) also has two vertices, so again we expect

$$\sigma \sim \alpha^2\left(\frac{\hbar c}{W}\right)^2$$

where W is now the photon–electron centre-of-mass energy. At very low energies (photon energy $E \ll m_e c^2$) we shall have $W \to m_e c^2$ and

$$\sigma \sim (\alpha\hbar c/m_e c^2)^2.$$

In fact the actual cross-section is $8\pi/3$ of this guess and is the classical Thomson cross-section of the electron.

This kind of guess–estimate runs into trouble when there is a virtual photon. Just above we noted that electron–nuclear scattering went as q^{-4} so if the kinematics allow $q^2 \to 0$ as in fact it does, the total cross-section becomes infinite. In fact the experimental cross-section does not because as $q^2 \to 0$ the electron impact parameter is becoming large and the incident electron is screened from the nucleus by the atomic electrons. See Problem 1.3. In the bremsstrahlung process, Fig. 9.4, the virtual photon gives the total cross-section the same properties.

9.6 Tests of QED

Measurements have been made of cross-sections for many processes for which QED can give a calculated cross-section. In general the theory is correct but the measurements cannot normally be made with great precision due to the presence of radiative effects and to experimental uncertainties. To understand the former, consider Bhabha scattering (Fig. 9.3a). We can attach a real photon line to one of the in- or out-going particles so that the final state contains a photon. Although this is one more vertex, it has a very large probability for the emission of low-energy photons: in fact the cross-section diverges and can only be tied to real physics by considering the experimental conditions. Thus measuring the real Bhabha cross-section may require corrections to the data which depend on the actual theory to be tested! In addition the experimental uncertainties limit the results to a precision that is normally worse than 1 in 10^3.

The real tests of QED come elsewhere and with astonishing precision. We will consider one and list some others. The Dirac theory of the electron predicted that the magnetic moment would be 1 Bohr magneton. QED requires some corrections due to higher-order Feynman diagrams—see Fig. 9.8. Now the difference from 1, foreseen in Section 9.3, is explained: in addition, the calculated and measured values now agree to a precision of better than 1 in 10^{10} In the hydrogen atom the $2s_{1/2}$–$2p_{1/2}$ levels are degenerate according to Dirac's theory, but are, in fact, split (the Lamb–Retherford shift) by an amount

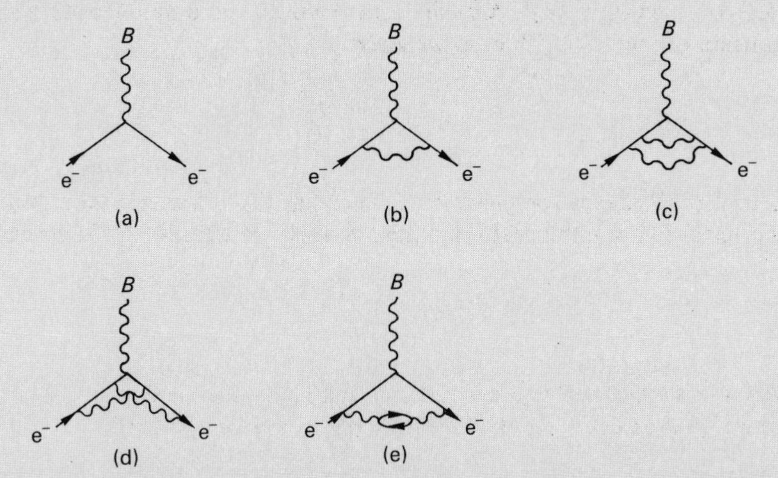

Fig. 9.8 The magnetic moment of the electron. The calculation obtains the energy of the electron in a magnetic field, here represented by B. Diagram (a) is that for the leading-order term: the remainder are the first few higher-order corrections. The theoretical result for the magnetic moment in Bohr magnetons is

$$1+\tfrac{1}{2}(\alpha/\pi)-0.328478966(\alpha/\pi)^2+1.1765(\alpha/\pi)^3-0.8(\alpha/\pi)^4$$
$$=1.001\,159\,652\,307(110).$$

The experimental result is

$$1.001\,159\,652\,193(10).$$

The figures in brackets give the uncertainties in the last digits. The theoretical calculation is uncertain due to errors in the estimates of the contributions of the highest-order terms expected to contribute at this level of precision and to the uncertainty in the known value of α.

predicted by the QED calculation. The other area of success is in the calculation of the fine and hyperfine structure (see Fig. 10.15) and lifetime of **positronium** (Ps), the atom consisting of an electron and positron bound by their Coulomb attraction. The total spin of the system (that is, the vector sum of the two intrinsic spins), quantum number s, can be 0 or 1 (para- and ortho-Ps respectively). The total angular momentum j is the vector sum of s and any relative orbital angular momentum l, as indicated by the usual spectroscopic notation, $n^{2s+1}L_j$. The two lowest-energy states 1^3S_1, and 1^3S_0 are split by about 10^{-3} eV (on a binding energy of 6.8 eV). The 2^3S_1 and 2^3P_2 state are split by 5×10^{-5} eV. In both cases the QED calculation and experiment agree. A further success for the theory is the calculation of the muon magnetic moment and the properties of muonium, the atom that consists of a μ^+ and an e^-. A summary of some of these successes is given in Table 9.3.

These results have given physicists great confidence in QED. However, the theory has been extended by unifying it with the theory of weak interactions into the electroweak theory (Section 9.13). This theory has a broader range of processes to which it can be applied and a larger energy range in which it can be expected to be valid than either of its parts. More speculative is the possibility that the electroweak theory is itself to be unified with the theory of other

DEFINITIONS AND KEYWORDS

Positronium The name given to the bound state of an electron and a positron. Symbol: Ps.

Table 9.3 Some of the successes of QED

Measure	Measurement	QED calculation
Hydrogen		
$2^2s_{1/2}-2^2p_{1/2}$ splitting (MHz)	1057.862(20)	1057.873(20)
Magnetic moments		
Electron (Bohr magnetons)	1.001 159 652 193(10)	1.001 159 652 307(110)
Mu-meson (μ magnetons $e\hbar/2m_\mu$)	1.001 165 923(8)	1.001 165 920 0(20)
Positronium		
$1^3S_1-1^1S_0$ splitting (GHz)	203.3887(7)	203.381(20)
$2^3S_1-2^3P_2$ splitting (GHz)	8.6196(28)	8.6252
1^3S_1 decay transition rate* (s^{-1})	$7.0314(70)\times10^6$	$7.0388(2)\times10^6$
1^1S_0 decay transition rate† (s^{-1})	$7.994(11)\times10^9$	$7.985\,251\,6(51)\times10^9$

* 1^3S_1 Ps decays to 3γ.
† 1^1S_0 Ps decays to 2γ.
The figure in parentheses in each case is the estimated uncertainty on the least significant figures of that number.

interactions (see Section 13.6). Also electrons and positrons might have structure and that certainty would demand a profound change. At the present, experimental results test for structure down to about 10^{-18} m: the agreement with the electroweak theory indicates that structure, if any, must be on a smaller scale.

9.7 Nuclear forces

The forces which must exist between the constituents of nuclei represent the existence of an interaction which we now have to investigate. Frank and Ernest (Fig. 9.9) have, of course, got it wrong so let us review what we have learnt about the nuclear forces from earlier chapters. The fact that nuclei are bound even in the presence of the Coulomb repulsion between the protons indicates that this force is stronger than the latter. In Sections 4.5 and 8.11 we made some qualitative statements about nature of nuclear forces from the success of semi-empirical mass formula. These were:

1. The nuclear force acts only to a distance of about 2 fermi, that is, it is short ranged (see Section 3.1).
2. The force saturates: this is believed to be the combined effects of a short-range ($\simeq 0.5$ fm) repulsive force, a tensor force (see Section 9.8), and the Pauli exclusion principle.

The success of the single-particle shell model indicates that:

3. In nuclei there is a spin–orbit interaction which splits the energy of single particle states of the same l but different j, the total angular momentum.

Let us now look at **mirror nuclei**: these are pairs of nuclei with the same odd A but the number of protons and neutrons in one nucleus (Z, $N = A - Z$) is the same as the number of neutrons (N') and protons (Z') in the other

Fig. 9.9 Frank and Ernest discuss some physics problems including how velcro binds neutrons and protons into nuclei.

($Z' = N, N' = Z$). The example in Fig. 9.10 shows the energy levels up to 8 MeV for the two nuclei ${}^{7}_{3}$Li and ${}^{7}_{4}$Be, which are mirror. The closeness, qualitative and quantitative, of the energy levels has an important consequence: both nuclei have the same number of neutron–proton (np) pairs but ${}^{7}_{3}$Li has 3 pp and 6 nn pairings and ${}^{7}_{4}$Be has 3 nn and 6 pp pairings. Allowing the Coulomb effects, the equality of the level structure means that nn forces are equal to pp forces (within the errors involved in correcting for Coulomb effects). This is **charge symmetry**. Note that it says nothing about the np force.

Now let us look at a triplet of nuclei ${}^{6}_{2}$He, ${}^{6}_{3}$Li, ${}^{6}_{4}$Be. The shell model suggests that these nuclei have a doubly-closed-shell core (${}^{4}_{2}$He) with two extra nucleons. Figure 9.11 shows the energy levels of these three nuclei and we note that levels formed in ${}^{6}_{2}$He are matched with those in ${}^{6}_{4}$Be and with some of those in ${}^{6}_{3}$Li in the same energy range. If we consider the space-spin wavefunction of the two extra-core nucleons, it can be symmetric or antisymmetric under exchange of the coordinates of these two nucleons. It must be antisymmetric for the pp, nn and for some of the np wavefunctions: the matching suggests that the nucleon–nucleon force is the same when the two nucleons have wavefunctions which are the same, as is the case for the antisymmetric wavefunctions for the two extra-core nucleon in these three nuclei. This is **charge independence**. The remaining states in ${}^{6}_{3}$Li have symmetric wavefunctions for the extra-core nucleon pair (np) which have no nn or pp analogues and so have a different pattern of energy levels. Clearly charge independence is a stronger statement than is charge symmetry and contains the latter.

DEFINITIONS AND KEYWORDS

Mirror nuclei Two isobars in which the number of neutrons in one is the same as the number of protons in the other, and vice versa.

Charge symmetry states that, apart from Coulomb effects, the force between two protons is the same as between two neutrons if they are in the same state of relative motion and total spin.

Charge independence states that, apart from Coulomb effects, the force between any two nucleons is the same if they are in the same state of relative motion and total spin.

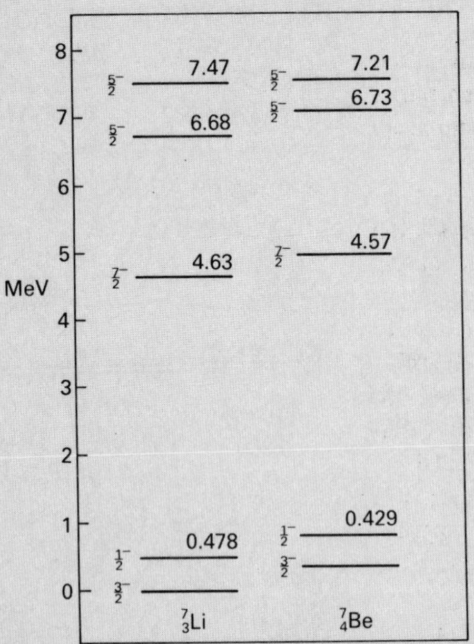

Fig. 9.10 The ground and first few excited states of the mirror nuclei $^7_4$Be and $^7_3$Li. Their energy levels differ only by the effects of the Coulomb term in the mass formula and of the neutron–proton mass difference. The mass scale gives the nuclear mass based on zero for the nucleus $^7_3$Li. The numbers on each level give the spin-parity j^P and the excitation above the ground state in mega-electron-volts. A pattern of analogous states is clear.

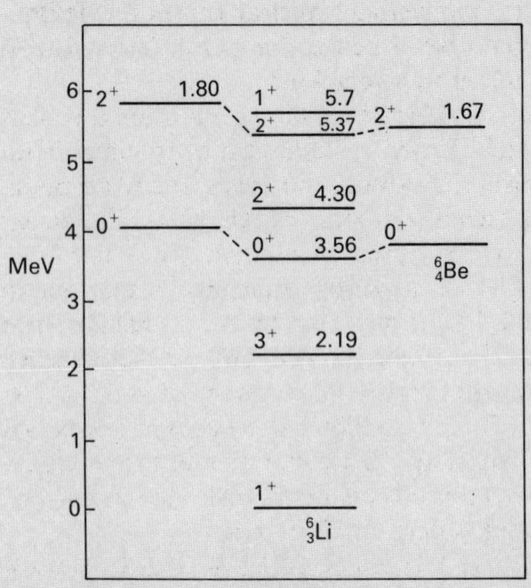

Fig. 9.11 The ground and first few excited states of three isobars with $A = 6$. The mass scale gives the nuclear mass based on zero for the nucleus $^6_3$Li. Two sets of analogous states are visible.

These conclusions appear to rest on a qualitative comparison of energy levels. However, there are more rigorous tests of charge independence that show it to be correct to the level of a few per cent. The physical origins of charge independence have now been traced to the flavour independence of the forces between quarks (see Section 10.11).

This concept of charge independence leads to the idea of isotopic spin: this subject opens a completely new perspective which we explore briefly in Section 10.11.

The next logical step is to look at the nuclear force as manifest in a system containing only two nucleons. Here we appear to have three choices of nucleon pairs

(a) proton–proton,

(b) neutron–proton,

(c) neutron–neutron,

with any pair either

(1) bound, or
(2) unbound.

However, of the six combinations only three are available: there is only one bound system, consisting of a neutron and proton (the deuteron, symbol $^2_1$H or d). The unbound systems must be investigated in neutron–proton and proton–proton scattering. Neutron–neutron scattering can only be observed indirectly or in circumstances not normally available.

9.8 The bound two-nucleon system

This means looking at the deuteron. The binding energy is 2.22 MeV and it has total angular momentum (nuclear spin) of 1. It is found to have even parity (conventionally we write the spin-parity as $j^{P}=1^{+}$). What do these facts mean for the deuteron wavefunction?

For a two particle system the wavefunction describes their motion about their mutual centre of gravity, so that the coordinates in this case are the separation of the neutron and proton, and the polar and azimuthal angle of the line joining them: thus there is one state of relative orbital angular momentum (quantum number l) to be specified. In addition, the spins of the two nucleons can add to give a singlet or a triplet total spin state (quantum number $s=0$ or 1, respectively). The orbital and total spin angular momenta add vectorially to give the total angular momentum (j). Any eigenstate can be described in spectroscopic notation by ${}^{2s+l}L_j$ where $L=S,P,D,F,G, \ldots$ for $l=0,1,2,3,4 \ldots$ respectively. Note carefully that we use the words *total spin* for the vector sum of the nucleon spins. The vector sum of the *total spin* and the orbital angular momentum is called the *total angular momentum* and is the *nuclear spin* of the whole nucleus, a deuteron in this case.

The deuteron spin-parity 1^{+} means that its wavefunction is ${}^3S_1$ or ${}^3D_1$ or a superposition of the two. This wavefunction is symmetric. The fact that the deuteron is bound makes us look at the pp and nn systems: if np will bind why not pp or nn? Firstly these pairs cannot be in a ${}^3S_1$ or a ${}^3D_1$ state (see Problem 9.5). The antisymmetric states ${}^1S_0$, ${}^3P_0$, ${}^3P_1$, ${}^3P_2$ and others are available but no bound states of nn or pp exist. Considering the S-states alone we see that the ${}^3S_1$ state of two nucleons binds but the ${}^1S_0$ does not. The immediate conclusion is that the nuclear force is spin dependent, that is, different from the singlet to the triplet state, less attractive in the former. If we recall the definition of charge independence we see that the nn, pp, and np force will be the same in the antisymmetric ${}^1S_0$ state: the existence of only one two-nucleon bound state, the deuteron, is consistent with what is implied in charge independence, that the force can be different in the symmetric ${}^3S_1$ state from that in the antisymmetric ${}^1S_0$ state.

The deuteron has a magnetic moment of 0.8574 nuclear magnetons. This is almost what is found (0.8797) from adding 2.7925 magnetons of a proton to -1.9128 of a neutron: this addition is what is expected in the ${}^3S_1$ state. The deuteron also has a positive electric quadrupole moment which must be the result of a prolate (see Section 3.9) deformation of the nuclear charge. Such a deformation cannot exist in a pure ${}^3S_1$ state, but requires a superposition of some ${}^3D_1$ state. Quantitative studies show that the probability amplitudes of the ${}^3S_1$ and ${}^3D_1$ states are about 0.98 and 0.20 respectively. Therefore the probability of finding the deuteron in the ${}^3D_1$ state is about 4%. This admixture predicts a magnetic moment closer to the measured value than that given by a calculation assuming that the deuteron is ${}^3S_1$ alone.

A pure central force in the deuteron ground state cannot mix ${}^3D_1$ into the dominant ${}^3S_1$ and so we are forced to find another component: this is the **tensor force.** In the caption to Fig. 9.12 we write the quantum-mechanical operator form for the potential and beside it a classical analogue: the analogue is the potential which exists between two electric dipoles. The force in both cases does not act along the line of centres and depends not only on the separation but also on the angles between the spins (electric dipole moments) and the line of centres.

There have been many studies of the deuteron and its wavefunction. Its

$$U = \frac{-p_1 p_2}{4\pi\varepsilon_0 r^3}\{2\cos\theta_1\cos\theta_2 - \sin\theta_1\sin\theta_2\cos\phi\}$$

$$= \frac{-1}{4\pi\varepsilon_0 r^3}\left\{\frac{3(\boldsymbol{p}_1.\boldsymbol{r})(\boldsymbol{p}_2.\boldsymbol{r})}{r^2} - \boldsymbol{p}_1.\boldsymbol{p}_2\right\} \qquad U = f(r)\left\{\frac{3(\boldsymbol{S}_1.\boldsymbol{r})(\boldsymbol{S}_2.\boldsymbol{r})}{r^2} - \boldsymbol{S}_1.\boldsymbol{S}_2\right\}$$

Fig. 9.12 (a) A possible disposition of two electric dipoles, $\boldsymbol{p}_1$ and $\boldsymbol{p}_2$, and the formulae for the potential energy in this circumstance. The angle φ is the azimuthal angle between the plane containing $\boldsymbol{p}_1$ and $\boldsymbol{r}$ and that containing $\boldsymbol{p}_2$ and $\boldsymbol{r}$. In the figure this angle is zero and is not depicted. (b) Two nucleon spins, $\boldsymbol{S}_1$ and $\boldsymbol{S}_2$, disposed in the same way as in (a) and the formula for the potential between them. The term $f(r)$ represents an unknown radial dependence, apart from which the angular dependences are identical. Note, however, that the nuclear tensor force is neither electric nor magnetic in origin: nucleons have no detectable electric dipole moment (see Section 8.8) and the magnetic moment they do have (and any reasonable electric dipole moment, if that were possible) gives a tensor contribution too small to be the origin of the observed nuclear tensor force.

physical properties do not uniquely determine the parameters of the potential that exists between the neutron and proton in this bound state.

9.9 The unbound two-nucleon system

This is essentially to do with scattering. Data exist on np elastic scattering for neutrons of kinetic energies of fractions of an electron-volt up to many giga-electron-volts. Low-energy protons (<100 keV) do not penetrate matter sufficiently to allow scattering experiments, so data on pp scattering exist only for kinetic energies greater than several hundred kilovolts. A complication in both cases is that at energies above about 240 MeV there is the possibility of inelastic scattering in which π-mesons are produced (see Section 10.2).

What does the data consist of? Below the threshold for π-meson production the scattering is elastic and the interesting quantities are

(1) the total cross-section,

(2) the differential elastic scattering cross-section,

(3) quantities associated with the scattering of spin polarized nucleons by nucleons.

We cannot pursue these subjects in this course. However, we can briefly indicate how the data are analysed. The unbound two-nucleon states in collision are a superposition of the states that the reader, we hope, found in Problem 9.5, and

DEFINITIONS AND KEYWORDS

Tensor force A force between two nucleons which does not act centrally. See Fig. 9.12.

Partial waves The wavefunction describing the collision of two unbound particles can be decomposed into a superposition of eigenfunctions of the operators $\boldsymbol{L}^2$, $\boldsymbol{J}^2$, and $\boldsymbol{S}^2$. These are the partial waves.

Exchange forces These are interactions between particles which have the effect of exchanging the identities or other properties, such as spin or charge, between the two. (Not to be confused with the Yukawa mechanism of boson exchange, although the latter can be the cause of exchange forces.)

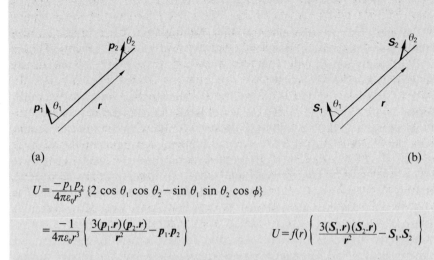

others of greater l. These states are called **partial waves**, and each wave is associated with a different set of values of l, s, and j. The probability of finding the two nucleons a distance apart less than the range of nuclear forces decreases rapidly with increasing l and at fixed l increases with increasing energy. Thus at a given energy only a limited number of partial waves from $l=0$ and up are significantly affected by the nuclear force; however, as the energy increases, the number of partial waves involved increases. This is really another aspect of the angular momentum barrier (Section 6.4): the greater l, the greater the momentum must be to bring the colliding particles close enough to be within the range of nuclear forces. The most sophisticated analyses propose potentials which are different for each partial wave but with guidance from theoretical considerations (see the next section). In addition there is evidence that the nuclear force has a hard repulsive core at separations of about 0.5 fm (Section 8.11). The parameters describing these potentials have to be adjusted to fit all the available scattering data and the properties of the deuteron. This is an ambitious

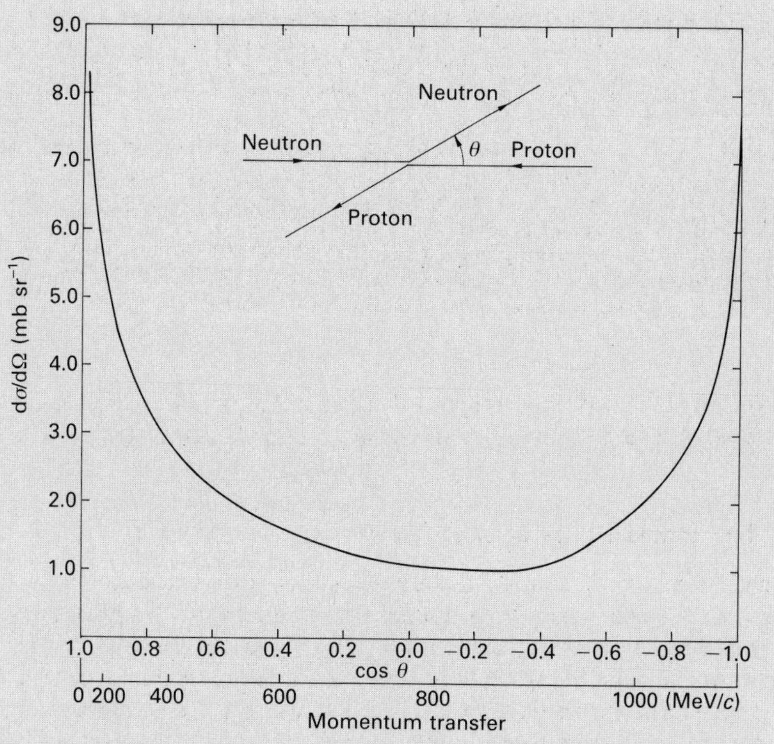

Fig. 9.13 The differential elastic scattering cross-section as a function of the cosine of the centre-of-mass scattering angle for elastic scattering of 630 MeV neutrons by protons. There is a strong forward peak with a rapid decrease in the scattering cross-section out to about 60°, which represents the increasing inability of the nuclear force to transfer increasing momentum. Beyond about 120° the cross-section increases again: here the momentum transfer is even larger if the neutron and proton retain their identity. However, if the nuclear force can exchange charge, then there can be an apparent large angle of scattering without a large transfer of momentum and the cross-section can be large. A momentum transfer scale is given below the abscissa.

The curve drawn is a smooth representation of data which has random and systematic errors; it must not be used as a source of accurate data. (Amaglobeli and Karzarinov, 1960).

programme and, in one sense, disappointing in its result: although describing the data, the result does not provide a fundamental description of the origin of nuclear forces.

At an elementary level elastic neutron–proton scattering at high energies provides one piece of evidence about nuclear forces. In Fig. 9.13 we show the angular distribution for the elastic scattering of 630 MeV neutrons by protons. The rapid decrease in differential cross-section as the angle of scattering increases (which is linked to increasing momentum transfer) from the large value at small angles of scatter tells us that the nuclear force, although strong, is not capable of transferring a large amount of momentum with large probability. However, there is also a backward peak which involves either a very large momentum transfer or another mechanism. The latter must be the case and the mechanism is provided by **exchange forces**: in one such force the neutron and proton exchange identity by charge exchange without a large momentum transfer. We shall see in the next section how this comes about.

In Table 9.4 we have reviewed the properties of nuclear forces which we have discovered from considering nuclei and the two nucleon systems, along with the major experimental fact which establishes each property.

Table 9.4 The properties of the nuclear force

Property	Evidence
1. It is strong.	Nuclei are bound in the presence of Coulomb repulsion.
2. It is spin dependent.	There is a $^3S_1$ but no $^1S_0$ bound state of two nucleons.
3. It has a tensor component.	The non-zero electric quadrupole moment of the deuteron.
4. It is short ranged.	Deviations from the Rutherford scattering formula.
5. It is charge independent. ($\supset$ charge symmetry)	The pattern of energy levels in the isobars of the light nuclei.
6. It has exchange properties	The backward peak in elastic neutron–proton scattering.
7. It saturates.	The almost constant nuclear density throughout the periodic table.

PROBLEMS

9.5 Consider a system consisting of two non-identical spin-$\frac{1}{2}$ fermions: list all the allowed states that are simultaneously eigenstates of the operators L^2, S^2, J^2 up to $l=3$ by their spectroscopic notation. For each find the value of j^P, where P is the parity, and the symmetry. Check that $^3D_1$ and $^3S_1$ are the only states with $j^P=1^+$.

All the states that you have listed are available to the neutron–proton system, but only half are available to the neutron–neutron and proton–proton systems: list these.

9.6 Use electrostatics to prove the formula given in Fig. 9.12 for the mutual potential energy of two electric dipoles. Choose a non-trivial set of values for θ_1, θ_2, and φ, and show that the force acting on one dipole does not act along the line of centres.

9.10 The Yukawa theory

In 1934 Hideki Yukawa recognized that the nucleon force could be the consequence of the exchange of a boson, with mass, between nucleons. The exchange of a massless boson between charges causes the Coulomb-like potential with $1/r$ dependence. The exchange of boson with mass m would give rise to a potential varying as $\exp(-mcr/\hbar)/r$. A very simplified argument for this result is given in Table 9.5. Yukawa knew that the range of nuclear forces was about 2 fm and so he predicted the existence of a boson of mass given by $\hbar/mc = 2$ fm; that is, $m \simeq 100$ MeV/c^2. He also predicted that

(1) this boson would be produced as a real particle in nucleon–nucleon collisions at sufficiently high energies,

(2) the boson would interact readily with nucleons and therefore have large cross-sections for scattering and absorption by nuclei,

and he suggested that

(3) the boson might be a virtual intermediary in neutron decay:

$$\mathrm{n} \to \mathrm{p} + \mathrm{B}^-$$
$$\qquad\quad \hookrightarrow \mathrm{e}^- + \bar{\nu}_\mathrm{e}.$$

This suggestion implies that if free, the boson will decay by this mode.

In 1934 the only place where sufficiently energetic collisions for boson production could be observed was in cosmic radiation impinging on the upper atmosphere. This radiation is mainly very energetic protons so that the bosons should be observable among the products of collisions between these protons and the nuclei of air in the top layers of the atmosphere. Studies at lower levels of the atmosphere indicated the existence of a particle of mass about 100 MeV/c^2, which was called the mesotron (prefix *meso* implying intermediate in mass between electron and proton). However, energetic mesotrons were able to penetrate too great a thickness of material (including the atmosphere) without significant absorption or scattering to be consistent with the properties predicted for Yukawa's boson. There were also other difficulties in reconciling the mesotron to Yukawa's boson.

This situation was not clarified until 1947 after Powell and co-workers had perfected techniques for observing energetic charged particles passing through photographic emulsions. Figure 9.14 shows four examples of the kind of event which they had observed in emulsions exposed to radiation at a high altitude and that enabled them to discover Yukawa's boson. They called it the π-mesotron (now π-meson or pion, symbol π). The cosmic ray mesotron was a decay product of the pion and they called it the μ-mesotron (now μ-meson or muon, symbol μ). These identifications were possible because the rate of observation of the pions was consistent with that expected for Yukawa's particle. Assuming that the muon (like the electron) experienced no nuclear interactions, and identifying it with the mesotron, the penetrating power of the latter could be explained and full reconciliation with Yukawa's predictions and with observations was possible. There was also indirect evidence for a neutral meson decaying to photons. In the years following, high-energy particle accelerators began to operate and the controlled production of these particles became possible. In Table 9.6 we give the intrinsic properties of these particles, including the neutral pion discovered in 1950.

Table 9.5 Yukawa's hypothesis and the range of nuclear forces.

Hideki Yukawa (1907–81)

A boson of energy E, momentum $\mathbf{P}$ and mass m satisfies the equation

$$E^2 - \mathbf{P}.\mathbf{P}c^2 = m^2c^4.$$

Making the replacements $E = +i\hbar\frac{\partial}{\partial t}$ and $\mathbf{P} = -i\hbar\nabla$ and making this equation an operator acting on a wave function ψ gives

$$-\hbar^2\frac{\partial^2\psi}{\partial t^2} + \hbar^2c^2\nabla^2\psi = m^2c^4\psi.$$

If $m = 0$ and in the right gauge, this is the wave equation satisfied by the electromagnetic potential φ (equation 9.2) and in the static case the equation is satisfied by the electrostatic potential of a point charge. If $m \neq 0$ the static equation is

$$\nabla^2\psi = \left(\frac{mc}{\hbar}\right)^2\psi.$$

and has a solution

$$\psi = (1/r)\exp(-r/a) \text{ with } a = \hbar/mc.$$

This result is interpreted as meaning that the bosonic potential of a point nucleon source will vary in this way with distance from the source and therefore that the internucleon potential will vary in the same way. Thus the potential is short ranged with

$$a = \text{range} = \hbar/mc.$$

If $a = 2$ fm, $m = 100$ MeV/c^2.

A more rigorous derivation of this result is given in Kane (1987).

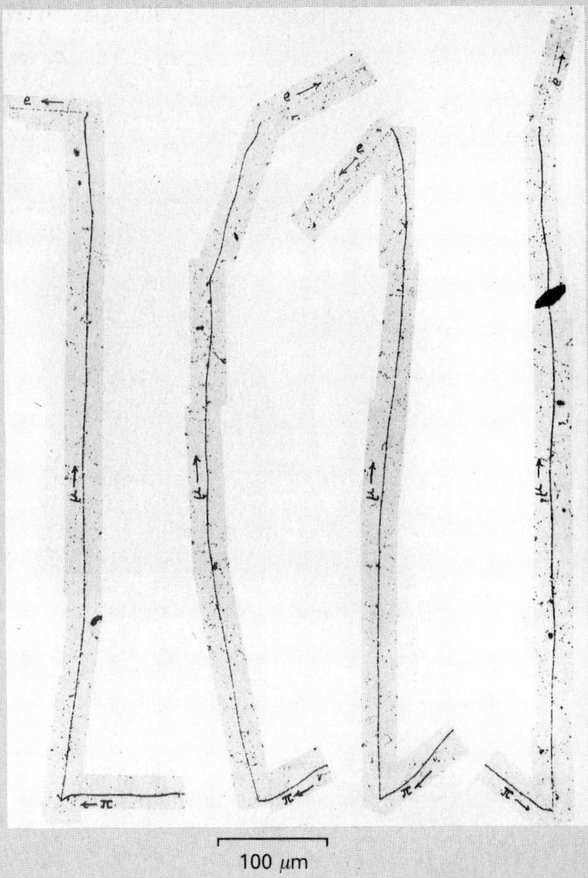

Fig. 9.14 The discovery of the Yukawa meson. Photomicrographs of four examples of one kind of event observed by Powell and colleagues in photographic emulsions exposed to cosmic rays at high altitudes, and interpreted as the successive decays $\pi^+ \to \mu^+ \to e^+$ (Powell, 1950.)

The passage of a charged particle renders a trail of silver halide grains developable. On normal photographic development a trail(track) of silver grains is left, each grain being of order 1–2 μm across. The density of grains along the trajectories of the particles is almost directly proportional to the rate of energy loss by ionization. That loss rate is greater the slower are the particles, so that the tracks become denser as the particle slows down and finally comes to rest. The low ionization by particles moving close to the velocity of light is witnessed by the sparseness of the tracks left by the positrons. The increase in density as a particle slows down is shown by the tracks left by the muons. These properties of the energy loss by ionization are described more fully in Section 11.2. Note that the lengths of the muon tracks are very nearly the same in all four examples: this is not an accident of choice but was true for all complete sequences observed. The length of the track is called its range and the near identical ranges indicated that all muons have the same energy when produced by the decay of the pion. It followed that the pion decay has a two-body final state.

Thus the full interpretation we can now give to such events is as follows: a positive pion enters the emulsion and after some distance comes to rest (the photographs show only the last few tens of microns of these tracks). The pion then decays $\pi^+ \to \mu^+ + \nu$, giving an unobserved neutrino of 29.8 MeV and a muon of 4.12 MeV. The muon travels about 596 μm and comes to rest. It decays $\mu^+ \to e^+ + \nu + \nu$. The neutrinos are not observed. The three-body decay means that the positron does not have a unique energy although this cannot be deduced from the photographs alone.

The effect of multiple scattering on the muons can be seen by the lack of straightness of their tracks.

Events of the chain $\pi^- \to \mu^- \to e^-$ are not seen in this form in photographic emulsion, or in any other material, since negative pions coming to rest in any material are attracted to a nucleus, captured (at a rate too great for decay to compete significantly) and lost in a nuclear disintegration which turns their rest mass energy into kinetic energy of ejected nuclear fragments. (See Fig. 11.1.)

Returning to the nuclear force problem we are now in possession of the idea that the nucleon–nucleon force is carried by the pions: some Feynman diagrams for the basic mechanism are shown in Fig. 9.15. The coupling of the pion to the nucleon is given by a constant g (equivalent to e in photon-charged particle coupling) so that the potential in the simple picture is

$$U(r) = (g^2/4\pi r)\exp(-mcr/\hbar).$$

Going beyond this potential, the exchange nature of the nucleon–nucleon force, which we discovered in np scattering, is now understood to be due to the exchange of a charged pion: the transfer changes $n \to p$ and $p \to n$ without a large transfer of momentum. Other properties of the nucleon–nucleon potential, for example its spin dependence, the tensor component, etc. could be understood in various versions of the theory, depending on the assumed nature of the pion field and on the form of the coupling to nucleons. Only a limited set of these theories was renormalizable, a property considered essential in view of its

Table 9.6 The major intrinsic properties of the pions and muons

	Pions			Muons	
	π^+	π^-	π^0	μ^+	μ^-
Charge	+1	−1	0	+1	−1
Rest mass	139.57 MeV/c^2		134.96 MeV/c^2	105.66 MeV/c^2	
Spin	0		0	$\frac{1}{2}$	
Mean life of free meson	2.60×10^{-8} s		8.7×10^{-17} s	2.20×10^{-6} s	
Major decay mode	$\mu^+ + \nu_\mu$	$\mu^- + \bar{\nu}_\mu$	$\gamma+\gamma$	$e^+ + \nu_e + \bar{\nu}_\mu$	$e^- + \bar{\nu}_e + \nu_\mu$
	Although the decay modes differ we think of the three pions as three different manifestations of Yukawa's boson.			The muon is, except for its decay, just like a heavy electron	

The overbars and subscripts which label different types of neutrinos emitted in the decay of pions and muons will be described in Sections 12.2 and 12.3. Of course, as we shall discover, all the decays involving neutrinos are related to the familiar β-decay.

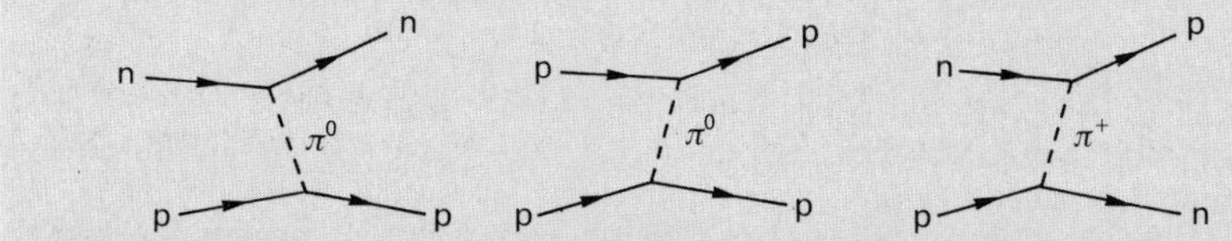

Fig. 9.15 Some Feynman diagrams for the exchange mechanisms involving the pions. Note again the appearance of the three-arm vertex with two fermions and one boson. In the charge exchange diagram, whether it is a π^+ one way or a π^- the other way is irrelevant, just as the time ordering of the vertices is irrelevant (see the caption to Fig. 9.3).

importance to QED. However, the important problem was that quantitative calculations were not successful. The reason lay partly in the value of $g^2/4\pi\hbar c$, which was about 15. As indicated in Section 9.4 the success of QED depends upon the smallness relative to 1 of α, the fine-structure constant, which makes the perturbation calculations manageable. The size of the pion–nucleon coupling constant prevents sensible use of perturbation theory; next-to-leading order has a bigger effect than leading order—and so on! Attempts to predict pion–nucleon scattering cross-sections were found to be hopelessly wrong as data from experiments using the new accelerators became available.

Yukawa's hypothesis has affected the analysis of the nucleon–nucleon scattering results in terms of potentials. The form and nature of the potentials that are used are very much influenced by those predicted from the exchange not only of pions but also by the exchange of heavier bosons such as the ρ and ω (Section 10.4). However, the nucleon–nucleon potential at separations greater than 1.5 fm appears to be well described by pion exchange alone. The small separations part of the potential is complicated by the exchange of these heavier bosons and is not well understood.

The way forward from the difficulties of applying Yukawa's ideas was through investigations into the properties of pions and, in particular, how they interact with nucleons and with other pions. This was done by measuring the cross-sections for elastic scattering of pions by nucleons and by investigating the properties of secondary particles produced in inelastic scattering of both pions and nucleons by nucleons. This led to the discovery of many new particles and to a beginning to the understanding of the structure of nucleons. This is the subject of elementary particle physics which is taken up in the following chapter. It leads to a new point of view: nucleons, pions, those heavier bosons and the new particles are now known to be bound systems of **quarks** and **antiquarks**, bound by a force due to the exchange of **gluons**. This relegates the nuclear force to a position with respect to the quark–gluon interaction similar to that held by inter-atomic and molecular forces with respect to the more fundamental electromagnetic interaction that dominates atomic structure. So nuclei may just be the molecules of particle physics. The theory of the quark–gluon interaction is called **quantum chromodynamics** (QCD) and it is based on what are believed to be very sound fundamental principles. However, no techniques presently exist for the precise prediction using QCD either of the properties of the quark-bound states or of the residual interactions between them. The next section gives a more detailed introduction to this aspect of nuclear and particle physics.

This shift of emphasis hinted by the introduction of quarks and gluons should not prevent us appreciating the importance of Yukawa's idea or realizing the huge effect it has had on our understanding of particle processes. The pion is not the fundamental object Yukawa envisaged but the concept of particle exchange lives on in modern theories of gauge particles such as the gluons (see Section 9.11).

DEFINITIONS AND KEYWORDS

Quarks and anti-quarks Spin $\frac{1}{2}$ fermions and anti-fermions now believed to be the constituents of hadrons. (See Section 10.3.)

Gluons Spin-1 bosons which, by exchange, are responsible for the forces among quarks and antiquarks. (See Section 10.7.)

Quantum chromodynamics The theory of the interaction of quarks and gluons.

9.11 Quarks, gluons, and QCD

We shall find in Chapter 10 that there are several different kinds of quark and that each kind, as specified by some conventional quantum numbers, can have one of three different values of a new quantum number called colour. The fundamental principle of quantum chromodynamics is that the fields describing the quarks are required to have a form of gauge invariance. The consequence is that there must exist, in this case eight, massless spin-1 bosons which interact with the quarks and are responsible for the forces between them. These are the gluons. In addition, colour is conserved and the theory is renormalizable. This is really an extension of a situation pre-existing in QED: a gauge invariance demands the existence of a neutral spin-1 boson (the photon, of course) and ensures renormalizability. Thus gauge invariance is now considered to be an essential element of all quantized field theories: the bosons required are called **gauge bosons**. We shall return to this subject in Section 13.4.

At this point we have to draw the Feynman diagrams for the quark–gluon interaction: quarks are spin-$\frac{1}{2}$ fermions and, as we have noted, gluons are spin-1 bosons. The basic vertex once again consists of the meeting of two fermion arms and a boson arm (Fig. 9.16a). But there are two other vertices, one with three gluon arms (Fig. 9.16b) and a second with four gluon arms (Fig. 9.16c). Thus a gluon can interact directly with another gluon in contrast to the situation in QED where a photon cannot interact directly with another. Thus no such 3- or 4-photon vertices exist in QED and their appearance in QCD is a consequence of a more complicated gauge invariance in the latter, called non-Abelian in the jargon of the trade, in contrast to the Abelian nature of gauge invariance in QED.

The quark–quark interaction is now by gluon exchange as in Fig. 9.17. It is believed that another consequence of the structure of QCD is that the central part of the potential between two quarks increases indefinitely as the distance between them increases, so that it is impossible to separate bound quarks: quarks are said to be **confined**. A poor analogy would be of a spring: the stored energy increases as the spring length is increased. The ends represent quarks, the spring represents the gluon field. If the spring is stretched so much that it breaks, two ends are created but no free ends! The material from which we and our environment are made appears to contain no free quarks and the fact that none has been observed even in the most energetic collisions observable is supposed to be due to quark confinement. More on this subject in Section 10.9.

The evidence for gluons as the messengers for quark interactions and that QCD is the theory which governs their behaviours is now strong. It will be given in Section 10.10. The quark structure of pions and nucleons, and of other hadrons, will be described in Section 10.3 and onwards.

DEFINITIONS AND KEYWORDS

Gauge bosons These particles are the carriers of the interaction between the fundamental fermions and are the modern version of Yukawa's boson as carrier of a force. We know twelve gauge bosons: 8 gluons, the photon, the W^+, W^-, and the Z^0.

Confinement This is the term describing the observation that it appears impossible to liberate quarks, or antiquarks, from hadrons.

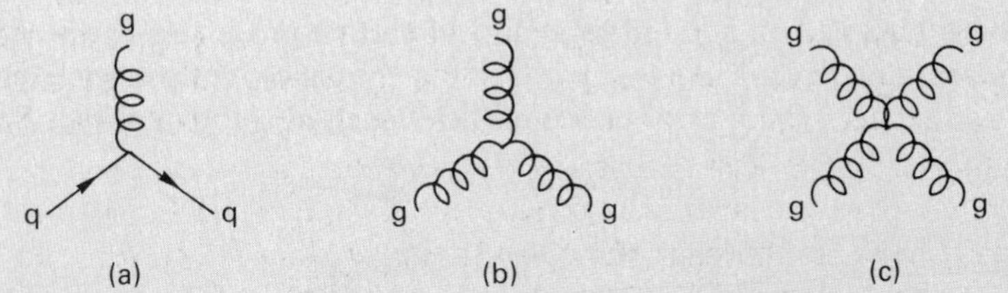

Fig. 9.16 The basic vertices of quantum chromodynamics: (a) quark–quark–gluon; (b) gluon–gluon–gluon; (c) four-gluon.

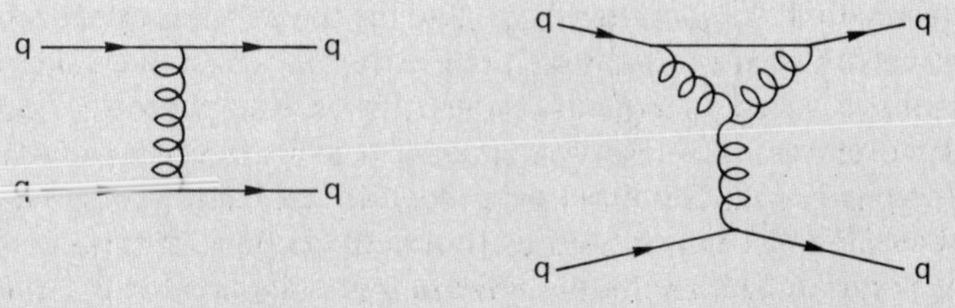

Fig. 9.17 Two Feynman diagrams representing the interaction between two quarks.

9.12 The strong interaction

We hope that the last section has widened the reader's view of the nuclear force: it introduced the pions and indicated that there are other heavier bosons related to the pions which could be involved in the nuclear force, and that all these particles could be readily produced in energetic collisions. Thus the nuclear force is a part of a wider interaction that exists between many particles. It is called the **strong interaction** and the particles which experience it are called **hadrons.** We expect and will find that the hadrons are composed of quarks and antiquarks. This compositeness of hadrons gives them a size of about 1 fm. In contrast we have noted (Section 3.2) that electrons, and of course positrons, muons, and neutrinos, do not experience the strong interaction: they are called **leptons.** They appear to be point-like.

The reader should note that the heavier bosons mentioned as being related to the pions and likewise involved in the nuclear force are not the gluons introduced in the last section.

DEFINITIONS AND KEYWORDS

Strong interaction This term covers the interactions that exist between members of a whole class of particles, including the pions and nucleons, that are believed to be composed of quarks and antiquarks.

Hadron The generic name for all particles that experience the strong interaction.

Lepton The generic name for all fermions, charged or uncharged, which do not experience the strong interaction (see Table 12.1).

9.13 The weak interaction

The reader has met β-decay in its simplest reaction

$$n \rightarrow p + e^- + \bar{\nu}_e.$$

This decay is a manifestation of the **weak interaction**. Called weak because, as we shall be able to show in Section 12.11, its coupling strength is effectively very small.

When Fermi in 1933 considered β-decay, he was guided by electromagnetism: if a nucleus could create and emit a photon, as it does in γ-decay, then why not the ability to create an electron and a neutrino, if energy permitted? In the language of Feynman diagrams, his theory of β-decay envisaged a neutron changing into a proton and emitting the $e^-\bar{\nu}_e$ system, all these changes happening at a point (see Fig. 9.18). The theory is consistent with β-decay data if the coupling of the particles is described by a constant $G_F = 1.436 \times 10^{-62}\ \mathrm{J\,m^3}$, the Fermi coupling constant. This theory is very successful but is unsound theoretically because, although leading-order calculations are correct, next-to-leading-order calculations are divergent: the theory was found to be non-renormalizable. (See Section 12.11 for another twist to this story.)

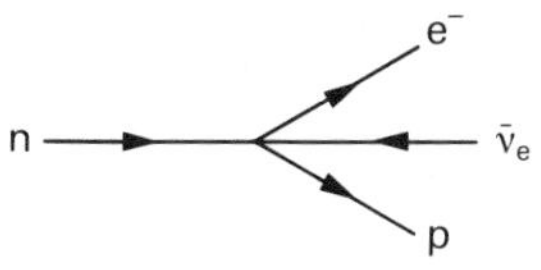

Fig. 9.18 The first β-decay theory had a four-fermion vertex.

In attempts to cure these difficulties it was proposed that there existed a carrier of the weak interaction. This was inspired by Yukawa's proposal concerning the role of his boson in the decay of the neutron, although the new carrier had to be very much more massive than the pion. This idea only displaced the onset of the symptoms of disease in the theory to higher energies and other circumstances, unless the fields describing the behaviour of the carriers and of the fermions with which they interacted were required to have a local gauge invariance similar to that possessed by QED and QCD. That cured the difficulties and the theory became renormalizable. So convincing was this solution that the Nobel prize was awarded to Glashow, Weinberg, and Salam, whose independent work culminated in this solution, even before the proposed carriers were observed as free particles. A Nobel prize later went to Rubbia and van der Meer for work which resulted in the observation of these carriers. So there are two stories here and in this chapter it is right to tell the outline of the theoretical proposal. The experimental observation is described and illustrated in Figs. 9.19 and 9.20. See also Fig. 12.8.

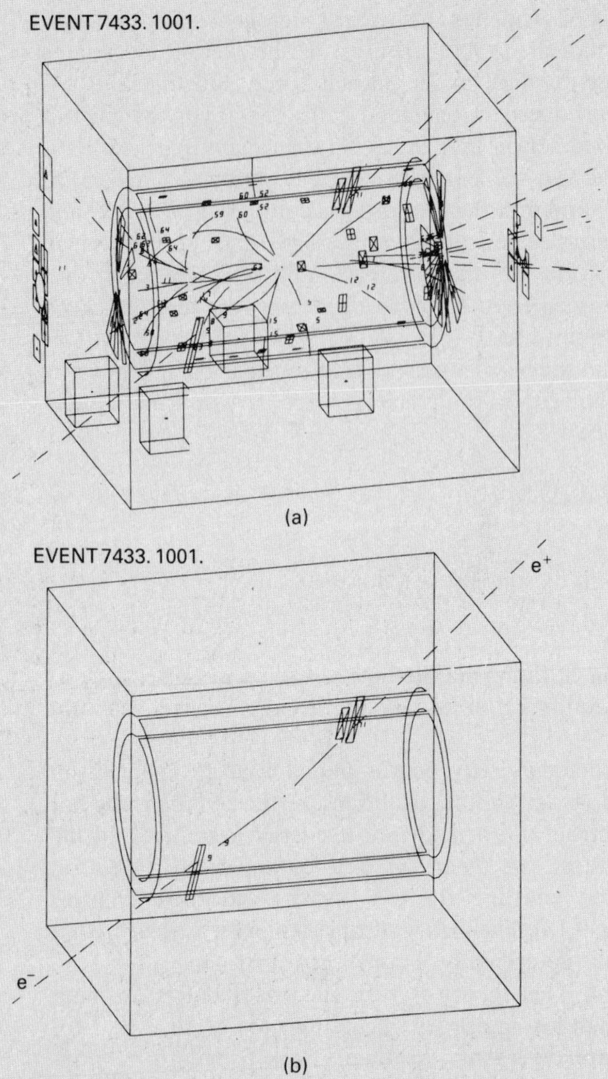

Fig. 9.19 The discovery of the Z^0. A computer reconstruction of an event in which a $Z^0 \rightarrow e^+e^-$ was observed in the UA1 apparatus at the p–p̄ collider at CERN. The 400 GeV/*c* proton synchrotron at the European Centre for Nuclear Research (CERN) was adapted so that it could act as a storage ring with protons of 270 GeV/*c* rotating in one sense and antiprotons of the same momentum rotating in the opposite sense. Two places on the ring were instrumented to observe collisions between p and p̄. The total centre of mass energy for these collisions was, of course, 540 GeV, well above threshold for real Z^0 or $W^\pm$ production. Most collisions give two jets of hadrons (mostly pions) collimated by low transverse momentun to the forward and backward direction with respect to the collision axis. These hadrons are the consequence of quark fragmentation (see Section 10.9) after small momentum transfer collisions between quarks and gluons from the proton with antiquarks and gluons from the antiproton. Occasionally a collision produces a real W or a real Z^0 which decays. One decay mode of the Z^0 is

$$Z^0 \rightarrow e^+ + e^-.$$

This yields two very energetic particles, frequently at large angles and well clear of the fragmentation hadrons. Electrons (and positrons) are easily identified and since their momenta are determined from the curvature of their trajectories in a magnetic field, it is possible to calculate the mass of their parent if they are the consequence of a two-body decay (see Fig. 2.7 for another example of this technique). In (a) there is a computer reconstruction of many of the charged tracks and of calorimeter (Section 11.2) hits of one event. If the tracks and calorimeter hits are each subject to a cut which eliminates all those with a transverse momentum less than 2 GeV/c, only one electron and one positron track survive, as shown in (b). This pair is consistent with the hypothesis that they originate from the decay of a particle of mass about 90 GeV/c^2. A number of such events allowed the unambiguous identification of the production and decay of the expected Z^0. (UA1 Collaboration, 1983b.)

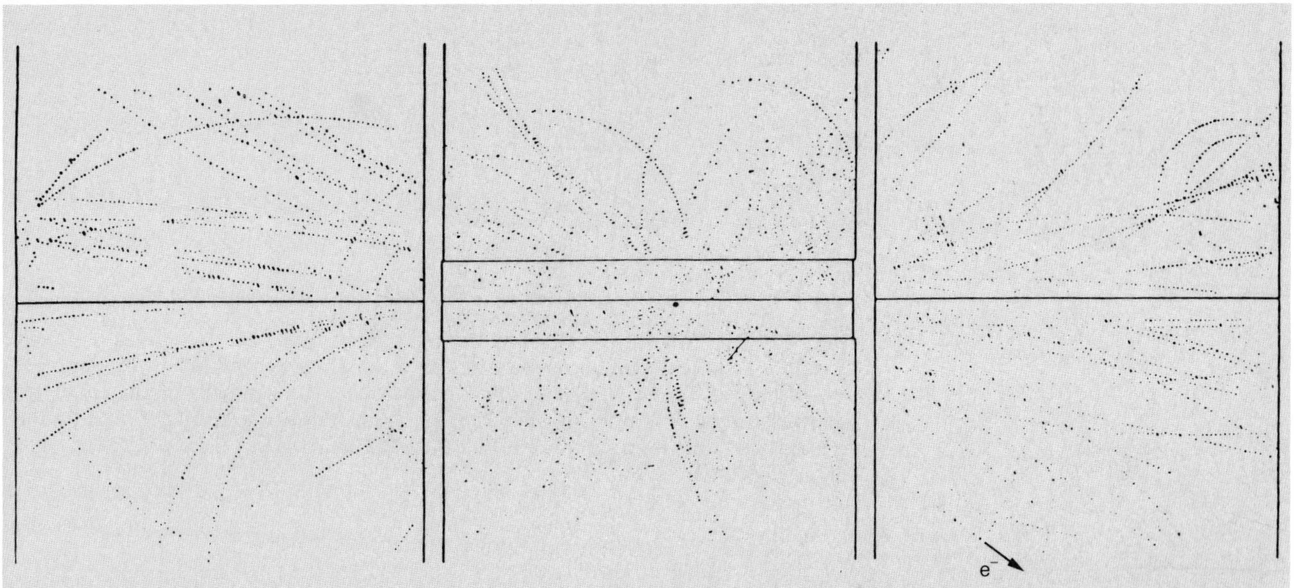

Fig. 9.20 This figure shows the computer reconstruction of another event from the UA1 apparatus. The reconstruction in Fig. 9.19 showed tracks seen in outer parts of the detector; this figure shows tracks seen in the inner detector.

The W was also expected to be produced at a rate comparable to that for the Z^0 but it is not expected to have a decay mode with as clear a signature as that of the e^+e^- mode of the Z^0. One decay is

$W^+ \to e^+ + \nu_e$ or, for the opposite charge, $W^- \to e^- + \bar{\nu}_e$.

This mode means an energetic e^+ or e^-, often at large angle. The event shown has one large transverse momentum electron, picked out by the arrow, which can be interpreted as due to the production and decay of a W^-.

In such events the e does not have a unique energy as the W itself may have considerable energy. Since the neutrino is not observable, it was impossible to find the mass of the W using the same techniques as was used to find the mass of the Z^0; a more subtle analysis had to be made but with results just as successful. (UA1 Collaboration, 1983a.)

The carriers are three-gauge bosons W^+, W^- and Z^0, and for their action we construct two basic vertices shown in Fig. 9.21. As for the QED vertex, there are two fermion arms and one boson arm. In the QED vertex it was implicitly assumed that the interacting fermion did not change, except in its energy and momentum. We note that the W vertex connects two different fermions: the Z^0 is more like a photon and its vertex connects identical fermions, or a fermion with its antifermion. Another difference: since both M_W and M_Z are much greater than both m_e and m_μ it is possible to have all three lines real (extend Problem 9.1 to confirm this).

Let us take a look at the diagrams for neutron decay in Figs. 9.18 and 9.22. In Fermi's theory the amplitude for the process of 9.18 contains a factor G_F; the amplitude for 9.22 contains a g_W for each vertex (see Table 9.7) and a factor $(M_w^2c^2 - q^2)^{-1}$ for the propagation of the virtual W from vertex to vertex. Here M_W is the W rest mass and q is the relativistic four-momentum carried by W (as in Fig. 9.6). In β-decay $|q^2c^2|$ is small ($\leqslant 1\,\text{MeV}^2$) so that if $M_w^2c^2 \gg |q^2|$ we would expect $G_F \approx g_W^2/M_W^2$. An interesting fact emerges: if $g_W^2 = e^2/\varepsilon_0$ (e is the electronic charge) then $M_W \approx 90\,\text{GeV}/c^2$ will give the experimentally known value of G_F. This situation is summarized in Table 9.7. Thus unification with electromagnetism is a possibility, and such a unification is the heart of the Glashow, Weinberg, and Salam model. That theory contains a parameter, the **Weinberg angle**, θ_W, which it cannot predict. However, the masses of the Z^0 and W, the strengths of their coupling to each kind of fermion are all predicted from the

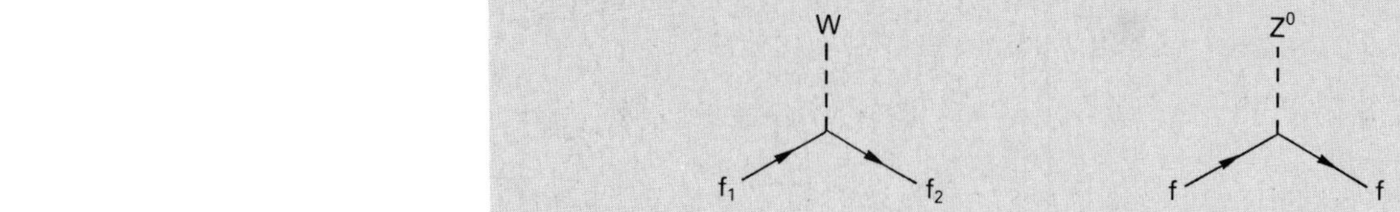

Fig. 9.21 The two basic vertices of the weak interaction. The solid lines represent fermions and the dotted lines represent one of the gauge bosons, $W^{\pm}$ or Z^0. Since the W is charged, f_1 and f_2 are not the same particle. The strengths of the couplings depends on the fermions, on whether those are coupled to a W or a Z^0 and on the electroweak coupling constant g (Table 9.8). We shall discuss these vertices in more detail in Section 12.4.

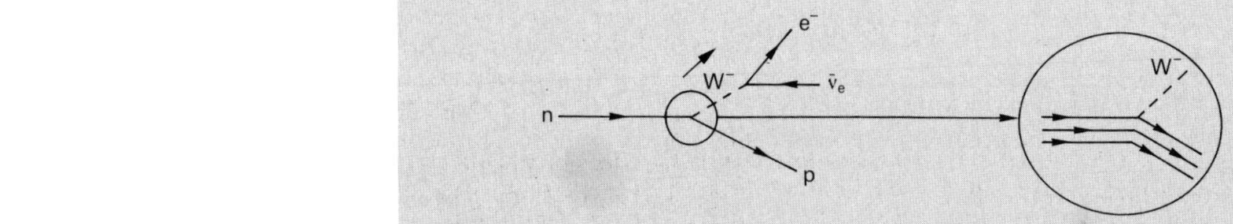

Fig. 9.22 The Feynman diagram for neutron decay in the electroweak theory. In fact the neutron has structure and the W vertex at the neutron end is attached to a quark line, as in the inset.

experimentally determined values of e, G_F, and θ_W. The predictions for the masses and widths are given in Table 9.8. The status of the measurements of these quantities is given in that table and in Table 12.5.

This is the so-called **electroweak theory**. The development of this theory, the observation of the $W^{\pm}$ and Z^0 bosons, and the discovery of processes due to the Z^0, will stand as a landmark in the history of particle physics, but the theory does have some features which must indicate that it is incomplete and cannot stand alone. The reader will have spotted one difficulty: the gauge invariance should require massless gauge bosons yet the W and Z^0 are the two heaviest elementary particles known. 'Where does the θ_W come from?' is an obvious question. This parameter is presumably connected with the mechanism which gives the $W^{\pm}$ and Z^0 mass. In the theory the mass-giving mechanism relies on the postulated existence of at least one heavy, spin 0 boson called a **Higgs boson.** It is named after P. W. Higgs, who discovered the mechanism by which such a particle can break the symmetry of a postulated triplet W^+.W^-,W^0 plus a neutral singlet particle we call V^0, all zero-mass gauge bosons, to give massive W^+ and W^-, a massive Z^0 and the massless photon **(the Higgs mechanism)**. There is no direct experimental evidence for the existence of any Higgs bosons—but, of course, the search is on.

Table 9.7 The weak interaction and an intermediate vector boson.

If the weak interaction is not

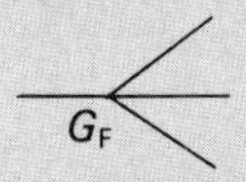

but

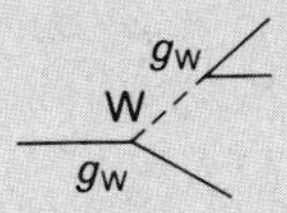

then what is the relation between g_W and G_F? Quantum field theory says

$$G_F/(\hbar c)^2 \approx g_W^2/(M_W^2c^4 - q^2c^2),$$

where q is the four-momentum transferred (Fig. 9.6) by the W, and M_W is the mass of the W (q is space-like in β-decay so that $q^2 < 0$). Now no dependence on q^2 is seen in β-decay so we assume $M_W c \gg |q|$. Therefore we have

$$G_F/(\hbar c)^2 \approx g_W^2/M_W^2c^4$$

so that

$$\frac{g_W^2}{4\pi\hbar c} \approx \frac{G_F M_W^2 c^4}{4\pi(\hbar c)^3}.$$

If $g_W^2/(4\pi\hbar c)$ is to have the same value as $e^2/(4\pi\varepsilon_0\hbar c)$ $[=(137)^{-1}]$, then

$$M_W c^2 \approx 89 \text{ GeV}.$$

Therefore the weak interaction could be weak, relative to electromagnetism, because of the large W mass: the 'range' of the weak force between fermions is then about 2×10^{-3} fm. We have used $G_F = 1.436\times10^{-62}$ J m^3 $= 0.896\times10^{-7}$ GeV fm^3 (see Table 12.8).

However, the above treatment oversimplifies the situation.

We shall return to the details of the weak interaction in Chapter 12, and to some aspects of the Higgs boson(s) in Section 13.4.

9.14 Conclusion

The object of this chapter has been to introduce a moderately unified and modern view of the interactions between particles. In the following chapters that view will be given more substance by our treatment of the physics of elementary particles. The experimental and theoretical development of this subject has been prodigious since 1945 and this fact makes the choice of strategy for the following chapters debatable. An historic approach would favour, for example, a tour through strong interactions which followed a route including the discovery of the hadron resonances, the study of scattering processes and deep inelastic scattering of leptons by nucleons to the unifying ideas of the quark–parton model and of quantum chromodynamics. An alternative approach to strong interactions is to start with quarks and some simple postulates and from that find that it is possible to explain a large body of experimental data. Since these data have more form within the context of the quark model we chose the latter approach and will develop our knowledge of the quark model in stages, each stage substantiated by sufficient but not excessive data. This is the program for Chapter 10.

Table 9.8 The Glashow–Weinberg–Salam electroweak theory.

A problem of notation exists. Since we have been using Standard International units, the fine structure constant α *is* $e^2/(4\pi\varepsilon_0\hbar c)$. Most books on electroweak theory have $\alpha = e^2/(4\pi\hbar c)$. So, where they write e we shall have to write $e/\sqrt{\varepsilon_0}$. This is a nuisance, as the business of units can be very confusing. In what follows, e is therefore in coulombs. The photon γ, the Z^0 and the W^+, W^- bosons are the observable consequence of a hypothetical massless four-component field. The proposed existence of one Higgs boson causes a mixing of the two electrically neutral components of the field into the massless photon and the massive Z^0, and gives mass to the charged components. The mixing is parametrized by an angle θ_W, called the **Weinberg angle**. Then the coupling g (Fig. 9.21) and the masses of these gauge bosons are related by

$$\frac{g^2}{4\pi\hbar c} = \frac{e^2}{4\pi\varepsilon_0\hbar c}\frac{1}{\sin^2\theta_W},$$

$$M_W^2 = \frac{g^2}{2^{5/2}G_F},$$

and

$$\frac{M_W}{M_Z} = \cos\theta_W,$$

Where G_F is the Fermi coupling constant (Tables 9.7 and 12.8, Section 12.8).

In 1988 the experimentally determined value of θ_W was

$$\sin^2\theta_W = 0.233 \pm 0.006,$$

which lead to the predictions

$$M_Z = 87.1 \pm 0.7\ \text{GeV}/c^2 \quad \text{and} \quad M_W = 75.9 \pm 1.0\ \text{GeV}/c^2.$$

However, there are some radiative corrections which change these values to

$$M_Z = 91.6 \pm 0.9\ \text{GeV}/c^2 \quad \text{and} \quad M_W = 80.2 \pm 1.1\ \text{GeV}/c^2.$$

These particles have many decay modes and although the full widths cannot be predicted with confidence in the absence of certain information about the quarks and neutrinos they were expected to be

$$\Gamma_Z \simeq 2.6\ \text{GeV} \quad \text{and} \quad \Gamma_W \simeq 2.1\ \text{GeV}.$$

The first observations of the Z^0 and $W^\pm$ were made with the proton-antiproton collider at the European Centre for Nuclear Research (CERN), Geneva (Figs. 9.19 and 19.20), and gave masses and widths which were, in 1988, given as

$$M_W = 81.0 \pm 1.3\ \text{GeV}/c^2,\ \Gamma_W < 6.5\ \text{GeV},$$
$$M_Z = 92.4 \pm 1.8\ \text{GeV}/c^2,\ \Gamma_Z < 5.6\ \text{GeV}.$$

In 1989, the electron-positron colliders capable of the formation of Z° particles commenced operation and within a few weeks gave

$$M_Z = 91.01 \pm 0.03\ \text{GeV}/c^2,\ \Gamma_Z = 2.60 \pm 0.10\ \text{GeV}.$$

These results are given in more detail in Table 12.5.

DEFINITIONS AND KEYWORDS

Weak interaction The interaction responsible for β-decay and other decay processes involving leptons and quarks (see Chapter 12).

Electroweak theory This is a unification of the quantized theories of the electromagnetic and weak interactions.

Weinberg angle This angle describes the degree of the mixing that generates from four gauge bosons, the observable bosons γ, $W^\pm$, and Z^0.

Higgs mechanism The mechanism by which the existence of a spin-0 particle can give a gauge boson mass without breaking the gauge symmetry.

Higgs boson The predicted spin-0 boson which must exist if the Higgs mechanism is responsbile for giving mass to the gauge bosons $W^\pm$, Z^0. There may be more than one Higgs boson.

In Chapter 11 we shall discuss the electromagnetic interaction and in Chapter 12 the weak interaction. This program should give the reader an opportunity to put the ideas introduced in this chapter into context. In Chapter 13 we shall summarize what is now called the standard model and look at the existing problems of particle physics and the direction future developments might take.

Have we omitted anything? Figure 9.23 reminds us that we have not discussed gravity. The gravitational potential is so small in all known basic atomic and

"THERE'S THE STRONG FORCE, THE WEAK FORCE, GRAVITY, ELECTROMAGNETISM, AND THEN THERE'S THAT CERTAIN SOMETHING I FEEL WHEN I SEE YOU."

ig. 9.23 A reminder that we have forgotten the gravitational interaction.

ubatomic processes that its effect is entirely negligible. However, it is worth oting that although the general theory of relativity correctly describes large-cale gravitational effects, a quantized version does not yet exist. Gravitational adiation is expected to exist although it has not yet been observed directly. The uantum of the radiation has been called the graviton before observation. The ature of the gravitational field implies that it will be a spin-2 particle.

Gravity will figure in our description of supernovae (Sections 14.9 and 14.10). The huge energy release in such events can be explained only if the assumption is nade that these stellar explosions are driven by stellar core collapse caused by ravity.

References

itchison, I. J. R. and Hey, A. J. G. (1982) *Gauge Theories in Particle Physics.* Adam Hilger, Bristol.

maglobeli, N. S. and Kazarinov, Yu. M. (1960) *Journal of Experimental and Theoretical Physics* (USSR), **37,** 1125–9.

ane, G., (1987), *Modern Elementary Particle Physics.* Addison-Wesley.

owell, C. F. (1950). *Reports on Progress in Physics,* **13,** 350–424.

JA1 Collaboration (1983a). *Phys. Lett.,* **122B,** 103–116.

JA1 Collaboration (1983b). *Phys. Lett.,* **126B,** 398–410.

10

Hadrons and the Quark–Parton Model

10.1 Introduction

There is a problem in discussing the hadrons and their spectroscopy. T introduce the problem we go back to the hydrogen atom. The radiativ transitions may be technically difficult to observe and to measure but concep tually they are simple and need only the classical idea of a wavelength t characterize each one. The first explanations of the energy levels by Bohr and b Schrödinger can be introduced to newcomers without a prerequisite detaile knowledge of the spectrum. And the quantum-mechanical explanation ca provide a foundation for understanding the more complicated aspects of atomi structure. In the case of hadrons, the spectrum of energy levels has propertie which include three different decay mechanisms and completely new quantur numbers that have no relation to anything in classical physics. Should th newcomer be given a tour of the zoological garden of partially classifie particles or taken straight to the description of a model which goes a long wa towards explaining their spectroscopy and properties? That is the problem. W have chosen to do a very brief review of what the hadrons are about and then t describe the model. We hope the reader will not be put off the subject by th sudden plethora of particle states that will appear. For the readers who woul like to have a guided tour through the discovery of the principal particles t complement this chapter, we refer them to Close *et al.* (1987), or to Cahn an Goldhaber (1989).

10.2 The hadrons

We recall that the hadrons are those particles which experience the stron interactions, the interactions which we associate with, among other things, th force which binds nucleons into nuclei. Thus the nucleons are hadrons and, sinc the exchange of pions is believed to be the major contributor to the inter-nucleo force, the pions must also be hadrons. We are interested in their static propertie such as mass, charge and spin, amongst others; however, to investigate th properties of the interactions it is necessary to measure and interpret th scattering of hadrons by hadrons. Experimentally, the direct investigation i limited to targets of protons or of neutrons (lightly bound in deuterium); target of complex nuclei may also be used although the bound state of the nucleon may complicate the interpretation. Incident beams are normally restricted t protons, neutrons and those hadrons, such as the charged pions, which are easily produced in the collisions of nucleons and which have mean lives sufficiently long to allow the formation of collimated, adequately monoenergetic beams. Antiprotons (symbol p̄) can also be produced and manipulated into useful beams. Recent developments in colliding beam machines have opened up the possibilities of investigating p–p and p̄–p collisions at energies very much higher than the equivalent centre-of-mass energies that can be attained with conventional accelerators and fixed targets. Measurements have been made of total and partial cross-sections, of differential angular cross-sections in elastic scattering, and of cross-sections and the properties of final states in inelastic collisions.

In Figs 10.1 and 10.2 we show plots of the total collision cross-sections for protons and antiprotons and for positive and negative pi-mesons incident on stationary protons as a function of momentum. An examination of the plot for protons shows that the total proton–proton cross-section falls from high values at low incident momentum to a mimimum; above the threshold for pion production ($P = 777$ MeV/c) the cross-section rises to a plateau where it varies slowly with momentum and has a magnitude of about 40 mb. This value would be expected for a classical proton of radius 1.8 fm. The anti-proton cross-section is large at low momentum due to the effect of annihilation into states of several pions, but as the incident momentum increases it falls and converges towards a common value with the proton–proton cross-section. The pion–proton total cross-section reaches a plateau above incident momentum of about 2 GeV/c at a cross-section of about 25 mb, suggesting a pion radius of about 1.0 fm. These radii are not properly defined and therefore must not be taken

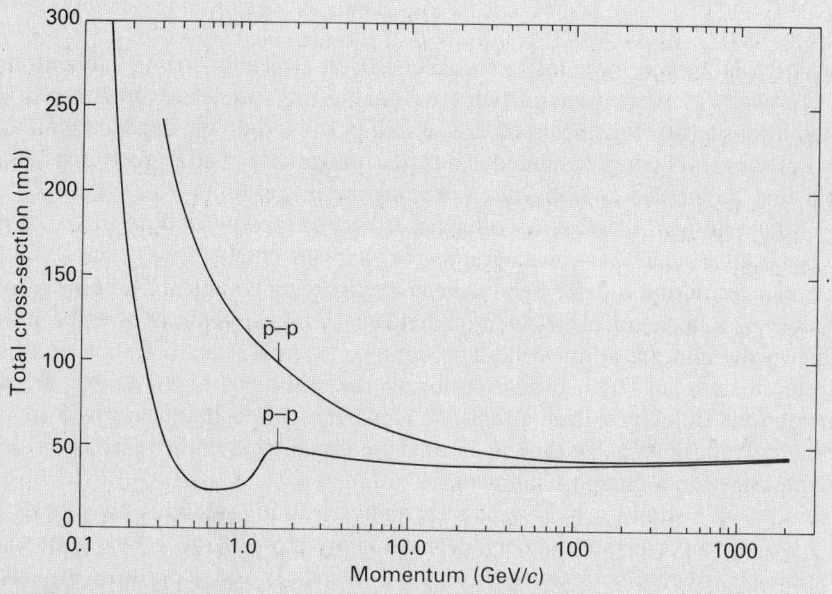

Fig 10.1 The total cross-section for protons and antiprotons on protons as a function of incident laboratory momentum. The p–p cross-section falls from high values at low momenta to a minimum at about the threshold for single pion production (777 MeV/c); at higher momenta the cross-section rises to a maximum and then levels at about 40 mb. The p̄–p cross-section at low momenta is dominated by the effect of annihilation (p̄p → pions); at higher momenta the cross-section converges with that for p–p collisions and at very high momenta both are rising slowly together.

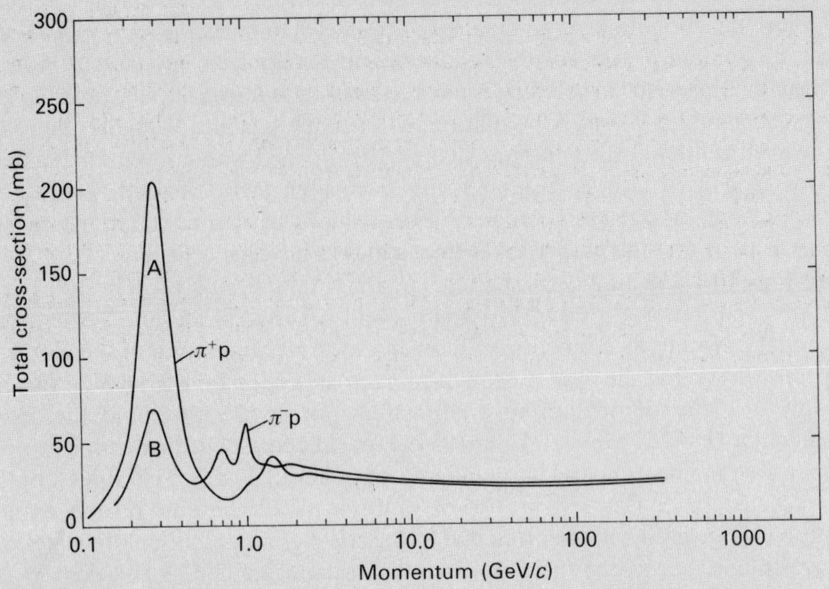

Fig. 10.2 The total cross-sections for positive and negative pions on protons as a function of incident laboratory momentum. As with protons and antiprotons, the two cross-sections converge and start a slow rise with increasing energy. At low momenta the cross-sections change rapidly with momentum showing peaks characteristic of compound state formation. These peaks are sometimes called resonances. Peak A is due to the formation of the $\Delta^{++}(1232)$ and peak B is due to the $\Delta^{0}(1232)$ (Section 10.3).

seriously. However, these total cross-sections are typical of strong interactions. It follows that we expect partial cross-sections of processes due to strong interactions to be of the order of tens of millibarns and down, depending on the part of the total being considered. Thus, this magnitude of cross-section labels a collision mechanism as being due to the strong interaction.

To give an idea of what is going on in hadron–proton collisions, we show some bubble chamber photographs of hadron interactions. The bubble chamber technique is described in the accompanying comment. A photograph of an event in a chamber has an immediate visual impact and can often be given a first-order interpretation without recourse to extensive measurements on a set of stereo images. These properties bring the reality of high-energy particle interactions quickly to our attention. However, the technique proved to be immensely more valuable than its immediate visual impact: the reasons for this are described in a second Comment.

Figure 10.3 shows a bubble chamber photograph containing two inelastic proton–proton collisions at an incident momentum of 24 GeV/c. One of these is a spectacular event with fourteen charged secondary particles, most of which are charged pions. This event was the one with the largest number of charged secondaries found among a large number of events photographed in this particular exposure of a chamber to energetic protons. Figure 10.4 shows an interaction of an incident 16 GeV/c charged pion with a proton in the same chamber. This is an untidier photograph than that of Fig. 10.3 because of a background of low-energy tracks. However, one of the incident pions has caused an event in which four charged secondary particles were produced, one

Comment The principles of bubble chamber operations

A liquid under pressure at a temperature which is its atmospheric pressure boiling point will not boil. If the excess pressure is reduced to zero the liquid will normally boil. However, boiling may not start immediately and it is then said to be in a superheated state. In this state boiling can be triggered by sharp edges, or by dust or ions in the liquid. A charged particle traversing the liquid leaves a trail of ions in the liquid that act as centres for boiling in the superheated state. The initial stages involve the growth of bubbles on the ions. These bubbles may be flash photographed; with correct relative timing of the reduction of the pressure (so-called expansion because the volume of the liquid is increased by the movement of a confining piston), of the passage of the particles, and of the flash, the bubbles may be caught when they are large enough to be photographed but before they have grown too large. The result is a photograph showing tiny bubbles along the trajectories of the charged particles which have traversed the liquid during the sensitive period. Stereo photography allows a three-dimensional reconstruction of these so-called tracks.

Most chambers have been used so that the liquid acts as both target and detector. A collision is then directly visible by the track generated by the incident particle (if charged) and by all the charged particles leaving the collision. Many liquids have been used. Hydrogen and deuterium provide targets of free protons and of almost free protons and neutrons respectively. Propane, neon–hydrogen mixtures have been used when a target density greater than that of liquid hydrogen or deuterium was essential as, for example, in the study of neutrino interactions (see Fig. 12.17).

Comment The contribution of the bubble chamber

The bubble chamber was invented by D. A. Glaser in 1952. The hydrogen bubble chamber was developed and put to use by L. Alvarez at the Lawrence Berkeley Laboratory in California. A small prototype was operating by 1954. By 1959 a chamber containing 500 litres of liquid hydrogen and a major length of 180 cm was operating ('the 72 inch'). This technique had a particularly profound effect on research into the spectroscopy of hadrons and into other properties of strong interactions in the energy range 100 MeV to 10 GeV.

Consider a hydrogen bubble chamber. Sizes have ranged from a few centimetres to two metres across. The mean free path for 10 GeV/c charged pions in liquid hydrogen (the density is 70.8 kg m^{-3}) is 9.4 m. Thus each photograph with ten pions of momentum 10 GeV/c entering the chamber one metre across will show, on the average over many photographs, just over one interaction. Thus at expansion rates of one every few seconds, several thousand events may be recorded per day of operation. Of course, the events cannot be of a chosen type, but this is a very useful yield and at ten incident particles per photograph the chamber will not be too crowded. These are very rough figures and will vary with the size, construction and mode of operation of the chamber.

The second thing is that the momentum of all changed particles may be measured from the curvature of their tracks in a magnetic field. All useful chambers were equipped with adequate magnetic fields. If a particle track may be identified as due to an electron, or pion, or proton, or . . . , then the energy may be calculated. Although identification is not always possible except at low momenta, it is often possible to go through a series of reaction hypotheses for a single event and to decide which is most likely to fit the constraints of energy and momentum conservation, thereby completely determining the nature of the event and its kinematics with some degree of confidence. In certain circumstances it is possible to determine the identity of an undetected, single neutral particle among the remaining charged particles produced in an interaction, as in the event of Fig. 10.5. From the totality of information obtained over many events it is possible to discover the existence of short-lived ($<10^{-20}$ s) states. For example, collisions in which a π^+ and a π^- are produced show evidence for a correlation between their momenta which indicates that they are frequently the decay products of a heavy (770 MeV/c^2) boson and not direct products of the collision. This effect is described in Fig. 2.7. However, the full physics of analyses of this kind is beyond the scope of this book.

The third important property of bubble chambers is their size and spatial resolution. Typically sizes are from a few centimetres to two metres across. The resolution, meaning the minimum distance in which gaps or details on tracks could be detected, was about 10 μm for small specialized chambers to about 150 μm for the largest chambers. By 1950 work with cloud chambers on cosmic ray interactions had shown the existence of 'long-lived' unstable particles: long-lived meaning lifetimes in the range 10^{-9} to 10^{-11} s. In that time such particles produced in high-energy collisions will move a few millimetres to tens of centimetres. These distances are easily observed in a bubble chamber and events involving the production of such particles are very striking. The two decays in Fig. 10.4 are examples. The discovery of charmed particles (Section 10.5) with mean lives of order 10^{-13} s required small, high-resolution chambers in order to observe the small distance moved between production and decay.

These properties made the bubble chamber one of the vital instruments in unravelling the spectrum of hadrons, in particular that of the strange particles, in the two decades following its invention. Later bubble chambers have been sized for other research goals; for example, the large-volume chambers used to investigate neutrino interactions.

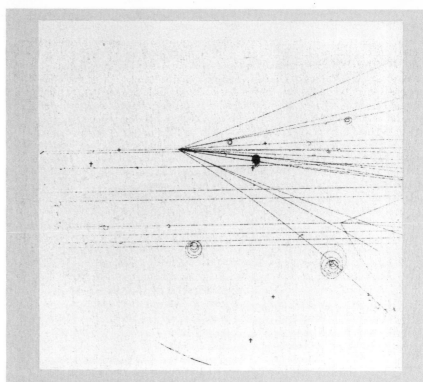

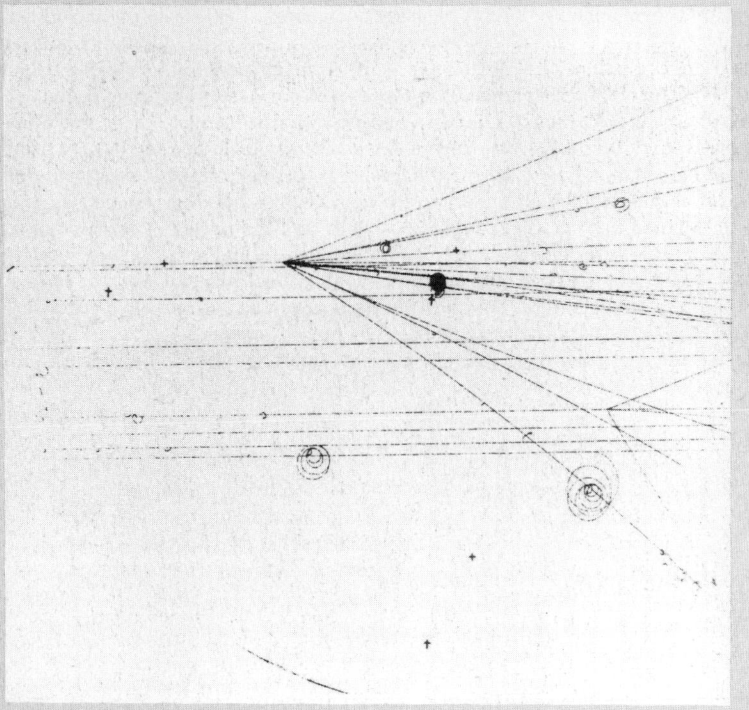

Fig. 10.3 A bubble chamber photograph of 24 GeV/c protons in liquid hydrogen. Nine protons enter the chamber from the left, of which two suffer inelastic collisions. The most spectacular has fourteen charged secondary particles and an unknown number of neutral secondaries. The incident momentum of 24 GeV/c means a total centre of mass energy of 6.84 GeV in these p–p collisions, plenty to provide the mass and kinetic energy for the two nucleons, which must be among the final state particles, and for a number of pions.

The chamber is cylindrical with an internal diameter of 32 cm. The boundary defined by this diameter can be seen on the photograph by the beginnings or ends of tracks. This gives a scale to the photograph. The magnetic field was 1.5 T along the axis of the cylinder and therefore perpendicular to the plane of the photograph. It is directed so as to bend negative particles to the right with respect to their direction of travel, as seen on this photograph. There is a δ-ray in this photograph, which gives this information about the direction of the field.

of which was involved in a second collision a few centimetres beyond the first. Among the products of these collisions is a neutral particle which decays into one positive and one negative particle after a free flight of several centimetres: this is an example of what was called a neutral V particle. Less easy to see is a positive particle produced in the second collision which has a kink in its trajectory. The most likely interpretation is that this secondary suffered decay in flight, producing one charged and one or more neutral particles: this would then be an example of what was called a charged V particle. These two photographs serve to show how secondary hadrons are copiously produced in energetic hadron–hadron collisions.

Figure 10.5 is a photograph, taken in the same chamber as those of Figs. 10.3 and 10.4, of 320 MeV positive pions in the liquid hydrogen filling. There is an event showing the classic $\pi^+ \to \mu^+ \to e^+$ decay sequence. The π^+ was the result of the reaction

$$\pi^+ + p \to \pi^+ + \pi^+ + n.$$

Fig. 10.4 A bubble chamber photograph of 16 GeV/c charged pions in liquid hydrogen. (This is the same chamber as in Fig. 10.3) This photograph is crowded with low-energy tracks due to electrons, positrons and pions. To see the 16 GeV/c pions look obliquely along the tracks from left to right and you will see ten incoming particles (from left) all of which have imperceptible curvature. One causes a reaction producing several secondaries, one of which causes a second reaction a few centimetres further. One of the charged tracks from this second interaction is kinked, implying an in-flight decay such as $\Sigma^+ \rightarrow p + \pi^0$. There is a neutral secondary which decays into one positive and one negative particle, but without a proper kinematic analysis it is not possible to decide whether it came from the first or the second interaction. The earliest observation of decays of this kind was made in cloud chambers exposed to cosmic rays. A neutral to two charged particle decay was called a neutral V, a name inspired by the topology. A charged particle decay in flight to one charged and one, or more, neutral particles was called a charged V. As seen in the plane of this photograph the magnetic field bends negative particles to the left of their direction of flight. Can you find a δ-ray which gives this information?

It is clear from these photographs that the strong interactions show a wide variety of phenomena, all of which have been the subject of intense research. However, in this chapter we have to concentrate on one aspect of the strong interactions, namely the spectroscopy of the hadrons. The neutral and charged V particles appeared to have mean lives of the order of 10^{-10} s. Figure 10.2 shows that the pion–proton cross-sections at low energies (<2 GeV/c laboratory incident momentum) have peaks, called resonances, which are interpreted as being due to the formation (Section 7.8) of compound states which in turn can be interpreted as excited states of the nucleons with mean lives of order 10^{-23} s. These observations opened the way to the discovery of many other hadronic states, some long-lived ($\sim 10^{-10}$ s), others very short-lived ($<10^{-22}$ s).

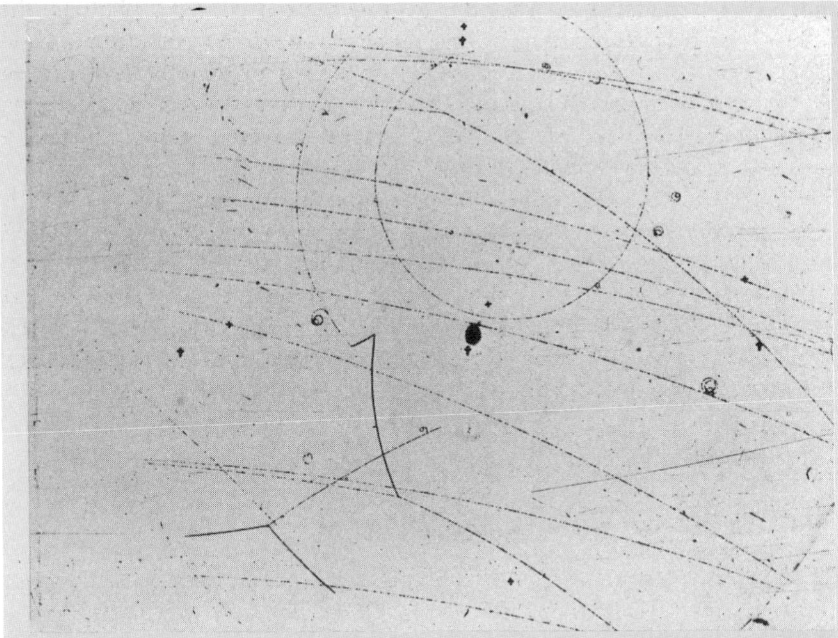

Fig. 10.5 A bubble chamber photograph of 320 MeV π^+ in liquid hydrogen. (This is again the same chamber as in Figs. 10.3 and 10.4; the magnetic field is bending negative particles to the left of their direction of flight.) The prominent event has a π^+ produced at about 120° to the direction (from the left) of the incident particle, coming to rest and decaying into a μ^+ and a neutrino, which leaves no track. The μ^+ decays to a positron and two neutrinos (see also Fig. 9.14). The positron makes one and a quarter turns of a tightening helix as it loses energy and before it leaves the chamber. The forward track from the original collision leaves the chamber. A measurement of track curvatures and range shows that the visible momentum and energy are not balanced in the collision and therefore that there is at least one neutral particle not directly observable. A proper kinematic analysis shows that this event is consistent with the reaction

$$\pi^+ + p \rightarrow \pi^+ + \pi^+ + n.$$

In addition to states produced by formation, hadron–hadron collisions produce many other short-lived hadronic states as well as pi-mesons and K-mesons (Section 10.4), either directly or as the decay products of the short-lived states. Any model of the hadrons must therefore take the existence of all these states into account.

10.3 The quark–parton model: Stage I

Feynman used the term **parton** to refer to the constituents of the hadrons. In Sections 9.11 and 9.12 we introduced the idea that the candidates for the role of partons are the spin-$\frac{1}{2}$ quarks (q) and antiquarks ($\bar{q}$) and the spin-1 gluons. The quarks and antiquarks, which give a hadron its properties such as spin, charge, and some other attributes that we are to meet later in this chapter, are called the **valence quarks**. The hadrons also contain ephemeral $q\bar{q}$ pairs which do not change the nature of the hadron but which do contribute a significant fraction of its rest energy: these are called the **sea quarks**. The hadrons also contain gluons, the quanta of the field which leads to the forces between quarks. This is analogous to the statement that the hydrogen atom contains photons, the quanta of the field responsible for the attractive force between electrons and protons. However, photons are a much less obvious constituent of the hydrogen atom than gluons are of the hadrons. This importance we will not refer to again until Section 10.10 but we will not wish to forget that the strong quark forces are due to gluon exchange.

Our stage one model requires two **flavours** (that is two types) of quark, symbols u and d, for up and down. They have spin-$\frac{1}{2}$ and are therefore fermions and have antiparticles, ū and d̄, for anti-up and anti-down respectively. The u and d have electric charges $+\frac{2}{3}$ and $-\frac{1}{3}$ of the proton charge respectively: the antiquarks have opposite charges. This is summarized in Table 10.1. Now it is easy to make some hadrons (the reader should check that the total hadron charge comes out correctly):

Table 10.1 The light quarks.

Quark	Charge	Spin
Up, u	+2/3	$\frac{1}{2}$
Down, d	−1/3	
Anti-up, ū	−2/3	
Anti-down, d̄	+1/3	

Note that antiquarks always have all quantum numbers of opposite sign to those of the quark, except spin; the masses are also identical. Therefore in future we will list quarks only and imply the anti-quarks.

The proton, p, has valence quarks uud.
The neutron, n has valence quarks udd.
The positive pion, π^+, has valence quarks ud̄.
The negative pion, π^-, has valence quarks ūd.

What about the neutral pion, π^0? The first thought is that it might be uū or dd̄, but of course it is a quantum-mechanical superposition which we write

$$\sqrt{\tfrac{1}{2}}\,|\mathrm{d\bar{d}}-\mathrm{u\bar{u}}\rangle.$$

(Beware! This is not a precisely defined state vector.)

The next step is to examine how the quarks can be put into states which give the observed hadron spin and parity. This is straightforward for the pions: if the qq̄ are put into a singlet spin state (total spin quantum number $s=0$) with zero relative orbital angular momentum (quantum number $l=0$) then the total angular momentum, that is the spin, of the pion is zero. This qq̄ state is represented in spectroscopic notation by $^1S_0$ ($^{2s+1}L_j$). A fermion and an antifermion have opposite intrinsic parity (a fact which we can only state and cannot even attempt to justify and which we must ask the reader to memorize); thus a qq̄ state will have parity $(-1)(-1)^l$ and ud̄, dū, or dd̄-uū in a $^1S_0$ state has total odd parity. Thus we can explain the pion spin-parity $j^P=0^-$.

There is no difficulty in seeing that quarks uud can be arranged so that their spin angular momenta add vectorially to a total angular momentum $\frac{1}{2}$ to make the known proton spin. Similarly for the neutron. In both cases there is no relative orbital angular momentum between the constituent quarks. The intrinsic parity of quarks is not defined: this does not matter as in circumstances in which parity is conserved so also is quark flavour (see below) and any intrinsic parity a given quark has will appear in both before and after assessment of the parity and therefore factor out. The result is that we are free to define the parity of the nucleons as we wish. As there is no internal orbital angular momentum to contribute any odd parity, it is sensible to choose the nucleon parity to be even. Thus the nucleons have $j^P=\frac{1}{2}^+$. As a matter of interest we give in Table 10.2 the state vectors for spin and quark flavour for the proton and neutron.

What predictions could we make based on this very simple beginning? Since antiquarks appear to exist, any state of quarks and antiquarks can be turned into another state by changing quarks into antiquarks and vice versa. Note that we expect q and q̄ to have the same mass but opposite charge, so this process of **charge conjugation**, as it is called, changes hadron into antihadron and these two

Table 10.2 Quark flavour–spin state vectors for the nucleons in a state of $j_z = +\frac{1}{2}$. The arrow $\uparrow(\downarrow)$ means that the quark has its spin z-component $+\frac{1}{2}(-\frac{1}{2})$.

Proton, p:

$$\frac{1}{\sqrt{18}}|2u_\uparrow d_\downarrow u_\uparrow + 2u_\uparrow u_\uparrow d_\downarrow + 2d_\downarrow u_\uparrow u_\uparrow - u_\uparrow u_\downarrow d_\uparrow - u_\uparrow d_\uparrow u_\downarrow - u_\downarrow d_\uparrow u_\uparrow - d_\uparrow u_\downarrow u_\uparrow - d_\uparrow u_\uparrow u_\downarrow - u_\downarrow u_\uparrow d_\uparrow\rangle.$$

Neutron, n:

Change u → d and d → u in the proton state vector.

How to interpret this? The probability amplitude for finding in a proton the first quark a u with spin ↑, the second a d with spin ↑, the third a u with spin ↓ is the coefficient of the fifth term, namely $-1\sqrt{18}$, so that the probability of finding this configuration is 1/18.

must have equal mass but opposite charge. Under this transformation $\pi^+ \to \pi^-$ and $\pi^- \to \pi^+$. However, $\pi^0 \to \pi^0$ and it is therefore its own antiparticle, that is, it is self-conjugate. Applying charge conjugation to proton and neutron we obtain the antiproton ($\bar{\text{p}}$) and antineutron ($\bar{\text{n}}$): thus $\bar{\text{p}} = \bar{\text{u}}\bar{\text{u}}\bar{\text{d}}$ and $\bar{\text{n}} = \bar{\text{u}}\bar{\text{d}}\bar{\text{d}}$ define the valence constituents and from our argument about parity we see that these particles will have $j^{\text{P}} = \frac{1}{2}^-$. These particles cannot have a prolonged existence naturally in our environment, but have been produced and used for particle research at various high-energy accelerators. The proton and neutron belong to a class of particles called **baryons**: all three-quark states are baryons and of course three-antiquark states are called **antibaryons**. From now on, whenever we find a baryon we imply the existence of the corresponding antibaryon with the same mass and spin but opposite parity and charge (and other attributes).

The next thing we can propose is states obtained by changing the spin state of the valence quarks. If we change the $\text{q}\bar{\text{q}}$ state of ${}^1S_0$ in each of the three pions to a ${}^3S_1$ state we expect three $(+, 0, -)$ particles having $j^{\text{P}} = 1^-$. These particles are known as the ρ-mesons (ρ^+, ρ^0, ρ^-) and in Fig. 2.7 we have shown one of their manifestations. If we do the same to the qqq state of the nucleons we might expect to generate one excited state for each of the proton and neutron in which the quark spins add vectorially to make a state having total spin $\frac{3}{2}$. However, what is observed is four baryon states believed to have a quark contents uuu, uud, udd, and ddd. These states are represented by the symbols Δ^{++}, Δ^+, Δ^0, and Δ^- and they all have a mass about 1232 MeV/c^2. This mass is above the threshold (938 + 140 MeV) for $\Delta \to \text{N} + \pi$; there are no selection rules to hinder this transition, or its reverse, by the action of the strong interactions. The consequence is that the decay transition rate is about $10^{23}\,\text{s}^{-1}$ and the intrinsic line width is about 120 MeV. A second consequence is that these states will appear as compound states in pion–nucleon scattering, and, in fact, this was where they were first observed. The peak A in the total cross section for π^+p collisions shown in Fig. 10.2 is due to formation of the Δ^{++}:

$$\pi^+ + \text{p} \to \Delta^{++} \to \pi^+ + \text{p}. \tag{10.1}$$

The Δ^0 is seen in π^-p scattering (peak B in Fig. 10.2), the Δ^+ in π^+n and the Δ^- in π^-n. These states can also be produced (see Section7.6), for example:

$$\pi^+ + \text{p} \to \Delta^{++} + \pi^0 \to \pi^+ + \text{p} + \pi^0.$$

The reader may well ask why there should be two $j^P = \frac{1}{2}^+$ nucleon states (the ground state of the qqq system) but four of these $j = \frac{3}{2}^+$ excited states. The reasons become available when it is noted that the observed states are just those to be expected if the qqq wavefunctions are required to have certain symmetry properties; however, we cannot pursue that subject here. What we must do is note that going from a state with some quark spins antiparallel to one in which some of these configurations are changed to spins parallel increases the energy, that is, the mass (for example, $\pi^+(0^-) \to \rho^+(1^-)$ or $p(\frac{1}{2}^+) \to \Delta^+(\frac{3}{2}^+)$). This spin dependence is analogous to an atomic hyperfine interaction $\mathbf{S}_1 . \mathbf{S}_2$ involving electron and nuclear spins, but its effect on the total energy is proportionately much greater than in the atomic case.

The next freedom we have to change the state of $q\bar{q}$ or qqq systems is to add orbital angular momentum (l) between the quarks. For every one unit of l we expect another factor of -1 on the parity. In addition, we expect the energy to increase. Thus for the $q\bar{q}$ system we expect the states, in spectroscopic notation

$$^1S_0, {}^3S_1, {}^1P_1, {}^3P_0, {}^3P_1, {}^3P_2, {}^1D_2, \ldots$$

with j equal to

$$0, 1, 1, 0, 1, 2, 2, \ldots \quad \text{respectively,}$$

where we expect the mass to increase but not necessarily in this order. Note that the q and $\bar{q}$ move about their mutual centre of gravity and there can only be one orbital angular momentum; for qqq there is freedom to have two orbital angular momenta but no more. Therefore for the qqq system the way the orbital angular momentum can be included is more complicated than in the case of $q\bar{q}$. However, we do expect that there will be two bands of excited states, one of doublets based on the band head of neutron and proton, the second of quartets based on the Δ baryons, again with increasing spin and mass and alternating parity; many of these states are known. The proper classification and understanding of these states are based on the theory of angular momentum and of the permutation group (Feynman, 1972).

There is one other thing we can do to point to the existence of other new states. Look at the π^0 for which the flavour state vector for the quark flavours is $\frac{1}{\sqrt{2}}|d\bar{d} - u\bar{u}\rangle$. The state $\frac{1}{\sqrt{2}}|d\bar{d} + u\bar{u}\rangle$ is orthogonal to this and represents a possible $q\bar{q}$ state; the corresponding $^1S_0$, $j^P = 0^-$ hadron is believed to be the η-meson which has a mass of 549 MeV/c^2. Of course the $^3S_1$ partner to the $^1S_0$ π is the ρ-meson (770 MeV/c^2): so what is the $^3S_1$ partner to the $^1S_0$ η-meson? It is called the ω-meson and has a mass of 783 MeV/c^2. Table 10.3 summarizes the mesons that we have mentioned.

We now look at the kind of Feynman diagram we can draw for strong interaction processes; let us look at π^+p elastic scattering at the Δ^{++} compound state as in equation (10.1). Figure 10.6 shows how the quarks in the incident particles rearrange into the Δ^{++} compound state and then into the final state of two hadrons. It is usual to omit all the gluon exchange lines from such a diagram even though those exchanges are essential to the process; the reason is that such exchanges are occurring in the hadrons before, during, and after the collision, and it is impossible to pick out and draw any particular order of exchange as the dominant mechanism. Note that these mechanisms can lead to the annihilation or creation of $u\bar{u}$ or $d\bar{d}$ pairs, as in Fig. 10.6. This is physically what we expect and is incorporated in the rule of **quark flavour conservation**

Table 10.3 Mesons containing light quarks

Quark constituents	Quark space-spin states ${}^1S_0$, $j^P=0^-$	${}^3S_1$, $j^P=1^-$
$\lvert u\bar{d}\rangle$	$\pi^+(140)$	$\rho^+(770)$
$\frac{1}{\sqrt{2}}\lvert d\bar{d}-u\bar{u}\rangle$	$\pi^0(135)$	$\rho^0(770)$
$\lvert u\bar{d}\rangle$	$\pi^-(140)$	$\rho^-(770)$
$\frac{1}{\sqrt{2}}\lvert d\bar{d}+u\bar{u}\rangle$	$\eta(549)$	$\omega(783)$

Masses are given in parentheses to the nearest MeV/c^2. The flavour state function is over-simplified for the η and ω: there is some admixture of $s\bar{s}$.

The π mean lives are given in Table 9.6. The η has a full width of about 1.1 keV and has many decay modes, the dominant being $\gamma\gamma$, $3\pi^0$, or $\pi^+\pi^-\pi^0$.

The ρ mesons decay to two pions and each has a total width of 153 MeV. The ω decays to $\pi^0\gamma$ or $\pi^+\pi^-\pi^0$ and has a total width of 8.5 MeV.

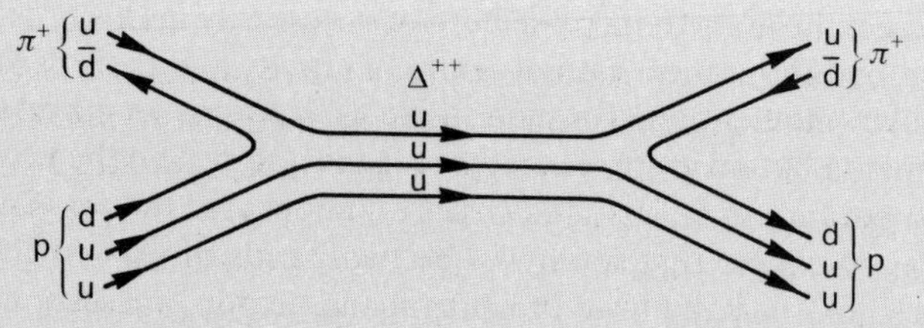

Fig. 10.6 The flow of valence quarks in the compound state scattering

$$\pi + p \rightarrow \Delta^{++} \rightarrow \pi^+ + p$$

or in quark terms

$$u\bar{d} + uud \rightarrow uuu \rightarrow u\bar{d} + uud.$$

The gluon exchange lines have been omitted.

which we can now formulate: In the strong interactions the number of quarks minus the number of aniquarks is a conserved quantity, separately for each flavour. Figure 10.6 obeys this rule. Of course, the creation of $q\bar{q}$ pairs which appear in produced hadrons in excess of those present initially requires energy.

We shall find (Section 12.4) that the weak interaction does not conserve quark flavour. However, it does obey the rule of **quark number conservation**: the number of quarks less the number of antiquarks is a conserved number. This last rule means that if any reaction starts with an excess of baryons over antibaryons, that excess will be conserved. Thus we can state the rule of **baryon number conservation**: the number of baryons less the number of antibaryons is a conserved quantity in all interactions. However, there are some theories which predict that the proton decays to mesons lighter than a proton and that process

DEFINITIONS AND KEYWORDS

Parton The generic name for the constituents of hadrons. These constituents are quarks, antiquarks, and gluons.

Valence quark The term for one of the constituent quarks or antiquarks that determine the external properties of a hadron.

Sea quarks In addition to the valence quarks every hadron contains a fluctuating population of $q\bar{q}$ pairs, mainly $u\bar{u}$ and $d\bar{d}$. These are the sea quarks. They are generated by quarks emitting gluons which in turn become $q\bar{q}$ pairs. The reverse steps lead to a loss of the pairs and an equilibrium is established which, on a very short time scale, is represented by a fluctuating number of pairs.

Flavour The term which describes the type of quark (up, down, etc.) or antiquark.

Charge conjugation The transformation that changes particle into antiparticle and vice versa. Mass, mean life and spin do not change. Additive quantum numbers such as charge, strangeness, charm, and so on, change sign. The parity of bosons does not change; fermion becomes antifermion and vice versa, and the parity changes.

Baryon The generic name for all the hadrons that are fermions and that have a mass equal to or greater than that of the proton.

Antibaryon The generic name for the hadrons that are the antiparticles to the baryons.

Quark flavour conservation Strong and electromagnetic interactions conserve flavour. That means that the number of quarks of a given flavour less the number of antiquarks of the same flavour is a conserved number in any process involving these interactions. The weak interaction does not conserve flavour.

Quark number conservation The number of quarks of all flavours less the number of antiquarks of all flavours is conserved in all interactions. This law will always hold unless proton decay is found.

Baryon number conservation The number of baryons less the number of antibaryons is conserved in all interactions. This follows from the nature of baryons and the conservation of quark number. This law will always hold unless proton decay is found.

must violate these two rules. If it occurs it is at a level that is totally insignificant in its effect on the strong interactions (see Sections 13.6 and 13.7).

The reader might now be wondering about combinations of quarks other than the $q\bar{q}$, qqq, and $\bar{q}\bar{q}\bar{q}$; why not qq, $qq\bar{q}$, or qqqq and so on? The answer to this lies with the existence of a new quantum number, colour. We will discuss this idea in Section 10.7. For the moment we note that if we restrict the number of valence quarks plus antiquarks to three or less and require the charge to be an integer, then we are restricted to the combinations we have used.

10.4 The quark–parton model: Stage II

But there is a third quark flavour: the quark is called **strange** with symbol s, and of course it has an antistrange partner, $\bar{s}$. The charge is $-\frac{1}{3}$ for the s and $+\frac{1}{3}$ for the $\bar{s}$.

Now let us see what we can build in the $q\bar{q}$ sector in a state of relative motion ${}^1S_0$, $j^P=0^-$. In addition to $u\bar{d}$, $\frac{1}{\sqrt{2}}|d\bar{d}-u\bar{u}\rangle$, $d\bar{d}$, $\frac{1}{\sqrt{2}}|d\bar{d}+u\bar{u}\rangle$ (π^+, π^0, π^-, η) we now have $u\bar{s}$, $d\bar{s}$, $\bar{u}s$, $\bar{d}s$, and $s\bar{s}$. These are called the K^+, K^0, K^-, $\bar{K}^0$, and η' mesons respectively. In the ${}^3S_1$, $j^P=1^-$ sector for the same valence quarks we already have ρ^+, ρ^0, ρ^-, ω to which we now add the K^{*+}, K^{*0}, K^{*-}, $\bar{K}^{*0}$, and φ (The * in superscript implies an excited state.) Masses are given Table 10.4 which also summarizes the quark content of mesons containing strange quarks. Readers will notice that we do not have a consistent notation: if the 1^- version of the 0^- K-meson is called the K* then the 1^-version of the 0^- π-meson should be called the π^*; in fact it is called the ρ! This and other anomalies are a consequence of a living subject where discovered particles were given names before the systematics were available to guide explorers to a more consistent notation. In fact, as far as notation is concerned, we shall try to keep in step with that employed in the *Review of Particle Properties* published every two years by the Particle Data Group (1988). Generally the long-lived particles which decay by weak or electromagnetic interactions have a simple symbol (π, K, η, Λ, Σ, Ξ,

Ω). Short-lived particles decaying by strong interactions normally have symbol followed by mass in parenthesis, for example Σ(1385) which is distinct from Σ. Thus strictly the φ, K* and ρ above should be $\varphi(1020)$. K*(892) and $\rho(770)$. We shall return to the distinction of short- and long-lived particles later in this section.

The name K-meson is frequently shortened to kaon just as π-meson is shortened to pion. The kaons have mean lives (see Table 10.4) sufficiently long to allow their production and use in controlled beams for experiments. In Fig. 10.7 we show the total cross-section for K^+ and K^- mesons incident on protons as a function of laboratory momentum. The K^+ cross-section is almost featureless apart from a rise to a plateau after the threshold for producing a single pion in a collision with a proton (707 MeV/c). The K^- cross-section shows evidence for compound states formation. As in the cases of π^+ and π^- and of p and $\bar{p}$, the kaon cross-sections are converging at high momenta and showing the beginnings of a slow rise with increasing momentum.

What about the 3q, $j^P = \frac{1}{2}^+$, sector? Starting with the n and p we can replace one u or d with an s. This gives four states:

$$|uus\rangle, \quad \tfrac{1}{\sqrt{2}}|(ud+du)s\rangle, \quad |dds\rangle, \quad \text{and} \quad \tfrac{1}{\sqrt{2}}|(ud-du)s\rangle,$$

which are known as

$$\Sigma^+, \quad \Sigma^0, \quad \Sigma^-, \quad \text{and} \quad \Lambda \qquad \text{respectively.}$$

If we replace two of u or d with 2s we get

$$|uss\rangle, \quad \text{and} \quad |dss\rangle,$$

known as

$$\Xi^0 \text{ and } \Xi^-.$$

Table 10.4 Mesons containing strange and light quarks.

Quark constituents	Quark space-spin states $^1S_0$, $j^P=0^-$	$^3S_1$, $j^P=1^-$	Strangeness
$\|u\bar{s}\rangle$	K^+ (494)	K^{*+} (892)	+1
$\|d\bar{s}\rangle$	K^0 (498)	K^{*0} (892)	+1
$\|\bar{u}s\rangle$	K^- (494)	K^{*-} (892)	−1
$\|\bar{d}s\rangle$	$\bar{K}^0$ (498)	$\bar{K}^{*0}$ (892)	−1
$\|s\bar{s}\rangle$	$\eta'(958)$	$\varphi(1020)$	0

Note that the η' is not pure $s\bar{s}$ but contains an admixture of lighter $q\bar{q}$.

The $K^\pm$ have a mean life of 1.24×10^{-8} s and many decay modes (see Problem 2.3). The K^0 and $\bar{K}^0$ do not have unique lifetimes because they mix under the effect of the weak interactions (see Section 13.3). The η' decays electromagnetically to $\eta\pi\pi$ or $\pi\gamma$ and has a width of 0.2 MeV.

The K* decay to Kπ and each has a full width of 51 MeV. The φ decays to $K\bar{K}$ or $\rho\pi$ and has a width of 4.4 MeV. (Where no charges are appended to particle symbols for decay modes all possibilities exist within the constraint of charge conservation. Only dominant decay modes given.)

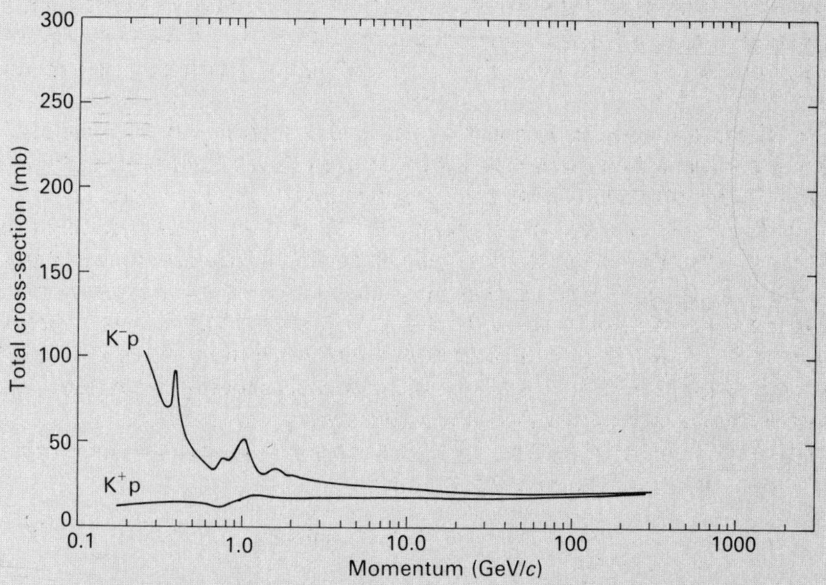

Fig. 10.7 The total cross-sections for positive and negative kaons on protons as a function of incident laboratory momentum. As with the charged pions, the two cross-sections converge and start a slow rise with increasing energy. At low momenta the K^-p cross-section shows peak characteristic of compound state formation. In contrast, the K^+p cross-section is featureless apart from a small rise above 700 MeV/c. To allow comparisons, the vertical scale is the same in this figure as it is in Figs. 10.1 and 10.2.

In the $j = \frac{1}{2}$ sector there is no state of three s-quarks, just as there is no 3u or 3d state; it is again the allowed symmetry properties of the spin-space wavefunction of the three quarks which explains this restriction. However, there is no such restriction in the 3q, $j^P = \frac{3}{2}^+$ sector. Building on the four Δ states (including uuu and ddd), there is a triplet of states uus, uds, and dds, known as $\Sigma^+(1385)$, $\Sigma^0(1385)$, and $\Sigma^-(1385)$. Then a doublet, uss and dss, known as the $\Xi^0(1530)$ and $\Xi^-(1530)$. Finally a singlet sss known as the Ω^-. All this is summarized in Table 10.5 (where we have put more precise values of the masses in the parentheses, if known).

Adding up the number of states we find there is a nonet (9) of $j^P = 0^-$ mesons, a second nonet of 1^- mesons, an octet (8) of $\frac{1}{2}^+$ baryons and a decuplet (10) of $\frac{3}{2}^+$ baryons (that is before considering any states containing internal orbital angular momentum). Historically, the ninth members (η' and φ) of the nonets were not associated with the other eight so that the occurrence of octets seemed to be a dominant property of the hadron spectrum. Feynman, Gell-Mann and others recognized this to be a pattern that could be explained in abstract group theory. Their version was called the **eightfold way** and allowed them to predict the existence of the Ω^- before it was observed. This story is told by Close *et al.* (1987) and by Cahn and Goldhaber (1989). Gell-Mann and, independently, Zweig proposed that it was all the consequence of the existence of what we now call quarks.

The masses of these baryons are also given in Table 10.5, and we can note some simple facts about them. Every time an s quark replaces a u or a d quark the mass increases by about 130 to 300 MeV/c^2. Also the spin dependence that

Table 10.5 Baryons containing strange and light quarks.

Quark constituents	Quark total spin states $j^P=\frac{1}{2}^+$	$j^P=\frac{3}{2}^+$	Strangeness
$\|uuu\rangle$	---	$\Delta^{++}(1230)$	0
$\|uud\rangle$	p(938)	$\Delta^{+}(1231)$	0
$\|udd\rangle$	n(940)	$\Delta^{0}(1232)$	0
$\|ddd\rangle$	---	$\Delta^{-}(1234)$	0
$\frac{1}{\sqrt{2}}\|(ud-du)s\rangle^{\dagger}$	$\Lambda(1116)$	---	−1
$\|uus\rangle$	$\Sigma^{+}(1189)$	$\Sigma^{+}(1383)$	−1
$\frac{1}{\sqrt{2}}\|(ud+du)s\rangle^{\dagger}$	$\Sigma^{0}(1192)$	$\Sigma^{0}(1384)$	−1
$\|dds\rangle$	$\Sigma^{-}(1197)$	$\Sigma^{-}(1387)$	−1
$\|uss\rangle$	$\Xi^{0}(1315)$	$\Xi^{0}(1532)$	−2
$\|dss\rangle$	$\Xi^{-}(1321)$	$\Xi^{-}(1535)$	−2
$\|sss\rangle$	---	$\Omega^{-}(1672)$	−3

†The quark contents are written in this way to distinguish the Λ from the Σ^0. This reminds us that strictly the spin-flavour state functions are different although both contain uds.

Note that within a given multiplet of the same spin (one of N, Δ, Σ, Ξ, see Section 10.11), which have members differing only in repeated $u \rightarrow d$, the masses increase by a few MeV/c^2 for each replacement of u by a d. This is due to a change of the contribution of the Coulomb interaction to the mass and to the fact that the d quark appears to be a few MeV/c^2 heavier than the u quark.

The boxes separate the long-lived particles which have weak interaction decays (n,Λ(1116), Σ(1189), ...) from the short-lived particles which have strong interaction decays (Δ, Σ(1383), Ξ(1532), ...). The Σ^0 decays electromagnetically to $\Lambda\gamma$ and the proton is stable.

we have already noted in the last section applies to the strange quarks; changing from $j=\frac{1}{2}^+$ to $j=\frac{3}{2}^+$ for the same valence quarks increases the mass. Note also that replacing a u by a d quark increases the mass by a few MeV/c^2: the actual mass change is due to a combination of the greater mass of the d quark relative to the u quark and a change in the contribution of the Coulomb interaction to the mass (analogous to the Coulomb term in the semi-empirical mass formula for nuclei, Section 4.4).

The baryon states of Table 10.5 become the band heads for a series of states of increasing angular momentum and mass as increasing steps of orbital angular momentum are added to the relative internal motion of the quark constituents. Those baryons with two light quarks and one s quark have strangeness −1 and baryon number +1; the K^-p system has the same quantum numbers and therefore these baryons which have mass greater than the K^-p channel mass ($m_K + M_p = 1432$ MeV/c^2) may appear as compound states in this system. This is the origin of the peaks in the K^-p total cross-section shown in Fig. 10.7. For example, the peak at 395 MeV/c is due to the reaction

$$K^- + p \rightarrow \Lambda(1520) \rightarrow \text{several channels}$$

Fig. 10.8 shows a liquid hydrogen filled bubble-chamber photograph chamber of an event in which a 1 GeV π^--meson strikes a stationary proton

The attached sketch shows the interpretation in terms of particles we have introduced. Note that it is

$$\pi^- + p \rightarrow K^0 + \Lambda,$$

or in quark terms

$$\bar{u}d + uud \rightarrow d\bar{s} + uds.$$

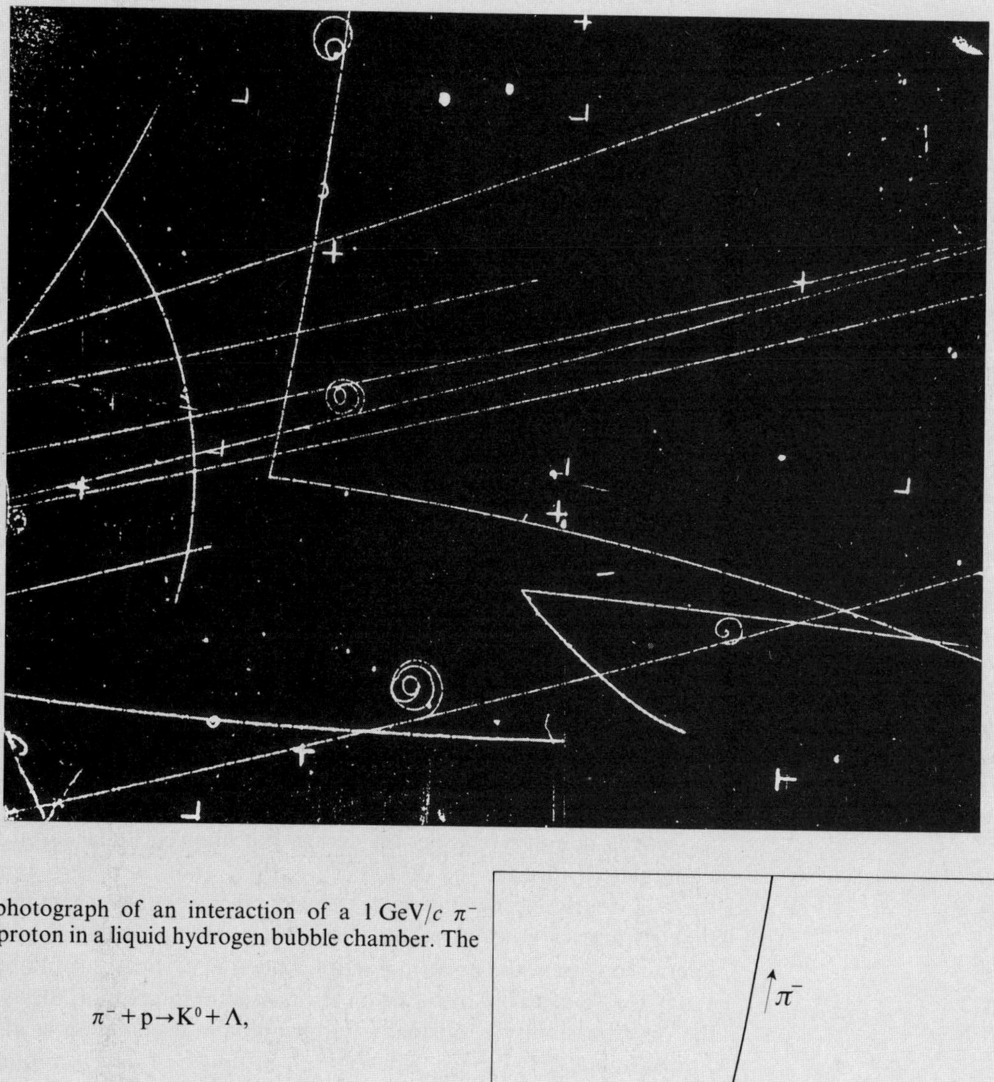

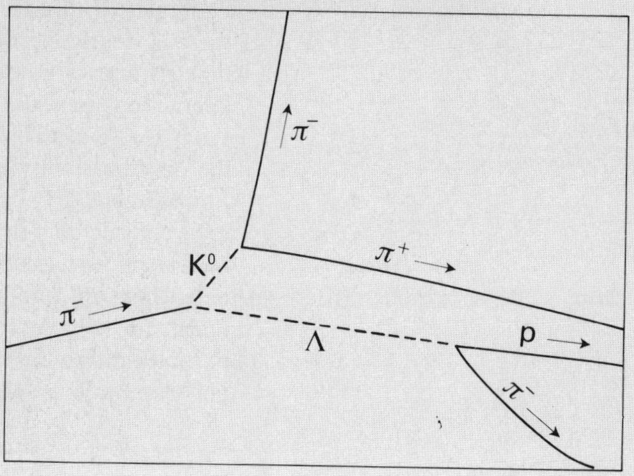

Fig. 10.8 A photograph of an interaction of a 1 GeV/c π^- meson with a proton in a liquid hydrogen bubble chamber. The reaction is

$$\pi^- + p \rightarrow K^0 + \Lambda,$$

followed by

$$K^0 \rightarrow \pi^+ + \pi^-,$$

and

$$\Lambda \rightarrow \pi^- + p.$$

This is an example of associated production of two strange particles.

DEFINITIONS AND KEYWORDS

Strange A quark flavour.

Eightfold way The importance of octets of particles was highlighted by this name given to a version of an abstract group theory applied to the classification of hadrons.

Strangeness The quantum number now associated with the presence of one or more strange quarks in a hadron. The strangeness of a hadron equals the number of $\bar{s}$ quarks less the number of s quarks that it contains. The total strangeness is conserved in strong and electromagnetic interactions.

Associated production In the collision of two hadrons containing no heavy (including strange) valence quarks, the production of hadrons containing heavy quarks, if proceeding by the strong interactions, must conserve flavour. The simplest final state may then consist of one hadron containing a heavy quark plus a second hadron containing the heavy antiquark. This is called associated production.

In this reaction a $u\bar{u}$ pair annihilates and, using some of the available kinetic energy, an $s\bar{s}$ pair is created and these two quarks appear in different hadrons. Before the recognition of the existence of quarks, particles which we now see as containing strange quarks were given a quantum number called **strangeness**: one s-quark contributed strangeness $S = -1$, an $\bar{s}$-quark contributed $S = +1$. This quantum number was additive and was conserved in strong interactions. In Fig. 10.7 the collision of two particles of zero strangeness leads to two hadrons of opposite strangeness, a process called **associated production**. We see now that this conservation law is a consequence of the strong interactions conserving quark flavour. But how do we know that it is the strong interactions causing this reaction? Observations showed that the cross-section for associated production of strange particles, although not a major part of the total, was sufficiently large (~ 1 mb) to be attributable solely to the effect of the strong interactions.

What of the fate of these strange particles?

$$\Lambda \to \pi^- + p,$$

$$K^0 \to \pi^+ + \pi^-.$$

These processees do not conserve quark flavour. There are no s or $\bar{s}$ in any of the final hadrons although both initial hadrons are strange. In addition these particles moved several centimetres before decaying and this corresponds to mean lives of approximately 10^{-10} s. That time is far too long for the decay mechanism to be attributed to the strong interactions for which decay mean lives are $\leqslant 10^{-20}$ s (see $\Delta^{++} \to \pi^+ + p$ in Section 10.2). So we have a flavour changing, not strong interaction; in fact in Section 9.13 we hinted that the weak interaction could change flavour and this is what is happening. We shall return to this subject in Chapter 12 but here do note some general principles. Excited states of hadrons will decay into lower mass combinations of hadrons under the action of the strong interactions with transition rates $10^{21}\,s^{-1}$ or greater if all quantum numbers are conserved, in particular quark flavour. However, once the lowest mass hadronic state has been reached under that restriction then further downward steps under the influence of the strong interactions are impossible. The electromagnetic interaction is weaker than the strong by a factor of approximately 10^{-2} but is not so restrictive and so may cause the next downward step: there will be no changes in quark flavour but changes occur in the wavefunction which permit decays not allowed under strong interactions. Examples of this are the decays $\Sigma^0 \to \Lambda + \gamma$, and $\eta \to \pi^+ + \pi^- + \pi^0$: the first of these is clearly electromagnetic, the second is also electromagnetic but to see that requires a careful analysis which we cannot undertake here. If strong interactions were the only interaction, then the Σ^0 and η would be stable. Once states stable against strong or electromagnetic interaction decays are reached the only possibility is normally decay by flavour-changing weak interactions. For example: $\Sigma^+ \to \pi^0 + p$, $K^0 \to \pi^0 + \pi^0$, $n \to p + e^- + \bar{\nu}_e$. These possibilities have to be considered if this next step of the decay is to be predicted or explained. The final result with established mechanisms is that all free hadrons decay to proton or antiproton, photons, electrons, positrons, and neutrinos. A typical decay scheme is shown in Fig. 10.9. In the presence of normal matter the final charged particles anti to this material have their individual death in annihilation. Thus a positron meets an electron and

$$e^+ + e^- \to \gamma + \gamma.$$

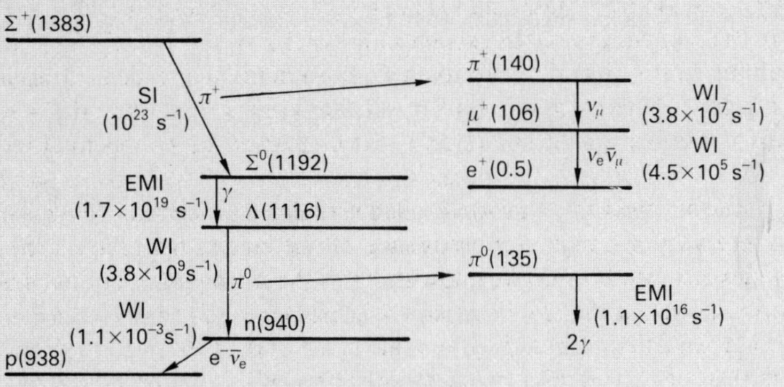

Fig. 10.9 The decay scheme which leads from $\Sigma^+(1383)$ to $pe^+e^-\nu_e\bar{\nu}_e\nu_\mu\bar{\nu}_\mu\bar{\nu}_\mu\gamma\gamma\gamma$! The interactions responsible for each step are given by the abbreviations SI, EMI, WI for strong, electromagnetic and weak interactions respectively. The transition rates are given in parentheses.

An antiproton meets a nucleus and annihilates with a nucleon, usually into several pions, each of which has its ultimate decay to photons, electrons, positrons and neutrinos.

The distinction between long-lived and short-lived states is that the latter cannot be seen directly in detectors such as a bubble chamber by leaving a visible track or a neutral or charged V (Fig. 10.4), whereas the long-lived can be so detected. This distinction separates, almost perfectly, particles decaying by the weak interaction from these decaying by the strong or by the electromagnetic interactions. However, the Particle Data Group (1988) includes the electromagnetic decays in the long-lived group and lists all of these under the heading 'stable'.

PROBLEMS

10.1 Use the information in Fig. 10.7 and the density of liquid hydrogen (71 kg m^{-3}) to calculate the mean free path for 5 GeV/c K^+ mesons in liquid hydrogen.

10.2 Look up the necessary masses and calculate the Q-values for each of the following reactions:

$$\pi^- + p \rightarrow K^0 + \Lambda,$$
$$K^- + p \rightarrow \pi^0 + \Sigma^0,$$
$$K^- + p \rightarrow K^0 + \Xi^0,$$
$$p + p \rightarrow K^+ + \Sigma^+ + n,$$
$$\Xi^- + p \rightarrow \Lambda + \Lambda,$$
$$\pi^- + p \rightarrow K^0 + \overline{K}^0 + n,$$
$$K^- + p \rightarrow K^+ + K^0 + \Omega^-.$$

Write down the valence quark content for each of the different particles and check that the conservation laws of electric charge, flavour, strangeness and baryon number are satisfied throughout. Draw a quark flow diagram for the last reaction.

10.3 Two of the following reactions cannot occur under any circumstances and a third cannot occur by strong interactions. Find these and indicate the reasons in each case:

(a) $K^- + p \to \bar{K}^0 + n$,
(b) $\pi^+ + p \to + K^+ + \Sigma^+$,
(c) $\pi^- + p \to K^+ + \Sigma^0 + \pi^-$,
(d) $\pi^- + p \to K^- + \Sigma^+$,
(e) $\bar{K}^0 + p \to K^- + p + \pi^+$,
(f) $\bar{p} + p \to \pi^+ + \pi^+ + \pi^- + \pi^- + \pi^+$,
(g) $\pi^+ + p \to K^0 + \Sigma^0 + \pi^+ + K^+ + \bar{K}^0$,
(h) $K^- + p \to \Sigma^+ + n + \pi^-$,
(i) $\pi^- + p \to \Sigma^+ + \Sigma^- + K^0 + \bar{p} + \bar{\Sigma}^+ + n$,
(j) $\pi^- + p \to \bar{\Sigma}^- + \Sigma^0 + p$.

The notation $\bar{\Sigma}$ means the antiparticle to the Σ and the sign convention is that $\bar{\Sigma}^+$ has charge +1 and is therefore the antiparticle to the Σ^-.

10.4 The terms *hadron, lepton, boson, fermion* are used in the classification of particles. Explain their meaning, giving examples of their use. Which can be applied to *quarks*?

The lowest mass (non-strange) baryon states are:

	Electric charge −1	0	+1	+2
nucleons:		n	p	
Δ-baryons:	Δ^-	Δ^0	Δ^+	Δ^{++}

(a) What are the quark constituents of each of these states?

(b) Assuming the quarks are in states of zero relative angular momentum, what fundamental difficulty appears to be associated with the Δ states, which have $j = \frac{3}{2}$, and how is it resolved?

(c) How do you explain the occurrence of excited states of the nucleon and Δ with higher values of j?

(d) What approximate sequence in the parity of these higher states would your simple model predict?

The Δ^0 and the Λ (strangeness −1) both decay to proton and π^--meson. Explain why the Δ^0 mean-life is $\sim 10^{-23}$ s while that of the Λ is 2.6×10^{-10} s.

(Adapted from 1984 examination of the Final Honours School of Natural Science, Physics, University of Oxford.)

10.5 Describe the experimental evidence which supports the quark model of hadron structure.

In a simple quark model the lightest hadrons are considered as bound states of u, d and s quarks or antiquarks. Give the composition of the mesons π^+, π^0, K^+, K^- ($S=+1$) and $\bar{K}^0$ ($S=1$) and the baryons n, p and Λ according to this model.

State, with reasons, which of the following reactions can proceed via the strong interaction:

$$K^- + p \to \bar{K}^0 + n$$
$$K^0 + n \to \Lambda + \pi^0$$
$$K^- + p \to \Lambda + \pi^0$$
$$K^0 + p \to K^+ + n.$$

In the additive quark model the total interaction cross-sections of high-energy hadrons are assumed to be due to the sums of the interaction cross-sections of the constituent quarks. Taking account of the differing cross-sections for different quark pairs (qq) and assuming that $\sigma(qq) = \sigma(q\bar{q})$, use this model to prove the relation

$$\sigma(\Lambda p) = \sigma(pp) + \sigma(\bar{K}^0 p) - \sigma(\pi^+ p).$$

At a laboratory beam momentum of 100 GeV/c the total interaction cross-sections are slowly varying functions of momentum. Given that

$$\sigma(\bar{K}^0 p) = 20 \times 10^{-3}\ \text{b}$$
$$\sigma(\pi^+ p) = 24 \times 10^{-3}\ \text{b}$$
$$\sigma(pp) = 39 \times 10^{-3}\ \text{b}$$

make an estimate of the interaction cross section of the Ξ^- hyperon (quark composition dss) with protons.

(Adapted from 1986 examination of the Final Honours School of Natural Science, Physics, University of Oxford.)

10.6 Outline the quark model description of the π meson, which has $j^P=0^-$. Give the quark model description and strangeness of other mesons with $j^P=0^-$ which can be constructed from u, d and s quarks. How do mesons with $j^P=1^-$ arise in this model? What j^P would you expect the first excited states of the 0^- and 1^- mesons to have?

The ρ meson has strangeness $S=0$, $j^P=1^-$ and a mass of 770 MeV. What would you expect to be the principal decay mode of the ρ^+? Why can the ρ^0 meson not decay to $\pi^0\pi^0$? What would you expect to be the principal decay model and approximate lifetime of the ρ^0? Describe briefly how the ρ^0 might be produced and indicate how its lifetime might be measured or inferred.

(Adapted from 1985 examination of the Final Honours School of Natural Science, Physics, University of Oxford.)

10.7 State what is meant by (*a*) charge symmetry (*b*) charge independence of nuclear forces. Illustrate charge symmetry by referring to the energy levels of light nuclei.

Explain qualitatively the following phenomena:

(i) At a certain momentum the cross-sections for π^+-neutron $\sigma(\pi^+ n)$ and π^--proton $\sigma(\pi^- p)$ interactions are equal whereas $\sigma(K^+ n) = 22$ mb and $\sigma(K^- p) = 55$ mb.

(ii) The ρ^0 and K^0 mesons both decay predominantly to $(\pi^+ + \pi^-)$. The mean life of the ρ^0 is 10^{-23} whereas that of the K^0 is 0.89×10^{-10} s.

(iii) The mean life of the π^+-meson is 2.6×10^{-8} s, whereas that of the π^0-meson is 0.8×10^{-16} s.

A beam consisting of π^+ and μ^+ mesons strikes an iron absorber. Given that the cross section for π^+ interactions with iron is 500 mb/nucleus calculate the thickness of iron necessary to attenuate the π^+ beam by a factor of 1000. Explain why the muons would be attenuated by a different factor.

[The density of iron is 7900 kg m^{-3}. The relative atomic mass of iron is 55.85.]

(Adapted from 1987 examination of the Final Honours School of Natural Science, Physics, University of Oxford.)

10.5 The quark–parton model: Stage III. Heavy flavours

The discovery of the strange particles was made during investigations into the effects of cosmic radiation but the elucidation of their properties occurred during the decade following 1954 when accelerators of sufficient energy became available to permit their production under controlled conditions. The quark model had been proposed and developed during the 1960s but it was not generally accepted until 1974. The three-flavour quark model presented certain theoretical difficulties which led to the prediction of the existence of a fourth flavour, given the name **charm** (symbol c). It was first observed experimentally in 1974. Theorists were soon discussing the possibility of two more flavours. The

DEFINITIONS AND KEYWORDS

Charm A heavy quark flavour. The charm quark has a charge $+\frac{2}{3}$ and a mass which is about 1.65 GeV/c^2.

Beauty A heavy quark flavour. Sometimes called bottom. The beauty quark has charge $-\frac{1}{3}$ and a mass which is about 4.5 GeV/c^2.

Truth A heavy quark flavour. Sometimes called top. The top quark is expected to have charge $+\frac{2}{3}$ and a mass greater than 70 GeV/c^2.

fifth, called **beauty** (symbol b, sometimes called bottom), was discovered in 1978. This makes a sixth flavour almost a certainty but at the time of writing experimental evidence for its existence is lacking, although the name **truth** (or top, symbol t) is ready waiting. The properties of these quarks are included in Table 10.6. Note that we have given a mass in each case; no free quarks have been observed so these assignments are based on an analysis of the masses of hadrons containing these quarks, using models of the forces between them based upon the nature of the gluon field. The large mass of the flavours above strange make it difficult to create hadrons containing these quarks. Of course starting with u and d, it is necessary to create $c\bar{c}$, $b\bar{b}$, or $t\bar{t}$ pairs in order to produce hadrons with these particular flavours. This is very expensive in energy and even in the most energetic collisions it is very unlikely; in addition, such states which decay into several ordinary mesons or baryons are hard to find in the large multiplicity of pions produced in high-energy hadron–hadron collisions. A method of producing these heavy flavours is described in Section 10.6.

Laying aside the t-quark for which there is no compelling experimental evidence, we can see that there are new branches of hadron spectroscopy based upon adding c or b quarks to the ingredients already available. The possibilities are numerous and perhaps not critically interesting unless particular states can tell us more about the nature of quarks and the forces between them and if such states can be sufficiently readily produced and studied. For every branch we expect to find excited states based upon ground states stable against all but the weak interaction and thus long-lived. For example, the combinations

$$c\bar{d},\ \bar{c}d,\ c\bar{u},\ \bar{c}u$$

in the $^1S_0$ state have been observed, studied, and named

$$D^+,\ D^-,\ D^0,\ \bar{D}^0.$$

The charged D have a mass of 1869 MeV/c^2, the uncharged 1865 MeV/c^2; they decay by the weak interaction and have mean lives of approximately 10^{-12} s and

Table 10.6 The quarks.

Quark name	Symbol	Charge	Mass MeV/c^{2}*	Flavour quantum number	
Up	u	$+\frac{2}{3}$	330	Isotopic spin $t=\frac{1}{2}$	$t_3=+\frac{1}{2}$
Down	d	$-\frac{1}{3}$	333		$t_3=-\frac{1}{2}$
Strange	s	$-\frac{1}{3}$	486	strangeness	$=-1$
Charm	c	$+\frac{2}{3}$	1 650	charm	$=+1$
Beauty	b	$-\frac{1}{3}$	4 500	beauty	$=-1$
Truth†	t	$+\frac{2}{3}$	>70 000?	truth	$=+1$

*Masses are derived from an analysis of hadron masses and are not directly observable.
† Predicted but awaiting experimental confirmation.

All quarks other than the u and d have isotopic spin 0. See Section 10.11 about the significance of isotopic spin.

4×10^{-13} s respectively. Although short even on the weak interaction scale, these times are that way because of the large energy released and hence the large phase space available to the decay products. The interesting thing about the change of flavour involved in their decay is that

$$c \to s \qquad (\text{or } \bar{c} \to \bar{s})$$

is favoured by about 20 to 1 over

$$c \to d \quad (\text{or } \bar{c} \to \bar{d}).$$

Thus we find an abundance of kaons in the decays of these charmed mesons and

$$D^+ \to K^- + \pi^+ + \pi^+$$

is about twenty times more probable than

$$D^+ \to \pi^- + \pi^+ + \pi^+.$$

The consistent picture which includes this result is described in Section 12.4. Another example of long-lived states is given by the $c\bar{s}$ (D_s^+) and $\bar{c}s$ (D_s^-) which have $j^P = 0^-$. The interesting thing here is that the first excited state ($^3S_1, j^P = 1^-$, $D_s(2113)$ is only 142 MeV/c^2 above the D_s. Decay by emission of a charged pion to the ground state is impossible and decay by emission of a neutral pion energetically very unfavoured, so that the decay is by γ-ray emission. (In fact, even if pion emission were energetically allowed, this is a forbidden transition for the strong interactions and the emission of a pion would be by the effect of the electromagnetic interaction; a virtual photon would be exchanged internally among the quarks allowing a change in quantum number not otherwise allowed. See Problem 10.12.)

An area which is now of interest to us is that provided by the $s\bar{s}$, $c\bar{c}$, and $b\bar{b}$ systems. The neutral mesons we met in Sections 10.3 and 10.4 already included the $u\bar{u}$, $d\bar{d}$, $s\bar{s}$ systems and we noted that they can mix quantum mechanically. Thus the η contains some $s\bar{s}$ in addition to $u\bar{u}$ and $d\bar{d}$. However, the degree of mixing is not what would be expected if u, d, and s all had equal masses (which they do not have) and equal interaction with gluons (which they do have). The φ (Table 10.4) should have the flavour state vector

$$\frac{1}{\sqrt{3}}|u\bar{u} + d\bar{d} + s\bar{s}\rangle$$

but in fact it is very close to

$$|s\bar{s}\rangle.$$

This absence of mixing becomes increasingly evident as we go to $c\bar{c}$ and then $b\bar{b}$.

In the next section we shall show how some of these states may be produced and discuss the $c\bar{c}$ system more fully in the section after that.

10.6 Producing heavy flavours

Several accelerators now exist capable of producing electron and positron beams of energies up to 50 GeV (and higher in the future) and of storing these beams in a circular beam chamber (storage ring) where bunches of each circulate in a counter rotating sense. Thus there are at least two points on the circumference where a bunch of e^+ repeatedly meets head-on a bunch of e^-. (There is one other scheme which does not involve a storage ring but arranges e^+e^- bunch crossovers in another way.) Collisions can occur at every crossover: what can happen?

1. Elastic e^+e^- scattering (Bhabbha scattering).
2. Inelastic e^+e^- scattering in which two virtual photons, one from each of the electric fields of the colliding particles, interact and produce hadrons (so-called two-photon physics).
3. Annihilation of the e^+e^- into two or three real photons.
4. Annihilation of the e^+e^- into a virtual photon followed by the materialization of this photon into a charged lepton plus antilepton pair or quark plus antiquark pair. The latter leads to hadronic final states.

Each of these is complicated by the possibility of either e^+ or e^-, or both, radiating one or even more real photons during the process (extra in the case of 3): this is an unavoidable radiative effect and one which experimentalists using electrons and positrons have to learn to live with and to correct for (see Section 9.6). We shall not discuss processes 1, 2, 3, or these radiative effects, but concentrate on 4. Now look back at Fig. 9.7. This is the Feynman diagram for the particular process of high-energy e^+e^- annihilation to a $\mu^+\mu^-$ pair. The first step is the production of a virtual photon; if the e^+ and e^- have equal energy E but opposite momentum (as in a storage ring) then this photon has energy $2E$ and zero momentum in the laboratory. This photon is virtual and time-like. It must therefore materialize and it can do that into any particle–antiparticle pair which is charged and for which the particle mass, m, satisfies $2mc^2 < 2E$. In Fig. 10.10 we analyse the contribution to the amplitude for high-energy e^+e^- annihilation to a fermion–antifermion pair. From this we show how to calculate, apart from a common factor, the total cross-section for any such process. To make the information correct a factor has been included which will be explained shortly; it is the colour factor, C, in Fig. 10.10.

Note that the photon creates particle pairs at a rate depending on their charge squared, however exotic these particles are relative to the more familiar ones. The only caveat is that there must be sufficient energy, which in the case of $q\bar{q}$ pairs means enough to produce the hadrons in which the quarks finally reside. Thus, through $c\bar{c}$ production, to obtain a c bound into a D^+ and a $\bar{c}$ in a D^- requires 2×1869 MeV. The high-energy annihilation of e^+e^- has the property that it puts all the collision energy into one virtual particle that, in turn, can put all that energy into one $q\bar{q}$ pair. At sufficiently high energy, it is only the charge squared which determines the relative rates with which each flavour is produced. In principle gluons couple equally to all flavours so that gluon–gluon collisions at high energy could produce heavy $q\bar{q}$ pairs democratically; however, there are no free gluons and such collisions can only be part of the general mess of parton–parton interactions occurring in hadron–hadron collisions. Under those circumstances there is very little chance of having more than a small fraction of the laboratory collision energy concentrated on a gluon–gluon

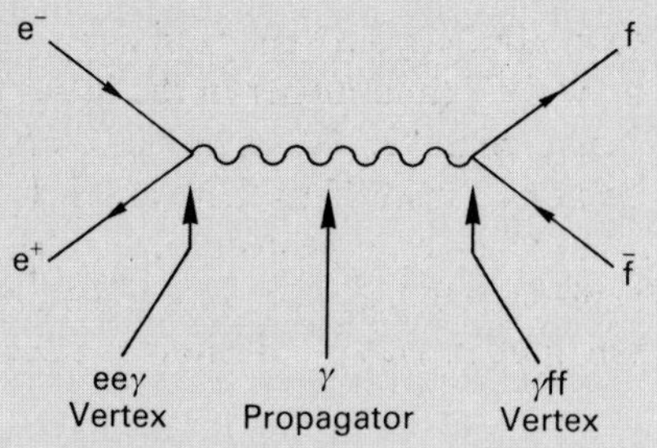

Fig. 10.10 This figure shows the Feynman diagram for high energy electron–positron annihilation to fermion plus antifermion

$$e^+ + e^- \rightarrow f + \bar{f}.$$

For a given choice of $f\bar{f}$, the amplitude for the transition is found by multiplying three factors; they are:

eeγ vertex	γ propagator	γff vertex
$-\sqrt{\alpha}$	$1/W$	$Q_f\sqrt{\alpha}$

where α is the fine structure constant, Q_f is the charge on f in units of the positron charge, and W is the total centre-of-mass energy. Then the total amplitude is proportional to their product $-\dfrac{Q_f\alpha}{W}$.

Then the total cross section, σ, is proportional to $\dfrac{Q^2\alpha^2}{W^2}$.

Missing factors: $\dfrac{4\pi}{3}(\hbar c)^2 C$ (*C* is the colour factor)

Then $$\sigma(e^+e^- \rightarrow \mu^+\mu^-) = \frac{4\pi}{3}(\hbar c)^2\frac{\alpha^2}{W^2}, \qquad (Q_f = -1, C = 1)$$

and $$\sigma(e^+e^- \rightarrow u\bar{u}) = \frac{16\pi}{9}(\hbar c)^2\frac{\alpha^2}{W^2}, \qquad (Q_f = +2/3. C = 3)$$

and $$\sigma(e^+e^- \rightarrow \text{all } q\bar{q}) = 4\pi(\hbar c)^2\frac{\alpha^2}{W^2}\sum_q Q_q^2. \qquad (C = 3)$$

If $f\bar{f}$ is a $q\bar{q}$ pair, then it is assumed that there is a probability of 1 for becoming hadrons after the production of the $q\bar{q}$ pair: then

$$R = \frac{\sigma(e^+e^- \rightarrow \text{hadrons})}{\sigma(e^+e^- \rightarrow \mu^+\mu^-)} = 3\sum_q Q_q^2.$$

Note that the cross-sections are scale invariant and that these formulae apply only if $W \gg 2m_fc^2$.

collision. Therefore high-energy e^+e^- annihilation at present wins easily as a method of producing heavy flavours.

We have to note that, except in specific circumstances, e^+e^- annihilation leads to the production of several hadrons, not just one or two containing the q and $\bar{q}$ generated by the photon. Thus well above threshold ($2E \gg 2 \times 1869$ MeV) charmed hadrons can be produced but usually along with other hadrons, most probably pions. The mechanism by which this happens is not clearly understood but in Section 10.9 we shall describe a simple model which has some element of truth in it. At this moment we make the following comment: after the step $e^+e^- \to q\bar{q}$ the production of hadrons occurs with unit probability so that the total cross-section for $e^+e^- \to$hadrons is the same as the total cross-section for $e^+e^- \to \Sigma q\bar{q}$ where the sum is over all flavours for which the energy is above threshold. With this in mind let us look at that part of the total annihilation cross-section which leads to hadrons. From the information given in Fig. 10.10 we expect all partial annihilation cross-sections to decrease as $1/E^2$ so it is convenient to remove this rapid variation by displaying and using not that cross-section but the ratio

$$R = \frac{\text{cross-section for } e^+e^- \to \text{hadrons}}{\text{cross-section for } e^+e^- \to \mu^+\mu^-}.$$

The muons ($\mu^+\mu^-$) are straightforward to detect in the same apparatus as that detecting hadrons (How is all this done? See Problem 10.8.) and determining the ratio avoids some systematic errors which would plague the measurement of the two cross-sections in separate experiments. What do we expect? From Fig. 10.10 we see that if the denominator is 1 (corresponding to unit charge squared for the muon) then the numerator is $C \times \Sigma\, Q_q^2$, where Q_q is a quark change and the sum is as above. Thus below $2E = 2m_\pi c^2$, where m_π is the mass of the lightest hadron

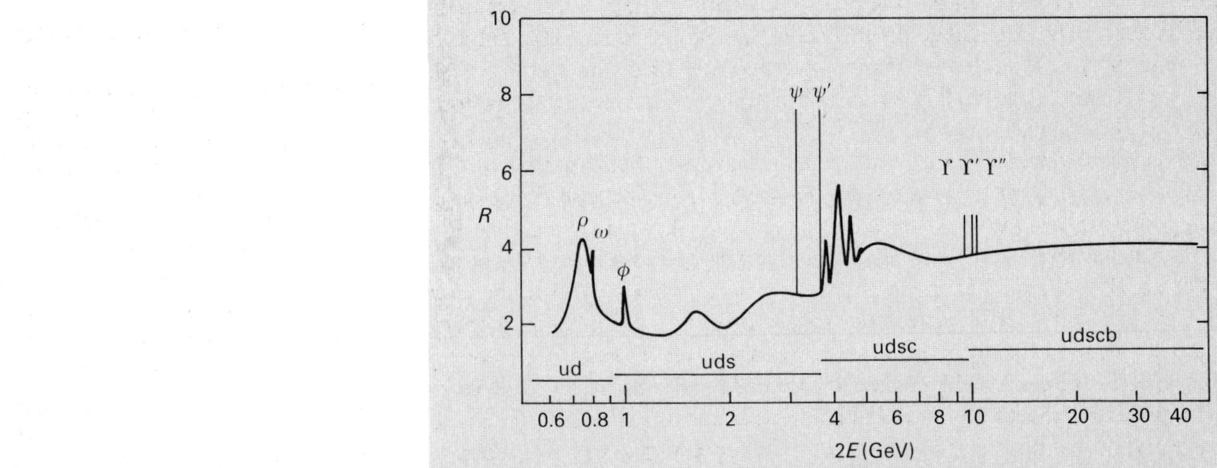

Fig. 10.11 The ratio

$$R = \frac{\text{cross-section for } e^+e^- \to \text{hadrons}}{\text{cross-section for } e^+e^- \to \mu^+\mu^-}$$

as a function of the total centre-of-mass energy $W = 2E$ (each beam has energy E). The curve is drawn through many measured points and represents only a rough average; however, it does reproduce the major features. The peaks are labelled by the conventional symbol representing that vector meson. The narrow peaks due to the ψ and Υ families are shown as vertical lines; the actual values of the peak cross-section are not represented. The above-$B\bar{B}$ threshold Υ states are omitted. The horizontal lines marked with quark flavours are the values expected away from resonances and in the absence of colour.

the π^0, no hadrons can be produced. As the energy increases from that point up to the threshold for K^+K^- production (the charged K mesons are the lightest strange hadrons) R should approach $C(\frac{4}{9}+\frac{1}{9})$, that is, C times the sum of squares of the charges of the u and d quarks. From that second threshold up, R should be $C(\frac{4}{9}+\frac{1}{9}+\frac{1}{9})=2C/3$ until that for $D^0\bar{D}^0$ production when it should become $C(\frac{4}{9}+\frac{1}{9}+\frac{1}{9}+\frac{4}{9})=\frac{10}{9}C$.

Fig. 10.11 shows the measured values of R. Two facts are immediately obvious:

1 There are resonances and other features changing rapidly with energy.

2 At high energies and away from resonances R is such that $C=3$. The next two sections deal with these two observations and their significance (but in the order 2,1).

10.7 The value of R and colour

That the value of C comes out at about 3 is one of the more weighty pieces of evidence for the quantum number of **colour** which we first mentioned in Section 9.11. Every flavour of quark has one of three possible values of this quantum number; these values are Red(R), Green(G), Blue(B). The anti-quarks are given one of the opposite colours which in conventional colour terminology are the complementary colours Cyan($\bar{R}$), Magenta($\bar{G}$), and Yellow($\bar{B}$). A meson containing a $R\bar{R}$ $q\bar{q}$ will be white, or colourless, just as the addition of red and cyan gives white light. Other ways of obtaining colourless systems are by the addition of green and magenta ($G\bar{G}$), or of blue and yellow ($B\bar{B}$), or of red and green and blue (RGB), or of cyan and magenta and yellow ($\bar{R}\bar{G}\bar{B}$). The factor of three in the e^+e^- annihilation to hadrons is now explained; there are three kinds of $q\bar{q}$ for each given flavour, namely $R\bar{R}$, $G\bar{G}$, or $B\bar{B}$, and so the cross-section is three times that expected without colour.

This may appear to be a complicated and arbitrary way of explaining a factor of three. However let us examine the consequence if we add another ingredient: it is the **hypothesis of colourless hadrons**, that is that all hadrons are colourless in the sense described above. The technical term for a colourless state is that it is a **colour singlet**. (Each quark is a member of a **colour triplet**). The immediate consequence is that the only observable combinations of quarks and antiquarks are those which are colour singlets and from above we see that those are $q\bar{q}$ (mesons), qqq (baryons) and $\bar{q}\bar{q}\bar{q}$ (antibaryons). Thus the problem exposed at the end of Section 10.3 as to why these are the only combinations of quarks to be observed has become one of colour and why observable states are colour singlets; more on this later.

Let us briefly look at the colour part of the state vectors for hadrons. The mesons have the colour state vector

$$\frac{1}{\sqrt{3}}|R\bar{R}+G\bar{G}+B\bar{B}\rangle. \tag{10.2}$$

So there are equal amplitudes for finding the q and $\bar{q}$ as $R\bar{R}$, or $G\bar{G}$, or $B\bar{B}$. There can be no cross terms such as $R\bar{B}$ or $G\bar{R}$ as they add to give a colour so cannot be parts of a colour singlet state function. What about the baryons? The colour singlet state vector for three coloured quarks is

$$\frac{1}{\sqrt{6}}|RGB-GRB+BRG-RBG+GBR-BGR\rangle.$$

DEFINITIONS AND KEYWORDS

Colour The name given to a quantum number associated with quarks and gluons. It is trichotomic with values conveniently called red, green and blue(but having nothing to do with optical colour). The quantum number can also take on its anti-(complementary) colour values cyan, magenta and yellow.

The hypothesis of colourless hadrons This hypothesis states that all observable hadrons are colour singlets.

Colour singlet This is the term that describes a state which is a combination of coloured particles such that the total colour effect is zero. Such a state is sometimes said to be colourless.

Colour triplet A system which must have one and only one of the three values of colour (or anticolour) is said to belong to a colour triplet. Quarks (or antiquarks) are in this class.

Colour octet A system in which the colour quantum number can have one of eight values chosen by combining one colour with one anticolour (a ninth choice is a colour singlet). The eight gluons are in this class.

This state vector is antisymmetric under the exchange of any pair of labels. This solves another problem which we have not revealed so far; consider the decuplet baryon states Δ^{++}, Δ^{-}, Ω^{-}. They have flavour contents uuu, ddd, sss respectively, the quark spins are all parallel to give $j=\frac{3}{2}$ and there is no orbital angular momentum. This state is manifestly symmetric under the exchange of any two quarks, a situation which violates the Pauli exclusion principle, a principal which must apply to a system of identical fermions. By including the colour singlet state function the whole state function becomes antisymmetric, as required by Pauli. In fact, all baryon quark space–spin–flavour state functions are symmetric and must be made totally antisymmetric by attaching this colour singlet function.

The gluons are coloured but in a way that is different from quarks: their state functions have the cross-terms. There are nine ways of choosing one colour and one complementary colour. One combination is that of equation (10.2) which is the colour singlet. The remaining eight are not singlets but members of a **colour octet** and, although some of their state vectors may appear to be colourless they are not singlets. So we expect eight coloured gluons and for interest we show their colour state vectors in Table 10.7. Referring to Fig. 10.12 we can see how gluon exchange can lead to colour exchange between interacting quarks. Colour is conserved at each step in the process.

The reader may wonder how the colour state functions can be derived in a theory that apparently leans heavily on the description of colour perception. In fact the correct mathematical structure is provided by that of the symmetry group $SU(3)$. The theory of colour perception provides a close and 'colourful'

Table 10.7 Coloured gluons

The eight gluons have colour state vectors:
$\lvert R\bar{G}\rangle, \lvert R\bar{B}\rangle, \lvert G\bar{R}\rangle, \lvert G\bar{B}\rangle, \lvert B\bar{R}\rangle, \lvert B\bar{G}\rangle, \frac{1}{\sqrt{2}}\lvert R\bar{R}-G\bar{G}\rangle, \frac{1}{\sqrt{6}}\lvert R\bar{R}+G\bar{G}-2B\bar{B}\rangle.$

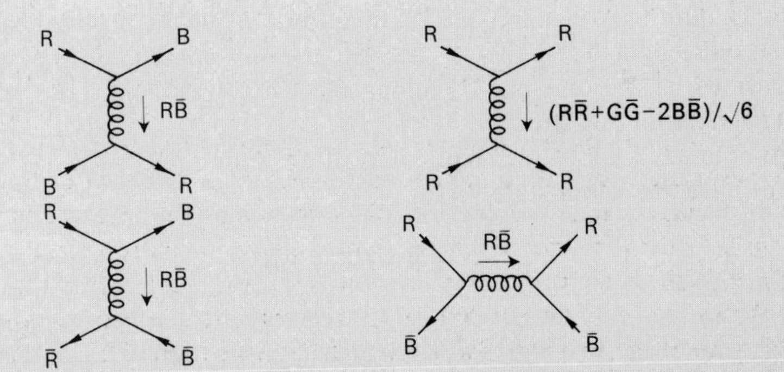

Fig. 10.12 Examples of gluon exchange between quarks and between quark and antiquark. Some exchanges lead to colour exchange, others do not. The apparently colourless exchange $R\bar{R}+G\bar{G}-2B\bar{B}$ is not so because the gluon with this state vector is not in a singlet colour state.

analogy which avoids the abstract nature of the group theory. However, the group theory was applied before the analogy gave the ideas a name. Clearly quarks do not have real colour and the meaning of this new quantum number remains a mystery. Readers should note that the success of applying $SU(3)$ theory implies something very important about quarks and gluons: the symmetry appears to be perfect, which is equivalent to saying that all gluons have the same mass, that quark mass is independent of colour and that one constant alone is needed to define the strength of the coupling of all quarks to any gluon and of gluons to gluons (see Fig. 9.16). A part of the story is the harmonious union of colour symmetry and of gauge theories in applications to particle physics; the theory of quantum chromodynamics (see Sections 9.11 and 13.4) embodies all these ideas. It is believed to be the correct theory of the strong interactions of quarks and gluons. However, in spite of that confidence, the theory has many difficulties of calculation which have made it impossible to obtain sound quantitative predictions except in the case of a few problems.

Returning to the level of observable hadrons we see that the concept of colour provides a consistent solution to three problems. In fact it also solves two other problems. The first concerns the mean life of the π^0-meson, which was about nine times shorter than predicted by what was considered to be a reliable theoretical calculation. The second concerned certain theoretical difficulties involving calculations of self-energy which gave infinity as the result. Both these problems disappeared when colour was introduced.

We have used the observed value of R to introduce the idea of colour. There are some subjects which follow on from this concerning the nature of the gluon field between quarks and the reason why free quarks are not observed; we shall return to these subjects in Section 10.9.

10.8 Resonances in e^+e^- annihilation and quarkonia

We promised at the end of Section 10.6 to return to look at the resonances seen in the value of R as a function of energy. Even before e^+e^- colliders were built resonant structure was expected for the following reason. The annihilation of e^+e^- through a single photon intermediate state followed by materialization as $q\bar{q}$ or $\mu^+\mu^-$ has certain properties. The photon has spin 1 (light can be polarized) and e^+e^- annihilation takes place at a point. The only way the e^+e^- can give angular momentum to the photon is by annihilating in the $^3S_1$ state. Since it appears quarks and muons are as point-like as electrons seem to be, it follows that the $q\bar{q}$ are produced in a $^3S_1$ state. But we already know of several hadrons which are $q\bar{q}$ in a $^3S_1$ state (ρ^0, ω, φ), the neutral **vector mesons**. Therefore each of these vector mesons is expected to appear as a compound state (resonance) in e^+e^- annihilation when the total energy in the collision centre of mass is equal to the rest mass energy of the vector meson. These mesons have decay modes which are predominantly to hadrons, so the ratio R is expected to show the resonant behaviour. The peaks in Fig. 10.11 are labelled with the symbol for the relevant vector meson. The e^+e^- colliders were built with several objectives, one of which was the investigation of higher-mass vector mesons; they proved to be a goldmine yielding far beyond their makers' dreams.

The break through came in 1974 when an e^+e^- collider operating at SLAC (Stanford Linear Accelerator Center) discovered the peaks marked ψ and ψ' in Fig. 10.11. Simultaneously a group at BNL (Brookhaven National Laboratory) observed the state ψ produced in collisions of 30 GeV protons with light nuclei:

they named it the J. The immediate difficulty was that, if these were vector mesons and hadrons, why were they such narrow resonances? The observed widths indicated mean lives of order 10^{-20} s, about 100 times longer than expected for a hadron decaying to hadrons. Many possibilities were examined but the interpretation that has survived is that these are the 1^3S_1 and 2^3S_1 states of $c\bar{c}$, where the principal quantum number $n = 1, 2$ has the same meaning as the principal quantum number in S states of the hydrogen atom. The ψ'' at a mass of 3770 MeV/c^2 is broad, about 25 MeV, in contrast to 63 and 215 keV for the ψ and ψ' respectively. The sudden change in width is now known to be due to the fact that the mass of the ψ'' is above the thresholds (3729, 3739 MeV) for decay into the $D\bar{D}$ system:

$$\psi'' \to D^0 + \bar{D}^0 \qquad \text{or} \qquad D^+ + D^-.$$

In quark terms

$$c\bar{c} \to c\bar{u} + \bar{c}u \qquad \text{or} \qquad c\bar{d} + \bar{c}d.$$

The $\psi(3097)$ and $\psi'(3685)$ are below these thresholds and can decay only to hadrons not containing c or $\bar{c}$. Such decays imply internal annihilation of the $c\bar{c}$; the simplest mechanism allowed by colour symmetry is that involving a three-gluon intermediate state, as shown in Fig. 10.13(a). This has a very low probability compared to the $\psi'' \to D\bar{D}$, Fig. 10.13(b), and leads to the relatively long mean lives of these states. This situation had not been anticipated but was explained once it had been observed.

The assignment of 1^3S_1 and 2^3S_1 to the ψ and ψ' lead immediately to the possibility that other states 1^1S_0, 2^1S_0, 1^3P_0, 1^3P_1, 1^3P_2 and 1^1P_1 (see Table 10.8

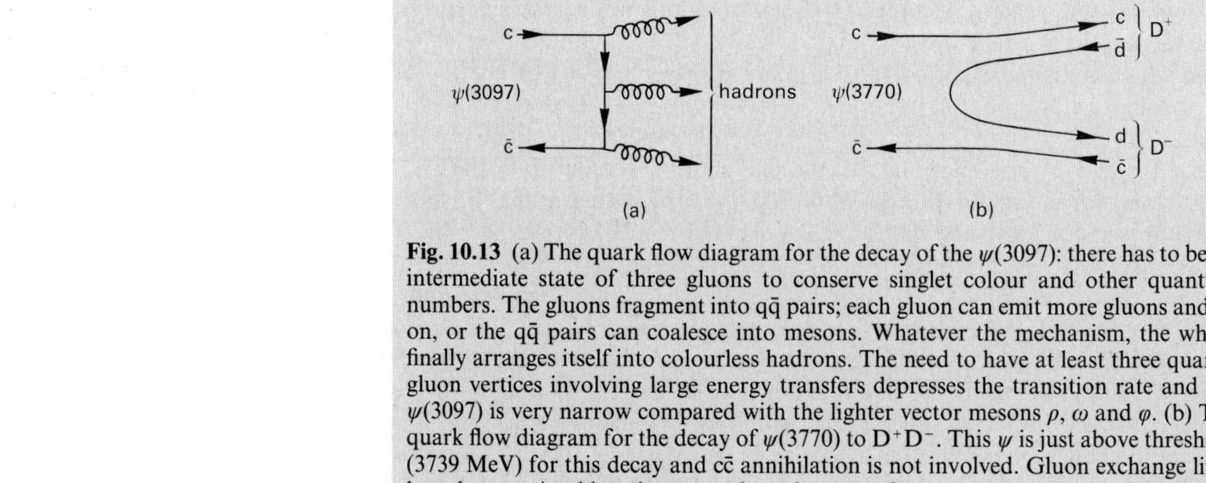

Fig. 10.13 (a) The quark flow diagram for the decay of the $\psi(3097)$: there has to be an intermediate state of three gluons to conserve singlet colour and other quantum numbers. The gluons fragment into $q\bar{q}$ pairs; each gluon can emit more gluons and so on, or the $q\bar{q}$ pairs can coalesce into mesons. Whatever the mechanism, the whole finally arranges itself into colourless hadrons. The need to have at least three quark–gluon vertices involving large energy transfers depresses the transition rate and the $\psi(3097)$ is very narrow compared with the lighter vector mesons ρ, ω and φ. (b) The quark flow diagram for the decay of $\psi(3770)$ to D^+D^-. This ψ is just above threshold (3739 MeV) for this decay and $c\bar{c}$ annihilation is not involved. Gluon exchange lines have been omitted but there must be at least one from c or $\bar{c}$ to a vertex where the $d\bar{d}$ pair materializes; however, the energy transfers are all small compared with those in figure (a) and consequently the quark–gluon coupling is much stronger and the transition rate is not depressed.

This situation points to a property of the quark–gluon coupling constant which we have not so far mentioned: it becomes weaker as the energy becomes greater (see Section 13.4).

Table 10.8 Notation

The lowest charmonium *P, D, F*, ... states are labelled 1 *P*, 1 *D*, 1 *F*, ... The next lowest 2*S*, 2*P*, 2*D*, ... and so on, as is done in the case of the nuclear shell model (see Section 8.3). This in contrast to the atomic notation where the lowest *P* state is labelled 2*P*, the lowest *D* state is 3*D*, and so on.

about notation) might also be below threshold for decay into $D\bar{D}$ and observable as narrow states (none of these has $j^P = 1^-$, so cannot be seen as compound states in e^+e^- single virtual photon annihilation). However, the ψ' is high enough in mass and narrow enough to have a significant branching fraction for decay by emission of a single real photon to all these states except the 1^1P_1. The single photon decay to this state is forbidden and it has evaded observation. Fig. 10.14 shows all the remaining states and the major or important transitions between them or away from them. This system is called **charmonium** by analogy with the name positronium (Section 9.6) given to the bound state of positron and electron. In fact there are many analogies between positronium and charmonium and we can take a look at some of them here. Most readers will be familiar with the hydrogen atom and could demand comparison with that more familiar object. However, the hyperfine interactions

DEFINITIONS AND KEYWORDS

Vector meson A generic name for spin-parity 1^- hadrons.

Charmonium This is the name given to the bound state of a charm quark plus a charm antiquark. Symbol: Ch.

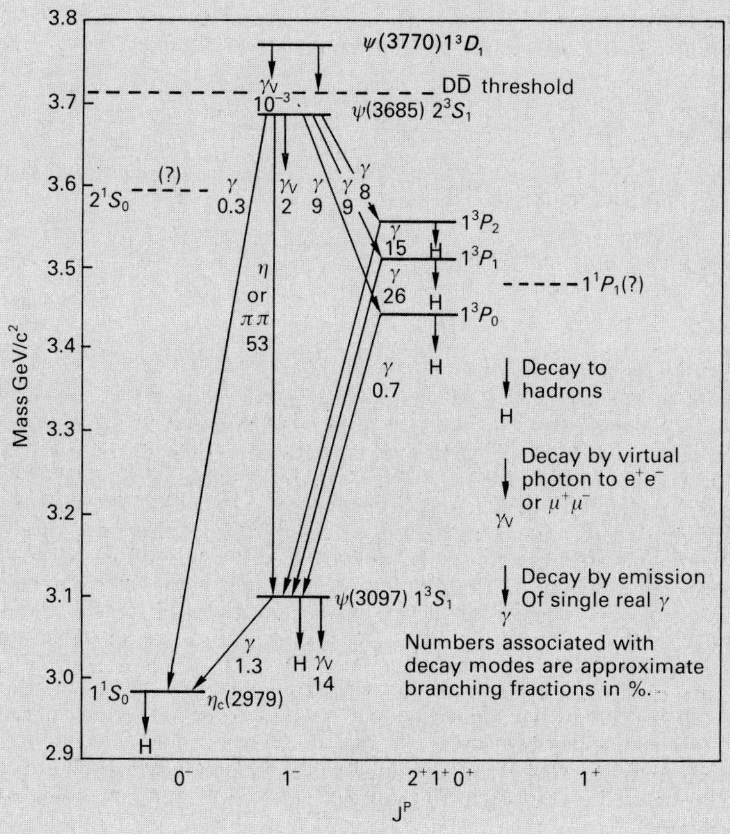

Fig. 10.14 The energy levels of charmonium below 3.8 GeV/c^2. Major or important decay modes are shown but no attempt has been made to cover 100% even in the broadest categories. Broken lines indicate expected position of unobserved levels and the $D\bar{D}$ threshold.

are frequently not included in discussions about the hydrogen atom because they have a very small effect, whereas in positronium and in charmonium they are very important. (Problem 10.10 asks you to discover why positronium hyperfine splitting of the ground state is nearly 150 times greater than it is for hydrogen.) Positronium (Ps, for short) and charmonium (Ch) are both bound states of two spin-$\frac{1}{2}$ fermions, so that the classification of angular momentum states are identical from the start and the same as those we are now accustomed to using in meson spectroscopy for $q\bar{q}$ states. Figure 10.15 compares the term diagrams for the two. Let us compare some interesting features:

1. The $2S$ and $1P$ levels of Ch are not near degenerate as they are in Ps. This means that the $q\bar{q}$ potential in Ch does not vary as $1/r$, the variation known to lead to this near degeneracy in the hydrogen atom and in Ps.
2. The Ch triplet levels 1^3P_0, 1^3P_1 1^3P_2 are, relative to the $2S$–$1S$ spacing (589 MeV in Ch, 5.1 eV in Ps) more split than the same levels in Ps. This indicates an important spin–orbit interaction in Ch.
3. The $1^3S_1 - 1^1S_0$ and the $2^3S_1 - 2^3S_1$ splitting in Ch are, again relative to the $2S$–$1S$ difference, larger than those in Ps.

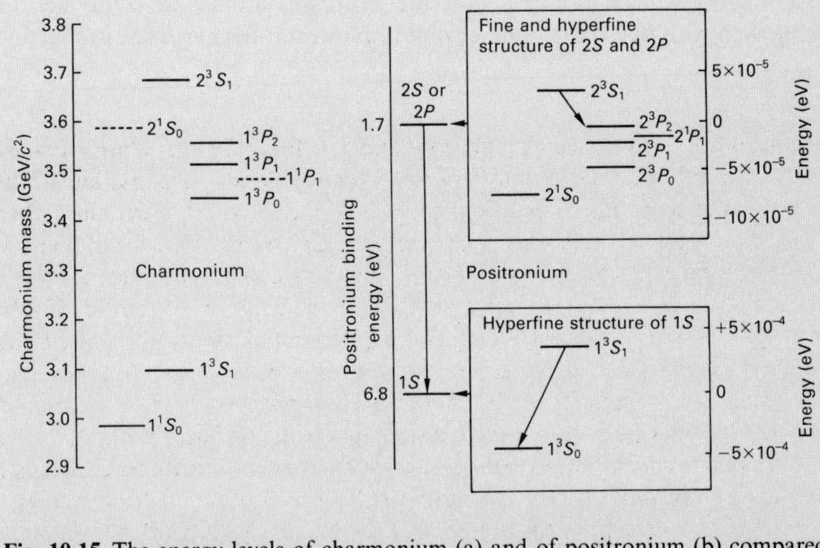

Fig. 10.15 The energy levels of charmonium (a) and of positronium (b) compared. Note that the details of the fine and hyperfine splitting of positronium have to be enlarged considerably compared with the scale set by the $2S$–$1S$, 5.1 eV splitting. Note also that we have used the atomic convention of labelling by $2P$ the lowest P states of positronium but have used the nuclear convention $1P$ for charmonium. The charmonium levels are given by broken lines when they have not been observed experimentally: their position is the result of calculation using a $q\bar{q}$ potential such as that of equation (10.3). No such modesty has restrained us in the case of the positronium levels: there is no doubt about the ground state hyperfine splitting but in the case of the $2S$ and $2P$ levels the level positions shown are the result of quantum electrodynamic calculations and only the $2^3S_1 \to 2^3P_2$ transition has been observed. None the less, great confidence can be placed in these calculations in view of their success in other systems (see Section 9.6). Here the prediction is that the $2^3S_1 - 2^3P_2$ splitting should be 8625 MHz and the observed value is 8628 ± 3 MHz, which reinforces that confidence. The main conclusion from the comparison is that the fine and hyperfine interactions are relatively much more important in charmonium than in positronium.

Let us see if we can make some sense of this. In spite of (1) suppose the potential between q and $\bar{q}$ does varies as $g^2/4\pi r$ where g is a quark–gluon coupling. Then just as the fine structure constant $\alpha = e^2/4\pi\varepsilon_0\hbar c$ we have a strong interaction fine structure constant $\alpha_s = g^2/4\pi\hbar c$. In Problem 10.11 you are asked to show that the energy levels of Ps are given by

$$E_n = -\frac{1}{4n^2}\,\alpha^2 m_e c^2\,,$$

where n is the principal quantum number, m_e is the mass of the electron and this is the simple non-relativistic Bohr–Schrödinger result. A few moments with a calculator shows this gives 5.1 eV for $E_2 - E_1$ of Ps. Rigorous QED calculations show that the ground state hyperfine splitting in Ps is given by

$$\Delta E(1^3S_1 - 1^1S_0) = m_e c^2\left(\frac{7}{12}\alpha^4 + O(\alpha^6)\right).$$

This gives 8.45×10^{-4} eV, as observed. In leading order the ratio of the second to the first splitting is $28\alpha^2/9 = 1.66 \times 10^{-4}$. Suppose we take the ratio of the corresponding energies in charmonium and put it equal to $28\alpha_s^2/9$:

$$\frac{116}{589} = \frac{28\alpha_s^2}{9}$$

which gives $\alpha_s = 0.25$. In fact for charmonium $\alpha_s \simeq 0.2$. What conclusion can we draw? If the q$\bar{q}$ potential is somewhat like the Coulomb potential and if the spin–orbit and spin–spin interactions scale as expected from positronium, then the potential is about 25 times stronger and the greater prominence of fine and hyperfine features of the charmonium energy levels can be understood.

Why did we do what is a rather outrageous calculation of this kind? The reason is that since gluons are massless the q$\bar{q}$ potential should vary as $1/r$ and have corresponding colour magnetic interactions which give rise to analogous spin–orbit and spin–spin forces; but that neglects a property that the gluon field has which is not to be found in the electromagnetic field. There is no attractive or repulsive force between photons in contrast to gluons which do attract one another; that self-attraction of the gluon field squeezes that field into a tube connecting the q and $\bar{q}$ so that the q$\bar{q}$ attractive force, instead of continuing to fall as the q$\bar{q}$ separation increases, remains constant independent of the length of the tube. This means that the q$\bar{q}$ potential, instead of increasing asymptotically to zero from negative as does the Coulomb potential between e^+ and e^-, continues to increase linearly with separation. This is a simplified picture for the situation which has motivated the use of the following potential in calculations attempting a quantitative explanation of the charmonium energy levels:

$$V(r) = -\frac{4\alpha_s \hbar c}{3r} + Kr. \tag{10.3}$$

The $\frac{4}{3}$ is there for technical reasons connected with the definition of the quark–gluon coupling constant and the $\hbar c$ is there to keep α_s dimensionless. This potential works rather well. There is also another consequence of this picture which will be discussed in the next section. And, of course, the simple calculation of α_s that we made earlier in this section will not be correct as it assumed the potential varied as r^{-1}.

We have not so far mentioned the other peaks near 10 GeV in R. There are three sharp peaks and three broad. These are interpreted as the 1^3S_1, 2^3S_1, 3^3S_1, 4^3S_1, 5^3S_1 and 6^3S_1 states of the $b\bar{b}$ system: they have symbol Υ (mass) and the masses are given in Table 10.9. The change of width indicates that there is at least one threshold between 3 and 4. The 4^3S_1 is above threshold for decay into a pair of the mesons $u\bar{b}$ and $\bar{u}b$ ($B^{\pm}$ at 5278 MeV/c^2) or $d\bar{b}$ and $\bar{d}b$ (B^0, $\bar{B}^0$ at 5279 MeV/c^2). These B mesons are the lowest mass states of the beauty spectroscopy and they will therefore be stable except against weak interactions. The potential of equation (10.3) has been applied successfully to the energy levels of this system, which is sometimes awkwardly called beauty-onium. Some of the expected P states have been observed and the experimental search for the remaining predicted levels is an ongoing research programme at the time of writing. Presumably ground states of the $b\bar{s}$, $\bar{b}s$, $b\bar{c}$, $\bar{b}c$ systems will also be stable except against decay by weak interactions: these states have not been observed.

The tendency for an increasing number of narrow $q\bar{q}$ $^3S_1$ states below threshold for $u\bar{q}$ plus $\bar{u}q$ production as the q mass increases in these **quarkonium** systems is an expected property of the potential of equation (10.3). If the truth (or top) quark mass is, for example, about 80 GeV/c^2, then truth-onium (toponium?) states will have masses near 160 GeV/c^2 and the number of narrow $t\bar{t}$ states below the threshold for $u\bar{t}$ plus $\bar{u}t$ production could be as many as twelve or more. In the other direction the 1^3S_1 state of $s\bar{s}$, which is known as $\varphi(1020)$ (and is strangeonium!) is just above threshold for the production of pairs of the lightest strange mesons, $K^0\bar{K}^0$ or K^+K^-. The 1^3S_1 states of $u\bar{u}$ and $d\bar{d}$ are mixed to give the observed $\rho^0(770)$ and $\omega(783)$ which are about 500 MeV above threshold for producing a pair of pions ($\pi^+\pi^-$).

The calculation of hadron masses using quantum chromodynamics or related models has not yielded good quantitative results. A phenomenological approach based upon a search for a consistent parametrization describing the separate contributions of quark masses, spin–orbit, spin–spin and Coulomb interactions has had a moderately good success. It can, for example, explain the ρ–π, the Σ–Λ, splittings and the mass differences inside the decuplet. However, a fuller discussion of this subject is outside the scope of this chapter.

The naming of the heavier quarks and of the quarkonium states has produced some awkward results: strangeness, strange particles and strange quarks are manageable but the b-quarks need a tidier name than beauty (quantum number beauty, beautiful quarks, and beautyonium?) or its alternative bottom (quantum number bottomness, bottom quarks, and bottomonium?). Topness, top quarks, and toponium, if they are found, are probably acceptable.

Let us revisit positronium. The states 1^3S_1 and 1^1S_0 decay to 3γ and to 2γ respectively, with mean lives of 0.14 μs and 0.125 ns. The first decay is

Table 10.9 The known upsilon states with $j^P = 1^-$.

State $b\bar{b}$	1^3S_1	2^3S_1	3^3S_1	4^3S_1	5^3S_1	6^3S_1
Mass MeV/c^2*	9460	10023	10355	10575	10860	11020
Width MeV*	0.052	0.044	0.026	24.	110.	79.

* Errors not shown, but considerable in the case of some widths.

reminiscent of $\psi(3097) \to$ three gluons and it is three in both cases for the same reason, a reason following from a conservation law connected with the operation of charge conjugation. The decay of ${}^3S_1$ Ps depends on α^6 (α^3 for three eeγ vertices and α^3 for the wavefunction squared at zero separation) and the ψ decay depends on α_s^6. This allows a calculation of α_s from the decay width of the $\psi(3097)$ (see Problem 10.9). The decay ${}^1S_0$ Ps $\to 2\gamma$ is analogous to that of π^0 (${}^1S_0$) $\to 2\gamma$.

PROBLEMS

10.8 Draw diagrams to illustrate the mechanisms by which leptons and hadrons are created in e^+e^- annihilation at high energy.

In an electron–positron collider the particles circulate in short cylindrical bunches of radius 1 mm (transverse to the direction of motion). The number of particles per bunch is 5×10^{11} and the bunches collide at a frequency of 1 MHz. The cross-section for $\mu^+\mu^-$ creation at 8 GeV total energy is 1.4×10^{-33} cm^2; how many $\mu^+\mu^-$ pairs are created per second? What is the rate of hadron production at this energy?

Indicate, with the help of a sketch, how μ^+ and μ^- from e^+e^- annihilation can be identified and suggest the principal sources of background.

(Adapted from the 1985 examination of the Finals Honours School of Natural Science, Physics, University of Oxford.)

10.9 Quantum electrodynamics predicts the decay rate for $1\,{}^3S_1$ positronium into three γ-rays to be given by

$$\omega = \frac{2\alpha^6(\pi^2-9)}{9\pi}\frac{m_e c^2}{\hbar}.$$

Assuming the same formula applies to the decay $\psi(3097) \to$ three gluons followed by fragmentation to hadrons, the fragmentation occurring with unit probability, calculate the effective α_s for the ccg vertices in this decay. The full width of the $\psi(3097)$ is 63 keV and 82% of decays are to hadronic final states. Assume the mass of the charm quark is 1.65 GeV/c^2

Now consider the radiative decay $\psi \to \gamma + g + g$. This, followed by fragmentation of the two gluons, would represent the mechanism for radiative decay to hadronic states not containing charmed quarks. In it one ccg vertex of $\psi \to 3g$ becomes a ccγ vertex. Modify the formula to give your prediction of the decay rate and estimate the branching fraction for this radiative decay mode.

10.10 The hyperfine splitting of the $1S$ state of positronium is about 150 times greater than the analogous splitting for the $1S$ state of the hydrogen atom. By considering the origin of the splitting, can you explain why this ratio is as great as this?

10.11 Show that the energy of the level of positronium with principal quantum number n is given by

$$E_n = -\frac{\alpha^2 m_e c^2}{4n^2}$$

where m_e is the mass of the electron.

10.9 Fragmentation

In the last section we introduced the idea of the gluon field between quarks squeezing itself down into a tube by self-attraction, sometimes called the colour flux tube since it conveys colour between the quarks. There are reasons to believe that the tube is a store of energy in the same way as a stretched spring contains energy and that that energy is given by the value of K in equation (10.3). A value which appears to fit the data is about 1 GeV per fermi of length. Increasing the $q\bar{q}$ separation means increasing the stored energy by doing work against the force tending to contract the spring. That force is, of course, 1 GeV fm^{-1} which is about 16 tonnes weight!

We now wish to use this picture to construct a model of how a $q\bar{q}$ pair once produced in e^+e^- annihilation becomes observed hadrons: this process that we are attempting to model is called **fragmentation**. As the pair separate after production, the 'spring' joining the pair will stretch and as it stretches it stores increasing energy until it becomes energetically possible for it to break with a new q attached to the end of the spring on one side of the break and a $\bar{q}$ of the same flavour at the other side of the break. The two segments of the spring may still be subject to stretching under the action of the possibly still rapidly moving quarks at the ends so that further breaking can occur. When breaking stops we are left with several spring segments each joining a q and a $\bar{q}$, not necessarily of the same flavour. Overall flavour must be conserved. Such segments consist of a q and a $\bar{q}$ with a binding field so they are mesons. Two of the segments, or mesons, each contain one quark from the original $q\bar{q}$. In Fig. 10.16 we have tried to illustrate this model by showing how the reaction

$$e^+ + e^- \rightarrow c + \bar{c} \rightarrow D^0 + D^- + \pi^0 + \pi^0 + \pi^+$$

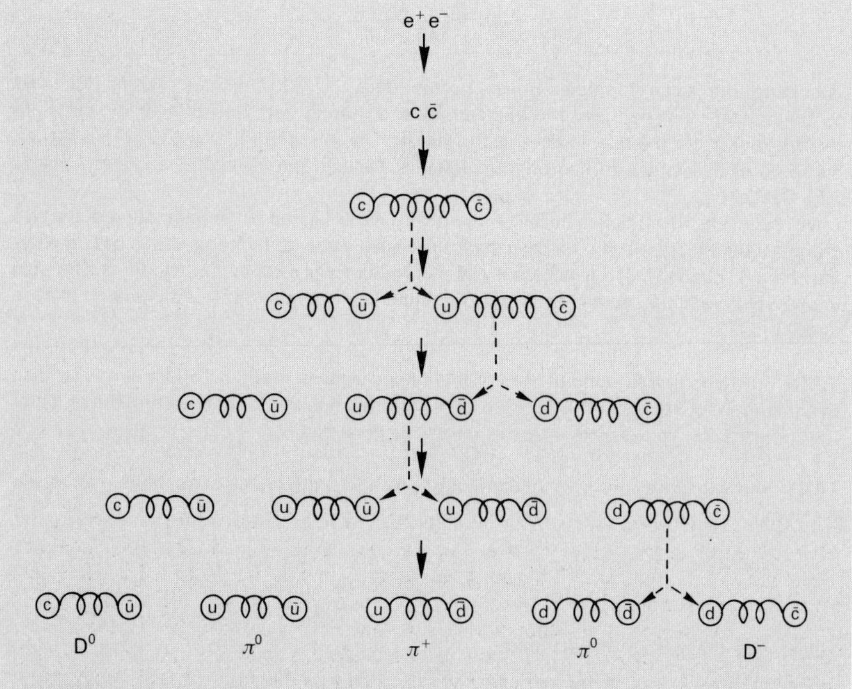

Fig. 10.16 The breaking spring model of fragmentation of a $c\bar{c}$ pair into $D^0D^-\pi^0\pi^0\pi^+$. The effect of the breaks in the gluon colour flux tubes are shown by dotted lines. The order of the mesons formed by the breaking of the spring into segments does not imply the same ordering in longitudinal momentum, although it is true that in the case of the fragmentation of heavy quarks, the meson with the heavy flavour does tend to have the largest fraction of that quark's original energy.

might occur. Sometimes baryons are produced; in that case the model sees a spring break producing not a $q\bar{q}$ pair but a di-quark pair, that is a qq on one side of the break and a $\bar{q}\bar{q}$ at the other. We leave it to the reader to draw a picture of that happening and developing so that one of the spring segments is a baryon and another an antibaryon.

There is an important property of the data which is explained by this model. The line of separation of the q and $\bar{q}$ is the line of stretching of the spring; when the spring breaks the forces acting on the mesons will be along this same direction and these mesons will therefore also be produced tending to move along that direction. This is just what is observed: the hadrons are produced in two back-to-back jets of particles. An example is shown in Fig. 10.17. However, the hadrons are not produced with all their momentum vectors in the $q\bar{q}$

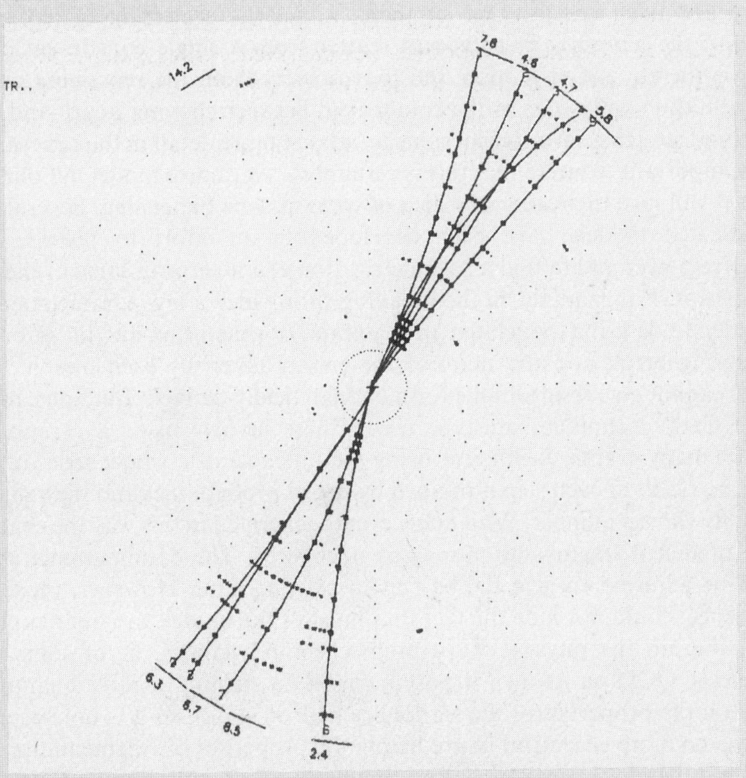

Fig. 10.17 A computer reconstruction of an e^+e^- annihilation event into two hadron jets as observed in the Tasso detector. The incident beam directions are perpendicular to the plane of the paper. The cross-over point occurs at the centre of a cylindrical detector which is coaxial with the beams. The detector has a solenoidal magnetic field parallel to the beams. The detector contains many elements but the important one here is a multiwire proportional chamber: it has the capability of determining many coordinates on the trajectories of charged particles in the gas of the chamber. This picture shows the projection of these coordinates and of the reconstructed trajectories onto the plane perpendicular to the beam axis. The magnetic field bends the trajectories of the charged particles into segments of a circle. The energy of the annihilation was 34.6 GeV.

The Tasso detector was operated at the e^+e^- collider PETRA at the Deutsches Elektronen-Synchrotron, Hamburg (DESY).

direction; each receives a random transverse component of momentum, averaging approximately 200 MeV/c, and distributed nearly exponentially. This magnitude of transverse momentum would be expected for a confinement distance of about 1 fm, presumably the transverse diameter of the colour flux tube. Longitudinally, many of the hadrons have much larger momenta, particularly at high energies of annihilation ($2E \gg 10$ GeV). The jet–jet direction, which has to be determined from the average momentum weighted direction of the particles, is assumed to be close to the original, unobservable, $q\bar{q}$ direction. That direction should be distributed in space with respect to the e^+e^- collision direction in the same way $[(I(\cos\theta) \propto 1+\cos^2\theta]$ as the final state in $e^+ + e^- \to \mu^+ + \mu^-$, since muons and quarks all have spin $\frac{1}{2}$. With the resolution permitted by the smearing effect of fragmentation, that is found to be the case. This is one of the stronger pieces of evidence that quarks do have spin $\frac{1}{2}$.

This model has been introduced in the context of hadron production in e^+e^- annihilation, but of course it can be applied to other circumstances in which quarks are given energy. One of these is inelastic scattering of leptons by nucleons: the lepton is presumed to scatter from a single constituent quark, which is thereby set in motion and moves away from the remaining quarks. Fragmentation will occur as the colour field between moving quark and those left behind stretches. We consider this process in more detail in the next section.

It is important to stress that this is certainly a very naïve model but our hope is that it will give the reader one idea of what may be happening. Several more sophisticated models have been developed in an effort to provide some predictive power and to find a parametrization characterizing some of the basic mechanisms. For example, in the breaking spring idea a few parameters would be needed to describe the relative probabilities of generating $u\bar{u}$, $d\bar{d}$, $s\bar{s}$, or $c\bar{c}$ at each spring break and the share of the energy taken by each meson. These models cannot give results in the form of analytically derived equations; instead Monte Carlo techniques must be used. These involve using a computer to generate many events, each event being the outcome of a whole series of steps where the result of each step is decided by preset probabilities and the value of a randomly chosen number. With many events generated in this way the character of the predicted fragmentation may be uncovered. The basic parametrization has to be adjusted to give the best fit to existing data. However, those bald statements should not hide the fact that most of the models are built as closely as possible on the physics of quantum chromodynamics, or of some other hypothesis. QCD on its own is not amenable to yielding reliable quantitative results on the properties of the stationary hadron states, so it is not surprising that it is no more successful in predicting the properties of fragmentation. The modelling therefore provides one of the means of making limited tests of the theory. One success for this approach was in the interpretation of data which gave direct experimental evidence for the existence of gluons; we will describe this in the next section.

These ideas may give a hint as to the reason for the confinement of quarks (Section 9.11). Starting with singlet colour state there is no way of removing a single quark without stretching the colour field indefinitely and that requires infinite energy. Conversely, an initially isolated quark may not be a sensible concept: we do consider an isolated electric charge but experience tells us that it has a conjugate charge somewhere. If the two charges are well separated the energy is small. An isolated quark may not be truly isolated, in which case, unless its conjugate is nearby, the system will have an immense energy! However, this kind of speculation raises more questions than it answers.

10.10 Further evidence for quarks and gluons

The description of the spectroscopy of hadrons in terms of constituent quarks could be an elaborate trap set by nature and the real explanation of these properties could lie elsewhere. It is essential therefore to look for support at other experimental evidence not concerned with spectroscopy. There are several areas but shall briefly discuss two; we can only do it briefly as a proper examination would take us beyond the scope of this book. The subjects are

(1) the **deep inelastic scattering of leptons** by nucleons;

(2) the properties of the final state in e^+e^- annihilation to hadrons.

Deep inelastic scattering means scattering in which large amounts of energy, ν, and momentum, $\boldsymbol{q}$, are transferred to the nucleon, where 'large' means $qc \gg Mc^2$ and $\nu \gg Mc^2$, where M is the nucleon mass, and 'inelastic' implies that the final state of hadrons is not a single nucleon but will consist of at least one baryon and one or more mesons. The lepton can be an electron, muon, or neutrino: here we will consider the scattering of electrons only, for the sake of simplicity. Figure 10.18 shows the quark–parton model picture of the process which is supposed to occur by the exchange of a virtual photon between the incident electron and one of the constituent charged partons, presumably a quark. The most convenient way to envisage the scattering is in the inertial frame which is the centre-of-mass of the virtual photon and the target nucleon. In that frame the approaching nucleon can be thought of as a bag of partons executing slow internal motions, slow because to an observer these motions are Lorentz time dilated. The collision takes place in a short time compared to the dilated period of these motions so that the partons can be considered as almost free. Thus the basic mechanism is e^- scattering at a free quark; this scattering is elastic and the cross-section can be calculated using quantum electrodynamics since both particles are assumed to be charged, point-like, and to have spin $\frac{1}{2}$. To assemble the basic cross-section into an effective cross-section for the whole nucleon means defining a function which gives the charge squared weighted probability of finding a quark having a certain fraction of the approaching nucleon's total momentum; these functions are called structure functions. All

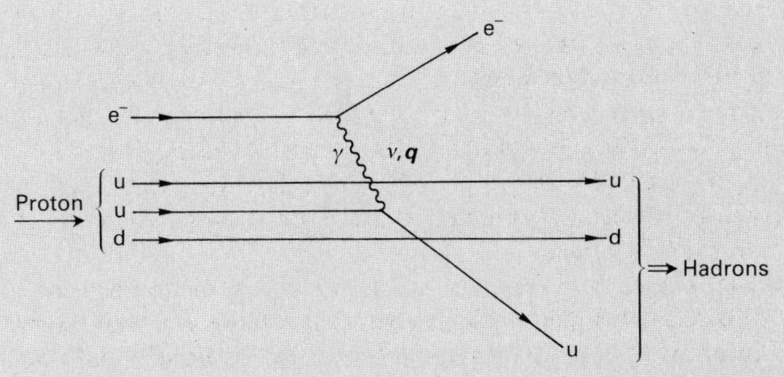

Fig. 10.18 The quark–parton model of the deep inelastic scattering of electrons by protons. The electron loses energy ν, and momentum $\boldsymbol{q}$, which are transferred by means of a virtual photon to one quark in the nucleon in what is effectively e^-q elastic scattering. The struck quark parts company from the spectator quarks and the stretching of the colour flux tubes between them initiates fragmentation into hadrons.

these functions have been determined from measurement and the results are consistent with the existence of three components of the nucleon, which are as follows.

1. Three valence quarks, uud or udd for proton or neutron, with the charges we have assigned earlier, and spin $\frac{1}{2}$.
2. A number of $q\bar{q}$ pairs, the sea quarks, which are mainly $u\bar{u}$ and $d\bar{d}$ with some $s\bar{s}$. These have no effect on the charge or flavour of the nucleon.
3. A number of neutral partons, which carry about half the nucleon's momentum and have spin 1. These are the gluons.

Items 2 and 3 cannot imply a fixed number of partons. All that can be determined is the fraction of the nucleon's momentum carried by each of these components on the average.

It was this interpretation of the first measurements of deep inelastic scattering of electrons by nucleons that generated further confidence in the quark–parton model. Indirectly, it provides evidence for the existence of gluons. In particular the discovery of the scale invariance (see Section 3.8) of the structure functions implied that the quarks were point-like on the scale determined by the momentum transfer (<0.1 fm). Later measurements showed that the scale invariance was not perfect and the deviation from scaling was just what was expected from applying QCD to deep inelastic scattering.

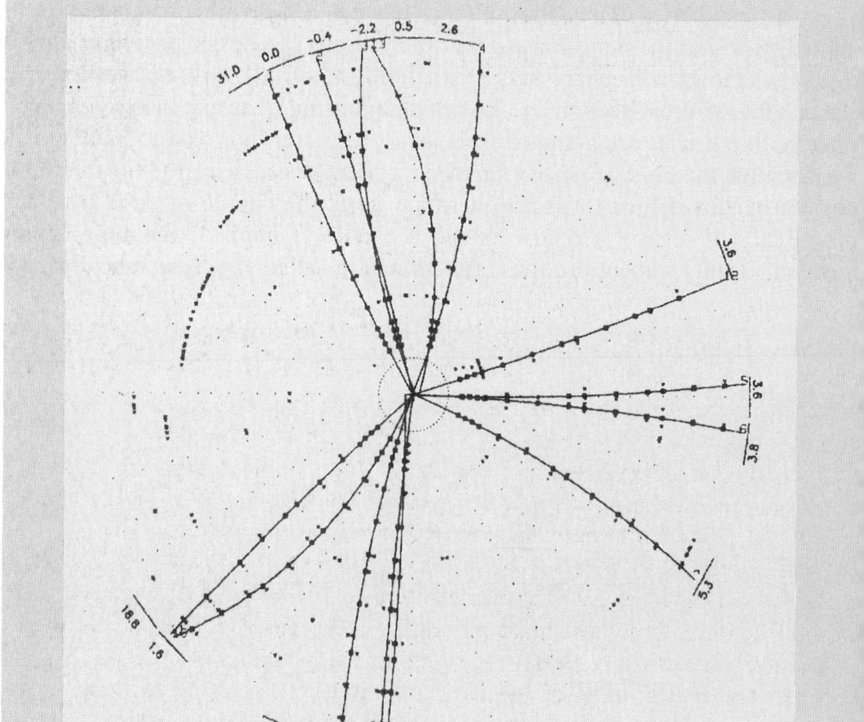

Fig. 10.19 The same as Fig. 10.17 except that this event is one of the rare, clearly separated, three jet events. The total energy is 35.16 GeV.

Contributions to this field were also made by measuring μ-meson and neutrino scattering. The latter involves different structure functions which provide information on nucleon structure that is complementary to that provided by measurements of charged lepton scattering. We shall discuss neutrino scattering in Section 12.11.

The more direct evidence for gluons comes from the study of the final hadrons in e^+e^- annihilation at high energies ($2E > 10$ GeV). We described in the last section how the hadrons are collimated into two back-to-back jets. However, it is possible for a quark to emit a gluon which can have a significant share of the energy. At high energies the gluon can have enough energy to form a separate jet of hadrons at an angle to the quark jets which makes the third jet clearly distinguishable. An example of this is shown in Fig. 10.19. Such clear three-jet events are rare and an event sample will show a whole spectrum of event shapes from the dominating two-jet through to the three-jet. The shape of the individual events changes from one where the hadrons' momentum vectors are collimated close to the jet–jet axis to one where these vectors are partially spread into a plane, occasionally separating into three jets. Each event has to be analysed to obtain the three parameters which describe its shape. One of these is greater the more obvious the three-jet structure becomes. The probability distribution of this parameter is compared with the prediction of QCD. The dominant difficulty is that the configuration of the quark, antiquark and gluon is seen through the blurring effect of the intervening fragmentation. Thus the analysis does depend on the model of fragmentation chosen. In spite of this difficulty, the results do require spin-1 gluons as an ingredient of QCD. (Of course, the possibility of three-jet events does not exclude the possibility of the emission of two energetic gluons giving four jets, and so on, but with rapidly decreasing probability.)

DEFINITIONS AND KEYWORDS

Fragmentation The process by which an energetically excited colour singlet assembly of quarks becomes observable hadrons.

Deep inelastic scattering of leptons The inelastic scattering of leptons by nucleons in which the energy and momentum transferred to the nucleon are large on a scale given by the rest mass energy of the nucleon.

10.11 Isotopic spin

The reader will not have failed to notice that many hadrons occur in groups each containing particles of almost equal mass: for example, the neutron and proton, the Σ^+, Σ^0, Σ^-, the π^+, π^0, π^-, the K^+, K^0, and so on. These groups of almost degenerate states are called **multiplets**. A careful examination of these multiplets shows that the changes in the quark structure in moving from one member to another of a multiplet involve up and down quarks or antiquarks alone. If the u and d quarks had equal mass and if electromagnetism were vanishingly weak compared with the strong quark–quark interaction, so that the Coulomb interaction contributions to the mass disappeared, we could expect the mass differences in each multiplet to be zero and there would exist an exact symmetry of nature. In fact, this is almost the case, and closely enough to make the symmetry very useful, but sufficiently imperfect to break the degeneracies and to allow the states to be individually identifiable.

The hypothesis to be developed is that it is impossible to tell the difference between the u and d quark except by their electric charge. An analogous situation exists in atomic physics where it is impossible to distinguish electron spin up (↑) from electron spin down (↓) except in the presence of a magnetic field. Therefore we invent **isotopic spin**, for which the basic states are isotopic spin-up (u quark) and isotopic spin-down (d quark) and we cannot tell the difference in the absence of electromagnetism. The electron has spin $s=\frac{1}{2}$ with z-components $s_z = +\frac{1}{2}$ (↑) or $-\frac{1}{2}$ (↓). The u and d quarks have isotopic spin $t=\frac{1}{2}$

Multiplet A set of degenerate or near degenerate quantum-mechanical states.

Isotopic spin The isotopic spin has components along an axis which are quantized with quantum numbers that distinguish the members of each multiplet of hadrons, in the same way that the z-component of an angular momentum distinguishes the magnetic substates of that angular momentum.

(t to represent the quantum number for total i-spin) with third components $t_3 = +\frac{1}{2}$ (u) and $-\frac{1}{2}$ (d). Now the algebra that we are all accustomed to use in angular momentum in order to construct states of higher spin from combinations of electron spin are available and it works in exactly the same way for isotopic spin. (To reduce the mouthful of isotopic spin we follow the lazy convention and say i-spin.) Two spins of $\frac{1}{2}$ each can be added to make a singlet ($s=0$) or a triplet ($s=1$) state of total spin angular momentum (as we have done in constructing $q\bar{q}$ states). Two i-spins of $\frac{1}{2}$ can make a singlet ($t=0$) or a triplet ($t=1$) state of total i-spin. Unfortunately we cannot have states of two quarks alone; however, the s quark (the lightest of the heavy quarks) has a mass too different from that of u and d quarks (the light quarks) to allow it to partake in the symmetry in a useful way so that 3q combinations of u and d and s quarks should show these multiplet structures. Thus we have the following.

1. The Ω^-(sss) contains no u or d quarks and is an isotopic singlet, $t=0$.
2. The Ξ^0(ssu) and Ξ^-(ssd) must be a doublet, $t=\frac{1}{2}$.
3. The suu, sud, sdd combinations contain two i-spins of $\frac{1}{2}$ so that there must be four states (two ways of choosing the third component, $t_3 = +\frac{1}{2}$ or $t_3 = -\frac{1}{2}$, for two quarks), which divide into three $t=1$ (the triplet Σ^+, Σ^0, Σ^-) and one of $t=0$ (the singlet Λ).
4. The uuu, uud, udd, ddd combinations contain three i-spins of $\frac{1}{2}$. Combining two as in (3) gives a triplet and a singlet; adding in the third gives a quadruplet ($t=\frac{3}{2}$) and two doublets ($t=\frac{1}{2}$). The quadruplet contains the Δ^{++}, Δ^+, Δ^0, Δ^-. One of the two doublets contains the two nucleons (n,p).

Of course, although that example uses the lowest 3q states, the i-spin multiplets occur throughout the baryon spectrum. In Table 10.10 we display the complete analogy that exists between the construction of states of higher spin and of higher i-spin from the basic doublets of spin $\frac{1}{2}$ and of i-spin $\frac{1}{2}$ respectively.

Turning to the meson sector ($q\bar{q}$ states) we expect to find i-spin singlets (η, ω, etc.) and triplets (π^+, π^0, π^- and ρ^+, ρ^0, ρ^-, and so on) when the constituent

Table 10.10 This table shows the close analogy between spin and isotopic spin in the construction of states of higher spin from the base states.

	Spin angular momentum: Quantum numbers	Spin angular momentum: State vectors	Isotopic spin: Quantum numbers	Isotopic spin: State vectors
Base states				
Doublet	$s=\frac{1}{2}$, $s_z=-\frac{1}{2}$	$\lvert\downarrow\rangle$	$t=\frac{1}{2}$, $t_3=-\frac{1}{2}$	$\lvert d\rangle$
	$s_z=+\frac{1}{2}$	$\lvert\uparrow\rangle$	$t_3=+\frac{1}{2}$	$\lvert u\rangle$
Vector addition of two spins				
Singlet	$s=0$ $s_z=0$	$\frac{1}{\sqrt{2}}\lvert\uparrow\downarrow-\downarrow\uparrow\rangle$	$t=0$ $t_3=0$	$\frac{1}{\sqrt{2}}\lvert ud-du\rangle$
Triplet	$s=1$, $s_z=-1$	$\lvert\downarrow\downarrow\rangle$	$t=1$, $t_3=-1$	$\lvert dd\rangle$
	$s_z=0$	$\frac{1}{\sqrt{2}}\lvert\uparrow\downarrow+\downarrow\uparrow\rangle$	$t_3=0$	$\frac{1}{\sqrt{2}}\lvert ud+du\rangle$
	$s_z=+1$	$\lvert\uparrow\uparrow\rangle$	$t_3=+1$	$\lvert uu\rangle$

uarks are chosen from u or d and ū or d̄ since again we are constructing higher spin states from two i-spins of $\frac{1}{2}$. The only difficulty is that the state vectors are ot immediately analogous to the ordinary spin state vectors as in Table 10.10 ecause antiquarks are involved; this need not worry us. Mesonic combinations f one light quark with a heavier quark (e.g. us̄ or ds̄) give doublets; ombinations of two heavy quarks (e.g. sc̄ or sb̄ or cb̄) give singlets.

The connection between i-spin and electric charge Q (in units of the proton harge) of a hadron made of u and d valence quarks is given by

$$Q = t_3 + B/2 \tag{10.4}$$

here B is the baryon number. For hadrons containing heavy valence quarks e formula becomes more complicated and it is easier to work out Q or t_3 from e quark content and the multiplet structure of the contained light quarks.

Since neutron and proton have i-spin $\frac{1}{2}$, nuclei will also have i-spin; however, e expect that, in all but lighter nuclei, the increasing effect of the nuclear harge and of the deviation from $Z = N$ will make it increasingly difficult to lassify the ground and excited states into i-spin multiplets. The i-spins of the ucleons pair up, as do the angular momentum vectors in the shell model, eaving the last unpaired nucleons to determine the total i-spin. Thus in mirror uclei (e.g. ${}^7_4$Be, ${}^7_3$Li) the core of nucleons (here 3p and 3n) has $t = 0$ and the last ucleon gives $t = \frac{1}{2}$ to the whole nucleus. Thus ${}^7_4$Be has $t = \frac{1}{2}$, $t_3 = +\frac{1}{2}$ and ${}^7_3$Li has $= +\frac{1}{2}$ and $t_3 = -\frac{1}{2}$. The ground and excited states of the two nuclei are i-spin oublets of equal energy (mass), apart from electromagnetic effects and the ffect of the neutron–proton mass difference. This is illustrated in Fig. 9.10. In he $A = 6$ isobars, the $Z = 2$, $N = 2$ core has $t = 0$ so that the two remaining ucleons can give the whole nucleus $t = 0$ or $t = 1$. Therefore ${}^6_3$Li will have a $t = 0$ round state and some $t = 0$ excited states which have no analogues in ${}^6_2$He or in Be. In addition, it will have some $t = 1$ excited states which do have analogues in hese neighbours. This is shown in Fig. 9.11; the pattern of $t = 1$ states are easily isible in spite of the mass splitting due to the effects mentioned. For the more eneral case the value of t_3 for the ground state of a nucleus is given by

$$t_3 = Z - A/2\ . \tag{10.5}$$

This formula follows directly from equation (10.4). In principle all that can be leduced about the nucleus is that $t \geqslant |t_3|$. However, ground states have the owest t possible consistent with their charge, therefore $t = |Z - A/2|$.

In Section 9.7 we introduced the concept of charge independence of nuclear orces. Charge independence is manifest in the near mass degeneracy of nembers of an isotopic spin multiplet. That is, that the force between two nucleons depends on t but not on t_3. In fact, the concept of isotopic spin was leveloped and applied to nuclear forces and nuclear structure long before quarks were discovered; the concept was based on the supposed indistinguishability of proton and neutron in the absence of electromagnetism. We have built it on the supposed indistinguishability of u and d quarks.

Apart from the classification of states, what are the other uses of isotopic spin? The 'symmetry of nature' slipped into the text above is, in this case, invariance under the simultaneous change u → d and d → u (plus likewise for the antiquarks). This situation is similar to the perfect symmetry associated with colour encountered in Section 10.7. Invariance means that interactions do not change and the properties derived from them such as transition rates, cross-sections and masses also do not change; in the specific case of isotopic spin the interaction implied is the strong interaction alone. The electromagnetic and weak interactions do not possess this symmetry. It can be proved quite generally that such a symmetry leads to conservation laws and in this case the law is that total isotopic spin of a system and its third component are conserved (they are good quantum numbers) during any processes inside that system that are caused by the strong interactions. (See Section 13.1 for more on these ideas.) Again there are very close analogies with angular momentum and its conservation. We expect to find selection rules and relations between reaction cross sections and between decay rates. Let us look briefly at one example of how selection rules can arise: consider the 2^3S_1 state of charmonium, $\psi(3685)$. It is massive enough to be energetically free to decay to the 1^3S_1 ground state $\psi(3097)$, by either of two channels:

$$\psi(3685) \to \psi(3097) + \pi^0, \tag{10.6}$$

$$\psi(3685) \to \psi(3097) + \eta. \tag{10.7}$$

Both states of charmonium have i-spin 0. The π^0 has i-spin 1 since it belongs to an i-spin triplet. The ψ has $t=0$. We can see immediately that reaction (10.6) is forbidden to proceed by strong interactions because the initial state has $t=0$ but the final state has $t=1$. This is not the case for (10.7) as both initial and final state have total i-spin zero. This prediction is verified by the data. Another, but less obvious example is that of the decay of the η; the strong interaction decay to pions is forbidden by a selection rule based on isotopic spin conservation and consequently the η decay is overtly electromagnetic ($\eta \to \gamma + \gamma$) or is covertly electromagnetic to pions ($\eta \to \pi^+ + \pi^- + \pi^0$), covertly because the electromagnetic mechanism is between the charges of the quarks and can only be inferred.

To exploit fully the theory of isotopic spin and the associated symmetry would take us beyond the scope of this book. For the present we hope the reader will note that the states of hadrons and of light nuclei belong to isotopic spin

Comment Isotopic spin symmetry

The strong interactions are invariant under any continuous rotation in isotopic spin space. The rotation through 180° around the second axis (equivalent to the y-axis in ordinary space) changes u→d, d→u, ū→d̄, d̄→ū. The analogy in ordinary space is the invariance of all known physical processes under continuous rotations in that space. The rotation through 180° around the y-axis changes spin ↑ to spin ↓ and vice versa. The latter invariance leads to the existence of eigenstates of angular momentum and to the conservation of angular momentum (see Section 13.1). The isotopic spin invariance also has analogous consequences and here are some of them:

1. The hadrons belong to isotopic spin multiplets.
2. The ground and excited states of light nuclei also belongs to isotopic spin multiplets.
3. Total isotopic spin of a system and its third component are conserved in strong interactions among the components of the system.

If the symmetry were perfect the members of any isotopic spin multiplet would degenerate. However, the electromagnetic interaction and the u–d quark mass difference spoil that symmetry. The consequences are that

(1) the hadron multiplet degeneracy is broken with splitting of order 1 to 10 MeV;
(2) the nuclear level degeneracies are also broken;
(3) processes involving electromagnetism can violate isotopic spin conservation.

The weak interaction also violates isotopic spin conservation. For example, in neutron β-decay the i-spin of the hadrons involved changes from $t=\frac{1}{2}$, $t_3=-\frac{1}{2}$ of the neutron to $t=\frac{1}{2}$, $t_3=+\frac{1}{2}$ of the proton (i-spin has nothing to do with the leptons). This change is a rotation of the i-spin vector. This suggests that it is better to think not of non-conservation but to give the selection rules that apply to the change from the initial to the final state hadrons in other interactions. In the case of electromagnetism these rules are that

$$\Delta t = 0, +1, \text{ or } -1.\ \Delta t_3 = 0.$$

For the weak interaction the rules cannot be given in a simple form and are not very helpful; we shall find in Section 12.4 that it is easier to think in terms of the changes in quark flavours.

ultiplets. If the symmetry were perfect, all members of a multiplet would have qual mass. Electromagnetic effects and the u–d mass difference break the ymmetry and give splittings of order 1 to 10 MeV. This situation is summarized the comment.

PROBLEMS

10.12 What is the isotopic spin of the D_s meson ($^1S_0$), $j^P=0^-$, $c\bar{s}$)? Why would the decay of the $D_s^* \rightarrow D_s + \pi^0$ be forbidden under the effect of the strong interactions? (D_s^* is the $^3S_1$ state of $c\bar{s}$.)

10.13 Explain how the isospin and strangeness quantum numbers arise from the existence of multiplets of elementary particles and from experimental observations on their production and decay. State the isospin selection rules observed in strong, electromagnetic, and weak interactions, and with reference to these selection rules discuss each of the following statements:

(a) the π^+ and π^- mesons are of equal mass, but the Σ^+ and Σ^- baryon masses differ by 8 MeV/c^2;

(b) the mean life of the Σ^0 baryon is many orders of magnitude smaller than those of the Λ^0 and Ξ^0 baryons;

(c) the reaction $K^+ + p \rightarrow \Sigma^+ + K^+$ has not been observed at any K^+ energy.

(Adapted from 1977 examination of the Final Honours School of Natural Science, Physics, University of Oxford.)

0.12 Conclusion

'his chapter has been a particularly dense one. We would like to round it off by oing two things. The first is to show in Fig. 10.20 the eightfold way diagrams of he octets of the $j^P=0^-$ mesons and of the $j^P=\frac{1}{2}^+$ baryons and the decuplet of $^P=\frac{3}{2}^+$ baryons. These patterns are moderately easy to memorize and may help ewcomers to remember the particles. When the quarks heavier than strange are dded the diagrams become more than two dimensional: in Fig. 10.21 we show s best as we can the pattern for the $j^P=0^-$ mesons when charm is added. The econd thing we wanted to do was set out a table of outstanding experimental acts which can be quoted as supporting the existence of quarks, or of spin $\frac{1}{2}$ for uarks, or of colour, and so on, each independently. The problem is that the vhole is an interlocking structure and it is impossible to order the facts into such eat categories. Instead we have made a table of the experimental facts (some oncatenated under rather vague titles such as 'hadron states') that in varying legrees contribute to each aspect of the model and that taken as a whole rovides overwhelming evidence for spin-$\frac{1}{2}$ quarks, spin-1 gluons, and colour. This is Table 10.11; by ticks it indicates the sources of particularly strong upport for particular details of the model.

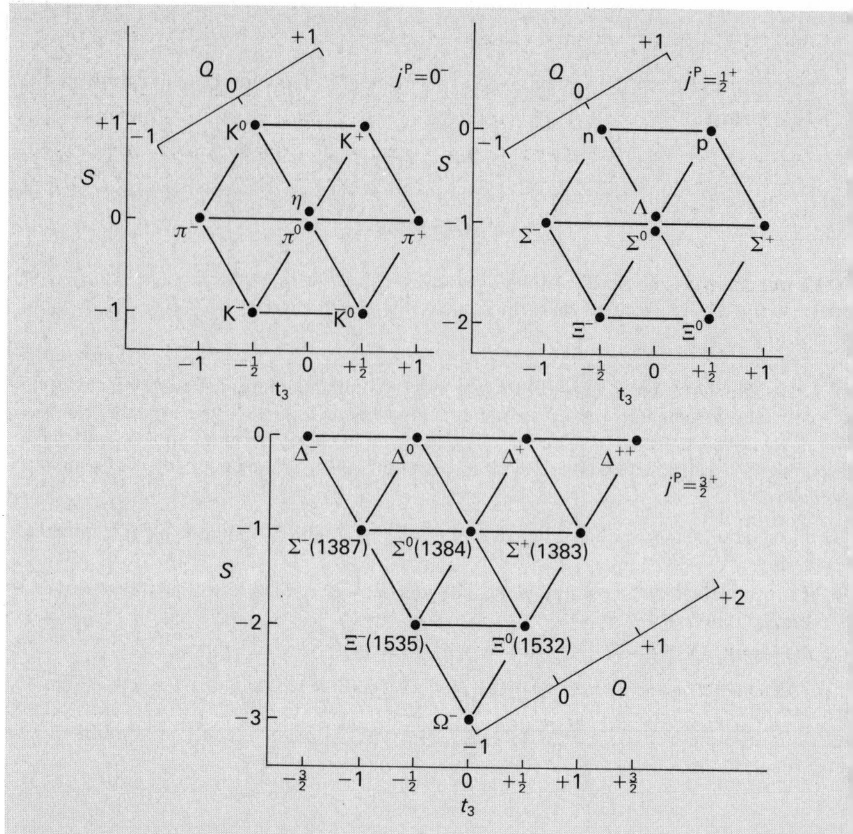

Fig. 10.20 The eightfold way diagrams for the $j^P=0^-$ octet of mesons, for the $j^P=\frac{1}{2}^+$ octet of baryons and for the decuplet of $j^P=\frac{3}{2}^+$ baryons. The symbols are S for strangeness, Q for charge and t_3 for the third component of isotopic spin. The particles have their conventional symbols. The technical term for such a diagram is *weight diagram*.

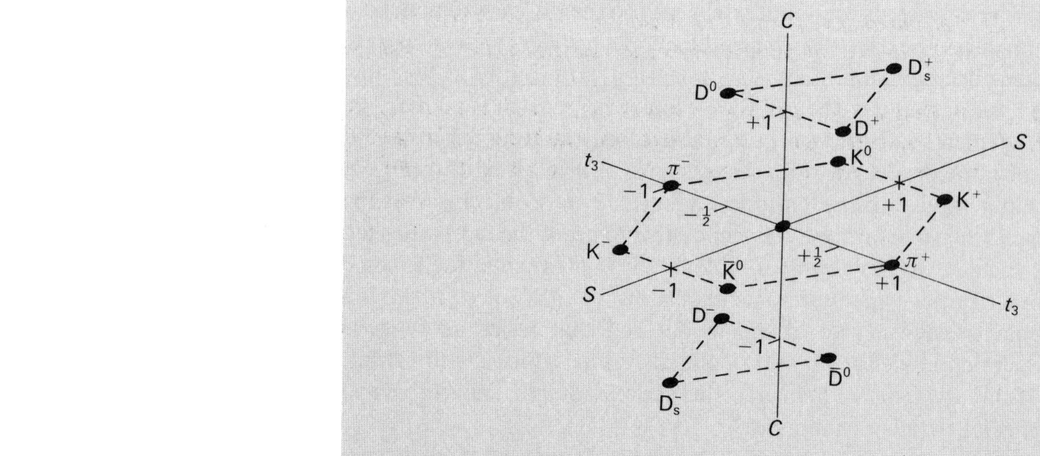

Fig. 10.21 The eightfold way weight diagram for the $j^P=0^-$ mesons including those with charm. The drawing is isometric with three orthogonal axes of third component of i-spin (t_3), of strangeness (S) and of charm (C) respectively. The dotted lines join particles on the periphery of the diagram sectioned at different values of charm, +1, 0, and −1. All the ${}^1S_0$ neutral, self-conjugate $q\bar{q}$ states are at the centre: they are the π^0, η (Table 10.3), η' (Table 10.4) and η_c (Fig. 10.14).

Table 10.11 The experimental evidence that supports aspects of the quark–parton model.

Experimental evidence	Quarks		Gluons		Colour	
	Existence	Spin $\frac{1}{2}$	Existence	Spin 1	Existence	Forces
Hadron states	√					
Hadron spins		√				
Quark statistics in Δ^{++}, Δ^-, Ω^-					√	
R for e^+e^- annihilation	√				√	
2-jet ang. distribution in $e^+e^- \to$ hadrons		√				
Inelastic lepton–nucleon scattering	√	√	√	√		
3-jets in $e^+e^- \to$ hadrons			√	√		
Mean lifetime of π^0.					√	
Only $q\bar{q}$, qqq, $\bar{q}\bar{q}\bar{q}$ hadrons observed					√	
Charmonium						√

References

Cahn, R. N., and Goldhaber, G. (1989). *The Experimental Foundations of Particle Physics*. Cambridge University Press.

Close, F., Marten, M., and Sutton, C. (1987). *The Particle Explosion*. Oxford University Press.

Feynman, R. P. (1972). *Photon-Hadron Interactions*. Benjamin, Reading, Massachusetts.

Particle Data Group (1988). Review of particle properties, *Physics Letters* B, **204**, 1–486.

Rochester, G. D. (1988). *Proceedings of the International Conference to Celebrate the 40th Anniversary of the Discoveries of the π and* V *Particles*, held at Bristol University, July 1987, pp 121–131 (ed. Foster, B. and Fowler, P. H.). Adam Hilger, Bristol.

11

The Electromagnetic Interaction

11.1 Introduction

As we indicated in Chapter 9, there now exists a theory that has unified the electromagnetic and the weak interaction. However, for us it is convenient to deal with them separately, since the area in which their effects overlap seriously is only now (1989) about to be investigated experimentally. In addition, many atomic and subatomic electromagnetic effects have been known and understood for many years and have played a large role in the experimental techniques of nuclear and particle physics, all without any need to connect them with the weak interaction.

This book was not planned to include much detail on experimental techniques. However, since these techniques almost exclusively depend on the electromagnetic interaction, we have three sections which describe how charged particles and photons interact electromagnetically with materials. All these processes lead to the production of free electrons and of positive and negative ions and many techniques depend on detecting these charged entities. We stop at their production and do not deal in detail with what happens thereafter in detectors making use of these phenomena. However, at least we can say that free electrons can be detected if produced in a suitable gas, liquid or solid. Atomic and molecular excited states are also produced: scintillation counters detect visible light emitted on their de-excitation when this occurs in certain transparent materials.

In Sections 11.2 and 11.3 we shall discuss the electromagnetic effects which cause kinetic energy loss from charged particles moving through materials. Clearly nuclear reactions can also occur and cause loss of energy but their significance depends on the circumstances. Photons are clearly absorbed or scattered by the effect of the electromagnetic interaction and in Section 11.4 we shall concentrate on those which are extra-nuclear: however, photons can be absorbed and scattered by nuclei; their absorption can cause the ejection of nucleons or, if the energy is high enough, the production of π-mesons or other short-lived particles. However, all these effects can be taken to be small compared with the extra-nuclear effects when considering the interaction of photons with materials.

From Section 11.5 on we shall deal with the nuclear and particle effects of electromagnetism. In Section 4.4 we discussed the Coulomb contribution to the mass of nuclei; in Sections 8.7 and 8.8 we discussed the moments of nuclei. We think of these as static effects and, in contrast, the effects we shall consider are dynamic: nuclei radiating photons or suffering changes engendered by the absorption of real or virtual photons.

11.2 The energy loss by ionization

A charged particle passing through ordinary material loses energy through many collisions with the bound electrons. We are in a position to make a rough attempt to find a simple version of the formula for the kinetic energy loss per unit distance of material traversed: this quantity is usually written as dT/dx and, if T is the kinetic energy of the particle, is a negative quantity. Its magnitude is called the **stopping power** and it is characteristic of the material. The collisions are caused by the electromagnetic interaction between the charge of the particle and that of the electron. Now our incident charged particle is normally a nucleus, for example, a proton or an α-particle, so the collisions that we are interested in are just like electron–nuclear scattering, only observed in an inertial frame in which the electron instead of the nucleus is at rest (neglecting for a moment the effects of electron binding). So we should be able to take over the Mott scattering formula (Table 3.1), which is just an extension of the Rutherford scattering formula, and use it to calculate the energy loss. This is done in Table 11.1. The result is the Bethe–Bloch formula for the stopping power. Of course the last steps are no more than a statement about the results of a proper calculation that takes into account the binding of the atomic electrons. However, it is interesting that we can go some way using ideas which had their beginning in Rutherford scattering.

Let us look at the major features of the formula for the stopping power:

$$-\frac{dT}{dx}=nZz^2\frac{4\pi\alpha^2\hbar^2}{m_e\beta^2}\left[\ln\frac{2m_ec^2\beta^2}{I(1-\beta^2)}-\beta^2\right]. \tag{11.1}$$

Table 11.1 Bethe–Bloch formula for the stopping power

The Bethe–Bloch formula for energy loss looks formidable and not transparent in its meaning. However, we can do half a derivation starting with a simple idea which helps to give the formula meaning. The reason for 'half' a derivation is that we will understand why our result is wrong and without serious justification just multiply our result by two!

We start with the Mott scattering formula (see Table 3.1) for the differential scattering cross-section for an electron of momentum P and velocity V by a heavy nucleus of charge $z|e|$

$$\frac{d\sigma}{d\Omega}=\frac{z^2\alpha^2(\hbar c)^2}{4P^2V^2}\operatorname{cosec}^4\frac{\theta}{2}\left(1-\frac{V^2}{c^2}\sin^2\frac{\theta}{2}\right).$$

The momentum transfer $|\boldsymbol{q}|=2P\sin(\theta/2)$, so we change the formula as in Problem 1.6 to a derivative in q^2:

$$\frac{d\sigma}{dq^2}=\frac{4\pi z^2\alpha^2(\hbar c)^2}{q^4V^2}\left(1-\frac{V^2}{c^2}\frac{q^2}{4P^2}\right).$$

Now we move from the frame in which the nucleus is at rest and the electron is moving to the frame in which the heavy nucleus of mass M is moving with velocity V towards a collision with a stationary electron. The momentum P in the formula is still the momentum of an electron of velocity V, that is $P=m_eV\gamma$, $\gamma=(1-V^2/c^2)^{-1/2}$. Now q^2 is the same in both frames. This is true non-relativistically, and, if q^2 is correctly defined, is also true relativistically. In both cases, for a free electron initially at rest, a transfer q^2 leads to an energy transfer ν to the electron given by

$$2m_e\nu = |q^2|.$$

Therefore

$$\frac{d\sigma}{dq^2} = \frac{1}{2m_e}\frac{d\sigma}{d\nu} = \frac{4\pi z^2\alpha^2(\hbar c)^2}{(2m_e\nu)^2V^2}\left\{1 - \frac{\nu}{2m_ec^2}\left(1-\frac{V^2}{c^2}\right)\right\}.$$

Thus the cross-section for energy loss in the energy range ν to $\nu+d\nu$ is

$$\frac{d\sigma}{d\nu}d\nu = \frac{2\pi z^2\alpha^2(\hbar c)^2}{m_eV^2}\frac{1}{\nu^2}\left\{1 - \frac{\nu}{2m_ec^2}\left(1-\frac{V^2}{c^2}\right)\right\}d\nu.$$

If this nucleus loses energy $-dT$ in a distance dx in a material containing n atoms of atomic number Z per unit volume, then

$$-dT = nZ\,dx\int_{\nu_{min}}^{\nu_{max}}\nu\frac{d\sigma}{d\nu}d\nu.$$

Therefore

$$-\frac{dT}{dx} = nZ\left[\frac{2\pi z^2\alpha^2(\hbar c)^2}{m_eV^2}\right]\left[\ln\left(\frac{\nu_{max}}{\nu_{min}}\right) - \left(1-\frac{V^2}{c^2}\right)\left(\frac{\nu_{max}-\nu_{min}}{2m_ec^2}\right)\right].$$

where ν_{max} and ν_{min} are the maximum and minimum values of the energy transfer and we have done the formal integration. Now for a heavy incident particle $\nu_{max}=2m_eV^2/(1-V^2/c^2)$ (see Problem 11.1). But, although ν_{max} is very much greater than ν_{min}, we do not know what the latter is. In fact, at large ν the bound electrons can be assumed to be free but at low ν that assumption cannot be made. It becomes possible to have an energy (ν) and momentum ($\boldsymbol{q}$) transfer that do not have to satisfy the constraint ($2m_e\nu=q^2$) imposed if those quantities impact on a free electron. The momentum and energy transfer can now lead not only to ionization but also to excitation of an atom. Thus at low ν we have not done the right integral; the correct one has to be done over the two variables ν and $\boldsymbol{q}$ for differential cross-sections that depend on the detailed atomic structure. This part of the integration gives an important contribution to the whole energy loss and we expect it to differ by some factor from the result that we so far have obtained: we can only give the final result. The Bethe–Bloch formula parametrizes these problems by defining a quantity I, the **mean excitation potential,** for all Z atomic electrons: this is an element-dependent parameter which has to be determined from experimental data and is approximately given by the formula $I=16(Z)^{0.9}$ eV (for $Z>2$). Now the correct result is found by replacing ν_{min} by I and doubling the result. That doubling cannot be justified here and must be taken on trust. What is important is that the formula is a good representation of the stopping power once I is known. In addition, we have established that it has a strong connection with the Rutherford scattering formula. The result is the Bethe–Bloch formula for the stopping power:

$$-\frac{dT}{dx} = nZ\left(\frac{4\pi z^2\alpha^2(\hbar c)^2}{m_eV^2}\right)\left[\ln\frac{2m_eV^2}{I(1-V^2/c^2)} - \frac{V^2}{c^2}\right].$$

Putting $\beta=V/c$, ρ=mass density of the material, A=atomic weight and N_A=Avogadro's number, we have

$$-\frac{1}{\rho}\frac{dT}{dx} = \frac{N_AZ}{A}\left(\frac{4\pi z^2\alpha^2(\hbar c)^2}{m_eV^2}\right)\left[\ln\frac{2m_ec^2\beta^2}{I(1-\beta^2)} - \beta^2\right].$$

This formula gives the energy loss in traversing normally a layer of unit mass per unit area; this is the usual way of giving numerical values of the stopping power.

The units are energy loss per metre if n is number of atoms of atomic number Z per cubic metre in the material; z is the charge on the incident particle. In the logarithm, I, the **mean excitation potential** (see Table 11.1), and the rest energy of the electron must be in identical units. Immediately we note that $\mathrm{d}T/\mathrm{d}x$ is independent of the mass of the moving particle but depends on its charge squared (z^2) and, at non-relativistic velocities V ($=\beta c$), on β^{-2}. These are the gross features. Their effect can be seen in the tracks of charged particles slowing down in photographic emulsion (Figs. 9.14 and 11.1) where the line density of developed silver grains is proportional to $|\mathrm{d}T/\mathrm{d}x|$.

However, it is clear that the formula cannot be correct at low velocities: the logarithm becomes negative when $\beta < \sqrt{(I/2m_e c^2)} \simeq 4 \times 10^{-3}$ in hydrogen and that is physical nonsense. In addition, moving positive particles at velocities less than about $7 \times 10^{-3}c$ can capture electrons and this reduces the value of its effective charge from $z|e|$ and thereby reduces $\mathrm{d}T/\mathrm{d}x$. That is not an irreversible event: one or more electrons may be captured and lost many times in the final stages of slowing down.

At the other end of the scale as $V \to c$ the $(1-\beta^2)^{-1}$ factor in the logarithm halts the decrease in the stopping power and there is a minimum at $\beta \simeq 0.96$ followed by a rise (Fig. 11.2). This **minimum ionization loss** is an important quantity: it and the stopping power are usually expressed as the energy loss in traversing normally a layer which has unit mass per unit area (see Section 2.9). That is $((1/\rho)\mathrm{d}T/\mathrm{d}x)_{\mathrm{min}}$, where ρ is the density of the material. Convenient units are energy loss in megaelectronvolts for a particle crossing normally a layer of material of thickness such that its superficial density is 1 gram per square centimetre ($\mathrm{g\,cm^{-2}}$). The minimum varies from 4.1 $\mathrm{MeV\,g^{-1}\,cm^2}$ in hydrogen, decreasing smoothly through the periodic table to 1.1 $\mathrm{MeV\,g^{-1}\,cm^2}$ in uranium. (For those wedded to SI units these figures become 0.41 and 0.11 $\mathrm{MeV\,kg^{-1}\,m^2}$; strictly you should also change megaelectronvolts to joules, but then you are straying far into numbers which no longer relate to what is needed for calculations with conceptually convenient magnitudes.)

Beyond the minimum, the Bethe–Bloch formula predicts that the stopping power increases indefinitely as $\ln(1-\beta^2)^{-1}$. At greater velocities of the incident particle, its electric field is Lorentz contracted and the transverse electric field is increased at large distances. The result is that the energy loss should also increase as more atoms come within the range of electric fields able to excite an atom. This is the origin of part of this rise of stopping power with velocity. But, in fact, the slope of the rise beyond the minimum is reduced by what is called the **density effect**. These long-distance effects are screened by the dielectric properties of any intervening molecules and the relativistic increase is reduced, as seen in Fig. 11.2 for the case of hydrogen. The effect is greater the greater the density of the material involved.

Measurements of the ionization deposited locally to the trajectory of a particle show that this quantity is less than that expected from the energy loss and does not continue to increase as the stopping power increases at greater energies of the particle; after a small relativistic rise from the minimum, this ionization reaches a value which remains constant at all greater energies. The data in Fig. 11.2 gives the average of the total energy loss by ionization. The equation for stopping power in Table 11.1 (and equation (11.1)) was derived by putting ν_{max} equal to the maximum energy that can be transferred to a stationary electron allowed by kinematics: thus many energy-losing collisions give secondary electrons (δ-rays) sufficiently energetic to deposit that energy decrement far or very far from the path of the particle losing energy. Many measurements of

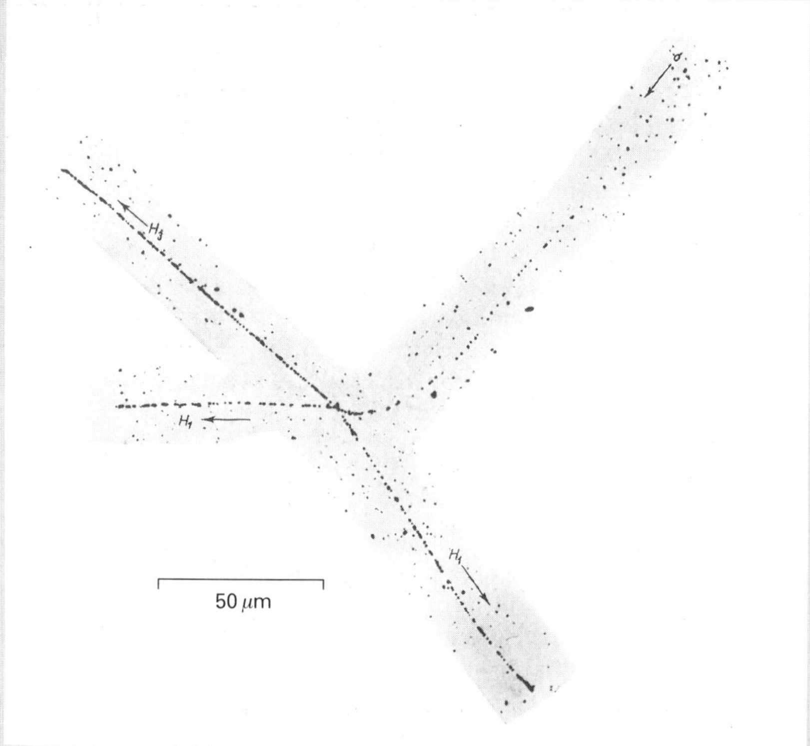

Fig. 11.1 This photograph shows tracks produced in a photographic emulsion exposed at a high altitude to cosmic radiation. The interpretation is that a π^--meson has produced the track marked σ, come to rest, been attracted into a Bohr-like orbit around a nucleus, and has then been absorbed by the nucleus. Its rest energy has disrupted the nucleus and three charged particles have been emitted. The two labelled H_1 are protons and that labelled H_3 is a triton. The shortest H_1 track is that of a proton which left the emulsion and had an initial energy greater than 3 MeV. The longer H_1 track is a proton of 3.7 MeV which came to rest in the emulsion. The triton, H_3, had an energy of 5.6 MeV and also stopped in the emulsion.

The ionization caused by energy loss makes silver halide grains of the emulsion developable into silver grains. The number of grains per unit length (grain density) of a track is a measure of $\mathrm{d}T/\mathrm{d}x$. We can therefore use this event to illustrate three properties of the process of slowing down charged particles which initially have a kinetic energy of a few mega-electron-volts, as follows.

1. The increase in ionization energy loss ($\sim V^{-2}$) as the particle slows down and approaches the end of its range. This property allows a determination of the original direction of motion of the particle and in this case a separation of the arriving particle from the products of the reaction.
2. The change of ionization energy loss along a track as a function of the residual range is greatest for the lightest particle (the π^-) because it loses more velocity in the same distance than does a heavier particle.
3. The multiple scattering in the last stages of slowing down to rest is greatest for the lightest particle.

The tracks in this photo should be compared to those of π^+ and μ^+ mesons in Fig. 9.14. The emulsion of Fig. 9.14 was much more sensitive than that of Fig. 11.1 and was capable of detecting electrons and other minimum ionizing particles (Section 11.2). This difference of sensitivity also shows up by giving much less dense tracks in Fig. 11.1.

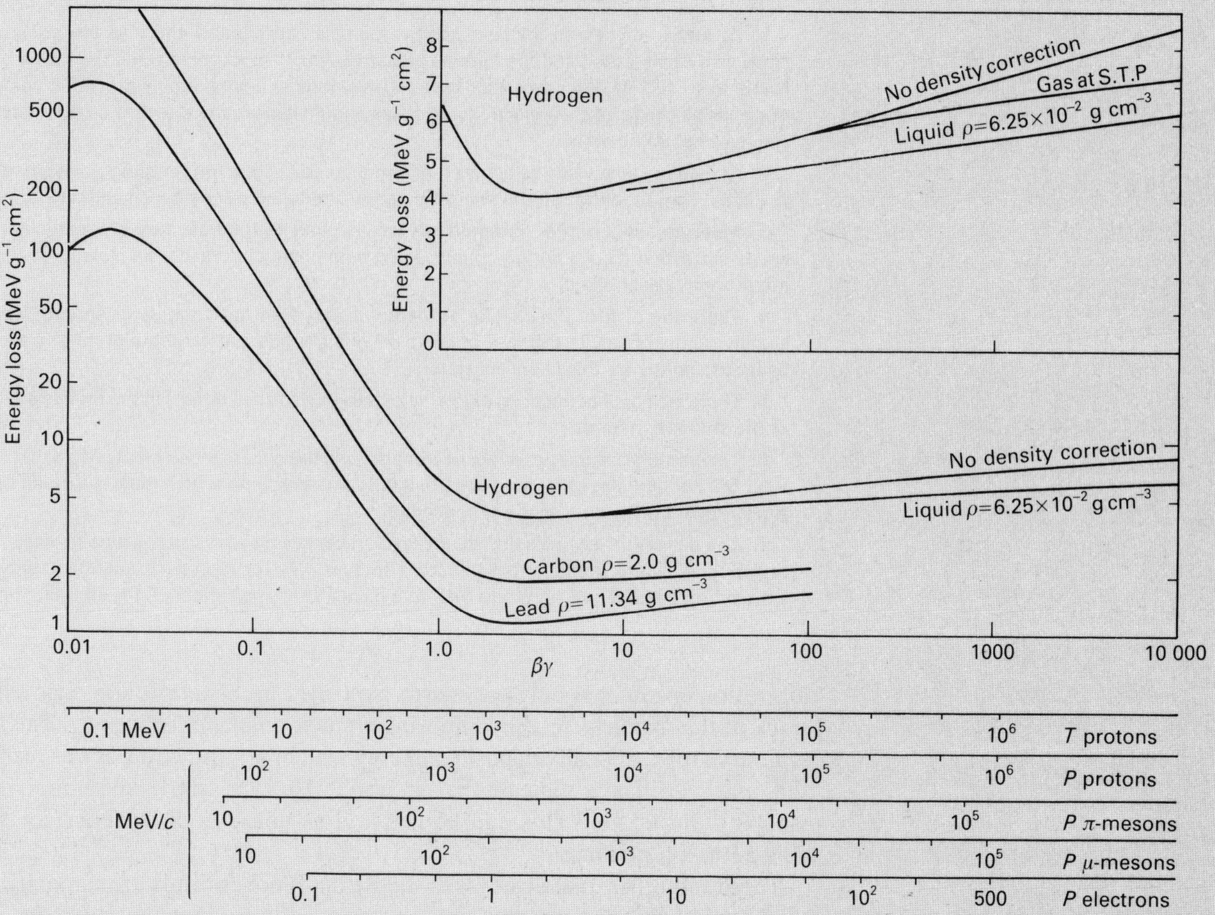

Fig. 11.2 The stopping power of hydrogen, carbon and lead as a function of $\beta\gamma = P/Mc$ for a singly positively charged particle of momentum P and mass $M \gg$ the mass of the electron. The curves are labelled with the material density ρ in g cm^{-3}. The stopping power is expressed as MeV g^{-1} cm^2; that is energy loss in MeV for a particle traversing normally a layer which has a surface density of one gram per centimetre squared. Thus at the minimum a proton loses 4 MeV in about $\rho^{-1} \approx 14$ cm of liquid hydrogen but in lead loses about 1.1 MeV in less than a one millimetre. The inset with a linear vertical scale shows the relativistic rise in hydrogen without the density effect correction and curves showing the result corrected for the gas and for the liquid. Horizontal scales are attached to show the kinetic energy (T) and momentum (P) of protons and the momentum of π-mesons, μ-mesons. There is also a momentum scale for the electron: although the curve has the same form the energy loss by ionization by electrons is slightly different from that by heavy particles.

At low values of $\beta\gamma$ the curves for carbon and lead show the fall away and turn over from the $1/V^2$ dependence as the moving particle (assumed to be positively charged) picks up and loses electrons, or becomes insufficiently energetic to knock out electrons from the inner atomic shells. The first of these low-velocity effects will depend on the charge of the particle and will, therefore, be different for π^+ from π^-. The second effect will depend on the energy of the particle and so on its mass for a given $\beta\gamma$. Therefore these curves, drawn for protons, do not apply immediately to the other particles at $\beta\gamma < 0.1$.

Comment The Bethe–Bloch formula

1. At low velocities an incident positive particle can capture and lose electrons many times as it slows down and the formula breaks down. This occurs at velocities below about 6×10^6 m s^{-1} (0.2 MeV for protons). Fast negative ions or neutral atoms can suffer complete or partial stripping of their electrons: in these circumstances the effective charge may also vary and be unknown.

2. At extreme relativistic velocities the formula predicts that the stopping power increases ($-dT/dx \sim \ln(\gamma)$), although the rate of increase is reduced by the density effect.

3. Measurements of the ionization left locally close to the trajectory of a particle losing enegy show that it does not go on increasing with energy but reaches a plateau, in spite of the property described in 2.

4. The value of I for a particular element will vary from the formula given and depends slightly on the phase. In the case of chemical compounds, the contribution of one element depends slightly on this chemical state.

5. There are small corrections at low velocities due to shell effects in the atomic structure of the stopping material.

6. The formula assumes that the incident particle has a mass much greater than that of the electron. For particles other than electrons (and positrons) for which there is a modified but similar formula, the error made is negligible.

7. The derivation assumes that the heavy particle has no spin and has a point-like charge distribution. These things affect the cross section only for the very rare large energy and momentum transfer collisions and therefore have a very small effect on the stopping power.

ionization encompass only the energy deposited in the immediate region of the track of that particle, thereby imposing a much smaller effective ν_{max}. The result is a curve for the locally deposited energy which, as a function of the energy of that particle, has a plateau that is about 10–15% above minimum in the stopping power for solids and liquids and as much as 50% above for gases, depending on pressure.

These and other qualifications to the Bethe–Bloch formula are summarized in the accompanying comment.

We cannot give all the details of the fine tuning of the formula for atomic and other effects: a great deal of effort has gone into determining the stopping power as a function of the properties of the moving particle and of the material. We end our discussion of the Bethe–Bloch formula by noting that we derived our formula for heavy incident particles. A more complicated but, in its essentials, a similar formula holds for electrons, and another for positrons. The differences from the formula for heavy particles are small and for the purposes of discussion we shall assume that they are the same. These light particles have much higher velocities for the same kinetic energy as protons, for example. This is brought out by an examination of Fig. 11.2. This is a plot of some stopping powers against $\beta\gamma$ ($\beta = V/c$, $\gamma = (1-\beta^2)^{-1/2}$) with added abscissa scales for the kinetic energy of protons and the momentum of several other unit charged particles. Thus the energy loss in hydrogen is the same (the minimum) for protons of kinetic energy 2.5 GeV as it is for electrons of momentum 1.74 MeV/c (kinetic energy 1.3 MeV). Such an electron will travel much less than another 4 cm in liquid hydrogen before coming to rest, even if ionization loss is the only mechanism of energy loss, whereas a proton of the same velocity has many tens of metres of potential travel in the unlikely event that it avoids inelastic or elastic scattering on a proton at rest. Thus we can see that light particles slowing down do not traverse as great a distance as do heavy particles of the same initial velocity. In addition light particles cover less distance in the last stages of slowing down when the energy loss per unit distance is large. Thus it is that measurement of the ionization loss of electrons or positrons almost always yields a value close to the minimum. On the other hand, a proton of a given kinetic energy loses energy more rapidly and comes to rest in a shorter distance than does an electron of the same kinetic energy.

This discussion indicates that a natural outcome of energy loss by ionization is the concept of **range.** Initially we will consider the distance, the **track length,** R, traversed by a particle of initial kinetic energy T_0 before it comes to rest in a stopping material. If we take ionization to be the only loss mechanism, then

$$R=\int_{T_0}^{0}\frac{\mathrm{d}T}{\mathrm{d}T/\mathrm{d}x}. \tag{11.2}$$

This integration cannot be performed analytically: however, we can do a rough job for a non-relativistic particle of mass M, charge z, putting

$$-\frac{\mathrm{d}T}{\mathrm{d}x}=\frac{az^2}{V^2}.$$

This neglects the slow variation of the logarithmic term. The quantity a is a constant which depends on the nature of the stopping material. It follows that

$$R=\frac{aT_0^2}{z^2M} \quad \text{or} \quad R=\frac{aMV_0^4}{4z^2}, \qquad \text{where} \quad V_0=\sqrt{(2T_0/M)}. \tag{11.3}$$

A more careful analysis shows that

$$R\propto T_0^{3/2}/M^{1/2} \qquad \text{or} \qquad R\propto MV_0^3$$

are a better but still only approximate representation of the variation of range. These formulae reinforce what we have said before: for the same T_0, the range is less the greater the mass but for the same V_0, $R\propto M$. Now try Problem 11.2. Figure 11.3 shows the range–kinetic energy curves for protons, pi-mesons and mu-mesons in carbon to illustrate some of these properties.

The loss of energy is not a smooth process and $\mathrm{d}T/\mathrm{d}x$ is really the average over very many collisions. In a finite distance the actual energy loss can fluctuate from the expected value. This effect is called **straggling**. Associated with the straggling of energy loss there is also **range straggling**. Consider the track length of particles of a fixed incident energy; the energy lost in each collision can fluctuate so that the number of collisions required to stop the particle fluctuates from particle to particle and hence the track length fluctuates: this is range straggling. These matters have been considered and quantified. Interested readers are referred to the book by Fernow (1986).

We emphasized that the range we have considered is the track length: many experimental determinations of range are made by determining the distance of transmission through absorbing material. Thus to determine the range for particles of a given energy the number surviving various thickness is measured. For α-particles of energy of up to tens of megaelectronvolts, the survival is very nearly 100% until a certain thickness, after which it drops to zero rapidly (see Fig. 11.4). In the case of electrons no such well-defined distance is found and survival sometimes follows an approximately exponential decrease. This difference can be blamed on the difference in scattering and multiple scattering for

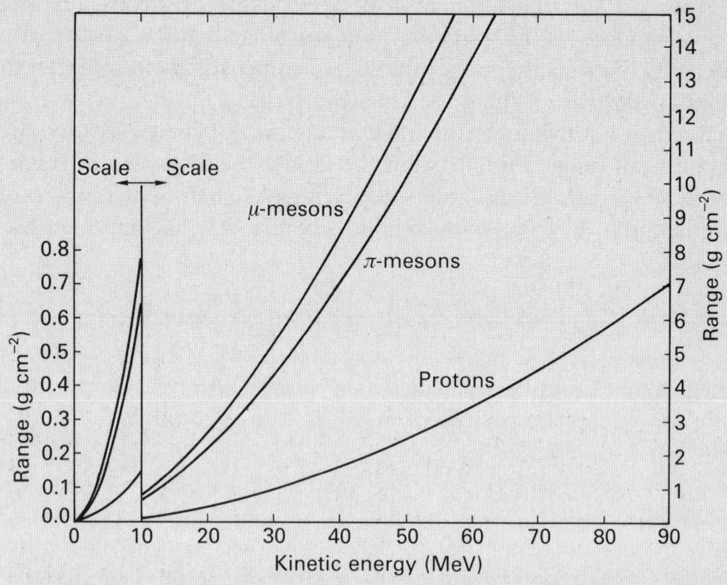

Fig. 11.3 The range–kinetic energy curves for protons, π-mesons and μ-mesons in carbon. Equation (11.1) tells us that at low energies the rate of energy loss is varying inversely as the velocity squared and therefore every increment in incident energy requires a disproportionate increase in range to remove that increase in energy; thus for all particles the range increases faster than linearly with the incident kinetic energy. In addition, Figure 11.2 shows that for the same kinetic energy lighter particles suffer less energy loss and therefore have greater ranges. Thus the features of this figure can be predicted from the properties of the stopping power.

the two particles. Light electrons of a given velocity have low momentum and are easily scattered or badly suffer the effect of multiple scattering, both of which cause an increasing deviation from a straight track, and that means that the track length has a projection on the original electron direction which is shorter or much shorter than the track length (see Fig. 11.5). For α-particles of the same velocity, the momentum is much greater and the track suffers much less deviation and the projected track length is in most cases only slightly less than the track length. For particles intermediate in mass these effects are intermediate between those for electrons and α-particles. For all but electrons the measured range is often the track length projected onto the incident direction of the particle. However, in techniques in which the trajectory of a particle may be seen (for example, photographic emulsions or cloud chambers), the real track length may be measurable and be called the range.

Multiple scattering was considered when we were thinking about Rutherford scattering (Section 1.1, Problem 1.7). It represents the effect of many small-angle Coulomb scatters, adding up statistically to give a deflection to the direction and a deviation from straight of the trajectory of a particle after a finite distance in a material (see Fig. 11.5). The theory is quite involved but we might guess that its effect will depend on the square of the charge (z^2) and inversely as some power of the momentum (P) of the moving particle, and on the square of the atomic number of the material (Z^2). However, the total effect is one of a random walk: for many trials, the frequency distribution of the space

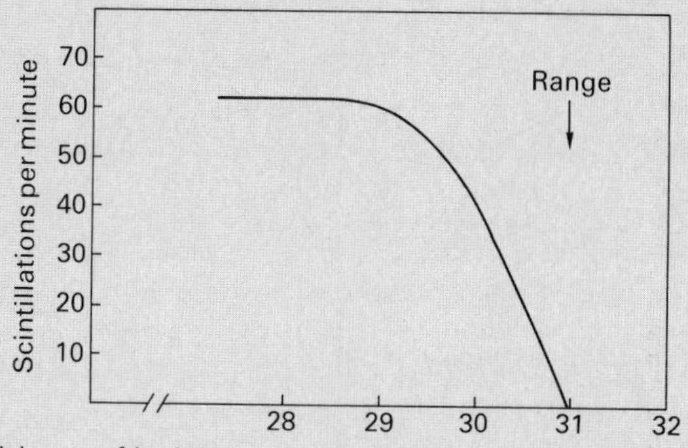

Fig. 11.4 The transmission of α-particles (7.68 MeV) as a function of thickness of hydrogen gas at 15°C and one atmosphere pressure. Taylor (1913) obtained this and other curves by visually counting the scintillations caused by individual α-particles from a source reaching a screen of zinc sulphide (see Table 1.3). The ordinate is scintillations per minute and the abscissa is the distance from source to screen. At this energy the chance of nuclear interaction before coming to rest is small. In addition, the loss of energy requires very many small energy transfers and the statistical fluctuation on the number required is a relatively small fraction: thus most of the α-particles travel the same distance before coming to rest. The transmission is 100% until a thickness close to the range is reached when it drops rapidly to zero. The definition of range is a matter of convention in view of the slight fluctuation in range and the fact that the multiple scattering gives a projected range less than the total path length (Fig. 11.5). In addition the source may have a significant thickness and the range could be cut short by the existence of a residual energy below which scintillations become visually undetectable. In this work the range was defined by the point where the curve intersected the axis.

Fig. 6.6 is a cloud chamber photograph of α-particle tracks which shows the almost unique range and small straggling of α-particles of the same energy.

angle deviation is approximately Gaussian, with a mean of zero and a standard deviation, θ_0; it is the variance θ_0^2 which has the behaviour that we guessed:

$$\theta_0^2 \propto \frac{z^2 Z^2}{P^2 \beta^2} L,$$

where L is the distance in which the deviation occurs. More usefully

$$\theta_0 = \frac{20\ \mathrm{MeV}/c}{P\beta} z \left(\frac{L}{L_\mathrm{R}}\right)^{1/2} \tag{11.4}$$

where P is the momentum in MeV/c. The length L_R is called the radiation length of the material; it has the property $L_\mathrm{R} \propto Z^{-2}$ and we will define it in Section 11.3 (see also Table 11.2). As an example, for lead $L_\mathrm{R} = 5.6$ mm: traversal of this distance by a proton at minimum ionizing power ($T = 2.4$ GeV, $P = 3.2$ GeV/c) causes an energy loss of 7.2 MeV and gives $\theta_0 = 6.3$ mrad. An electron of the same velocity ($T = 1.3$ MeV, $P = 1.73$ MeV) crossing $10^{-3}\ L_\mathrm{R}$ of lead loses about 7 keV by ionization but suffers more severe multiple scattering with $\theta_0 \simeq 360$ mrad (calculated assuming the formula for θ_0 is right in this regime,

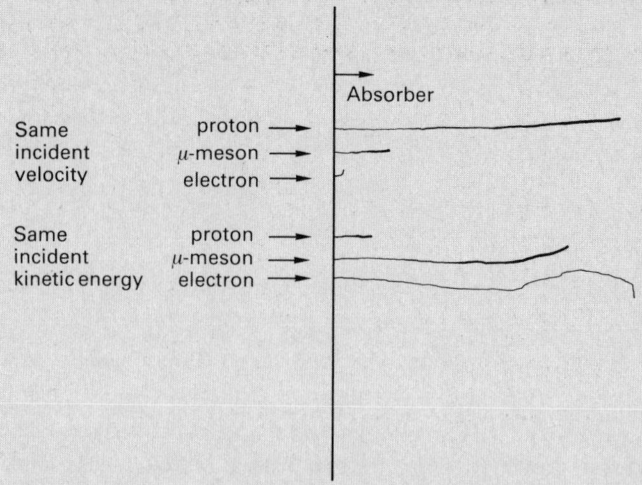

Fig. 11.5 Sketches showing typical paths for protons, μ-mesons and electrons coming to rest in an absorber (kinetic energies of a few megaelectronvolts). The electrons are lightest and suffer the greatest multiple scattering. As described in the text, for the same incident velocity the stopping power is initially the same and the most energetic particles, the protons, have the greatest range. For the same incident kinetic energy the stopping power is least for the electrons and therefore their path length is greatest. These remarks apply only at energies low enough that there are no energy-loss mechanisms other than loss by ionization. The length of the last electron path is under-represented.

which is very doubtful). An electron of the same momentum as that of the proton ($T = 3.2$ GeV), crossing L_R would suffer ionization loss and scattering almost identical to that of the proton: however, it suffers another very important energy-loss mechanism of an electromagnetic origin: as we shall find in the next section this expected radiative energy loss is about 2.0 GeV.

At low energies all particles can sometimes come to rest behind where they started, if they suffer large-scale scatters. Low-energy electrons are particularly prone to back scattering of this kind.

At the beginning of this section we called dT/dx the stopping power of a material. Since it varies with velocity it is an inconvenient measure. Better is the value of $(1/\rho)dT/dx|_{min}$, which is almost independent of the density of the stopping material and of the type of the particle losing energy. Table 11.2 gives values of this quantity for some materials. Another term frequently used is **minimum ionizing particle**: this means a particle moving so that in a given material the energy it is losing by ionization is at or near the minimum; this loosely means particles with velocities $\geqslant 0.9c$.

At this point we can ask how the kinetic energy lost by the originally incident particle is finally dissipated in a material which it traverses or in which it stops. The electrons generated by ionization are called δ-rays (Section 1.1) and can have energies up to ν_{max}, the maximum energy that can be transferred to a stationary electron by the moving particle (see Problem 11.1), although the maximum energy transfer is rare. These δ-rays can in turn cause further ionization and further δ-rays. At energies below ionization threshold the electrons will lose energy in collisions with atoms. Those last stages cause thermal excitations of the absorbing material. Thus all the energy lost by the

Table 11.2 Properties of some elements that are relevant to the electromagnetic processes leading to the stopping, slowing down, absorption or degradation of energetic radiation. The data in this table show how these properties change through the periodic table of the elements.

Element	Z	Atomic weight	ρ ($g\,cm^{-3}$)	I/Z (eV)	$\frac{1}{\rho}\left(\frac{dT}{dx}\right)_{min}$ ($MeV\,g^{-1}\,cm^2$)	L_R ($g\,cm^{-2}$)	L_R (cm)	T_{crit} (MeV)
H_2	1	1.008	0.0708*	20.4	4.09	61.3	866*	340
He	2	4.003	0.125*	19.6	1.95	94.3	754*	220
C	6	12.01	2.1–2.3†	12.3	1.85	42.7	~19.0	103
Al	13	26.98	2.70	12.3	1.63	24.0	8.9	47
Fe	26	55.85	7.87	10.7	1.49	13.84	1.76	24
Pb	82	207.19	11.35	10.0	1.14	6.37	0.56	6.9
U	92	238.03	18.7	9.56	1.10	6.00	0.32	6.2

*Liquid at boiling point.
†Commercial graphite.

Symbols: ρ=mass density, I/Z=ionization energy parameter per unit Z, $\frac{1}{\rho}\left(\frac{dT}{dx}\right)_{min}$ =minimum ionization energy loss, L_R=radiation length, T_{crit}=critical energy. We have used grams and centimetres in this table rather than the rather unwieldy SI units. For other elements and materials see Tsai (1974) and Janni (1982). The values of I are found by fitting the Bethe–Bloch formula to experimental data on the stopping power: the results have errors and all the figures given for each I/Z may not be significant. In addition, I will depend on the phase and molecular state of the element, particularly for hydrogen. The values of the minimum ionization loss are calculated using the I-values obtained as indicated. The values in the last three columns are calculated from theory. For more comprehensive data see the *Review of Particle Properties* (Particle Data Group, 1988).

incident energetic particle goes finally to heating the material, apart from a small amount of energy that may be tied up in damage to the material in which atoms are displaced from their original position or in which chemical bonds are rearranged. (Such effects may be important in situations in which the radiation dose is large.) In certain transparent materials, the molecular structure can become excited and then radiate in the visible or near visible part of the spectrum: such materials are called scintillators and the light is emitted with a mean lifetime of a few nanoseconds or longer, or very much longer, depending on the material. Coupling to a photomultiplier allows the light pulse to be converted into an electrical pulse and we have the basis of what is called a scintillation counter. Devices of this kind have a wide variety of uses in the detection of energetic particles and radiation. We also note that after the passage or stopping of a charged particle, a stage is reached in which there are ions and free electrons, having, presumably, thermal energies if free to move. It is at this stage that other experimental techniques intervene by collecting some of these electrons or ions and transforming them into electrical signals; if that is not done the ions and electrons undergo recombination. Devices using this method include, for example, the gas proportional counter, the liquid argon ionization chamber, the germanium (or silicon) solid state detector.

Many devices for detecting energetic particles rely on totally absorbing, or as near as possible totally absorbing, the energy of the incident particle and all its secondaries. Such devices are organized to give an analogue electrical signal which depends on the energy deposited and are called calorimeters; however, they do not normally rely on detecting the increment in temperature as that is

DEFINITIONS AND KEYWORDS

Stopping power The kinetic energy loss per unit length of path by a charged particle moving through a given material.

Mean excitation potential The parameter I in the Bethe–Bloch formula for the stopping power of a material. Normally found by fitting the formula to data for the stopping power. It is the geometric mean of the ionization potentials for all Z electrons of the stopping material.

Minimum ionization loss The minimum value of the stopping power for a given material. It occurs at a velocity of about $0.96c$ for a heavy charged particle.

Minimum ionizing particle A charged particle with a velocity such that its energy loss is at or near the minimum.

Density effect The reduction of stopping power at velocities approaching c due to the screening of long-range Coulomb effects by the electric polarizability of the material.

Range The distance moved by a charged particle through a material before coming to rest. Sometimes meaning distance projected onto incident direction of particle, sometimes meaning the track length.

Track length The distance along the trajectory of a particle before it comes to rest in a given material.

Straggling The fluctuations in the energy loss for a population of identical charged particles of the same kinetic energy traversing the same thickness of the same material.

Range straggling The fluctuations in range for a population of identical charged particles all of the same kinetic energy coming to rest in the same material.

too small to be easily detected, but on sampling with scintillation counters or proportional counters the ionization loss and appropriately summing the samples. These techniques are described more fully elsewhere: see, for example, England (1974), Kleinknecht (1986) and Fernow (1986).

There is one energy loss phenomenon that we must mention. The density effect is associated with the electric polarization of the medium in which the charged particle is moving. This polarization is varying and in a transparent medium this variation leads to the emission of visible photons. This is the Cerenkov effect: its properties are described in Fig. 13.4.

PROBLEMS

11.1 Show that the maximum energy that can be transferred to an electron, initially at rest, by a particle of rest mass M and total energy E is given by

$$\nu_{\max} = \frac{2m_e c^2(E^2 - M^2c^4)}{M^2c^4 + m_e^2c^4 + 2Em_ec^2} = \frac{2m_ec^2\beta^2\gamma^2}{1+\left(\frac{m_e}{M}\right)^2 + 2\gamma\left(\frac{m_e}{M}\right)^2}$$

where βc is the velocity of the particle and $\gamma = (1-\beta^2)^{-1}$, as usual. This is the relativistically exact formula. Show that for the extreme relativistic case $E \gg M^2c^2/m_e$

$$\nu_{\max} = E,$$

and for a less extreme case when $M \gg m_e$ and $E \ll M^2c^2/m_e$

$$\nu_{\max} = 2m_e\beta^2\gamma^2c^2.$$

Non-relativistically ($E = Mc^2 + T$, $T \ll Mc^2$) the maximum energy transfer is given by

$$\nu_{\max} = \frac{4m_eMT}{(m_e + M)^2}.$$

Calculate the value of $\nu_{\max}$ for μ-mesons with $E = 5$ GeV.

11.2 Show that if energy $\nu(\ll m_e c^2)$ and momentum $\boldsymbol{q}$ are transferred to a free, stationary electron, then $\boldsymbol{q}.\boldsymbol{q} = 2m_e\nu$. If $\nu > m_e c^2$ the problem is relativistic: in this case q is the four-momentum transfer. Look at Fig. 9.6 to find out what that means and show that $q^2 = -2m_e\nu$.

11.3 A particle a, of a mass $M_a(\gg m_e)$, charge $Z_a|e|$, and initial kinetic energy T_a has a range R_a. Show that the range in the same material for particle b is given by the scaling law

$$R_b(M_b, Z_b, T_b) = \frac{M_b}{M_a}\frac{Z_a^2}{Z_b^2}R_a(M_a, Z_a, T_a = M_a T_b/M_b).$$

(This does not apply to electrons or positrons.) What assumptions are implicit in the use of this scaling law? Is it valid for relativistic, massive particles?

11.4 Using the results given in Table 11.1, show that the kinetic energy (T) spectrum of δ-rays knocked out of atoms by a fast heavy particle is given approximately by

$$\frac{dN}{dT} = \frac{K}{T^2}$$

for $I \ll T \ll \nu_{max}$, where K is a constant.

11.3 The bremsstrahlung process

In Fig. 9.4 we showed the Feynman diagram for the emission of a photon by an electron scattered by a nucleus (it could equally have been a positron being scattered) and in Section 9.5 we discussed how to take into account some of the factors that go into the cross-section. Each vertex contributes a charge squared factor so we get for the cross-section per nucleus

$$\sigma \propto Z^2\alpha^3,$$

where Z is the atomic number of the nucleus. Without the full apparatus of QED we cannot make much more progress so we use some plausibility arguments. In classical theory, the electron radiates because it is accelerated or decelerated as it is deflected by the nuclear electric field. A simple view is that the electron suffers a violent deceleration as it radiates and this led to the German name for the radiation: **brems**(=brake) **strahlung**(=ray). In fact, the process is not limited to electrons. Any charged particle radiates, but the lighter it is, the greater is the acceleration or deceleration; thus the amplitude of the emitted radiation is inversely proportional to the mass and the intensity to the inverse mass squared. Bremsstrahlung is clearly most important for electrons and until later in this section we will concentrate the discussion on these particles. So we now have

$$\sigma \propto \frac{Z^2\alpha^3}{m_e^2 c^4}. \qquad (11.5)$$

If ν is the energy of the emitted photon we are interested in $\frac{d\sigma}{d\nu}$. So, developing equation (11.5) we guess that

$$\frac{d\sigma}{d\nu} \propto \frac{Z^2\alpha^3(\hbar c)^2}{m_e^2 c^4 \nu} \tag{11.6}$$

since it is dimensionally correct. We now turn this into energy loss per unit distance moved by the electron:

$$-\left(\frac{dT}{dx}\right)_{rad} = n\int_{\nu_{min}}^{\nu_{max}} \nu \frac{d\sigma}{d\nu} = \frac{nZ^2\alpha^3(\hbar c)^2}{m_e^2 c^4}(\nu_{max} - \nu_{min}). \tag{11.7}$$

The subscript 'rad' means energy lost by radiation. The n is again the number per unit volume of the nuclei, atomic number Z. Now ν_{max}, the maximum energy loss per encounter, is T, the kinetic energy of the electron, and the minimum loss, ν_{min}, is essentially zero. In addition we have not picked up some numerical factors. In this case it is $4\ln(183/Z^{1/3})$. So we now have

$$-\left(\frac{dT}{dx}\right)_{rad} = \frac{4nZ^2\alpha^3(\hbar c)^2}{m_e^2 c^4} T \ln\frac{183}{Z^{1/3}}. \tag{11.8}$$

In fact, that numerical factor is very important because the logarithm expresses the effect of the range of impact parameters of the incident electron in which radiation is expected to occur during the traversal of an atom. At large impact parameters the nucleus is screened by the atomic electrons and thus the cross-section is limited. We are missing some very important physics in our rough approach.

Since $-\dfrac{dT}{dx} \propto T$ we have

$$\frac{dT}{T} = -A\, dx,$$

where A is a constant, so that

$$T = T_0 e^{-Ax}.$$

Thus there is a distance in which, on the average for many trials, the electron energy is reduced e-fold. That distance is called the **radiation length** L_R and it is given by

$$L_R = \left[\frac{4nZ^2\alpha^3(\hbar c)^2}{(m_e c^2)^2} \ln\frac{183}{Z^{1/3}}\right]^{-1}. \tag{11.9}$$

For lead, $Z=82$ and $n=3.30\times10^{28}\,m^{-3}$ gives $L_R = 5.3$ mm—close to 5.6 mm, the value found if all the small corrections for atomic effects which we have omitted from our formula are included. Table 11.2 gives the radiation lengths for lead and some other elements.

The spectrum of emitted photons is given by equation (11.6):

$$d\sigma \sim \frac{d\nu}{\nu}.$$

up to the maximum possible, $\nu_{max} = T$. This means that approximately equal energy is radiated into equal intervals of $d\nu$. Of course our procedure has assumed a sharp cut-off at $\nu = T$. In fact there is a roll-off which depends on the kinematics and the details of the screening.

We would like to compare $(\mathrm{d}T/\mathrm{d}x)_{\mathrm{rad}}$ with $(\mathrm{d}T/\mathrm{d}x)_{\mathrm{ion}}$, the energy loss by ionization at relativistic velocities. We do not know the latter at extreme relativistic velocities so we will take its minimum value, which is given by equation (11.1) with $\beta=0.96$ and $z=1$. This is not quite correct for electrons but will do well enough. Using equation (11.8) we find (for electrons or positrons only)

$$\frac{(\mathrm{d}T/\mathrm{d}x)_{\mathrm{rad}}}{(\mathrm{d}T/\mathrm{d}x)_{\mathrm{ion}}} = \frac{Z\alpha}{\pi m_e c^2} T\beta^2 \left[\frac{\ln(183/Z^{1/3})}{\ln[2m_e c^2\beta^2/I(1-\beta^2)]-\beta^2}\right], \quad \beta=0.96.$$

Putting $I \simeq 16Z^{0.9}$ eV this ratio is approximately

$$\frac{ZT(\mathrm{MeV})}{560}.$$

This ratio is 1 for electrons at what is called the **critical energy,** T_{crit}. Table 11.2 gives some values of T_{crit} calculated with more precision than is possible using this simple ratio. A comparison will show that $T_{\mathrm{crit}}=560/Z$ MeV is correct to about 10% for all but the lowest Z. Below T_{crit} the energy loss by ionization dominates; above T_{crit} the radiative energy loss wins and grows rapidly with T. (Note, however, that medical X-ray sources and television picture tubes operate at electron energies well below the critical energy. The first are designed to make use of the bremsstrahlung, which is still an important energy-loss mechanism. The television tubes have to have thick lead glass walls and faces designed to absorb the X-rays in order to protect the viewer.)

The formula we have used for $(\mathrm{d}T/\mathrm{d}x)_{\mathrm{rad}}$ depends inversely on the electron mass squared. Therefore for a charged particle of mass m the radiative loss is down by the factor $(m_e/m)^2$. The next heavier charged particle is the μ-meson with $(m_e/m_\mu)^2=2.34\times10^{-5}$. Thus the muon radiation length and its critical energy are greater by the reciprocal of this factor, 4.28×10^4. Thus for μ-mesons and the heavier charged particles energy loss by radiation is not important until very high energies are reached, for example, $T\simeq 300$ GeV for muons in lead.

Again we stress that the formula for radiative energy loss gives an average over many trials. One charged particle will suffer radiative energy losses which can fluctuate hugely from the average. And the number and energy of the radiated photons will also fluctuate. And, of course, the behaviour of positrons is nearly identical to that of electrons: the differences are due to the different wave functions in the nuclear Coulomb field, an effect significant only at very low energies (<1 MeV).

There is nothing so far to tell us about the angular distribution of the emitted photon or of the degraded electron: with respect to the direction of the incident electron both are emitted close to the incident direction with a root mean square angle of order $m_e c^2/E'$, where E' is the total energy of the secondary particle. Thus 500 MeV photons emitted by 1 GeV electrons have an r.m.s. angle of order 1 mrad ($=0.06°$).

We now see an additional problem to associating range with electrons, except well below the electron critical energy. An electron can lose, by radiating one photon, a large fraction of its energy with much higher probability than applies to heavier particles. Such abrupt losses shorten the track length. The same applies to heavier charged particles except that even at energies well below the critical energy the ranges can be so great that the probability of a nuclear interaction is large enough for inelastic collisions with nuclei to become the main cause of abrupt energy loss.

DEFINITIONS AND KEYWORDS

Bremsstrahlung The radiation (photons) emitted by a charged particles suffering scattering in the electric field of a nucleus.

Critical energy The kinetic energy of a charged particle at which the rate of energy loss by the bremsstrahlung process is equal to that by ionization. At higher energies, loss by the former process is the greater.

PROBLEM

11.5 Show that the average radiative energy loss of electrons of $P=3.2$ GeV/c crossing one radiation length of lead is about 2.0 GeV.

A beam of electrons of energy 1 GeV traverses normally a foil of lead $\frac{1}{20}$th of a radiatron length thick. Show that the angular distribution of bremsstrahlung photons of energy 800 MeV is determined more by multiple scattering of the electrons than by the angular distribution in the basic radiation process.

11.4 Photon absorption and scattering

Clearly the interaction of photons in material is an electromagnetic phenomenon. Photons can be scattered or completely absorbed. The processes that we shall consider are:

(1) **Rayleigh scattering**;

(2) **Compton scattering**;

(3) **photoelectric effect**;

(4) **pair production**.

Coherent scattering (elastic) by the whole atom is Rayleigh scattering. Compton scattering (Section 9.5) is strictly photon scattering by an initially free electron: in normal circumstances electrons are bound and for photons of energy much greater than the atomic binding energy the electron is essentially free. However, for photon energies comparable with the electron binding energy the electrons cannot be assumed to be free. We include under the umbrella of **incoherent scattering** all the photon–atom inelastic scattering processes. The photoelectric effect involves the ejection of a single electron from an atom on absorption of the photon: the electron receives the photon energy less the binding energy and the atom recoils with the small out-of-balance momentum. Pair production is the conversion of an energetic photon in the electric field of a nucleus into an electron–positron pair, which share the photon energy.

The business of physicists involves a huge range of photon wavelengths. For us, it is sensible to consider only photons of energy greater than about 100 keV. Below that energy the photoelectric effect dominates absorption and is characterised by the existence of the K, L, M, ... absorption edges; we wish to avoid becoming too involved in such atomic physics. Even above 100 keV the atomic binding of the electrons makes the photoelectric effect strongly dependent on the atomic number; in addition, it is ejection of an electron from the K-shell which gives by far the largest part of the total atomic cross-section. For each K-shell electron the contribution to the cross-section is given by

$$\sigma_{\mathrm{PE}} \sim Z^5 \alpha^4 \left(\frac{m_e c^2}{E_\gamma}\right)^n$$

where E_γ is the photon energy and $n=\frac{7}{2}$ at $E_\gamma < m_e c^2$ and changes to 1 at $E_\gamma \gg m_e c^2$ (subscript PE for photoelectric effect). There is no simple physical argument that leads to this result: the fifth power of Z is possibly attributable to the fact that increasing the probability density of the K-shell electrons near the nucleus increases the probability of the photoelectric ejection (Z^3) and the nucleus has to take up some momentum which it does by the absorption of a virtual photon emitted by the electron (Z^2).

Compton scattering was briefly discussed in Section 9.5 where we guessed that the cross-section (σ_C, C for Compton) was given by

$$\sigma_C \simeq \frac{\alpha^2(\hbar c)^2}{W^2}$$

where W is the centre-of-mass energy of the photon–electron system and is given by

$$W = \sqrt{[(m_e c^2)^2 + 2E_\gamma m_e c^2]}.$$

The complete calculation was first performed by Klein and Nishina; their result gives the two extreme energy results:

$$(1)\ E \ll m_e c^2\ (W \approx m_e c^2);\quad \sigma_C = \textbf{Thomson cross-section}\ (\sigma_T) = \frac{8\pi}{3}\frac{\alpha^2(\hbar c)^2}{(m_e c^2)^2}, \tag{11.10a}$$

$$(2)\ E \gg m_e c^2\ (W \approx \sqrt{(2E_\gamma m_e c^2)});\quad \sigma_C \approx \frac{4\pi\alpha^2(\hbar c)^2}{W^2}\left(\ln\frac{W}{m_e c^2} + \frac{1}{4}\right). \tag{11.10b}$$

Our original guess in Section 9.5 about the form of cross-section was correct at low energies but we missed the logarithm, which has a dominant effect at high energies. The Klein–Nishina cross-section decreases from its value at $E_\gamma = 0$ of 0.665 b (the Thomson cross-section) as the photon energy increases. However, at low energies the incoherent scattering cross-section per atom falls below the value $Z \times$ the Klein–Nishina cross-section. This can be seen in Fig. 11.6, where we see that the incoherent scattering cross-section in carbon and lead are not going to reach the low-energy limit of $Z\sigma_T$. In fact the incoherent cross-section must tend to zero as $E_\gamma \to 0$ because the inelastic processes have a finite threshold energy. An examination of Fig. 11.6 shows that Rayleigh (coherent) scattering cross-section increases and the incoherent decreases as the incident photon energy decreases.

We note that, at high energies,

$$\sigma_{PE} \sim \frac{1}{E_\gamma} \quad \text{and} \quad \sigma_C \sim \frac{\ln E_\gamma}{E_\gamma}.$$

The fourth process is pair production. One Feynman diagram is shown in Fig. 11.7: compare it with the diagram of Fig. 9.4. The two have the same structure, the difference being that two arms are interchanged: therefore we can expect the cross-section will contain factors close to those we started with in bremsstrahlung and displayed in equation (11.5), that is

$$\sigma_{PP} \simeq \frac{Z^2\alpha^3}{m_e^2 c^4}$$

(subscript PP for pair production). Clearly the pair production involves a momentum transfer (q) to the nucleus (see Problem 9.4). This q decreases as the incident energy becomes greater and the electron mass becomes less significant.

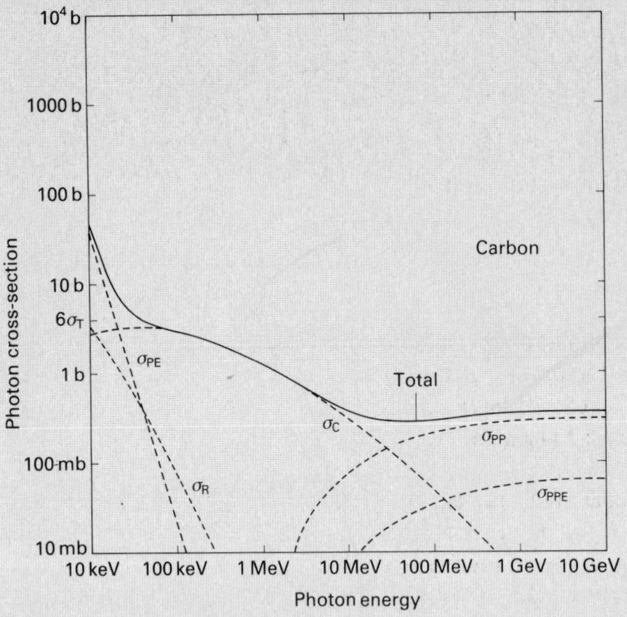

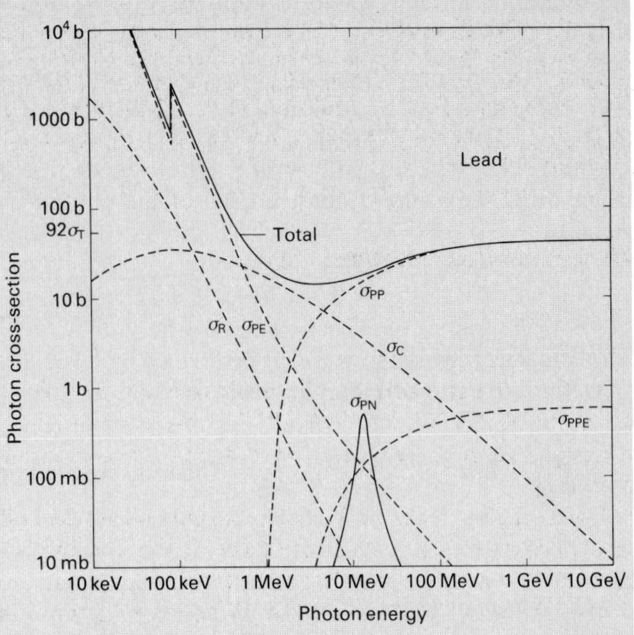

Fig. 11.6 The partial and total cross-sections per atom for photon interactions in (a) carbon and (b) lead as a function of photon energy (Storm and Israel, 1970). Subscripts mean:

PE = photoelectric effect,
PP = pair production in the electric field of the nucleus,
PPE= pair production in the electric field of Z electrons,
C = incoherent photon (Compton) scattering,
R = Rayleigh scattering,
PN = photonuclear cross-section.

Low energy is dominated by the photoelectric effect, the Compton effect is important at energies near 1 MeV and pair production dominates at high energies. Note that the latter includes not only pair production in the nuclear field but also production in the separate fields of the Z atomic electrons. The scattering processes, Compton and Rayleigh, lead to a scattered photon of degraded energy, whereas pair production and the photoelectron effect lead to complete absorption of the photon. The effect of the nuclear giant dipole resonance (Section 11.11) in carbon is a peak reaching about 7 mb and is not tall enough to show; the peak of this resonance in lead is about 500 mb and is shown. (Harvey *et al.*, 1964.) However, in both, and in other elements, the resonance contributes no more than a few percent to the total atomic cross-section for photons, and then only over a very limited energy range.

As q decreases the electric field needed to take up q decreases and so the photon–nucleus impact parameter can increase. And that implies an increasing cross-section. That increase would continue indefinitely but for the screening effect of the atomic electrons. Once again a full development requires an understanding of atomic effects. We are not surprised to find that the σ_{PP} starts from zero at threshold, $E_\gamma = 2m_ec^2$ ($=1.02$ MeV), rises and then reaches a plateau by about 1000 MeV. That plateau depends on the material and is, for $E_\gamma \gg m_ec^2$,

$$\sigma_{\mathrm{PP}} = \frac{28}{9}\frac{Z^2\alpha^3(\hbar c)^2}{(m_ec^2)^2}\left[\ln\frac{183}{Z^{1/3}} - \frac{2}{7}\right].$$

Of course, the photon disappears on producing a pair, so we can define a mean free path L_{PP} for the process:

$$L_{\mathrm{PP}} = \frac{1}{n\sigma_{\mathrm{PP}}}$$

$$= \left[\frac{28}{9}\frac{nZ^2\alpha^3(\hbar c)^2}{(m_ec^2)^2}\left\{\ln\frac{183}{Z^{1/3}} - \frac{2}{7}\right\}\right]^{-1}$$

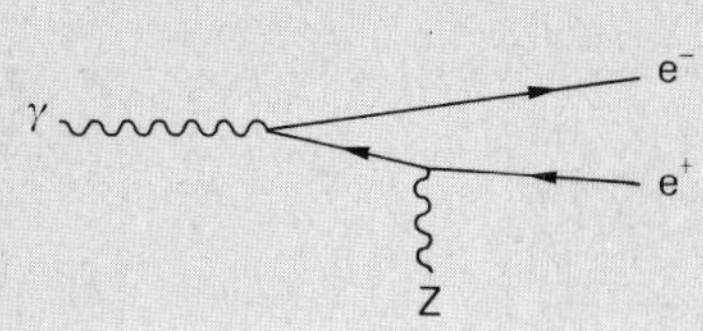

Fig. 11.7 One of the two Feynman diagrams for the production of an electron–positron pair by a photon in the electric field of a nucleus. Compare this with the diagram for the radiation of a photon by an electron (or positron) in the field of a nucleus, Fig. 9.4. (The reader is invited to find the second diagram for pair production.)

which gives, from equation (11.9)

$$L_{\text{PP}} \simeq \frac{9}{7} L_{\text{R}}.$$

Therefore, not surprisingly, the mean free path for pair production by a very energetic photon ($E_\gamma > 1000$ MeV) is closely related to the mean distance in which an energetic electron ($E_e \gg$ critical energy) loses all but a fraction e^{-1} of its energy.

Classical optics tells us that an absorbing object also diffracts light; the same must apply to the present circumstances in that the photon absorption processes must be accompanied by elastic scattering (Rayleigh). However, the contribution of the Rayleigh scattering to photon attenuation is small compared both with the photoelectric effect and with (except at low energies) the incoherent scattering, and it decreases to become negligibly small for photons of energy greater than about 10 MeV.

We can now put all this together to show how these processes determine the total cross-section for a photon. In Fig. 11.6 we show this cross-section for photons of energy 10 keV to 1000 GeV in carbon and lead. These two elements provide a contrast in the relative importance of the major processes. In carbon the photoelectric effect is just giving way to Compton scattering at 20 keV as the most important; pair production begins to dominate at about 100 MeV. In lead Compton scattering is a minor effect except in range from $\frac{1}{2}$–5 MeV. We note that there is a minimum in the total cross-sections: it is at about 4 MeV in lead,

DEFINITIONS AND KEYWORDS

Rayleigh scattering The elastic scattering of photons by atoms.

Compton scattering The scattering of photons by free electrons.

Photoelectric effect The ejection of an atomic electron by the absorption of a photon.

Incoherent scattering of photons by atoms The inelastic scattering of photons by atoms.

Thomson cross-section The total cross-section for photon–electron scattering in the low frequency limit.

Pair production The materialization of a photon into an electron–positron pair in the electric field of a nucleus or of an electron.

Radiation length The distance in a given material in which members of a population of electrons moving at an extreme relativistic velocity lose, on the average, all but a fraction e^{-1} of their initial kinetic energy.

Electromagnetic shower The name given to an event in which an energetic photon, electron or positron shares its energy among many such particles by the bremsstrahlung and pair-production processes, finally dissipating its energy by absorption and degradation of these secondary particles.

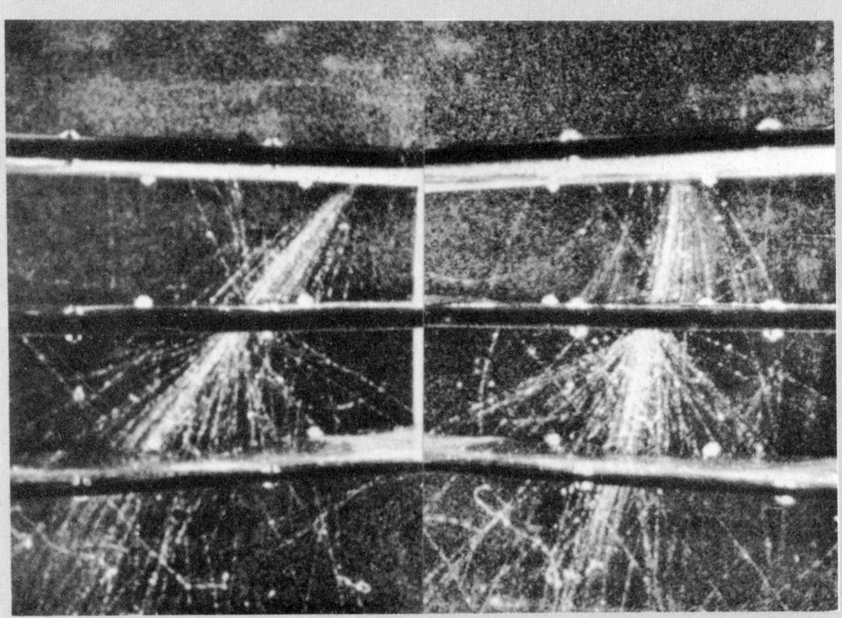

Fig. 11.8 Two cloud-chamber photographs of electromagnetic showers developing in lead plates (thicknesses from top down 1.1, 1.1, and 0.13 radiation lengths) placed within the chamber which was exposed to cosmic radiation at sea level. There are two views in each photo, obtained by aiming the camera so that it sees simultaneously directly and via a mirror into the chamber. The lower photograph shows a shower which was either started by a photon or missed the sensitive region above the top plate. In both, the growth and attenuation of the shower is evident. Note that many of the tracks have systematic curvature or distortion, probably due to some difficulty in maintaining stable and uniform temperature conditions in the chamber. These photographs were obtained by L. Fussel and used in a talk by J. C. Street which was published in 1939.

0 MeV in carbon, and 250 MeV in hydrogen; it corresponds to maximum mean ree paths of about 250, 700, and 900 kg m^{-2} respectively. (In real length 22, 310 nd 12 700 mm in the solids and the liquid, respectively.)

Of course the discussion so far in this section concerns the first event that appens to a photon. Such an event transfers some or all of the photon energy o a recoiling electron or atom or to an electron–positron pair. An interesting uestion is 'what finally happens to the photon's total energy?' If it can all be ransferred to low-energy electrons, then it will finally be dissipated as thermal nergy in the material, as described in Section 11.2 for all charged particles. We an now discern that before that process is complete an event can occur that can ave many branches. An incident energetic photon can create an e^+e^- pair: ach member of the pair can radiate photons which in turn can create further airs or photoelectrons, which in their turn radiate photons. Thus the energy of ne photon becomes the energy of many photons, electrons and positrons. ositrons can annihilate, moving or at rest, in an encounter with an electron, reating two or three photons. As the energy is spread among many particles, he photoelectric effect absorbs the photons, and low energy electrons lose nergy which becomes thermal energy. The thickness in which the process is ssentially complete is, for example, about 20 radiation lengths at 10 GeV, ncreasing by 1 radiation length for every e-fold increase in energy. The name iven to such an event is an **electromagnetic shower**. Clearly a shower can also be nitiated by an incident electron or positron. Figure 11.8 shows cloud chamber hotographs of electromagnetic showers observed in an exposure to cosmic adiation.

11.5 The radiation of photons by nuclei and particles

Rutherford gave the name γ-rays to the electrically neutral radiation emitted by he naturally occurring radioactive materials. It soon became clear that this adiation had the properties of very energetic X-rays and was a part of the lectromagnetic spectrum. So usage has given *γ-ray* the meaning of a photon associated with a nuclear or a particle process. There is no sharp energy boundary: X-rays which we associate with atomic processes extend up to 100 keV, whereas there are rotational bands in very heavy nuclei which involve transitions emitting photons of less than 10 keV. However, that is the extreme and we associate photon energies of the order of 1 MeV with the de-excitation of light nuclei. In Table 11.3 we describe one of the methods of detecting and measuring the energy of photons of order 1 MeV.

The factor of 10^4 in the ratio of the size of atoms (10^{-10} m) to the size of nuclei (10^{-15} m) means that the energies of excited levels in hydrogen (10 eV) become 10^6 eV for the excited levels of light nuclei. Electric dipole transitions have rates proportional to the cube of the energy and the square of the dimension of the radiating system; this has the result that radiative transition rates of nuclei are 10^5 times greater than those of atoms. Even as order-of-magnitude statements they are very approximate, and as we shall see almost immediately we have to understand a huge range of radiative transition rates. However, we hope that this comparison will remind the reader of the connection between size and energy and will help us to place γ-rays in the context of the electromagnetic spectrum.

We shall discuss radiation from both nuclei and particles: let us consider such a system which has an excitation or mass that makes it unstable: if the Q-value is

Table 11.3 The detection and measurement of the energy of γ-rays. Gamma-rays are photons and, in principle, can have any energy. However, usage gives 'γ-ray' the meaning of a photon created in a nuclear or elementary particle process. Here we are concerned with the former, for which the energy range is from a few kiloelectronvolts to several megaelectron-volts. (The latter have energies from tens of megaelectronvolts and up.)

One of the most powerful detectors of nuclear gamma rays is the lithium-drifted germanium detector, Ge(Li). It consists of a block of germanium with two thin electrodes on opposite faces and an intervening junction between *n*- and *p*-type sections. The depletion layer at the junction is extended by the presence of the lithium and by the application of a large reverse voltage. The generation of an electron–hole pair in the depletion layer is followed by their separation and drift to anode and cathode respectively. If the cathode is maintained at zero potential and the anode isolated from the applied voltage by a suitable resistance, the processes of drift and collection lead to a negative change of potential at the anode. The *RC* of the circuit restores the potential to its original value. This instantaneous generation and collection of a large number of electron–hole pairs leads to the appearance of an electrical pulse at the anode of amplitude proportional to the number of pairs. A sequence of such events may be analysed to give a spectrum of the frequency of events having different numbers of pairs.

A charged particle loses energy by ionization (see Section 11.2), and in germanium this energy is transformed with high efficiency into electron–hole pairs. The number of pairs is directly proportional to the energy loss. Thus a Ge(Li) detector which is almost all depletion layer is an excellent device for measuring the energy of charged particles which enter and stop in the germanium or for measuring the energy loss of particles which traverse it. How is this used to measure γ-ray energies? Clearly the γ must deposit all its energy, in the form of moving charged particles, in the depletion layer although that cannot be arranged to happen with every event. The processes by which a γ-ray produces such particles are (see Section 11.4):

(1) the photoelectric effect, which produces an electron with a kinetic energy less than the γ-ray energy;

(2) the Compton effect, which produces a degraded photon and a recoil electron;

(3) electron–positron pair production (threshold 1.02 MeV).

We leave to the reader to see how, in a sufficiently large piece of germanium, all these processes, and others, can lead to complete containment of the energy. Problem 11.6 is concerned with the interpretation of results obtained when the containment is incomplete. Figure 7.8 shows another example of a spectrum obtained with a detector of this kind.

large enough, decay may occur by emission of a strongly interacting particle. Examples are

$$\begin{aligned}
\Delta^{++}(1232) &\rightarrow \pi^{+} + \mathrm{p} + 154\ \mathrm{MeV}, & \omega &= 1.7\times 10^{23}\ \mathrm{s}^{-1},\\
{}^{15}_{7}\mathrm{N}^{*}(12.15) &\rightarrow {}^{14}_{6}\mathrm{C} + \mathrm{p} + 1.94\ \mathrm{MeV}, & \omega &= 2.5\times 10^{19}\ \mathrm{s}^{-1},\\
&\rightarrow {}^{14}_{7}\mathrm{N} + \mathrm{n} + 2.31\ \mathrm{MeV}, & \omega &= 4.5\times 10^{19}\ \mathrm{s}^{-1},
\end{aligned}$$

where $\Delta^{++}(1232)$ is the three-quark baryon we met in Section 10.3 (see Table 10.5) and ${}^{15}\mathrm{N}^{*}$ is an excited state of ${}^{15}\mathrm{N}$ at 12.15 MeV. In general, such decays have transition rates of $\sim 10^{19}\ \mathrm{s}^{-1}$ and greater. However, if the mass is insufficient for emission of a strongly interacting particle, then the next possibility may be de-excitation by photon emission:

$$\Sigma^{0}(1192) \rightarrow \Lambda(1116) + \gamma + 76.9\ \mathrm{MeV}, \quad \omega = 1.7\times 10^{19}\ \mathrm{s}^{-1}, \tag{11.11}$$

$${}^{137}_{56}\mathrm{Ba}^{*} \rightarrow {}^{137}_{56}\mathrm{Ba} + \gamma + 662\ \mathrm{keV}, \quad \omega = 4.5\times 10^{-3}\ \mathrm{s}^{-1}, \tag{11.12}$$

where Σ^0 and Λ are the two lightest neutral strange baryons (Section 10.4 and Table 10.5). These examples are chosen from almost the extremes of transition rate and indicate the large range which we have to try to understand. We note that it overlaps with the range of transition rates for the emission of strongly interacting particles.

Classically an object with charge structure can be represented by a charge distribution with a spherical dependence which can be decomposed into a sum of spherical harmonics (Section 8.7). Each harmonic gives rise to a contribution to the potential at large distances. The first term represents the component which is spherically symmetric, is called the monopole moment and is, of course, equal to the total charge of the object. Charge conservation requires that this total charge remains constant, but its radial distribution can oscillate, the so-called breathing mode: that oscillation causes no change to the electric field outside the object and no energy can be radiated. The next three terms correspond to the components of the electric dipole moment: oscillations of this vector do give rise to EM radiation, the electric dipole (E1) radiation. The next five terms correspond to the possible arrangements of an electric quadrupole (see Sections 3.9 and 8.8): oscillations of this quadrupole give rise to electric quadrupole (E2) radiation. The reader will recognize that this series can be continued. The classical situation is that each class of polarity gives radiation with an angular distribution characteristic of the class: for example E1 has $\sin^2\theta$ distribution with respect to the polar axis aligned along the direction of the oscillating dipole. The angular distribution becomes more complicated for increasing polarity: an oscillating 2^L-pole charge distribution gives radiation classed as EL which has an angular distribution which can be as complicated as $\cos^{2L}\theta$. We call L the **multipole order** or **multipolarity**.

How does this transform into quantum-mechanical terms? Radiation is now quantized as photons and the emitting object changes from one state (A), total angular momentum quantum number j_A, to a state of lower energy (B) and angular momentum j_B. The change is simple: L now represents the change in the total angular momentum of the object emitting the photon. Since angular momentum is a vector which is conserved we must have

$$|j_A - j_B| \leqslant L \leqslant j_A + j_B.$$

(This is a quantum-number formulation of a vector requirement, symbolically written $\boldsymbol{j}_A - \boldsymbol{j}_B = \boldsymbol{L}$. See Table 11.5.) Thus j_A and j_B determine the permitted values of L in the transition which takes A$\rightarrow$B and emits a photon. Normally L has the lowest possible value allowed, which is $|j_A - j_B|$, except that $L=0$ is forbidden; the reasons for this we shall investigate shortly. Thus $j_A = 1 \rightarrow j_B = 0$ must have $L=1$ and will occur by an E1 (electric dipole or, as we shall see later, M1, magnetic dipole) transition. A transition $j_A = 2 \rightarrow j_B = 1$ could be $L = 1$, 2, or 3 but will almost certainly have $L=1$. The L-value determines the angular distribution of the radiation, as it does classically: however, that distribution can be detected only if the radiating state is polarized or the polarization state of the final state is detected. We cannot explore that subject in any detail but a few words are in order. Obviously a $j_A = 0 \rightarrow j_B = 1$ has no direction associated with the state A so the probability of emission of the photon is the same in all directions. For the transition $j_A = 1 \rightarrow j_B = 0$, if the initial state is polarized with $j_{Az} = +1$, then the radiation has a distribution, $1+\cos^2\theta$. In the same transition, if the initial assembly is one of equal and incoherent population of the three magnetic substates, then there is no direction associated with that assembly and the photons must be emitted isotropically.

We must now consider why $L=0$ is not allowed: this implies that photons cannot carry away from a transition zero angular momentum and we must understand why. Photons can be polarized which follows from the vector nature of the electromagnetic field and that in turn means they have intrinsic spin 1. A massive particle of spin 1 has three magnetic substates, $s_z = +1$, 0, and -1. The real photon is massless and has only two substates, corresponding to the two possible states of polarization of light. (That this is a consequence of the zero mass cannot be proved or justified here and the reader must take it on trust.) The most illuminating basis is that of the right and left circular polarization; for a photon propagating along the z-axis, these states correspond to $L_z = +1$ and -1. This means that there must be a change of j_z of the source of -1 or $+1$. None of this can happen if $L=0$. The reader may well ask if the spin of the photon is 1 why L is not restricted to 1? An answer is that the photon and the recoiling nucleus can have relative orbital angular momentum so that the photon appears to carry away more angular momentum than its own intrinsic spin. The orbital angular momentum is a vector perpendicular to the direction of the photon (the z-direction, remember) and so contributes nothing to L_z, and the argument based on the z-components is not disturbed. However, this argument is dangerously over-simplified since relativistically the distinction of orbital angular momentum as a part of the total angular momentum does not exist.

At this point we are going to postpone the consideration of parity changes in the transitions and of the role of the magnetic transitions, although to some extent these are the logical next subjects. Instead we want to proceed to consider briefly the actual rates in electric transitions, and return later to these subjects.

PROBLEM

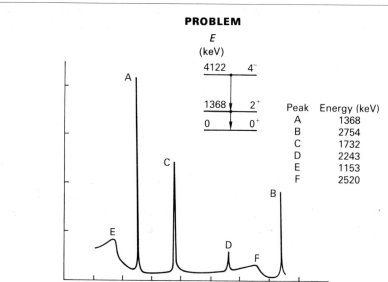

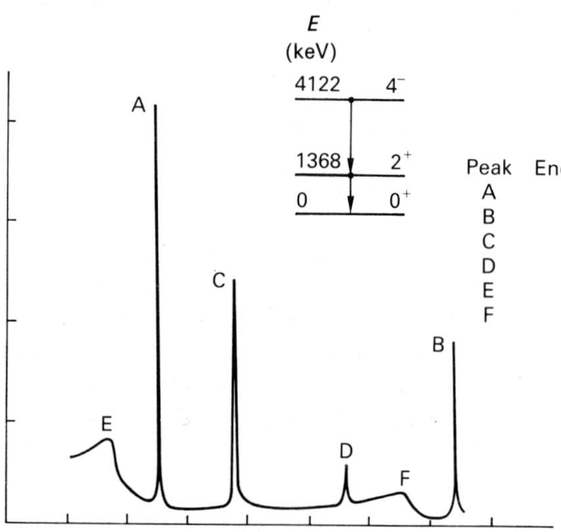

Peak	Energy (keV)
A	1368
B	2754
C	1732
D	2243
E	1153
F	2520

11.6 Find a formula for the maximum energy that a photon of energy E_γ can transfer to a free electron.

The figure shows the energy spectrum obtained when a small solid state detector is exposed to γ-radiation from $^{24}_{12}Mg$. Account for the structures at E and F. The peaks D and C are 511 keV and 1022 keV lower in energy, respectively, than B. Explain how they arise.

(Adapted from the 1984 examination of the Final Honours School of Natural Science, Physics, University of Oxford.)

11.6 Rates for electric transitions

There is no simple way of deriving the transition rate: the problem is connected with the fact that the electromagnetic field has quanta, the photons, which are relativistic from the beginning. The problem is treated in many books and at various levels. Corney (1977) provides a semiclassical derivation for the E1 transition probability, A_{ki}, from an initial state k to a final state i of the radiating system (Corney, equation (4.18)):

$$A_{ki} = \frac{\omega_{ki}^3}{3\pi\varepsilon_0 \hbar c^3} \left| \int \psi_i^* e\boldsymbol{r} \psi_k \, dV \right|^2, \tag{11.13}$$

where $\hbar\omega_{ki}$ is the energy of the emitted photon and the integral is the matrix element of the electric dipole operator $e\boldsymbol{r}$ between initial and final states of the emitting system. Let us rewrite this in a notation closer to our style. We hijack ω to represent the transition rate and put the integral as $e\langle \boldsymbol{r} \rangle$. Then

$$\omega(\text{E1}) = \frac{4\alpha}{3}\left(\frac{E_\gamma}{\hbar c}\right)^3 c |\langle \boldsymbol{r} \rangle|^2 \tag{11.14}$$

where E_γ is now the energy of the emitted photon and α the fine structure constant. Dimensionally this is correct (see Problem 11.7). It has the number of powers of α that we expect from emission of a single photon (Section 9.4) although the third power of E_γ is not obvious. The matrix element $\langle \boldsymbol{r} \rangle$ is worth one length dimensionally: note that it is the matrix element between the initial and final state of the radiating system and does not contain the wavefunction of the photon. The transition rate, ω, cannot be easily evaluated as the wavefunction of the charged particles (protons in nuclei, quarks in particles) are not sufficiently well known to permit a precise calculation of $\langle \boldsymbol{r} \rangle$. However, we can easily obtain a rough order of magnitude result: clearly for radiating nuclei $|\langle \boldsymbol{r} \rangle|$ will be approximately the nuclear radius, $R = 1.2A^{1/3}$ fm so for a 1 MeV γ-ray we have

$$\begin{aligned} \omega(\text{E1}) &\approx \frac{4}{3 \times 137}\left(\frac{1}{197}\right)^3 3 \times 10^{23} (1.2A^{1/3})^2 \\ &= 5.5 \times 10^{14} A^{2/3} \text{s}^{-1}. \end{aligned} \tag{11.15}$$

For $A = 100$ this gives a mean life of about 8×10^{-17} s.

How do we do the step to the electric quadrupole transition rate? A static electric quadrupole has the dimension of charge times area. The quantum-mechanical operator, Q, will have matrix elements of the same dimension. The latter must appear squared in the transition rate, so to keep the overall dimension of $(\text{time})^{-1}$ we must bring in a $(\text{length})^{-2}$ and that is done with another factor $(E_\gamma/\hbar c)^2$. So we have

$$\omega(\text{E2}) \sim \left(\frac{E_\gamma}{\hbar c}\right)^5 c |\langle Q \rangle|^2.$$

Since $|\langle Q \rangle| \sim eR^2$ we have

$$\omega(\text{E2}) \sim \alpha \left(\frac{E_\gamma}{\hbar c}\right)^5 cR^4.$$

For example, in the case of $A = 100$ and $E_\gamma = 1\,\text{MeV}$ this gives $\omega(\text{E2}) = 7 \times 10^{12}\,\text{s}^{-1}$. However, we have missed some factors depending on L. Before quoting those we can generalize: every increase in multipolarity L by one increases the power of both R and $(E_\gamma/\hbar c)$ by two. In a simple-minded approach this is like an angular momentum barrier reducing the rate as L goes up: the relevant factor in the formula for $\omega(\text{E}L)$ is $(kR)^{2L+1}$ where $\hbar k$ $(= E_\gamma/c)$ is the photon's momentum. If the wavelength of the photon, $2\pi/k$, is long compared with the radius, the more angular momentum it must carry away the lower is the transition rate. So, for the general electric transition and building on $\omega(\text{E1})$, we get

$$\omega(\text{E}L) \sim \alpha \left(\frac{E_\gamma}{\hbar c}\right)^{2L+1} R^{2L} c \tag{11.16}$$

The constant missing is

$$\frac{2(L+1)}{L[(2L+1)!!]^2}\left(\frac{3}{L+3}\right)^2, \qquad \text{where } (2L+1)!! = (2L+1)(2L-1)(2L-3)\cdots 1.$$

Where does this come from? Most is from a proper definition and evaluation of the matrix elements of the multipole operators. However, a part comes from a better treatment of the replacements that we made above:

$$\text{E1: } \langle er \rangle \to eR$$

$$\text{E2: } \langle Q \rangle \to eR^2$$

$$\text{E3: } \langle \text{E3 transition operator} \rangle \to eR^3$$

This better treatment is due to Weisskopf and the result is called the **Weisskopf single-particle rates**. The *single-particle* because his treatment assumed one proton alone makes a change of orbit inside the nucleus: this is an over-simplification. In Table 11.4 we have summarized our own crude formulae and have quoted those of Weisskopf.

Returning to the change in $\omega(\text{E}L)$, we noted that it was by a factor of order $(E_\gamma R/\hbar c)^2 = (kR)^2 = (2\pi R/\lambda)^2$ as L is increased by 1. At $A = 100$ and $E_\gamma = 1\,\text{MeV}$ this factor is about 8×10^{-4}. So for every increase in multipolarity by 1 we expect a decrease of the order of 10^{-3} in the transition rate, other things being kept the same. Thus for given j_A and j_B, the multipolarity of the radiation will be by the lowest L allowed, $|j_\text{A} - j_\text{B}|$ (not 0), unless a parity selection rule intervenes, of which more shortly.

How well do these single-particle formulae fit the date? The answer is poorly! The formulae tend to over-estimate the transition rates by several orders or magnitude. In addition, the observed rates of a given multipolarity do not scale as might be expected. For example, from equation (11.13) we could expect the quantity $\omega(\text{E1})/(E_\gamma R)^3$ to be a constant over many E1 transitions. It in fact varies by a factor of about 10^4. This should be put in the context of a huge total range of transition rates; the Weisskopf formula predicts rates varying from about $10^{17}\,\text{s}^{-1}$ for a 2 MeV E1 to about $10^{-6}\,\text{s}^{-1}$ for a 20 keV, E3 transition. The conclusion to be drawn is that the calculation of the matrix element in each case is impossible to do with any precision: this is not surprising in view of the complexity of the nucleus and our lack of knowledge of its wavefunctions. In

Table 11.4 Transition rates for photon emission by nuclei and other hadrons.

	Electric $\omega(EL)$	Magnetic $\omega(ML)$
Rough estimate	$\alpha\left(\frac{E_\gamma}{\hbar c}\right)^{2L+1} cR^{2L}$	$\alpha\left(\frac{E_\gamma}{\hbar c}\right)^{2L+1} cR^{2L-2}\left(\frac{\hbar}{M_p c}\right)^2$
To obtain the Weisskopf single particle rate multiply by	$\frac{2(L+1)}{L[(2L+1)!!]^2}\left(\frac{3}{L+3}\right)^2$	$\frac{20(L+1)}{L[(2L+1)!!]^2}\left(\frac{3}{L+3}\right)^2$

To obtain ω in s^{-1} use:

R	$\approx$radius of emitting system$=1.2A^{1/3}$ fm for nuclei.
$M_p c^2$	=proton rest mass energy=938 MeV.
α	=fine-structure constant$=(137)^{-1}$
$\hbar c$	=197 MeV fm.
c	=speed of light$=3\times10^{23}$ fm s^{-1}.
$(2L+1)!!$	$=(2L+1)(2L-1)(2L-3)\cdots$ 5.3.1

To obtain the width $\Gamma=\hbar\omega$, use

Γ(eV) $=6.57\times10^{-16}\times\omega(s^{-1})$.

Weisskopf's rates for nuclei are often given as line widths in the form

$\Gamma(E1) = 0.07\ E_\gamma^3\ A^{2/3}$,

$\Gamma(M1) = 0.021\ E_\gamma^3$,

$\Gamma(E2) = 4.9\times10^{-8}\ A^{4/3}\ E_\gamma^5$,

where Γ is in electronvolts if E_γ is in megaelectronvolts.

spite of that the Weisskopf formulae can be used to help to decide on the multipolarity L of a transition if its mean life is known or to give a first guide to a rate to be expected assuming a multipolarity.

There is one exception to the Weisskopf formulae over-estimating the transition rates. It occurs in connection with deformed nuclei and their rotational levels (Section 8.10). Many of their rotational excited states are based on the angular quantum number series 0^+, 2^+, 4^+, ... De-excitation occurs by successive E2 transitions. The motion of rotation is collective and so the transitions are far from single particle. Many proton charges may be moving collectively and this enhances the transition rates by about 100 over the Weisskopf ω(E2). Figure 7.8 shows the spectrum of photons emitted after the Coulomb excitation of $^{234}_{92}$U: the lines corresponding to the E2 transitions between rotational levels are labelled.

11.7 Rates for magnetic transitions

The magnetic transitions have been waiting for attention. Our discussion of the electric transitions was based upon the classical idea of changing electric moments. However, classically, charge distributions can be constant yet have currents flowing which give rise to magnetic moments: therefore changing magnetic moments will radiate. Let us consider the simplest case in a quantum-mechanical context. A nucleon has spin $s=\frac{1}{2}$, and magnetic moment; a transition in which the spin changes direction $s_z=+\frac{1}{2}\rightarrow s_z=-\frac{1}{2}$ and the energy changes will cause the radiation of a photon or be caused by the absorption of a photon. Such a transition is the simplest form of a magnetic dipole (M1) transition. If this occurs in an odd-A nucleus for the unpaired nucleon which is in a shell model $s_{\frac{1}{2}}$ state, then $|j_{Az}-j_{Bz}|=1$ and $L=1$. Can we adopt the E1 transition rate formula, equation (11.14), to this case? The $\langle e\boldsymbol{r}\rangle$ cannot apply so we have to find a replacement charge times length. In formal terms $e\boldsymbol{r}\rightarrow\boldsymbol{\mu}$, electric to magnetic dipole moment. Now the magnetic moment of a nucleon is $\boldsymbol{\mu}=g_s\,(e\hbar/2M)\,\boldsymbol{s}$, where g_s is the gyromagnetic ratio for the nucleon spin ($+5.59$ for p, -3.82 for n), M is the nucleon mass and $s=\frac{1}{2}$ (Section 8.6). The quantity $\hbar/Mc$ is a length, the nucleon reduced Compton wavelength, which is equal to 0.21 fm. This suggests that the replacement of the matrix element $\langle e\boldsymbol{r}\rangle$ by eR that we made in getting the E1 Weisskopf formula becomes the replacement

$$\langle\boldsymbol{\mu}\rangle \rightarrow e\frac{\hbar}{4Mc}g_s,$$

so that we get

$$\omega(\mathrm{M1}) = \frac{4}{3}\alpha\left(\frac{E_\gamma}{\hbar c}\right)^3 c\left(\frac{g_s\hbar}{2Mc}\right)^2. \tag{11.17}$$

We can see immediately that, for the same energy, the M1 transition rate is smaller than the E1 rate by about

$$\left(\frac{g_s\hbar}{4McR}\right)^2,$$

which for a proton ($g_s=5.586$) and $A=100$ is of the order of 10^{-3}.

The higher magnetic transitions are less simple to understand: as in the case of electric transitions they can change the nuclear spin by L when the transition is ML. The magnetic multipole that is changing is due to the current of both the spin and the orbital motion of the proton involved: the former with $s=\frac{1}{2}$ can only contribute one unit of multipolarity so the rest must be contributed from the orbital motion which in turn brings in the size of the nucleus. Thus from M1→M2 will add to the transition rate a factor of $(kR)^2=(E_\gamma R/\hbar c)^2$ as in E1→E2. Building on M1 we expect therefore

$$\omega(\mathrm{M}L) \sim \alpha\left(\frac{E_\gamma}{\hbar c}\right)^{2L+1} cR^{2L-2}\left(\frac{\hbar}{Mc}\right)^2. \tag{11.18}$$

This is just like $\omega(\mathrm{E}L)$ except one $R^2\rightarrow(\hbar/2Mc)^2$. (We have dropped $g_s/4$ in our approximation.) The full Weisskopf single particle estimate is

$$\omega(\mathrm{M}L) = \frac{20(L+1)}{L[(2L+1)!!]^2}\left(\frac{3}{L+3}\right)^2\alpha\left(\frac{E_\gamma}{\hbar c}\right)^{2L+1} cR^{2L-2}\left(\frac{\hbar}{Mc}\right)^2.$$

Once again this over-estimates the actual transition rates found. We note that for $L>1$ the change in orbital motion of the single particle is assumed to be a part of the changing current that radiates; therefore the particle must be charged and the formula can apply to protons only. Table 11.4 summarizes these formulae along with those for the electric transitions.

Just as in the case of EL transitions, the ratio $\omega(\mathrm{M}L+1)/\omega(\mathrm{M}L)$ is of order $(kR)^2$. And we note again that the $\omega(\mathrm{M}L)/\omega(\mathrm{E}L)$ ratio is of order $(\hbar/McR)^2$.

11.8 Selection rules in γ-ray emission

As we said in Section 11.5 and can now extend by including magnetic transitions, the multipolarity L satisfies

$$|j_A - j_B| \leqslant L \leqslant j_A + j_B.$$

The other relevant quantum numbers are the parities of states A and B.

Now all the evidence is that parity is conserved in electromagnetic transitions. The photon itself behaves like a spin-1 particle with odd intrinsic parity but, since it has zero mass and is relativistic, it turns out we cannot easily do a reckoning of spin and orbital angular momentum and the related parities involving the emission (or absorption) of a photon. However, we can discover whether the parity of the emitting (or absorbing) system changes or not. A change in this parity does not imply parity non-conservation: the parity of the initial and final states, including the photon, are the same. To find what happens to the parity of the system we must look at the matrix element. As parity is conserved the transition rate cannot change under the change of coordinates $\boldsymbol{r} \rightarrow -\boldsymbol{r}$ (see Section 13.1). This in turn means that the matrix element must also remain unchanged under this parity transformation (which we symbolize by P). In E1 transitions the matrix element is $\langle \boldsymbol{r} \rangle \equiv \langle \psi_B | \boldsymbol{r} | \psi_A \rangle$. We therefore must have

$$P\langle \psi_B | \boldsymbol{r} | \psi_A \rangle = \langle \psi_B | \boldsymbol{r} | \psi_A \rangle.$$

Since the polar vector $\boldsymbol{r}$ changes sign, we require the states ψ_A and ψ_B to have opposite parity, and we conclude that the radiating source parity changes in E1 transitions. Electric quadrupole transitions have two lengths in the operator Q (equation (11.14)) and so A and B have the same parity. Electric octupole E3 has three lengths in the operator so that A and B have the opposite parity. Thus we have alternating change/no change of parity in the series E1, E2, E3, ..., and, with an obvious notation:

$$\mathrm{E}L\text{: } P_B = (-1)^L P_A.$$

The magnetic transitions start with the M1 matrix element

$$\langle \psi_A | \boldsymbol{\mu} | \psi_B \rangle.$$

Now $\boldsymbol{\mu}$ is an axial vector and does not change sign under P: therefore A and B have the same parity and M1 transitions do not change parity. Building upwards, an extra factor $\boldsymbol{r}$ comes into the operator with each increase of L by 1. Thus M2, M3, M4, ... transitions alternately do, do not, do ... change the parity:

$$\mathrm{M}L\text{: } P_B = -(-1)^L P_A.$$

All this is summarized in Table 11.5.

Table 11.5 Angular momentum and parity selection rules in the absorption or emission of real photons.

L	Polarity 2^L	Name	Parity change Electric	Parity change Magnetic
1	2	Dipole	yes	no
2	4	Quadrupole	no	yes
3	8	Octupole	yes	no
.	.	....	...	..
L	2^L	2^L-pole	$(-1)^L$	$-(-1)^L$

If we draw a triangle with sides of length equal to the quantum number L, to the initial, and to the final state total angular momentum (spin) quantum numbers (j_A and j_B respectively) then this triangle must close to satisfy angular momentum conservation. Normally, the polarity is determined by the least L to satisfy this condition.

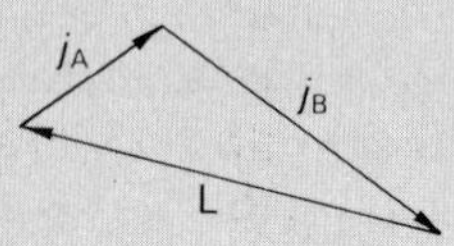

What relevance do these selection rules have? Clearly they can encompass any change in spin and parity (except two, see later in this section). However, since the transition rate decreases rapidly with increasing L, the nature of the transition is determined by the lowest L value possible consistent with the parity change rules. (See Problem 11.7). Only occasionally is there any uncertainty: for example

$$\Delta^+(1232) \rightarrow p + \gamma + 294 \text{ MeV},$$
$$j^P = \tfrac{3}{2}^+ \quad \rightarrow j^P = \tfrac{1}{2}^+$$

can be E2 or M1 by the selection rules. Since both of these transition amplitudes are down by factors like 30 from an E1 amplitude it could be that there are amplitudes due to both present in the final state and there is a possibility of interference effects which would then appear in the angular distribution of γ-rays from a sample of polarized Δ-particles.

The reader will recall that there is one transition which is not included in the rules given above; L cannot be zero. If $j_A = j_B = 0$ then since $L \geqslant 1$ no transition involving the emission of a photon is possible. This rule is absolute: $0 \rightarrow 0$ transitions are impossible involving the emission or absorption of a real photon. Note the inclusion of *real* qualifying photon. However, if the photon is virtual it can cause or be emitted in a $0 \rightarrow 0$ transition if there is no parity change. So how can a virtual photon be involved? In the first possibility the virtual photon can be exchanged with one of the atomic electrons so that the transition energy is used to eject an electron from the atom: this is **internal conversion**. An example is the first excited state of $^{72}_{32}$Ge at 691 keV and spin-parity $j^P = 0^+$: it decays to the ground state which also has $j^P = 0^+$ with a transition rate of $1.7 \times 10^6 \text{ s}^{-1}$ ejecting an atomic electron as it does so. The second possibility is **internal pair production**: the transition energy is used to create an electron–positron pair (threshold 1.02 MeV). A well-known example is the first excited state of $^{16}_{8}$O at

6.05 MeV which has $j^P = 0^+$: it decays by internal pair production to the ground state. These processes are not, in fact restricted to $j_A = 0 \rightarrow j_B = 0$ transitions, but they do provide the only mechanism for electromagnetic de-excitation for a nucleus with a first excited state with $j^P = 0^+$ and nowhere to go except to the lower $j^P = 0^+$ ground state. There is one other process by which $0 \rightarrow 0$ decays can occur: it is by the emission of two photons. However, it is a higher-order electromagnetic effect and is significant only if first-order transitions are forbidden. The $0^+ \rightarrow 0^+$ decay of ${}^{16}_{8}O^*(6.05)$ to the ground state by the 2γ mode has a branching fraction of about 2.5×10^{-4}.

Now what about $0 \rightarrow 0$ transitions with parity change? They require the emission of two real photons, or one real photon with one virtual photon, or two virtual photons, each virtual photon leading to internal conversion or pair production. No examples are known among nuclei. However, an example is

$$\pi^0 \rightarrow \gamma + \gamma.$$

The neutral π-meson, π^0 has $j^P = 0^-$ and it makes a transition to a state of no hadrons, a state which has $j^P = 0^+$! So the transition is $0^- \rightarrow 0^+$. One or both of these photons can be virtual so that there are two other (rare) decay modes

$$\pi^0 \rightarrow \gamma + e^+ + e^-,$$

and

$$\pi^0 \rightarrow e^+ + e^- + e^+ + e^-.$$

Internal conversion is a process which also competes with normal γ-ray decay: the conversion can be from any of the shells K, L, M, ... in X-ray nomenclature. It occurs predominantly from the $1s$-state as this is the one with the greatest overlap of the electron wavefunction with the nucleus. The probability density at the nucleus for a $1s$ electron is $Z^3\alpha^3(m_e/\hbar)^3$. The virtual photon is coupled to charge at each end so that brings in another e^2 into the amplitude, so we expect the internal conversion transition rate, ω_{IC}, to be proportional to $Z^3\alpha^5$. In fact the actual transition rate is not as important as the **internal conversion coefficient**: this is the ratio of the probability of internal conversion to real γ-ray emission, $\omega_{IC}/\omega_\gamma$. This ratio can be broken into parts corresponding to conversion of a K, or an L, or an M, ... shell electron respectively. Since all ω_γ for whatever multipolarity contains one power of α, the fine-structure constant, we expect

$$\omega_{IC}/\omega_\gamma \propto Z^3\alpha^4.$$

(Beware, reader, because most books use α, α_K, α_L, α_M, ... for the internal conversion coefficients! We have reserved α for the fine-structure constant so cannot follow this notation!) Evidently, internal conversion becomes relatively more important as Z increases.

Theoretically the value of $\omega_{IC}/\omega_\gamma$ also depends on the energy released and on the multipolarity, electric or magnetic. It does not depend on the nuclear part of the matrix element which is common to both rates and therefore cancels. Thus the measurement of the ratio leads, via tabulations of calculated values, to the multipolarity of the transition. This is one of the methods of determining the multipolarity.

Returning to selection rules in γ-ray emission or absorption, we remind readers that in a comment with Section 10.11 we gave the selection rules that applied to isotopic spin. However, that subject takes us beyond the scope of this book and we cannot usefully discuss it further.

11.9 Nuclear isomerism

Isotope, isobar and isotone were defined in Section 1.6: briefly, same Z, same A, and same N respectively! So what is an **isomer**? The isomer of a stable nuclueus has the same Z and A but is heavier and is, therefore, an excited state. There is nothing special about excited states so how did this term come to be used? In the early days of investigations into radioactivity many excited states of heavy nuclei were recognized and observed to decay by γ-ray emission. However, the lifetimes were too short to be susceptible to measurement. Later, states which are unstable and emit a γ-ray were found with mean lifetimes which could be measured, usually fractions of a second but some much longer. For example, the excited state at 80 keV of $^{234}_{91}$Pa, which is the product of the β^- decay of $^{234}_{90}$Th and has a mean life of 101 s, was the first to be found. Such states with measurable lifetimes were called isomers. Of course, what is measurable has changed and the boundary to isomers is no longer clear. Roughly any excited state with a lifetime greater than about 10^{-6} s may be called an isomer. Looking back we can see that a long lifetime will be associated with a large change in $j_A \rightarrow j_B$ or a low-energy γ-ray, or both.

In certain parts of the periodic table, the systematics of the shell model give an energy ordering of levels which has, for a single unpaired nucleon, a high angular momentum neighbour to a ground state of low angular momentum, or vice versa (for example, $1h_{11/2}$ and $3s_{1/2}$ in $^{113}_{48}$Cd). The transition, E5, has a mean life which makes the excited state an isomer. Such circumstances are frequent where the spin–orbit splitting brings high-j states down close to low-j states in the shell of lower principal quantum number. The consequence is regions in the periodic table called **islands of isomerism** where isomers are a common occurrence. These islands are given as evidence for the shell model of nuclei (see Section 8.2).

DEFINITIONS AND KEYWORDS

Multipole order (multipolarity) The quantum number ($L \geqslant 1$) describing the change in total angular momentum of a system which radiates or absorbs a photon.

Weisskopf single-particle rates Formulae for the transition rates for the emission of photons by nuclei, derived assuming that only one proton within the nucleus changes its state of motion.

Internal conversion The process by which a nuclear excited state decays by the direct ejection of one of its atomic electrons; it occurs normally in competition with photon emission.

Internal conversion coefficient The ratio of the probability of internal conversion to that of photon emission.

Internal pair production The process by which a nuclear excited state decays by the direct production of an electron–positron pair.

Isomer An excited state of a nucleus which decays electromagnetically and has a mean life greater than about 10^{-6} s.

Islands of isomerism Regions of the periodic table particularly rich in isomers.

The reader should now look at Problem 11.8 which asks for an estimate of the decay transition rates for the decays given in equations (11.11) and (11.12) as examples in Section 11.5. The last is an isomer and the transition is from an excited state of $\frac{11}{2}^-$ to the ground state of $\frac{3}{2}^+$. Problems 11.9 and 11.10 are also concerned with estimating transition rates and determining multipolarities.

PROBLEMS

11.7 Show that equation (11.14) is dimensionally correct.

11.8 Consider the decay of equation (11.12): given that the spin-parity of the $^{137}_{56}\text{Ba}^*$ excited state is $\frac{11}{2}^-$ and that of the ground state is $\frac{3}{2}^+$, decide on the multipolarity of this photon-emitting transition and calculate the expected transition rate using a formula from Table 11.4.

Determine the multipolarity of the radiative decay $\Sigma^0 \to \Lambda\gamma$ (equation (11.11)) and estimate the transition rate using a formula from Table 11.4. Do you see any modifications that should be made to the formula that you wish to use?

11.9 Classify the following transitions by their multipolarity and calculate the expected transition rates:

$$^{175}_{71}\text{Lu}^*(\tfrac{9}{2}^+) \to {}^{175}_{71}\text{Lu}(\tfrac{7}{2}^+) + \gamma + 114\,\text{keV}, \qquad \omega = 5\times10^{9}\,\text{s}^{-1},$$
$$^{60}_{27}\text{Co}^*(2^+) \to {}^{60}_{27}\text{Co}(5^+) + \gamma + 58.6\,\text{keV}, \qquad \omega = 4\times10^{-5}\,\text{s}^{-1},$$
$$^{60}_{28}\text{Ni}^*(2^+) \to {}^{60}_{28}\text{Ni}(0^+) + \gamma + 1.33\,\text{MeV}, \qquad \omega = 9.5\times10^{11}\,\text{s}^{-1},$$
$$^{44}_{21}\text{Sc}^*(1^-) \to {}^{44}_{21}\text{Sc}(2^+) + \gamma + 68\,\text{keV}, \qquad \omega = 4.5\times10^{6}\,\text{s}^{-1},$$
$$^{47}_{21}\text{Sc}^*(\tfrac{3}{2}^+) \to {}^{47}_{21}\text{Sc}(\tfrac{7}{2}^-) + \gamma + 767\,\text{keV}, \qquad \omega = 2.6\times10^{6}\,\text{s}^{-1}.$$

Comment on any discrepancies between expected and actual transition rates wherever you can.

11.10 Use the given transition rates for the following decays to attempt a determination of the multipolarity:

$$^{86}_{37}\text{Rb}^* \to {}^{86}_{37}\text{Rb} + \gamma + 556\,\text{keV}, \qquad \omega = 1.1\times10^{-2}\,\text{s}^{-1},$$
$$^{95}_{40}\text{Nb}^* \to {}^{95}_{40}\text{Nb} + \gamma + 236\,\text{keV}, \qquad \omega = 5.8\times10^{-7}\,\text{s}^{-1},$$

Look up the transitions (try Lederer *et al.*, 1978) to find how near you are to the correct classification.

11.10 Other electromagnetic processes

The objective of this section and the one following is to widen our horizons with respect to the electromagnetic interaction. In Sections 11.5–11.9 we discussed the radiation of photons by nuclei and particles. But of course the reverse is possible. Nuclei can absorb photons, resulting in an excited nucleus or in the ejection of nucleons (**photonuclear reactions**); at high energies the **photoproduction** of mesons and baryon–antibaryon pairs is possible. The selection rules and the multipole classifications are the same as in photon radiation. Thus the electromagnetic part of the meson photoproduction reaction

$$\gamma + \text{p} \to \Delta^+(1232) \to \pi^0 + \text{p}$$

is $j^{\text{P}} = \frac{1}{2}^+ \to j^{\text{P}} = \frac{3}{2}^+$ and is therefore M1 or E2. A complete analysis of data shows that it is dominantly M1 and that this observation has a physical explanation.

The p and Δ^+ have the same valence quark content (uud), no relative orbital angular momentum between the quarks and therefore a quark total spin state which is $\frac{1}{2}$ and $\frac{3}{2}$ respectively: clearly, flipping one quark spin in the proton can change the total spin as required and such a spin flip changes the magnetic moment, just as is expected in an M1 transition.

This reaction has, in fact, the features of a compound state reaction. Of course, the intermediate Δ^+ is not a compound of the incident photon and the target proton but the reaction cross-section has a resonance shape as expected from the formation of an intermediate state of well-defined quantum numbers and of relatively long lifetime. This cross-section is shown in Fig. 11.9(a). This example does not mean that all meson photoproduction reactions go by the compound state mechanism. The equivalent of direct reactions also occur: a photon, or a virtual photon, absorbed by a nucleon has interacted with one of the quarks and this can lead, if the energy is sufficient, to the production of one or more hadrons by the quark fragmentation process (Section 10.9), without the formation of any intermediate (compound) state.

These features are also seen in nuclear reactions caused by photons. Any reaction conserving energy, momentum, charge, and the number of baryons

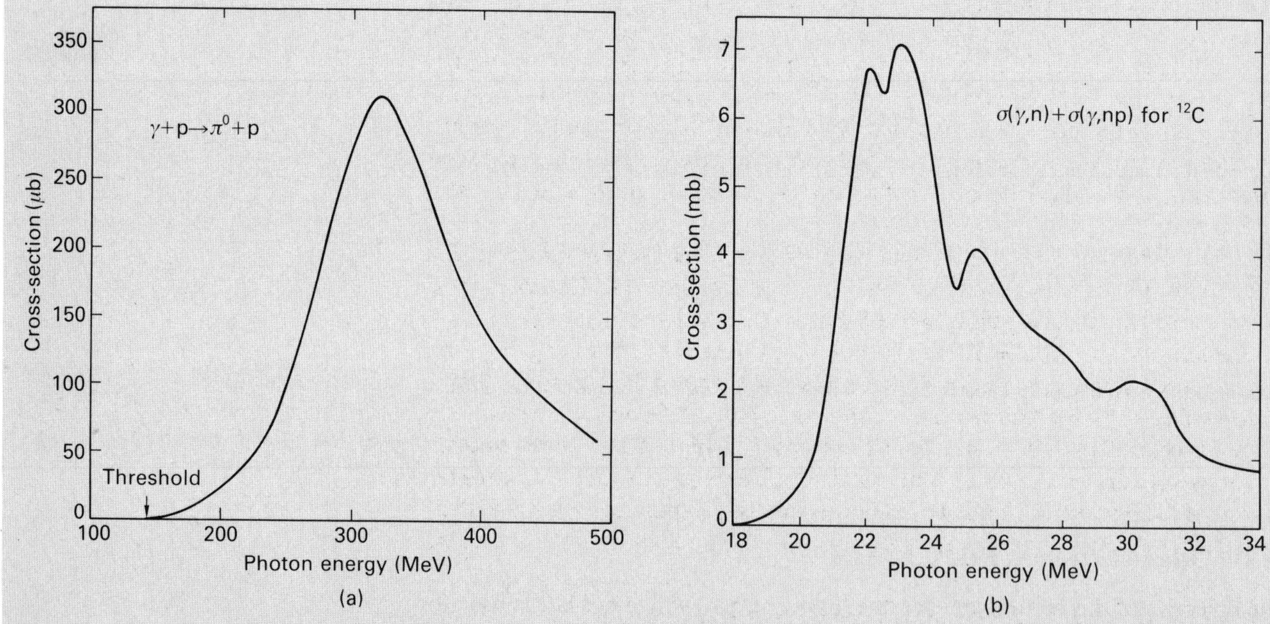

Fig. 11.9 (a) The cross-section for the reaction

$$\gamma + \text{p} \rightarrow \pi^0 + \text{p}$$

as a function of incident photon energy. This reaction shows resonant behaviour at centre-of-mass energies corresponding to the formation of the state $\Delta^+(1232)$. Note that the cross-section reaches about 300 μb to be compared with about 200 mb in π^+p scattering at the same total centre-of-mass energy. (b) The sum of the cross-sections for the ${}^{12}_{6}$C(γ,n) and the ${}^{12}_{6}$C(γ,np) reactions as a function of photon energy. The broad peak is due to the **giant dipole resonance** in which at one instant all the protons are displaced in one direction and all the neutrons are displaced in the opposite direction. This resonance decays predominantly by the emission of one or more nucleons and is a feature of all nuclei. These total giant dipole cross-sections reach maxima that are about 3% of geometrical for carbon and increase to about 50% for lead.

The curves in these figures have been drawn to represent data which has errors: therefore they cannot be taken to be an accurate representation of the real cross-section. The data of figure (a) is based on the compilation by Alekhin *et al.* (1987) and figure (b) is based on data published by Fultz *et al.* (1966).

(Section 10.3) is possible: these include reaction types such as (γ,n), (γ,p), (γ,α), (γ,2n), (γ,np) and so on. Both the direct and compound state mechanisms occur but there is in all nuclei a very prominent occurrence of the latter called the **giant dipole resonance**. It is seen at photon energies of about 25 MeV in light nuclei, decreasing to about 12 MeV in the heaviest stable nuclei, with widths of several mega-electron-volts. This type of resonance is due to the mode of nuclear vibration in which at any instant the protons are collectively displaced one way while the neutrons are displaced in the opposite direction, with the centre of gravity remaining stationary. The excitation of this mode is clearly by E1 photon absorption: the resonance decays by the emission of one or more nucleons of which one at least is normally a neutron. Figure 11.9(b) shows the giant dipole resonance cross-section as a function of photon energy for carbon; there is some structure indicating that it is not just one simple mode of vibration, but nonetheless the model of the resonance is, except in detail, correct.

Resonant absorption of photons by nuclei can also occur to one of the excited states below the energy at which nucleon emission is possible. The only mode of decay will be by emission of a photon and we have what is essentially resonant photon scattering. This is sometimes called **resonance fluorescence**, a term taken over from atomic and molecular physics. However, this process is hard to observe since these excited states are narrow (of the order of milli-electron-volts) and it is difficult by conventional means to tune a beam of photons to have sufficient number in an energy range which overlaps the required excitation energy without an excessive number of unwanted photons at other energies, all giving rise to effects which dilute the effects of resonant scattering. However, there are a couple of techniques that enable the process to be observed and we shall devote Section 11.11 to brief descriptions of their principles.

So far in this section we have dealt with effects caused by real photons. But virtual photons can also cause nuclear and particle reactions. The source of photons can be either energetic charged leptons (electron, positron, the muons) or energetic charged nuclei. This almost closes one circle by reminding us of the elastic scattering of electrons by nuclei (Section 3.2) and of the elastic scattering of α-articles by nuclei (Rutherford scattering, Section 1.2). However, inelastic scattering is not excluded: energetic electrons and muons can cause nuclear reactions and the production of mesons from nuclei and nucleons (**electroproduction**). The important thing about charged leptons in this context is that they do not experience the nuclear, strong force, so that, at all but very large

DEFINITIONS AND KEYWORDS

Photonuclear reactions Nuclear reactions initiated by the absorption of a photon.

Photoproduction reactions Photon–nucleon or photon–nuclear collisions in which hadrons not previously present as the target or part of the target are produced.

Giant dipole resonance A collective excitation that can occur in all nuclei in which at any instant all the protons are moving one way and all the neutrons the other way.

Resonance fluorescence The elastic or inelastic nuclear scattering of photons by the compound state mechanism: the compound state is, by definition, an excited state of the nucleus normally below the threshold for nucleon emission.

Electroproduction The inelastic scattering of a charged lepton by a nucleon or a nucleus in which one or more hadrons are produced.

momentum transfers, charged leptons interact electromagnetically when scattered elastically or inelastically from nuclei or nucleons (see Section 10.10). At high momentum transfers, (>10 GeV/c) the weak interaction will become as important as the electromagnetic and, in fact, the interaction must be discussed in terms of the unified electroweak theory (Section 9.13). We say *will become* because charged lepton scattering experiments and measurements at momentum transfers of that magnitude have not yet been done.

The use of energetic charged nuclei as a source of virtual photons requires that the electromagnetic scattering is not masked by the strong nuclear force effects. This requirement can be satisfied by having the Z of the target and incident projectile sufficiently large that Coulomb barrier penetration is small. Then the large values of Z and the long range of the Coulomb force are the ingredients that cause significant electromagnetic elastic scattering (Rutherford), or inelastic scattering (Coulomb excitation, Section 7.6). Readers will be reminded that Coulomb excitation is a particularly fruitful way of producing rotational excited states of nuclei.

A question can be raised as to whether or not these nuclear, Coulomb effects of charged particles add to the energy loss and to the photon absorption effects discussed in Sections 11.1–11.4. The answer is that they do but only at a relatively very low level. The cross-section for a photonuclear reaction is normally very small compared to the cross-sections for pair production (see Fig. 11.6). The inelastic scattering of charged leptons by nuclei causes large energy losses but, as the cross-section is relatively very small, the average contribution to the energy loss is small.

We remind readers that one process involving virtual photons and particles

Table 11.6 A summary of electromagnetic effects of nuclei and hadrons according to whether they involve real or virtual photons.

Real photons	(Approximate total cross-section)
Emission of photons in radiative de-excitation of nuclei and of hadrons	—
Photonuclear reactions	(10–1000 μb)
Giant dipole resonance	(at resonance maximum $\sigma \simeq 1$–500 mb)
Photoproduction of hadrons	(1–100 μb)
Resonance fluorescence	(at resonance maximum $\sigma \simeq \lambda\!\!\!^{-2}$)
Nuclear Rayleigh scattering	($\sim 0.2(Z/A)^2$ μb)
Virtual photons	
Internal conversion	—
Internal pair production	—
Rutherford and Mott scattering	(see Section 1.3)
Coulomb excitation	(comparable to Rutherford scattering cross-sections)
Inelastic nuclear scattering of charged leptons	(0.1–10 μb)
Electroproduction of hadrons (which includes deep inelastic scattering) by charged leptons	(~ 1 μb)
Electron–positron annihilation into hadrons	(see Section 10.6)

The total cross-sections given are no more than a guide to the range of magnitudes to be found: note that those involving real photons are about 10^{-3} of those for similar reactions initiated by nucleons or other hadrons.

has been discussed before this chapter. It is the production of hadrons following electron–positron annihilation at high energies: see Section 10.5.

In Table 11.6 we summarize the processes we have discussed, albeit most of them very briefly.

We end this section by examining the cross-sections for photon-induced reactions as compared to those induced by hadrons. There are no simple rules which allow us to estimate cross-sections, but we can say that the former are expected to be smaller by a factor of 100–1000 than a comparable reaction induced by a nucleon or a π-meson. This is because the photon couples to charge with a strength characterized by α, the fine structure constant, which is smaller by this magnitude than an effective strong interaction coupling. We illustrate this in Fig. 11.10, which shows on the same graph the total cross-sections for negative π-mesons and photons on protons as a function of the centre-of-mass energy. The logarithmic plot shows the ratio of over 100 to 1 in favour of the strong interaction π^-p to the electromagnetic γp cross-section.

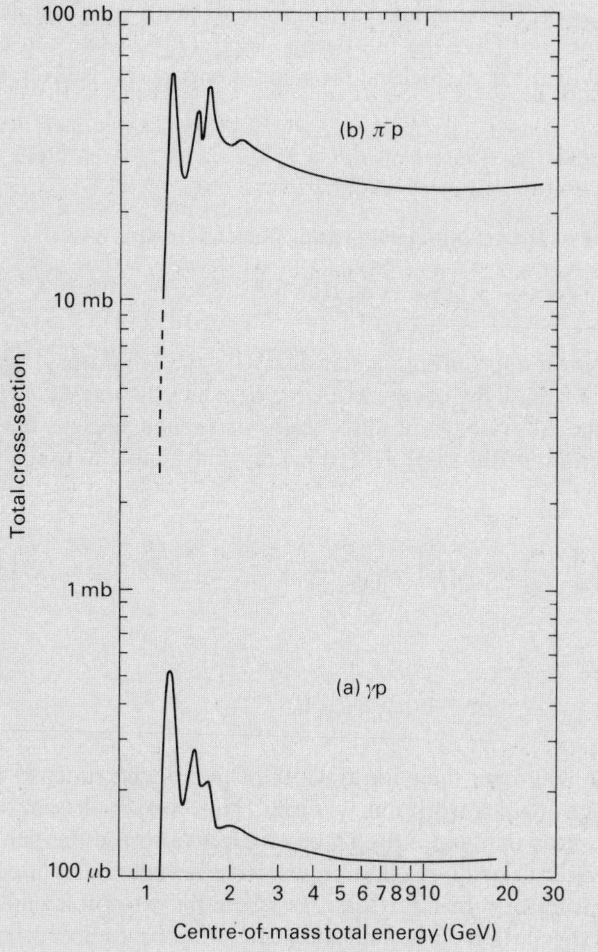

Fig. 11.10 A plot of the two total cross-sections (with a common vertical scale) as functions of the total centre of mass energy: (a) photon–proton, (b) negative π-meson–proton. Note that the structure is similar as both systems are channels to the same resonant states; however, the relative magnitudes are about 1 to 300 due to the strength of the electromagnetic coupling relative to the strong interaction coupling. Data from Flaminio *et al.* (1983) and Alekhin *et al.* (1987). Note that the γp cross-section does not include pair production.

Virtual photon cross-sections are more complicated: energy and momentum ($\times c$) are no longer equal, which means that the total cross-section for incident mono-energetic charged particles can involve a range of energy transfers to the target not directly comparable to the unique energy when mono-energetic photons cause reactions. However, since a virtual photon has to be radiated another α-factor is to be expected and the total cross-section will be smaller by a factor of 10^4 to 10^6 than a comparable strong interaction cross-section.

11.11 Resonance fluorescence and absorption of photons

The first process was described in the last section and we now describe how it may be observed in nuclei. The resonant absorption of a photon is clearly only the first step in resonance fluorescence and would be observed by a loss of transmission of the photons through a target when the photons had the resonance energy.

Let us first consider a specific example, that of $^{152}_{62}$Sm, which has an excited state at 963 keV with $j^P = 1^-$ (which we label $^{152}_{62}$Sm*) with a mean life of about 3.5×10^{-14} s (line width $\simeq 0.02$ eV) (see Fig. 11.11a). Under what conditions can the γ-ray from the decay of this samarium excited state scatter from a sample of $^{152}_{62}$Sm? This photon does not have the full decay energy ($E_0 = 963$ keV in this case) because of recoil of the final nucleus: the photon energy, E_γ, is given by

$$E_\gamma = E_0 \left(1 - \frac{E_0}{2M_{\mathrm{Sm}}c^2}\right), \tag{11.19}$$

where M_{Sm} is the mass of the recoiling samarium nucleus. In this case

$$E_\gamma = E_0 - 3.26 \text{ eV}.$$

Now consider this photon approaching a stationary (in the laboratory) $^{152}_{62}$Sm nucleus. In the centre-of-mass the energy available to excite the nucleus is less than E_γ because, in the laboratory, on absorption the excited nucleus has to recoil with the momentum of the photon. The energy E available to excite the nucleus is given by

$$E = E_\gamma \left(1 - \frac{E_\gamma}{2M_{\mathrm{Sm}}c^2}\right). \tag{11.20}$$

In this case

$$E_0 - E = 6.52 \text{ eV}.$$

This shortfall is about 300 times the natural width of the excited state, so E is too far from E_0 for resonant absorption to occur. However, if the emitting nucleus is moving towards the target, the Doppler effect can raise the photon energy, and with the right velocity, the photon will have the energy required to allow photon-resonant absorption. For the case where the photon is emitted along the direction of the motion of the source ($^{152}_{62}$Sm*), the required velocity βc is given by

$$E_0 \sqrt{\left(\frac{1+\beta}{1-\beta}\right)} = E_0 + 6.52 \text{ eV}.$$

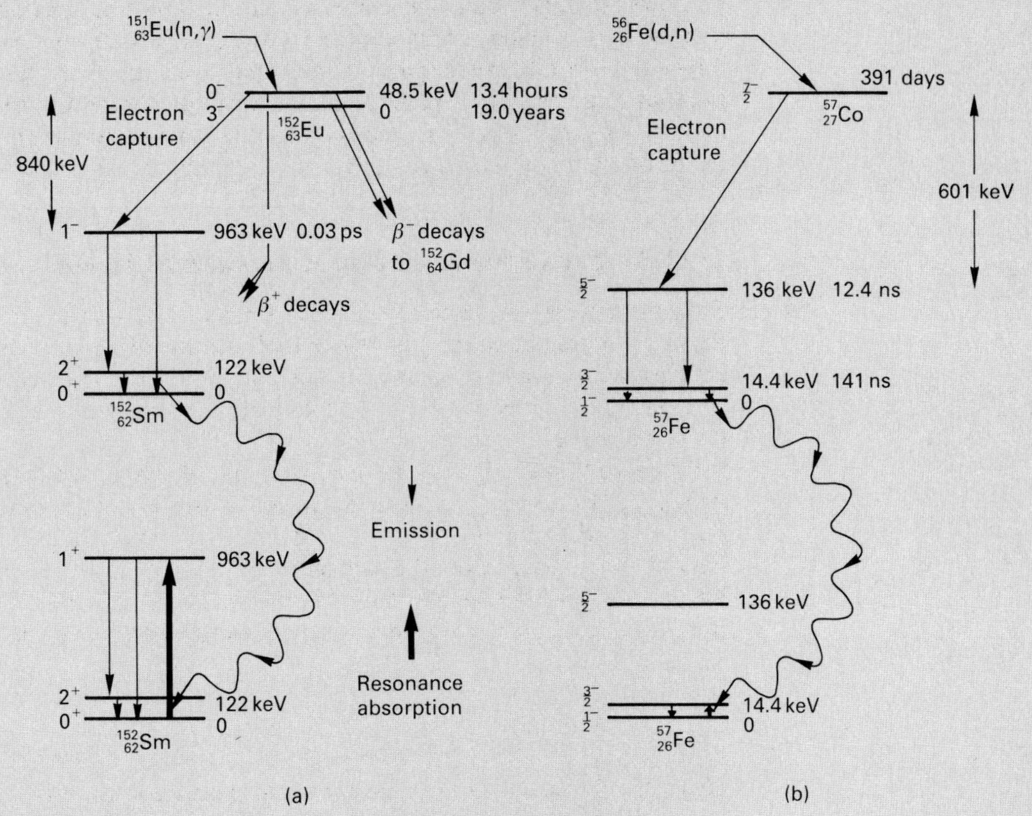

Fig. 11.11 The nuclear energy levels for the two examples of resonance fluorescence discussed in the text. (a) $^{152}_{62}$Sm: the transition from the 1^-, 963 keV excited state to the ground state can be observed to be resonantly scattered by a target of $^{152}_{62}$Sm if the radiating nucleus is moving with sufficient velocity towards the target. (b) $^{57}_{26}$Fe: The transition from the 14.4 keV excited state to the ground state can be observed to be resonantly absorbed by a target of $^{57}_{26}$Fe if both emission and absorption can be made recoil-less by using the Mössbauer effect.

In this figure the wavy line represents the real photon emitted by the radiating nucleus propagating to be resonantly absorbed by a target.

The result is $\beta c = 2040\ \mathrm{m\,s^{-1}}$. How can this velocity be obtained? The state $^{152}_{62}\mathrm{Sm}^*$ is produced in the electron capture decay of an isomer of $^{152}_{63}$Eu (Fig. 11.11a);

$$^{152}_{63}\mathrm{Eu} + e^- \rightarrow {}^{152}_{62}\mathrm{Sm}^* + \nu_e.$$

The Q of the decay is 840 keV, which gives the final state nucleus a recoil velocity of $1800\ \mathrm{m\,s^{-1}}$. This is almost enough; thermal agitation broadens the intrinsic line widths sufficiently that there is now overlap of the line widths and resonant absorption occurs. Note that this requires that the mean life of the $^{152}_{62}\mathrm{Sm}^*$ is much shorter than the time taken for it to come to rest after recoiling from the neutrino emission in electron capture. Once resonant absorption has occurred the excited nucleus will decay, giving, in this case, two γ-rays of 841 and 122 keV or one of 963 keV: thus we have resonance fluorescence. This circumstance was used in an experiment to measure neutrino helicity (Section 12.9) in order to pick out photons emitted forward in the decay of $^{152}_{62}\mathrm{Sm}^*$

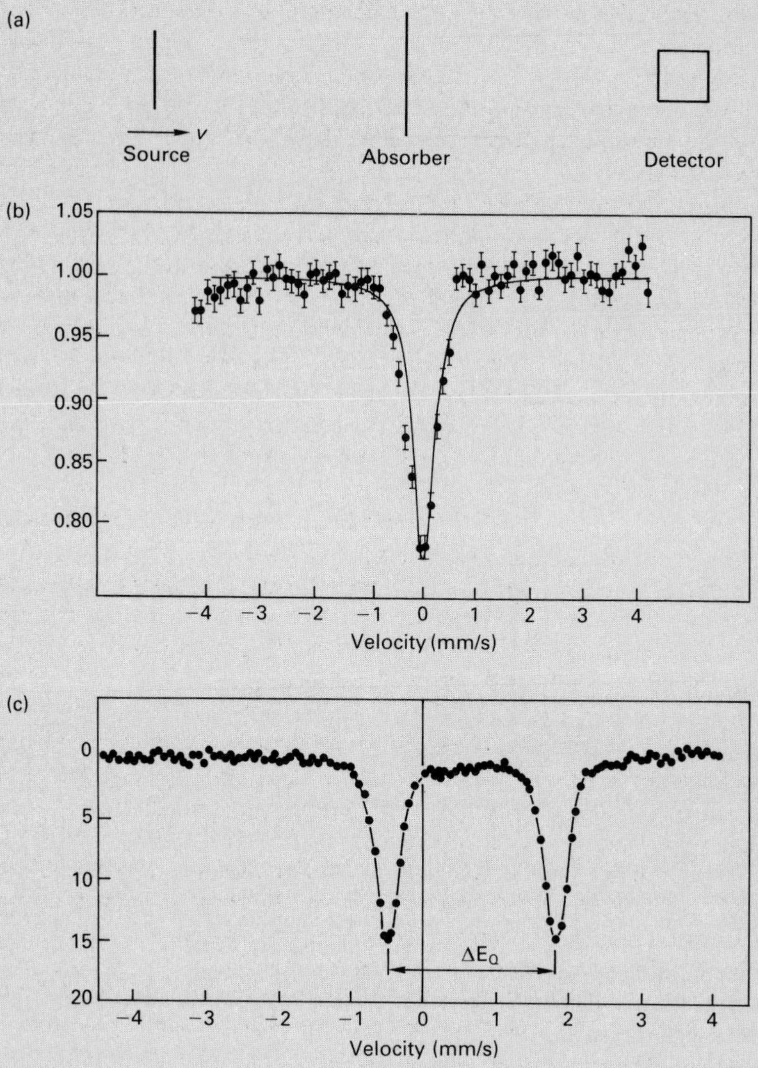

Fig. 11.12 (a) The arrangement of source, absorber and detector employed to observe resonant absorption as required for the observation and use of the Mössbauer effect. In some cases source and absorber will be maintained at a low temperature. (b) The detector relative counting rate as a function of the source velocity. In this case the source was $^{57}_{27}Co$ diffused into a stainless steel foil. The absorber was a similar stainless steel foil but without the radioactive cobalt. The solid line is the expected detector response and is the convolution of the source and absorber (identical) line shapes (see Problem 11.12). The points are the observed rates with their errors (Ruby and Bolef, 1960). (c) Shows the absorption when the source is $^{57}_{27}Co$ diffused into a chromium foil and the absorber is $^{57}_{26}Fe$ in a compound called biferrocenyl. (Wertheim and Herber, 1963.) The two inverted peaks correspond to the hyperfine splitting of the excited state of $^{57}_{26}Fe$ in the biferrocenyl due to the interaction of its electric quadrupole moment with the gradient of the electric fields generated by its environment. The centre of the splitting is shifted by what is called the isomer effect. The energy difference between ground and excited states of $^{57}_{26}Fe$ is slightly shifted by the difference in nuclear radii and the density of the atomic electron wavefunctions at the nucleus. The shift is different in the source from that in the absorber in the case of (c). This last example indicates the power of the Mössbauer effect in investigating certain aspects of condensed matter.

recoiling from electron capture. Of course, this example is a very special case where the properties conspire to allow resonant absorption and was selected for that reason.

Nuclear reactions may also produce excited states of nuclei with a velocity that allows compensation for the effect of recoil. Now the kinematics of the reaction may permit the choice of an angle of observation that selects the required velocity.

The formula for the energy of a photon emitted from a source suffering recoil (equation (11.19)), gives the clue to the next method of observing resonant absorption. The recoil energy loss decreases as the mass of the radiating system increases: if the excited nucleus of ground state mass M is locked into a crystal lattice which recoils without excitation, then the effective mass is not M but many orders larger and the recoil loss of energy becomes negligible. Similarly, if the identical target nucleus is likewise situated, there is no recoil on absorption and all the photon energy is available to excite the target. Overall there is no mismatch and resonant absorption can occur. This is the **Mössbauer effect**, named after the discoverer.

DEFINITIONS AND KEYWORDS

Mössbauer effect The phenomenon in which a nucleus emits a photon without energy loss due to recoil if the host atom is securely located in a crystal lattice.

The probability of recoil-less emission decreases with increasing photon energy and with increasing lattice temperature. The best example is the 14.4 keV transition in $^{57}_{26}$Fe (see Fig. 11.11b); this energy is so low that recoil-less emission occurs readily at room temperature. The half width of the excited state is 4.7×10^{-9} eV and it is produced in the decay by electron capture of $^{57}_{27}$Co. Figure 11.12(a) shows a schematic layout for the detection of resonant absorption. In the case of $^{57}_{26}$Fe the source could be a stainless steel foil into which had been diffused some $^{57}_{27}$Co so that the excited iron nucleus is in an environment identical to that for all the iron nuclei: the absorber could be an identical stainless steel absorbing foil. The energy match may be spoilt by giving the source foil a constant velocity: thus a mismatch may be achieved in the case of $^{57}_{26}$Fe by a source velocity of 1 mm s^{-1}. The convolution of the line shape with itself may be explored by measuring the transmission of the 14.4 keV photons by the absorber as a function of source velocity, as shown in Fig. 11.12(b).

If the absorber does not provide the same environment for the iron nuclei as does the source foil, then the energy of the ground and excited states may be shifted or the spin degeneracy lifted by hyperfine interactions with this environment. The absorption versus velocity curves then become more complicated but now give information about the physical and chemical state of that environment. A simple example of such a situation is shown in Fig. 11.12(c). The Mössbauer effect therefore has applications in the investigation of condensed matter, although wide use is prevented by the limited number of suitable sources. The effect has also been used to measure the decrease in energy of photons emitted at one height and detected at a greater height (the gravitational red shift). These and other applications are described by Wertheim (1964) and by Herber (1971).

Problems 11.11 to 11.14 will give the reader an opportunity to calculate the velocities and energy shifts typical of Mössbauer spectroscopy.

11.12 Summary

Electromagnetism is really very well understood unless one asks questions at a level which involves its role in the nature and behaviour of matter and their place in space–time. Given the existing understanding, the electromagnetic

PROBLEMS

11.11 Find the formula for the relativistic Doppler effect and compare it to that for non-relativistic velocities.

11.12 Given that the width of the Mössbauer absorption line in Fig. 11.12(b) is entirely due to the convolution of the identical line widths at emission and absorption, estimate that line width in eV.

11.13 Use the information given in Fig. 11.12 on the $^{57}_{26}Fe$ Mössbauer transition to calculate the quadrupole splitting in biferrocenyl and the isomer shift between this nucleus in a chromium foil and in biferrocenyl. Give the results in electronvolts. Why is it that the energy of the excited state of $^{57}_{26}Fe$ is split and not that of the ground state? Why is the result of the splitting a doublet?

11.14 A 14.4 keV photon from $^{57}_{26}Fe$ is red-shifted as it rises from a source at ground level to an absorber foil at a height of 10 m. What velocity of the absorber foil is required to compensate the red shift, and in which direction?

interaction of nuclei and particles gives information on the structure of those systems. For example, the determination of the multipolarity of a γ-ray transition and its rate for a given nucleus may give important information on the wavefunctions of the nuclear states involved. Again the rate of production of hadrons in electron–positron annihilation at high energies provides convincing evidence of the role of the colour quantum number in the physics of quarks.

In Section 13.2 we present a guide to recognizing the controlling interaction (strong, weak, or electromagnetic) in nuclear and particle decay and reaction processes.

Of course, electromagnetism was the result of the first unification which physicists are seeking for all the interactions of matter. Maxwell unified the physics of electrostatics and of magnetism. That particular unification involves an interaction between fields and charges of strength given by the fine structure constant. In modern quantum mechanics the interaction is that of a relativistic four-vector of the current j $(=\boldsymbol{j},\rho c)$, where $\boldsymbol{j}$ is the current density and ρ the charge density, with the four-vector potential A $(=\boldsymbol{A},\varphi/c)$, where $\boldsymbol{A}$ is the vector potential and φ the scalar potential. The interaction then contributes to the energy density of the system an amount

$$jA=\rho\varphi-\boldsymbol{j}.\boldsymbol{A}.$$

This is all that is required to describe completely the electromagnetic interaction.

The unification of electromagnetism with the weak interaction, which we introduced in Section 9.13 and which was referred to briefly at the beginning of this chapter, has no significant impact on the material discussed in this chapter, except to the case of charged lepton scattering at very high momentum transfers and to e^+e^- annihilation at energies comparable to and above the rest mass energy of the Z^0 boson (see Section 12.4). We shall discuss this unification again in Section 13.4.

References

Alekhin, S. I. *et al.* (1987). *Compilation of Cross Sections IV.* HERA and COMPAS Groups, CERN-HERA 87–01, European Organisation for Nuclear Research, Geneva.

Corney, A. (1977). *Atomic and Laser Spectroscopy.* Oxford University Press.

England, J. B. A. (1974). *Techniques in Nuclear Structure Physics*, Parts 1 and 2. Macmillan, London.

Fernow, R. C. (1986). *Introduction to Experimental Particle Physics.* Cambridge University Press.

Flaminio, V., Moorhead, W. G., Morrison, D. R. O., and Rivoire, N. (1983). Compilation of Cross Sections I. HERA Group, CERN-HERA 83-01, European Organisation for Nuclear Research, Geneva.

Fultz, S. C., Caldwell, J. T., Berman, B. L., Bramblett, R. L., and Harvey, R. R. (1966). *Physical Review*, **143,** 790–6.

Harvey, R. R., Caldwell, J. T., Bramblett, R. L., and Fultz, S. C. (1964). *Physical Review*, **136B,** 126–31.

Herber, R. H. (1971). *Scientific American*, **225,** 86–95.

Janni, J. F. (1982). *Atomic Data and Nuclear Data Tables*, **27,** 147–529.

Kleinknecht, K. (1986). *Detectors for Particle Radiation.* Cambridge University Press.

Lederer, C. M., Shirley, V. S., Browne, E., Dairiki, J. M., and Doebler, R. E. (1978). *Table of Isotopes* (7th edn.). John Wiley, New York.

Particle Data Group (1988). Review of particle properties. *Physics Letters B*, **204,** 1–486.

Ruby, S. L. and Bolef, D. I. (1960). *Physical Review Letters*, **5,** 5–7.

Storm, E. and Israel, H. I. (1970). *Nuclear Data Tables*, **7A,** 565–681.

Street, J. C. (1929). *Journal of the Franklin Institute*, **227,** 765–88.

Taylor, T. S. (1913). *Philosophical Magazine*, 6th Series, **26,** 402–10.

Tsai, Y-S. (1974). *Review of Modern Physics*, **46,** 815–51.

Wertheim, G. K. (1964). *Mössbauer Effect: Principles and Applications.* Academic Press, London and New York.

Wertheim, G. K. and Herber, R. H. (1963). *Journal of Chemical Physics*, **38,** 2106–11.

12

The Weak Interaction

12.1 A review

Up to this chapter we have been very free with references to the β-decay of nuclei, to the weak interaction decays of hadrons and to the basic mechanisms involving Z^0 and $W^{\pm}$ particles. It is now time to attempt to pull all this into a coherent and whole picture. However, for readers who may be taking liberties with the order in which they are reading chapters, here is a short reminder:

1. Beta decay is the process by which complex nuclei return towards the line of stability by emitting electrons or positrons or by electron capture (Chapter 4):

$$(Z,A) \rightarrow (Z+1,A) + e^- + \bar{\nu}_e, \tag{12.1}$$

$$(Z,A) \rightarrow (Z-1,A) + e^- + \nu_e, \tag{12.2}$$

or $$(Z,A) + e^- \rightarrow (Z-1,A) + \nu_e \text{ (electron capture).} \tag{12.3}$$

2. Weak interaction decays of hadrons. In addition to some β-decays such as $K^+ \rightarrow \pi^0 + e^+ + \nu_e$, many decays occur which are also due to weak interactions (Sections 10.3 to 5):

$$K^+ \rightarrow \pi^+ + \pi^0,$$

$$K^+ \rightarrow \pi^+ + \pi^+ + \pi^-, \text{ etc.}$$

3. A few decays do not involve hadrons: for example

$$\mu^+ \rightarrow e^+ + \nu_e + \bar{\nu}_\mu \tag{12.4}$$

4. In Section 9.13 we prepared readers for the idea that the gauge, spin-1 bosons, $W^{\pm}$, Z^0) are involved in the weak interactions.

In Section 9.12 we defined the leptons. These are spin-$\frac{1}{2}$ particles, some charged and some uncharged, which do not experience the strong interactions: they are very frequently involved in weak interactions. Thus the weak interaction decays of hadrons that involve only leptons in the final state (for example, $\pi^+ \rightarrow \mu^+ + \nu_\mu$) are called **leptonic hadron decays.** Those that have both leptons and hadrons are called **semi-leptonic hadron decays** (for example, $\Sigma^- \rightarrow n + \mu^- + \bar{\nu}_\mu$). Those that involve only hadrons are called **non-leptonic hadron decays** (for example, $K^+ \rightarrow \pi^+ + \pi^0$). β-decay is clearly a semi-leptonic deay. Decays or interactions involving leptons alone (for example, equation (12.4)) are called **pure leptonic interactions**.

All the leptons believed to exist are listed in Table 12.1. Since they have spin $\frac{1}{2}$, they obey Fermi–Dirac statistics and each lepton (particle) has an anti-lepton (anti-particle) partner. Let us consider the entries: electron and positron should be well known to the reader. The negative and positive μ-meson (or muons) were discovered in the cosmic radiation (Section 9.10): apart from their instability, they have all the properties of heavy electrons and positrons. Their existence is a puzzle which was increased by the discovery of the τ-mesons in 1975 (τ^+,τ^-), an even heavier electron–positron-like partnership. To each of these pairs we associate a neutrino–antineutrino pair labelled with a subscript which ties each such pair to a pair of charged leptons. So we now have three **lepton generations,** each of four particles:

1 e^-, e^+, ν_e, $\bar{\nu}_e$.

2 μ^-, μ^+, ν_μ, $\bar{\nu}_\mu$.

3 τ^-, τ^+, ν_τ, $\bar{\nu}_\tau$.

The masses of these neutrinos are known only within upper limits. They cannot easily be detected, so how can we be sure there are so many kinds? We attempt to answer that question in the next two sections.

DEFINITIONS AND KEYWORDS

Leptonic hadron decays Weak decays of hadrons that have leptons alone in the final state.

Semi-leptonic hadron decays Weak decays of hadrons that have leptons and hadrons in the final state.

Non-leptonic hadron decays Weak decays of hadrons that have hadrons alone in the final state.

Pure leptonic interactions Weak interactions, decays or scattering, involving leptons alone.

Lepton generations The three generations of leptons associated with the electron μ-meson and τ-meson respectively.

Table 12.1 The three generations of charged and neutral leptons.

This table gives the lepton numbers, masses or upper limits assuming particle and antiparticle masses are equal.

Generation	Leptons		Antileptons		Masses	
	Charged	Neutral	Neutral	Charged	Charged	Neutral
Electron	e^-	ν_e	$\bar{\nu}_e$	e^+	0.511 MeV/c^2	<14 eV/c^2
	$L_e=+1$		$L_e=-1$			
Muon	μ^-	ν_μ	$\bar{\nu}_\mu$	μ^+	105.66 MeV/c^2	<0.25 MeV/c^2
	$L_\mu=+1$		$L_\mu=-1$			
Tau	τ^-	ν_τ	$\bar{\nu}_\tau$	τ^+	1784 MeV/c^2	<35 MeV/c^2
	$L_\tau=+1$		$L_\tau=-1$			

The determination of the upper limit on the electron neutrino mass is described in Section 12.12. The mass difference between π^+ and μ^+ is known to about 1 part in 3×10^4; energy and momentum conservation connects this difference to the mass of the neutrino and the kinetic energy of the μ^+ in $\pi^+\rightarrow\mu^++\nu_\mu$. Thus the mass is given by a measurement of this kinetic energy but that cannot be done with a precision sufficient to do better than the upper limit given in the table.

A sample of τ-mesons of well-defined energy may be produced in high-energy electron–positron annihilation $e^++e^-\rightarrow\tau^++\tau^-$. The limit on the momentum spectrum of charged particles produced in decays such as $\tau^+\rightarrow\mu^++\nu_\mu+\bar{\nu}_\tau$ gives the upper limit in the mass of ν_τ. But again the precision is poor.

It is important to be able to recognize weak interaction decays. One or more of the following will label a decay, or interaction, as weak:

(1) neutrinos are involved;

(2) if hadrons are involved, there is a change of quark flavour;

(3) the transition rate, allowing for energy release, is very small compared with rates due to the electromagnetic or strong interactions.

In Section 13.2 we shall elaborate on these indicators and place them in the context of recognizing the different interactions.

12.2 Neutrino and antineutrino

Pauli proposed in 1932 that neutrinos existed (see Chapter 5) and Fermi validated their existence by his successful quantitative theory of β-decay in 1934. We give a very simplified version of this theory in Section 12.5.

Given that, why do we complicate matters by differentiating the neutrinos in β^+- and β^--decay?

$$n \rightarrow p + e^- + \bar{\nu}_e, \tag{12.5}$$

$$\text{`p'} \rightarrow \text{`n'} + e^+ + \nu_e \text{ (only for `bound protons')}. \tag{12.6}$$

It is tidy because with the lepton e^-, is emitted the antilepton $\bar{\nu}_e$, and with the antilepton e^+ is emitted the lepton ν_e. So we envisage the conservation of an additive lepton number. But perhaps ν_e and $\bar{\nu}_e$ are the same: what would be the consequence? If (12.5) represents a valid reaction then we can rearrange this reaction by taking the $\bar{\nu}_e$ across the $\rightarrow$ sign and changing it into its antiparticle and get another valid reaction:

$$\nu_e + n \rightarrow p + e^-. \tag{12.7}$$

This would be the forced decay of the neutron. Now if $\nu_e = \bar{\nu}_e$ and if the energy is right we can have the following sequence of events:

$$\text{`n'} \rightarrow \text{`p'} + e^- + \nu_e \tag{12.8}$$

$$\nu_e + \text{`n'} \rightarrow \text{`p'} + e^-.$$

The ν_e emitted in decay (12.8) forces the decay of a second neutron in the same nucleus and is never emitted. This is neutrino-less double β-decay. All that happens is that two electrons are emitted with total kinetic energy equal to the Q of the decay. $(Z,A) \rightarrow (Z+2,A) + 2e^-$. Because of stability conditions a candidate nucleus for this as a unique decay must be an even–even nucleus separated from its $Z+2$ partner by an odd–odd nucleus which exceeds the former in atomic mass. There are several such candidates but we will base our discussion on the proposed double β-decay (see Fig. 12.1):

$$^{82}_{34}\text{Se} \rightarrow {}^{82}_{36}\text{Kr} + 2e^- + 3.03 \text{ MeV}. \tag{12.9}$$

We must distinguish two possibilities:

1. No neutrinos emitted (symbolically $0\nu 2\beta$). This would be proof that $\nu_e = \bar{\nu}_e$, or that other as yet undetected neutrino properties exist.
2. Neutrinos are emitted (symbolically $2\nu 2\beta$), for example:

$$^{82}_{34}\text{Se} \rightarrow ^{82}_{36}\text{Kr} + 2\text{e}^- + 2\bar{\nu}_e + 3.03 \text{ MeV}. \qquad (12.10)$$

It is natural to assume that the neutrinos will be the kind we have assigned to β-decay. This transition is expected to occur through an intermediate virtual state of the nucleus $^{82}_{35}$Br. It cannot be by two successive β-decays through a real nucleus $^{82}_{35}$Br as energy cannot be conserved in the first step, as can be seen in Fig. 12.1.

The rate for the second mechanism (equation (12.10)) can be calculated with some degree of confidence: the rate for the first (equation (12.9)) can be calculated assuming $\nu_e = \bar{\nu}_e$. Experimentally it is a difficult matter of detecting the simultaneous (within 10^{-21} s) emission of two electrons. If their kinetic energy sums to 3.03 MeV, then it is $0\nu 2\beta$. If it does not, then it is presumably $2\nu 2\beta$. Until 1987 measurements on all practical candidates had yielded only upper limits. Such upper limits were well below the rate expected for $0\nu 2\beta$ due to $\nu_e = \bar{\nu}_e$, which therefore appears to be excluded. In 1987 events of the kind which are consistent with the decay 12.10 were reported.

In Fig. 12.2(a) we show the mechanism of neutron decay involving the W^- boson introduced in Section 9.13. In Fig. 12.2(b) we show this mechanism duplicated for $2\nu 2\beta$ decay of $^{82}_{34}$Se. If double β-decay without neutrino emission is

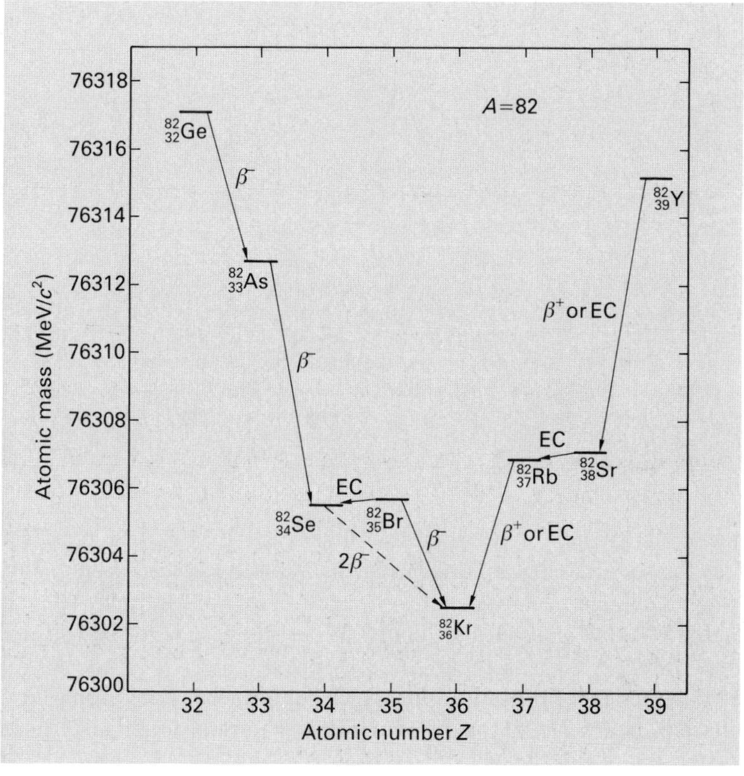

Fig. 12.1 A plot of the atomic mass against Z the atomic number for $A = 82$ nuclei. The possible 2β-decay is marked by the broken arrow. There are other atomic numbers at which the masses are also suitably arranged. The vital thing in this case is that $^{82}_{34}$Se is below $^{82}_{35}$Br in mass. The amount by which its mass is above that of $^{82}_{36}$Kr is important: the greater that excess is, the greater will be the transition rate, if the decay exists.

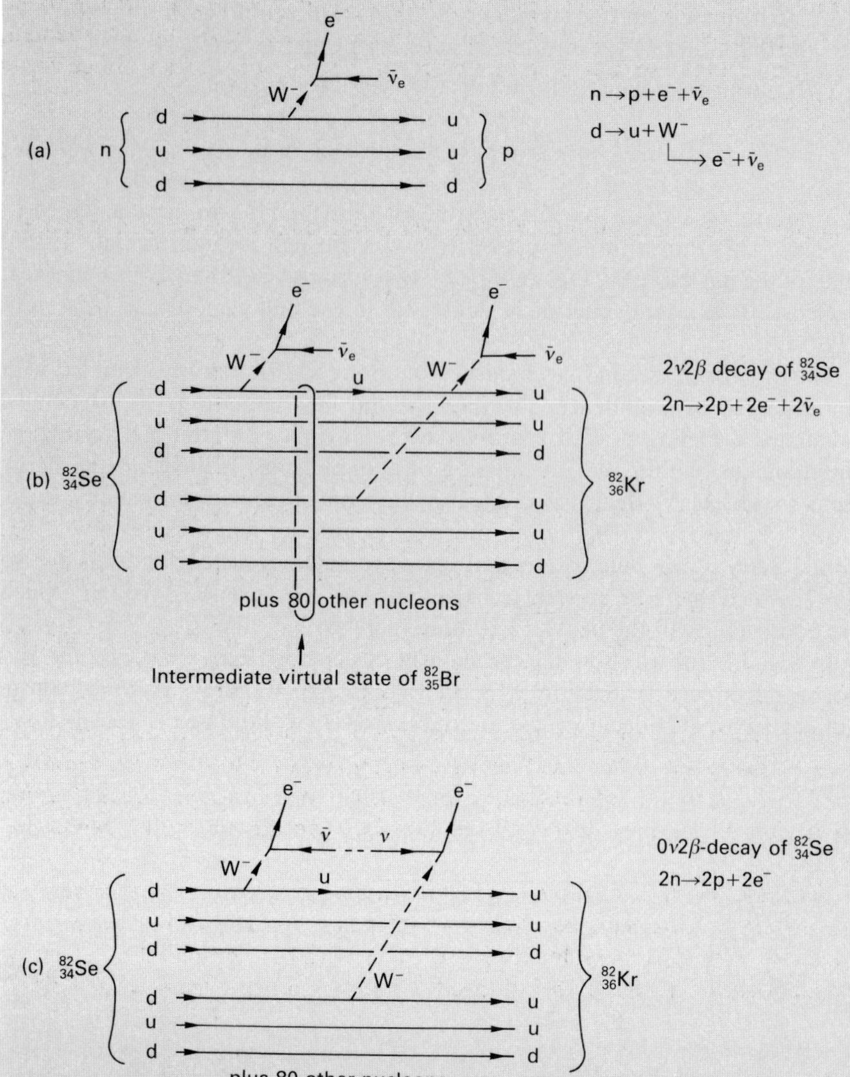

Fig. 12.2 (a) The mechanism of neutron β-decay involving the exchange of a virtual W boson. (b) $2\nu2\beta$-decay of ${}^{82}_{34}$Se. it must involve $2n \rightarrow 2p$ via an intermediate virtual state of the nucleus ${}^{82}_{35}$Br. It cannot involve a real state of ${}^{82}_{35}$Br because the mass of that atom is greater than that of ${}^{82}_{34}$Se. (c) The $0\nu2\beta$ decay requires that the $\bar{\nu}_e$ produced after the exchange of a first W^- becomes a ν_e (or $\nu_e \equiv \bar{\nu}_e$) to allow $\nu_e + W^- \rightarrow e^-$ when the second W^- is exchanged.

to occur, the antineutrino emitted in the first $W^- \rightarrow e^- + \bar{\nu}_e$ has to cause $\nu + W^- \rightarrow e^-$ at the second W emission. In Fig. 12.2(c) the second $\bar{\nu}_e$ line of Fig. 12.2(b) is turned back and becomes a a ν_e propagating forward in time. There is then a mismatch of the $\bar{\nu}_e$ emitted by the first W^- and the ν_e required to force $W^- \rightarrow e^-$. Thus only if $\nu_e = \bar{\nu}_e$ can the $0\nu2\beta$ decay occur.

There are two other experimental results which support the conclusion that ν_e and $\bar{\nu}_e$ are different. In describing the experiments we assume our assignments of equation (12.5) and (12.6) are correct. Nuclear reactors produce about 10^{17} neutron-rich fission products per second for each megawatt of operating power. These products β^--decay rapidly to longer-lived isotopes so that an operating reactor is an intense source of antineutrinos. These antineutrinos can cause the reaction.

$$\bar{\nu}_e + p \rightarrow n + e^+$$

but should not cause the reaction (written with the neutrino which can cause it)

$$\nu_e + {}^{37}_{17}\mathrm{Cl} \rightarrow {}^{37}_{18}\mathrm{Ar} + e^-,$$

which is essentially

$$\nu_e + \text{'n'} \rightarrow \text{'p'} + e^-.$$

The first reaction (proton to neutron) was observed at the expected rate in a detector operating near a high-power nuclear reactor (Reines and Cowan, 1959). In a separate experiment the second reaction (${}^{37}_{17}\mathrm{Cl}$ to ${}^{37}_{18}\mathrm{Ar}$) was not detected. The upper limit on the cross-section was 5% of that expected if $\nu_e = \bar{\nu}_e$ (Davis, 1955; Davis and Harmer, 1959).

The absence of both the $0\nu2\beta$ decay and the ${}^{37}_{17}\mathrm{Cl} \rightarrow {}^{37}_{18}\mathrm{Ar}$ reaction are negative experimental results which do not exclude their occurrence at a very low rate. For all but the last two sections of this chapter we assume that ν_e and $\bar{\nu}_e$ are distinct particles. However, there is one property of the weak interactions that will be discussed in later sections. This property, that of interacting preferentially with fermions of prescribed longitudinal polarization, forces a reinterpretation of these experimental results which we discuss in the penultimate section, 12.13.

Throughout this chapter and later, the word neutrinos can cover both neutrinos and antineutrinos. When we refer to antineutrinos it is for a specific purpose. If we have to refer specifically to neutrinos, we hope the context will make it clear that we are being specific rather than using the label generically.

12.3 Neutrinos galore

Even if we have decided that $\nu_e \neq \bar{\nu}_e$, why do we need to associate distinct $\nu\bar{\nu}$ pairs with each of the other charged lepton pairs, as listed in Table 12.1? This problem was raised by the absence of the decay

$$\mu^+ \rightarrow e^+ + \gamma, \tag{12.11}$$

which is apparently allowed by energy, momentum, angular momentum, and charge conservation. The proposed solution was that there was an additively conserved quantum number (L = number of leptons − number of antileptons) which applied separately to each lepton generation: electron, muon (and now to τ-meson) (L_e, L_μ, L_τ). In the decay of equation (12.11) the initial state has $L_\mu = -1$ and $L_e = 0$; the final state has $L_\mu = 0$ and $L_e = -1$. Thus if it occurred, there would be a change in both muon and electron lepton number: the absence of this decay is a property of this empirical conservation law. The neutrinos

DEFINITIONS AND KEYWORDS

Lepton generation number conservation The number of leptons of a given generation less the number of antileptons of the same generation is a conserved number. (L in Table 12.1.)

must be included in the game of counting leptons: as discussed in the last section we assign names so that in β^--decay the electron (lepton) is emitted in the company of an antineutrino (antilepton), and so on. It follows there must also be neutrinos for each generation of charged leptons. Thus, in addition to the uncharged electron neutrino and antineutrino pair $\nu_e\bar{\nu}_e$, we must also have muon neutrinos $\nu_\mu\bar{\nu}_\mu$ and tau neutrinos $\nu_\tau\bar{\nu}_\tau$. The conservation of separate lepton numbers for each generation is the **law of the conservation of lepton generation number**.

The existence of muon type neutrinos ν_μ and $\bar{\nu}_\mu$ as distinct from electron type neutrinos ν_e and $\bar{\nu}_e$ was confirmed experimentally. The basis of the method was the following: the dominant π^+ meson decay is

$$\pi^+ \rightarrow \mu^+ + \nu_\mu, \tag{12.12}$$

where we have used labels consistent with what we are hoping is confirmed.

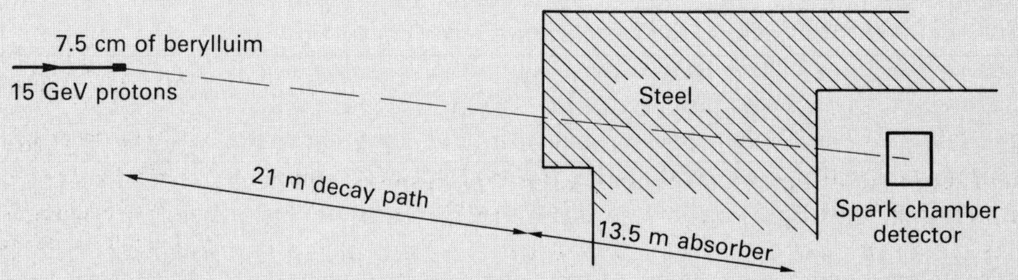

Fig. 12.3 The essentials of the first high-energy neutrino experiment. A 7.5 cm beryllium target was bombarded by 15 GeV protons: in the direction of 7.5° to the beam direction was a free flight path of 21 m followed by 13.5 m of steel. Behind the steel was a large 10 tonne spark chamber which acted both as target for neutrino interactions and detector of the products of the reactions. These are

$$\nu + N \rightarrow \mu^\pm + \cdots,$$

$$\nu + N \rightarrow e^\pm + \cdots.$$

The collisions of protons with nucleons in the beryllium nucleus produce a large number of π^+- and π^--mesons of which a fraction decay in the 21 m flight path in the direction of the detector:

$$\pi^\pm \rightarrow \mu^\pm + \nu.$$

The steel serves to remove the remaining π-mesons, protons or other strongly interacting particles (they and each of their secondaries have a mean free path of less than $\frac{1}{4}$ m in the steel). In addition it is thick enough to stop μ-mesons of energy up to 17 GeV, by the process of energy loss by ionization. Thus only neutrinos plus any products of neutrino interactions in the last part of the steel should emerge from the steel and reach the detector. In an exposure amounting to about 3.5×10^{17} protons striking the target, 29 events were detected and attributed to neutrino interactions producing muons and no events attributable to interactions producing high-energy electrons or positrons. The conclusion was drawn that the neutrinos from π-meson decay were different from those from β-decay which are known to be able to cause $\bar{\nu}_e + p \rightarrow e^+ + n$.

[The spark chamber: The passage of an ionizing particle through a suitable gas generates free electrons. The application of a large electric field can cause acceleration and multiplication of the electrons and lead to an electric spark discharge which can be photographed. The chamber used in this experiment consisted of 10 tonnes of aluminium plates each 25 mm thick and separated by 9 mm of a suitable gas. The traversal of all or a part of the chamber by a charged particle was detected by scintillation counters and this triggered the application of a large voltage across the gaps between the plates. The track of the particle would then be visible as a series of sparks. Mu-mesons of momentum greater than 0.3 GeV/c left a trail of well-aligned sparks. Electrons or positrons generated electromagnetic showers which produce a broken and irregular spark pattern. Events due to neutrinos occurred in the aluminium of the plates, producing spark patterns starting at a point within the chamber.]

Neutrino–quark scattering can occur: some typical reactions are:

$$\nu_\mu + \mathrm{d} \to \mu^- + \mathrm{u}, \tag{12.13}$$

$$\nu_\mathrm{e} + \mathrm{d} \to \mathrm{e}^- + \mathrm{u}. \tag{12.14}$$

In these two reactions we have assumed the lepton conservation law and quark targets (of course, these quarks have to be bound into the nucleons of real targets). As formulated, the neutrinos from π^+-decay can cause reaction (12.13) and produce μ^--mesons but cannot cause reaction (12.14) and produce electrons. Similarly antineutrinos from π^--decay can produce μ^+ but not positrons. The experiment in which this was confirmed as the situation was performed in 1962 (Danby *et al.*) and is briefly described in Fig. 12.3. Since that time neutrino and antineutrino beams have been improved in intensity and energy and used in many experiments (see Section 12.11). Thus there is now a great deal of evidence for the assignment of generation (e, μ) distinguishing lepton numbers to neutrinos. However, there is no corresponding evidence on the τ-meson generation. The τ was discovered in

$$\mathrm{e}^+ + \mathrm{e}^- \to \tau^+ + \tau^-$$

and has decays such as

$$\tau^+ \to \pi^+ + \bar{\nu}_\tau,$$
$$\tau^+ \to \mu^+ + \nu_\mu + \bar{\nu}_\tau.$$

These are all written assuming this generation is repeating the features of lepton number conservation established by the muon and electron generations.

This lepton number conservation is arbitrary and has no basis in more fundamental ideas. In fact, the reader will notice that we appear to have three generations of quark (du, sc, bt(?)) and three generations of leptons (e, μ, τ). The weak interactions do not behave in the same way for these two sets of three generations, as we shall see in the next section.

We shall, from now on, assume the correctness of the lepton generation number conservation law. Since it involves leptons, charged and neutral, it is relevant only in the operation of weak and electromagnetic interactions; it is not relevant to strong interactions and therefore no strong reaction should be proposed which violates this law.

Of course at some level this law may be violated. The decay of equation (12.11) cannot be proved experimentally to be forbidden. Measurements show that it has a branching fraction less than 5×10^{-11}. In Section 12.11 we shall discuss the possibility of neutrino oscillations which may occur if neutrinos have mass. They involve the oscillatory change of one neutrino type into another: that, if it occurs, clearly violates the law of the conservation of lepton number.

The evidence that these different neutrinos and antineutrinos exist is itemized in Tables 12.2 and 12.3 for convenience. We note that we assume the existence of the tau neutrino and antineutrino although there is no experimental evidence.

Table 12.2 The evidence that $\nu_\mathrm{e} \neq \bar{\nu}_\mathrm{e}$

1. Absence of neutrino-less double β-decay.
2. Antineutrinos ($\bar{\nu}_\mathrm{e}$) from bound neutron decay cause

   $$\bar{\nu}_\mathrm{e} + \mathrm{p} \to \mathrm{n} + \mathrm{e}^+$$

   but do not cause

   $$\nu_\mathrm{e} + \text{‘n’} \to \text{‘p’} + \mathrm{e}^-.$$

   (‘n’, ‘p’ mean nuclear bound neutron and proton respectively.)

But see Section 12.13.

Table 12.3 The evidence that charged and neutral leptons of each generation are distinct in that $\nu_\mathrm{e} \neq \nu_\mu$ and $\bar{\nu}_\mathrm{e} \neq \bar{\nu}_\mu$.

1. Absence of $\mu^+ \to \mathrm{e}^+ + \gamma$.
2. Neutrinos or antineutrinos from charged pion decay interact with nucleons to produce muons but not positrons or electrons.

PROBLEMS

12.1 Assign the lepton generation subscript and distinguish antineutrinos from neutrinos in the following reactions and decays. Use the symbols ν_e, $\bar{\nu}_e$, ν_μ, $\bar{\nu}_\mu$, ν_τ, $\bar{\nu}_\tau$.

$\pi^+ \to \pi^0 + e^+ + \nu$

$\mu^+ \to e^+ + \nu + \nu$

$\mu^- \to e^- + \nu + \nu$

$K^+ \to \pi^0 + e^+ + \nu$

$\bar{K}^0 \to \pi^0 + e^- + \nu$

$\Sigma^- \to n + \mu^- + \nu$

$\Sigma^+ \to \Lambda^0 + e^+ + \nu$

$D^0 \to K^- + \pi^0 + e^+ + \nu$

$\nu + p \to n + e^+$

$\nu + {}^{37}_{17}Cl \to {}^{37}_{18}Ar + e^-$

$\nu + p \to \mu^- + p + \pi^+$

$\nu + n \to e^- + p$

${}^3_1H \to {}^3_2He + e^- + \nu$

$\pi^+ \to \mu^+ + \nu$

$\pi^- \to e^- + \nu$

$\tau^- \to \pi^- + \pi^0 + \nu.$

12.2 Draw Feynman diagrams for the following decays: do this at the quark level for those involving hadrons.

$\tau^- \to e^- + \bar{\nu}_e + \nu_\tau$

$K^0 \to \pi^- + e^+ + \nu_e$

$D^+ \to \bar{K}^0 + \mu^+ + \nu_\mu$

$\tau^+ \to \pi^+ + \nu_\tau$

$\Lambda \to p + e^- + \bar{\nu}_e$

$\Xi^- \to \Sigma^0 + \pi^-$

$K^+ \to \pi^+ + \pi^- + \pi^+.$

12.3 Draw Feynman diagrams for the following reactions: do this at the quark level for those involving hadrons.

$\nu_e + e^- \to \nu_e + e^-$

$e^- + p \to n + \nu_e$

$\mu^+ + e^- \to \bar{\nu}_\mu + \nu_e$

$\nu_\mu + p \to \mu^- + \Delta^{++}.$

12.4 Check that the conservation laws given in the box of Fig. 12.4 are obeyed by all the examples in this figure.

12.5 Show that the following decays cannot occur by first order weak interactions:

$D^0 \to K^+ + \pi^- + \pi^0 + \pi^0$

$D^+ \to K^0 + \pi^+$

$D^0 \to K^+ + \mu^- + \bar{\nu}_\mu$

$K^0 \to \pi^0 + e^- + \bar{\nu}_\mu.$

12.4 The W and Z gauge bosons

It is now time to consider in more detail the role of the $W^\pm$ and Z^0 bosons. The reader may wish to make a brief re-visit to Section 9.13 and, in particular, to Figs 9.18 to 9.22. We now have to learn how to draw the diagrams for any weak interactions using the vertices of Fig. 9.21 and we will do that first for reactions involving the W^+ or W^-. So in Fig. 12.4 we have redrawn the simple vertex, and then established a time direction and drawn some examples: at each vertex we must satisfy electric charge conservation, lepton number conservation (assuming that the W^+ and W^- bear no lepton number, the only assumption consistent

(a)

Rules for the Wf$_1$f$_2$ vertex
1. Electric charge conserved
   and, from f$_1$ to f$_2$
2. Lepton (generation) number conserved
3. Quark number conserved
4. Quark flavour is not conserved.
5. Quark colour is conserved.

Examples follow in (b) to (f). Time from left to right

(b)

$\mu^- \to e^- + \bar{\nu}_e + \nu_\mu$
(μ^--decay)

(Electron antineutrino scattering
by electron)

(c)

$\pi^+ \to \mu^+ + \nu_\mu$

(The dominant decay mode of π^+)

$K^- \to \mu^- + \bar{\nu}_\mu$

(A decay mode of the K^+)

(d)

$n \to p + e^- + \nu_e$

$K^+ \to \pi^0 + \mu^+ + \nu_\mu$

(e)

$D^0 \to K^- + \pi^+$

(A decay mode of the D^0)

$D^+ \to \bar{K}^0 + \pi^+$

(A decay mode of the D^+)

(f)

Lepton $l = e, \mu$ or τ. The $\pm\frac{2}{3}, \pm\frac{1}{3}$ in superscript refers to the charge of the quark.
The $q\bar{q}$ final states will fragment into hadron jets.

Gluon exchange is omitted in the examples involving hadrons. The $u\bar{u}$
pair in $D^0 \to K^-\pi^+$ is created by gluon exchange with the other quarks.

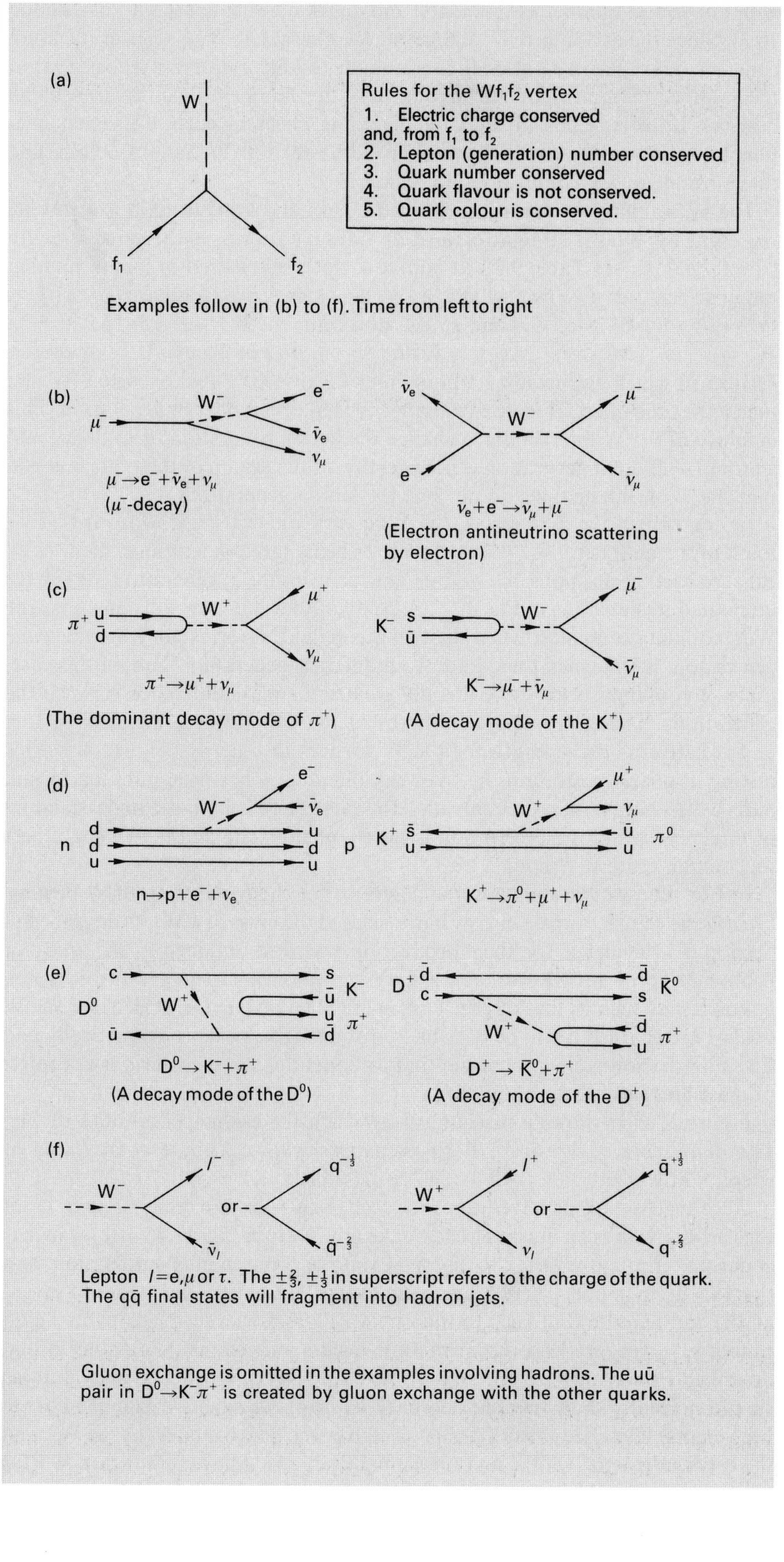

Fig. 12.4 (a) The Feynman diagram
for the Wf$_1$f$_2$ vertex (f = fermion),
with the rules for vertex construction
where fermions are, of course,
selected from quarks or leptons. The
three arms can be orientated in any
way with respect to the direction of
time. Remember that fermions with
their arrows pointing back in time
represent antifermions moving
forward in time. The W arrows can
point either way but, of course, a
W$^+$ moving back is the same as a
W$^-$ moving forward.

The rest of the figure shows
examples of how W vertices can be
arranged. (b) Pure leptonic
interactions. (c) Quark weak
interactions in leptonic hadron
decays. (d) Quark weak interactions
in semi-leptonic hadron decays. (e)
Quark weak interactions in
non-leptonic hadron decays. (f) Free
W decays.

The box encloses the rules that
apply to the vertex.

with our lepton number assignments) and quark number conservation (Section 10.3). Since the exchanged W changes quark charges at Wqq vertices it cannot conserve quark flavour but does conserve the (additive) quark number: that is the number of quarks of all flavours less the number of antiquarks of all flavours. Look carefully through the diagrams, checking each of the conservation laws in turn at each diagram. In addition, the W has no colour and therefore the quark colour does not change.

The W coupling to leptons (a charged lepton and a neutrino at a vertex we represent by W$l\nu$) is straightforward and always occurs at the same strength (strictly $g/2\sqrt{2}$, see Table 9.8). In addition, the conservation of lepton number requires that at a vertex where a W becomes leptons then $W^+ \to e^+ + \nu_e$, $W^+ \to \mu^+ + \nu_\mu$, $W^- \to e^- + \bar{\nu}_e$ and so on, do occur but $W^+ \to e^+ + \bar{\nu}_e$ ($\Delta L_e = -2$), $W^- \to \mu^- + \bar{\nu}_e$ ($\Delta L_\mu = +1$, $\Delta L_e = -1$) and so on, do not occur. Also, of course, vertices in which an incoming lepton emits or absorbs a real or virtual W and then leaves the interaction do not change the lepton number or generation. This inability of the W interaction to change the lepton generation or to decay into leptons of different generation is noteworthy: in the jargon of the trade it is said that the weak interactions do not mix the lepton generations.

In contrast to the W$l\nu$ vertex, the Wqq vertex is not straightforward. If we think of the quarks u + d, c + s, and t + b, as being three generations, then the W can connect generations as well as the pair inside a generation. We have attempted to show this in Fig. 12.5. A study of that diagram will show that the Wqq coupling is greatest when the two quarks at the vertex are from the same generation, is less when the qq are from different generations and, in this case, that the coupling seems to be less the greater the mass difference between the generations. The approximate strengths of the coupling are included in Fig. 12.5, relative to the strength for the W vertices in μ-decay (Fig. 12.4b). This mixing of quark generations by the weak interaction has been put on a formal basis which parametrizes the mixing. However, there is no basic understanding of this mixing and the description is purely phenomenological. We shall meet this mixing again in Section 12.8.

In Fig. 12.6 we have given some diagrams for the production of W bosons. Clearly the centre-of-mass energy has to exceed the mass of a W. As described in Section 9.13 evidence for their production was first obtained in the study of proton–antiproton collisions at 540 GeV centre-of-mass energy, in this case a basic process such as that of Fig. 12.6(a) or (b) will have operated. Since the W bosons are charged it will be possible to produce them electromagnetically and Fig. 12.6(c) shows one mechanism that will operate in e^+e^- collisions at centre-of-mass energies greater than $2M_W c^2$.

Figure 12.6(d) shows a mechanism in which the exchanged photon of Fig. 12.6(c) is replaced by a Z^0. Both exchanges will contribute comparably at energies at which $e^+ + e^- \to W^+ + W^-$ is possible.

Once real W bosons have been produced, how will they decay? Examples of the vertices have been drawn already: take any W from Fig. 12.4(b)–(e), which is connected through a vertex to any final state of fermion plus antifermion, and that is a decay mode. There are enough examples to allow the reader to invent all the basic modes (but keep to the vertex rules). However, those involving $q\bar{q}$ cannot give free quarks so fragmentation occurs in a manner analogous to that described in Section 10.8 for $e^+ + e^- \to q + \bar{q}$. Thus a W^+ decay into hadrons should normally show two back-to-back hadron jets: such a decay has not yet been clearly identified (1989). Table 12.4 gives the decay modes for the W^+ and the expected partial widths and branching fractions. Allowing a factor of 3 for coloured quarks, the partial width for $W \to f_1\bar{f}_2$ is expected to be the same for all

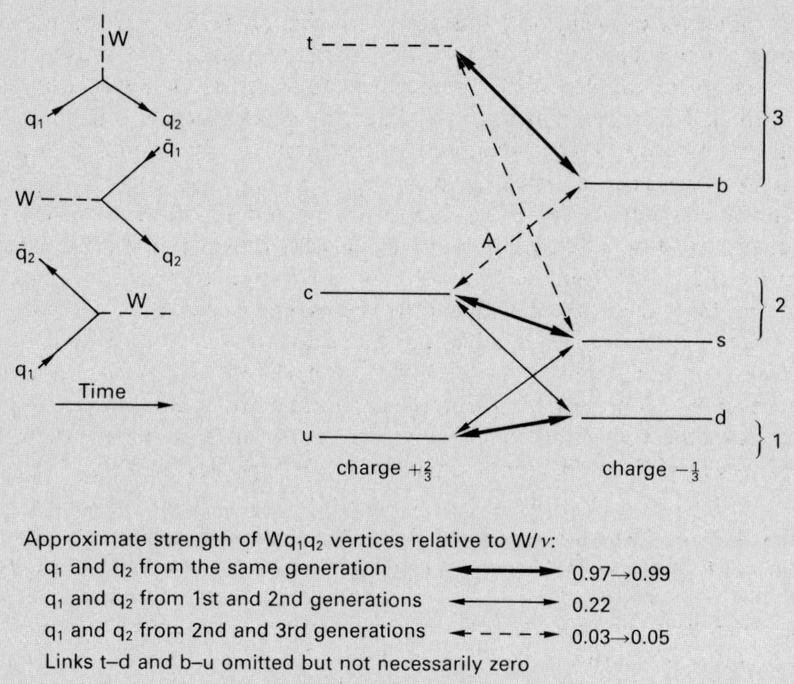

Fig. 12.5 The emission or absorption of a W will change a q_1 to q_2. The quarks or antiquarks can be chosen within the rules boxed in Fig. 12.4. The first interpretation of the above figure allows us to choose an allowed pairing by picking a double-headed arrow (for example arrow A pairing a charmed quark c with a beauty quark b). Then various time orderings give the following vertex effects:

$$c \to b + W^+, \quad c + W^- \to b, \quad \bar{c} \to \bar{b} + W^-, \quad \bar{c} + W^+ \to \bar{b}, \quad W^+ \to c + \bar{b}, \quad W^- \to \bar{c} + b,$$

and their time reverses.

The second interpretation of the figure gives an idea of the strength of the coupling constant at the vertex and hence a factor in the amplitude of any reaction involving that vertex. This strength is indicated by the thickness of the arrows: thus the strong vertices involve Wtb, Wcs and Wud. The figure gives approximate numbers in terms of $g/2\sqrt{2}$. This is the strength of The $Wl\nu$ vertex, where g is the coupling constant of the electroweak theory (Section 9.3).

The quarks are arranged vertically in relative order of mass as known approximately (Table 10.6). Thus decays are expected down the table. For example, the basic vertex steps in the decay from a b quark might be

$$\begin{aligned} &b \to c + W^- \\ &\quad \hookrightarrow s + W^+ \\ &\qquad\quad \hookrightarrow u + W^-. \end{aligned}$$

Transitions upwards can occur in, for example, neutrino scattering, if there is sufficient energy available:

$$\nu_\mu + d \to c + \mu^-.$$

The truth (top) quark t has not been observed but its couplings can be inferred from the known strengths. The b to u coupling is known to be very small and it follows that t to d will also be very small. Their omission from this diagram does not mean that they are absent.

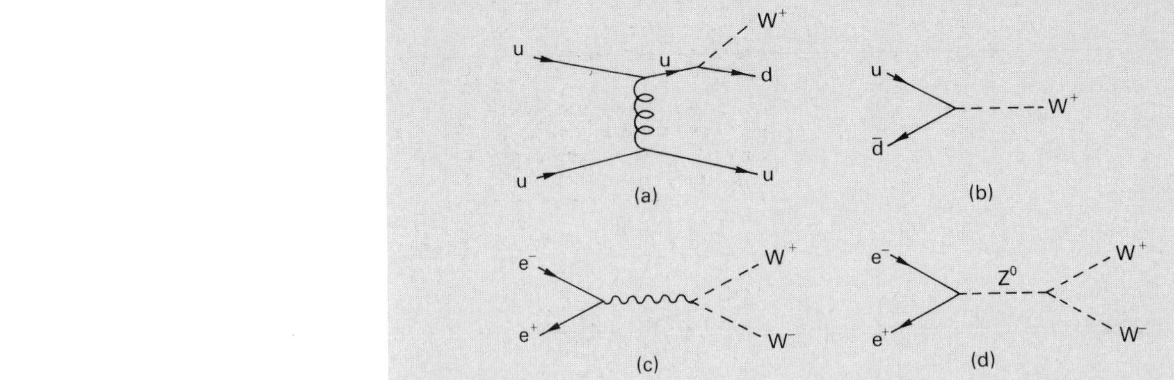

Fig. 12.6 Some contributions to real W boson production. In collisions between two hadrons a quark or antiquark from one interacts with a quark or antiquark from the other. (a) Shows an exchanged gluon providing sufficient energy and momentum transfer to enable a u quark to emit a real W^+. (b) Shows a u and a $\bar{d}$ annihilating at high energy to produce a real W^+. A virtual photon from e^+e^- annihilation materializes into a W^+W^- pair in (c), and in (d) it is the same reaction but involving an intermediate Z^0.

In case (a) the uu centre of mass energy must exceed $M_w c^2$. In (b) the $u\bar{d}$ centre of mass energy must equal $M_w c^2$. In (c) and (d) the e^+e^- centre of mass energy must exceed $2M_w c^2$.

f_1 and f_2 when both are from the same generation, apart from a small effect due to the mixing of quark generations and the effect of a decrease in final state phase space for the heavier fermions. This allows us to estimate branching fractions and lifetimes in some cases: Table 12.7 gives an example.

Table 12.4 The decay modes and values of the predicted partial widths and branching fractions for the W^+ gauge boson. (The W^- will have charge-conjugated decays but identical numbers.) Calculated using the Fermi coupling constant, $G_F = 8.96 \times 10^{-8}$ GeV fm^3 and $M_W = 80.9$ GeV/c^2.

Decay mode $W^+ \to$	Partial widths* GeV	Branching fractions	Branching fraction if top quark mass $\gtrsim$75 GeV
$e^+\nu_e$	0.23	0.083	0.11
$\mu^+\nu_\mu$	0.23	0.083	0.11
$\tau^+\nu_\tau$	0.23	0.083	0.11
$u\bar{d}$	0.72	0.25	0.34
$c\bar{s}$	0.72	0.25	0.34
$t\bar{b}$	(0.72→0)†	<0.25	0

Total width: 2.52→2.13 GeV, depending on the t-quark mass.
Mean life of W^+: ~3×10^{-25} s.

* Neglecting the rest mass of fermions in the final state and the mixing of quark generations in Wqq coupling. The cross-generation channels $W^+ \to u\bar{s}$, $u\bar{b}$, $c\bar{d}$, $c\bar{b}$, $t\bar{s}$, $t\bar{d}$ have small partial widths and whatever they add to the total width is compensated by a linked, small decrease in the $q\bar{q}$ partial widths tabulated.

† The mass of the t quark is not known but it is probably >70 GeV, in which case the lightest observable hadronic final state will be $(t\bar{u})+(\bar{b}u)$ with a threshold of >75 GeV. Thus the $(t\bar{b})$ partial width will be less than given here. If the t-quark mass is 75 GeV or greater it, cannot be a decay product.

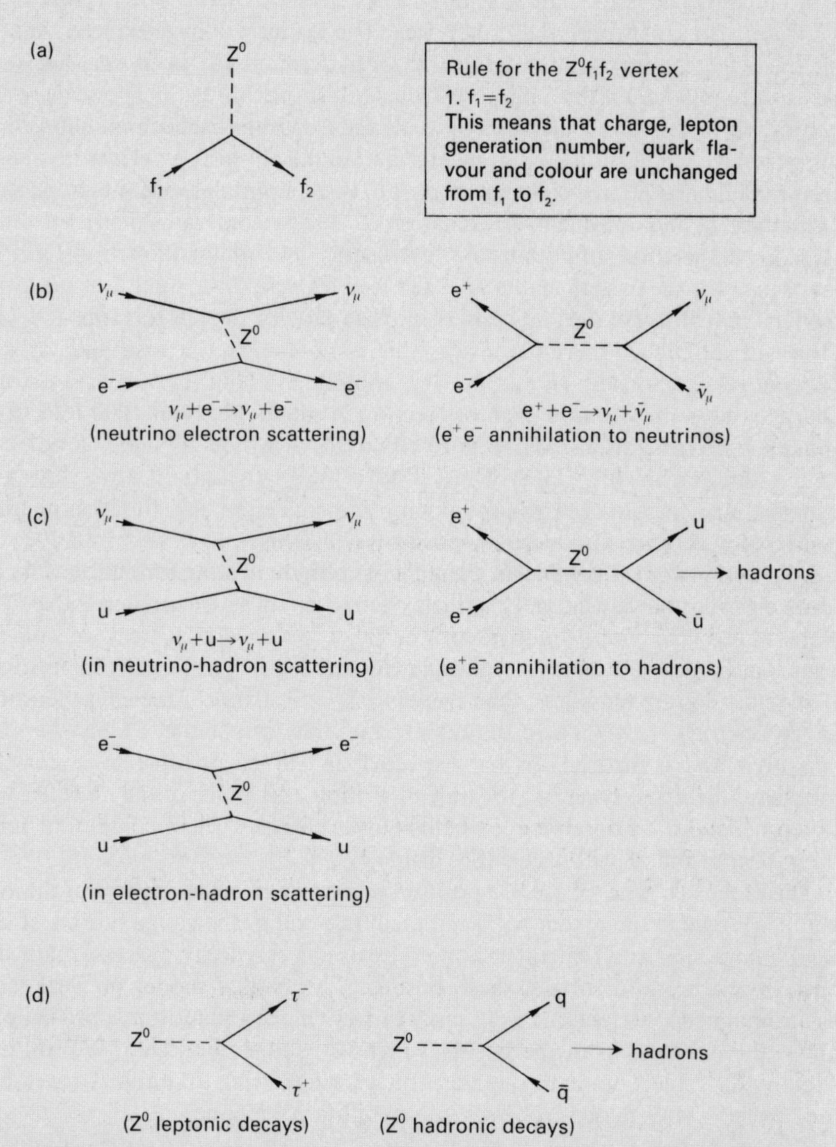

Fig. 12.7 (a) The Feynman diagram for the $Z^0f_1f_2$ (f = fermion) vertex with the box around the one rule that applies to the choice of f_1 and f_2, namely $f_1=f_2$. The rest of the figure shows examples of how the Z^0 vertices can be arranged. (b) Pure leptonic interactions. (c) Quark neutral current interactions. (d) Real Z^0 decays. The strength of the coupling at the vertex is predicted by the electroweak theory. It depends on the fermion but is comparable in magnitude for all.

Weak interactions that involve $W^\pm$ exchange are called **charged current interactions**. This statement brings us to the role of the Z^0. Weak interactions which involve Z^0 exchange are called **neutral current interactions**. In Fig. 12.7 we do for the Z^0 what we did in Fig. 12.4 for the $W^\pm$. This time the vertex rule is a single one and very simple. It is that the Z^0 is attached to a fermion current in which the fermion does not change, so that in $f_1\leftrightarrow Zf_2$ or $Z\leftrightarrow f_1\bar{f}_2$ we have $f_1=f_2$.

DEFINITIONS AND KEYWORDS

Charged current Weak interactions in which the mechanism is W^+ or W^- exchange.

Neutral current Weak interaction in which the mechanism is Z^0 exchange.

A consequence is that the first three rules satisfied by the Wf_1f_2 vertex (Fig. 12.4) are also satisfied by the Zf_1f_2 vertex. The reader will have noticed that, in all the Z^0 exchanges which involve charged fermions at both vertices, it is possible to replace a Z^0 by a photon. This is not surprising as the Z^0 and γ are the neutral partners in the unification of weak and electromagnetic interactions that we introduced in Section 9.13. It means that, in the physical processes involving the possibility of Z^0 or γ exchange, there are two amplitudes, one due to photon exchange, A_γ, and one due to Z^0 exchange, A_Z: the two diagrams then contribute $A_\gamma + A_Z$ to the total amplitude. At low energy and momentum transfers the amplitude for the exchange of a Z^0 is reduced because the virtual Z^0 is far away from its real mass (91 GeV/c^2). This is not the case for the photon and $A_\gamma \gg A_Z$. However, at higher-energy transfers this is no longer the case and A_Z can become very important. In fact in e^+e^- annihilation (Fig. 12.6(c)), if the total energy is near the Z^0 mass the cross-section is enhanced about 1000-fold from that proposed and discussed in Fig. 10.10. In this case the Z^0 is a real particle but with a finite mean life (3.0×10^{-25} s, $\Gamma \approx 2.3$ GeV). In Fig. 12.8 we show the expected and measured cross-sections for Z^0 production as a function of total centre-of-mass energy in electron–positron collisions.

This case just mentioned is an example of compound state formation. The Z^0 formed is free to decay into all channels open to it. These are listed in Table 12.5 with the expected branching fractions. Those which are $q\bar{q}$ will, of course, appear as hadron jets. Note that the neutrino modes are particularly interesting. If all neutrinos are mass-less, then there is a $Z^0 \rightarrow \nu\bar{\nu}$ channel for each generation of leptons (e, μ, τ, . .?) These decays are invisible: but if the Z^0 line-shape is measured and compared to that expected, it will be possible to count the number of neutrino types. At the time of writing, this is one of the objectives of experiments due to observe e^+e^- collisions at energies where compound state formation of the Z^0 will occur. (See Table 12.5.)

This is the last time we are in a position to discuss at length the role of the $W^\pm$ and Z^0. We hope the reader will keep them in mind as the gauge bosons of the weak interaction and remember how to analyse weak decay processes into the basic mechanisms involving these bosons. The unified model of weak and electromagnetic interactions as it exists in 1989 makes quantitative predictions about the interactions of the bosons which will shortly be tested. We will not explore this subject but in the next section we will go back to simple β-decay and look at such low-energy processes in a quantitative manner.

12.5 The Fermi theory of β-decay

In 1933 Sargent published an analysis of the decay rates of the naturally occurring β-emitters and showed that a large group obeyed an approximate rule that the decay rate, $\omega \propto T_{max}^5$, where T_{max} was the greatest electron kinetic energy in a given decay. Fermi in 1934 published the first quantitative theory of β-decay. Apart from the success of the theory it was a remarkable feat for two other reasons.

1. This was the first serious extension of the concept of creating photons to the concept of creating particles, electron plus neutrino in this case. This solved the then existing problem of 'where do the β-decay electrons come from?' since they could not exist in nuclei (see Section 1.6).

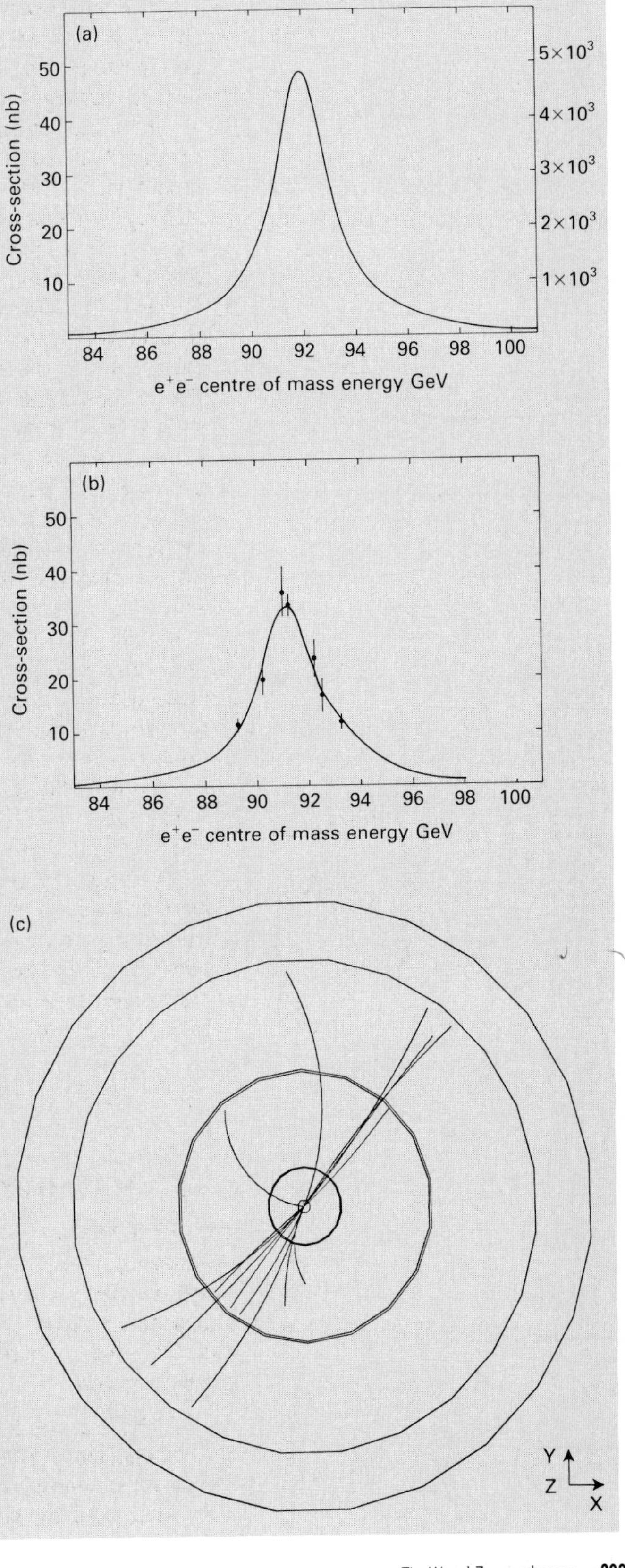

Fig. 12.8 (a) This plot shows the predicted (before 1989) cross-section for the compound state formation of the Z^0 in electron–positron collisions

$$e^+ + e^- \to Z^0,$$

as a function of total centre-of-mass energy and as it would be in the absence of radiative effects (see Section 9.6). The values assumed are $M_Z = 91.9\ \mathrm{GeV}/c^2$, which is the weighted average of results from the proton–antiproton collider at CERN (Fig. 9.19), and $\Gamma_Z = 2.56$ GeV (see Table 12.5). (b) The actual cross-section (points with errors) as observed in the Delphi detector at the LEP collider in November 1989 for the reaction

$$e^+ + e^- \to Z^0 \to \text{hadrons}.$$

The curve is a best fit (including radiative corrections) and yields $M_Z = 91.06 \pm 0.09\ \mathrm{GeV}/c^2$ and $\Gamma_Z = 2.42 \pm 0.21$ GeV. The measured peak cross-section is 33 ± 2 nb and is expected to be 33 nb.

(c) A computer reconstruction of one of the events

$$e^+ + e^- \to Z^0 \to \text{hadrons}$$

as observed in the Delphi detector. This shows the charged particle tracks observed in the inner detectors of this apparatus, projected onto a plane perpendicular to the e^+e^- collision axis. The jet structure expected for $Z^0 \to q\bar{q}$ followed by fragmentation is clearly visible in this event (compare with Fig. 10.17).

In (a) the cross-section scale on the left is in nanobarns. Using the scale on the right of this figure, the curve which gives values for the cross-section ratio

$$\frac{\sigma(e^+e^- \to Z^0)}{\sigma(e^+e^- \xrightarrow{\text{QED}} \mu^+\mu^-)}$$

is almost indistinguishable from that for the cross-section and is omitted. The reader will recall from Section 10.6 that the cross-section $\sigma(e^+e^- \xrightarrow{\text{QED}} \mu^+\mu^-)$ is the standard against which hadron production cross-section is measured with

$$R = \frac{\sigma(e^+e^- \to \text{hadrons})}{\sigma(e^+e^- \xrightarrow{\text{QED}} \mu^+\mu^-)}$$

At energies where five quark generations are produced R is about $\frac{11}{3}$. The QED over the → means the quantum electrodynamic process in which a photon is exchanged and which, for $\mu^+\mu^-$ production, has the cross-section

$$\frac{4\pi}{3}\left(\frac{\alpha\hbar c}{W}\right)^2$$

at a total centre of mass energy W (see Fig. 10.10). In fact, since the Z^0 decays to $\mu^+\mu^-$ with a branching fraction of about 3.4%, the total production of $\mu^+\mu^-$ pairs increases about 160 times at the Z^0 maximum over that expected from QED, one-photon exchange alone. Looking at the branching ratios for Z^0 hadronic decay (see Table 12.5) we see that hadron yield increases by even greater factors.

The cross-section for Z^0 production is such that when the LEP collider reaches its design intensity, the yield of Z^0 bosons will be of the order of 1 per second, which is bountiful by the standards of high energy particle physics.

Table 12.5 The predicted decay modes and values of the partial widths and branching fractions for the Z^0 gauge boson. Calculated assuming $\sin^2\theta_W = 0.230$, $M_Z = 91.9\ \text{GeV}/c^2$.

Decay mode $Z^0 \rightarrow$		Partial width* (MeV)	Branching fractions†
e^+e^-		88	0.034
$\mu^+\mu^-$		88	0.034
$\tau^+\tau^-$		88	0.034
$\nu_e\bar{\nu}_e$		175	0.068
$\nu_\mu\bar{\nu}_\mu$		175	0.068
$\nu_\tau\bar{\nu}_\tau$		175	0.068
$u\bar{u}$	hadrons	302	0.118
$d\bar{d}$	hadrons	388	0.152
$s\bar{s}$	hadrons	388	0.152
$c\bar{c}$	hadrons	302	0.118
$b\bar{b}$	hadrons	388	0.152
$t\bar{t}$	hadrons	0?	

Total width: ~ 2.56 GeV†
Mean life of Z^0: $\sim 2.6 \times 10^{-25}$ s.

* Neglecting rest mass of fermions in final state, except for $Z^0 \rightarrow t\bar{t}$ which may be energetically impossible.

† Assuming $Z^0 \rightarrow t\bar{t}$ is energetically forbidden and only three generations of leptons.

At the time of proof reading measured values of the mass, the width, and of some partial widths of the Z^0 have become available from the detectors operating at the Stanford Linear Collider (SLC) and at the Large Electron–Positron Collider (LEP at the European Centre for Nuclear Research, Geneva.) The average values are:

$$
\begin{aligned}
M_Z &= 91.01 \pm 0.03\ \text{GeV}/c^2,\\
\Gamma &= 2.60 \pm 0.10\ \text{GeV},\\
\Gamma_{ee} &= 0.089 \pm 0.004\ \text{GeV},\\
\Gamma_{\mu\mu} &= 0.085 \pm 0.009\ \text{GeV},\\
\Gamma_{\tau\tau} &= 0.087 \pm 0.010\ \text{GeV},\\
\Gamma_{\text{had}} &= 1.787 \pm 0.013\ \text{GeV},\\
\Gamma_{\text{inv}} &= 0.552 \pm 0.110\ \text{GeV}.
\end{aligned}
$$

The subscript 'had' refers to hadronic final states and therefore corresponds to the sum of all $q\bar{q}$ final states. The subscript 'inv' refers to invisible final states and therefore corresponds to the sum of all $\nu\bar{\nu}$ final states. These results are significant for three reasons:

1. They are consistent with the values predicted by the standard model.
2. They show that, if all neutrinos are very much lighter than the Z^0, there are only three varieties, with a confidence level of about 98%.
3. They confirm the equality, within errors, of the weak coupling of all three generations of leptons.

2. By its success it validated Pauli's hypothesis of the neutrino (Section 5.3).

We give here a very simplified version: it will enable us to understand how β-decay rates can vary over many orders or magnitude when we have asserted that the weak coupling constant is universal (Section 9.13).

For simplicity we will consider neutron decay:

$$\mathrm{n} \to \mathrm{p} + \mathrm{e}^- + \bar{\nu}_\mathrm{e} + Q(=782\ \mathrm{keV}).$$

We know that the W boson is involved in this decay as in Fig. 12.1(a), where the quark change is $\mathrm{d} \to \mathrm{u} + \mathrm{e}^- + \bar{\nu}_\mathrm{e}$. The matrix element for the decay contains a $g/2\sqrt{2}$ for each vertex of the W exchange and a factor $(M_\mathrm{W}^2c^4 + q^2c^2)^{-1}$ for the propagation of the W. Since the four-momentum squared q^2c^2 carried by the W is a few MeV^2, which is tiny compared with $M_W^2c^4$ (6.7×10^9 MeV^2) we can neglect q^2c^2 in the propagator and replace $g^2/8M_\mathrm{W}^2c^4$ by $G_\mathrm{F}/\sqrt{2}$; this is the correct form of the connection that we established approximately in Section 9.13. Fermi did not know about the W bosons so he thought of the decay in a way which we would now represent by a four-arm vertex, as in Fig. 9.18. He did not know about quarks so his vertex had a neutron in and a proton out instead of quarks. The strength of his four-fermion interaction was G_F and the difference by a factor of 2 is due to the fact that the interaction of gauge bosons with fermions in the electroweak theory is that of a vector field interacting with a fermion current which contains two parts, a vector and an axial vector current (see Section 12.7). We neglect these complications and take the coupling to be G_F. We use what Fermi called the Golden Rule number 2, the familiar formula for the transition rate:

$$\omega = \frac{2\pi}{\hbar}|M|^2\frac{\mathrm{d}N}{\mathrm{d}E}. \tag{12.15}$$

This formula is described in Table 12.6. The reader has probably met it before. In this case M is the matrix element between the initial state (n) and the final state ($\mathrm{pe}^-\bar{\nu}_\mathrm{e}$) of an operator describing the change of $\mathrm{n} \to \mathrm{p}$, and the creation of the electron and antineutrino, by the effect of the weak interaction. We do not define that operator here but state a result:

1. That $M = G_\mathrm{F}/V$, where V is the volume in which the particle wavefunctions are normalized (see also Table 12.9).

We also need to make some other simplifications, so we assume:

2. That the effects of angular momentum conservation and parity non-conservation can be neglected (Section 12.9).
3. That the neutron and proton are very massive relative to the decay energies so that the proton does not recoil and the electron and neutrino momenta are uncorrelated except in that the sum of their total (rest + kinetic) energies is equal to the total available energy. That energy is the $Q + m_\mathrm{e}c^2$ of the decay but also by necessity it is the E in $\mathrm{d}N/\mathrm{d}E$ of equation (12.15). So we have $E = E_\mathrm{e} + E_\nu$ where E_e and E_ν are the total relativistic energies of electron and neutrino respectively.
4. That the neutrinos are mass-less.

The quantity ω is the transition rate for the decay of the neutron. However, we know the electron momentum spectrum is continuous, so we really need a

Table 12.6 The Fermi Golden Rule number 2.

The transition rate ω for the transition from state A to state B is

$$\omega = \frac{2\pi}{\hbar}|M|^2\frac{dN}{dE},$$

where M is the matrix element of the operator (the interaction Hamiltonian) representing the interactions which causes the change from A to B, taken between the state functions representing these states. The quantity E is the total energy available in the centre of mass of the final state, N is the number of final states in momentum space available to B which have energy less than or equal to E.

For a single particle of momentum P emitted by a non-recoiling source, N is the volume of the sphere of momentum $P(=\frac{4}{3}\pi P^3)$ divided by $(2\pi\hbar)^3/V$, which is the volume in momentum space occupied by one quantum mechanical state. V is the volume in ordinary space in which the initial and final state wavefunctions are normalized. The final result for a single particle's decay transition rate is independent of V, so it must cancel from the calculation.

If there are two particles in the final state, then P is the momentum of one in the centre of mass (the other has equal and opposite momentum) and E is the sum of the two kinetic energies. Non-relativistically:

$$E = \frac{P^2}{2}\left(\frac{1}{m_1}+\frac{1}{m_2}\right)$$

and N is again

$$\frac{4\pi P^3 V}{[3(2\pi\hbar)^3]}.$$

To calculate a cross-section, divide by the sum of the velocities of the two colliding particles (v_a and v_b) in the centre of mass and multiply by V:

$$\sigma = \frac{\omega V}{v_a + v_b}.$$

The extra particle in the initial state brings in another V^{-1} into ω so that again a result, in this case σ, is independent of V.

differential decay rate, $d\omega/dP_e$, where P_e is the electron momentum. As M is constant that means we need

$$\frac{d^2N}{dP_e dE}.$$

If the electron momentum is fixed (P_e) the number of electron states N_e having momentum $< P_e$ is given by

$$N_e = \frac{4\pi P_e^3 V}{3(2\pi\hbar)^3}. \tag{12.16}$$

Remember, this comes from counting the number of discrete wavefunctions of wavelength greater than $2\pi\hbar/P_e$ that you can fit into a box of volume V. Similarly the number of neutrino states is

$$N_\nu = \frac{4\pi P_\nu^3 V}{3(2\pi\hbar)^3}.$$

Now $N = N_e N_\nu$ because to begin we have a complete freedom to imagine putting the e^- and the $\bar{\nu}_e$ independently into any momentum state we wish. Then

$$N = N_e N_\nu = \frac{16\pi^2 P_e^3 P_\nu^3 V^2}{9(2\pi\hbar)^6}.$$

Now P_e is fixed because we agreed to find the value of $d\omega/dP_e$ at a given P_e. That means, since $E = E_e + E_\nu$, we have

$$dE = dE_\nu = c\,dP_\nu \tag{12.17}$$

and we obtain

$$\frac{d^2N}{dP_e dE} = \frac{16\pi^2 P_e^2 P_\nu^2}{(2\pi\hbar)^6 c} V^2.$$

Putting everything together in the Golden Rule we have

$$\frac{d\omega}{dP_e} = \frac{G_F^2 P_e^2 P_\nu^2}{2\pi^3 \hbar^7 c}. \tag{12.18}$$

This differential transition rate gives the intensity spectrum of the electrons emitted. In Fig. 12.9 we show an example of this spectrum. At low electron momentum the intensity varies as P_e^2. As the electron momentum reaches its maximum value the neutrino momentum approaches zero and the spectrum varies as P_ν^2. Using the usual relation between E and P for a particle $E^2 = P^2c^2 + M^2c^4$ it is easy to rewrite the formula in its usual form:

$$\frac{d\omega}{dE_e} = \frac{G_F^2}{2\pi^3 \hbar^7 c^6} (E_e^2 - m_e^2 c^4)^{1/2} E_e (E - E_e)^2, \tag{12.19}$$

where $E = E_e + E_\nu = Q + m_e c^2$. We can see that this spectrum in electron energy

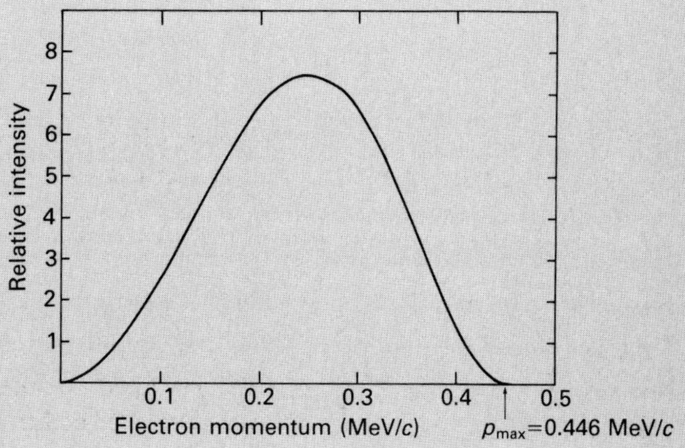

Fig. 12.9 The expected momentum spectrum of electrons emitted in the decay

$${}^{35}_{16}S \rightarrow {}^{35}_{17}Cl + e^- + \bar{\nu}_e + 0.1675\ \text{MeV}.$$

The spectrum has been calculated using the simple Fermi theory (equation 12.18) without correction for Coulomb effects and assuming that the antineutrino mass is zero.

extends from $E_e = m_ec^2$ to $E_e = Q + m_ec^2$ and on an energy scale there is some distortion from the bell shape of Fig. 12.9.

Let us now go backwards through our assumptions to see what has to be done to remove them, although we do not go into great detail. We do this extending our discussion beyond neutron decay to nuclear β-decay in general. Assumption 4: If the neutrino mass is not zero, the electron total energy spectrum extends only to the end point $Q + m_ec^2 - m_\nu c^2$ and the shape of this spectrum is modified: this is discussed in Section 12.12. Assumption 3: If the recoil of the daughter nucleus is taken into account, again there is shift in the end point and the spectrum shape. Assumption 2: As derived, equation (12.19) applies to what are called **allowed transitions**. Essentially these are transitions in which the electron and neutrino carry away a total of either one or zero units of angular momentum. The $e^- \bar{\nu}_e$ system can do this by being in a spin triplet state (**Gamow–Teller transitions**) or in a spin singlet state (**Fermi transitions**) respectively. If the decay occurs between nuclear states such that there is a change in nuclear spin greater than 0 or 1, or if the nuclear parity changes from parent to daughter, the angular momentum barrier depresses the rate and the e^- spectrum shape is altered. These are the **forbidden transitions**.

Note that, in spite of the name, these transitions are not absolutely forbidden but have degrees of forbiddenness which increase with the increase in the change of nuclear spin: with each increase in degree there is corresponding decrease in the transition rate, by a factor of about 10^2 or greater, if all other things remain the same. The **first forbidden transitions** have 0, 1, or 2 units change of nuclear spin and a change of nuclear parity. The second forbidden have 1, 2, or 3 units change of nuclear spin and no change of the nuclear parity. This continues and is rightly reminiscent of the multipole classification of electromagnetic transitions (Section 11.8).

Looking at assumption 1, we can guess rightly that there are many details hidden. One of the simpler of these is the effect of the Coulomb interaction. In β^--decay, the outgoing electron's wavefunction is greater at the charged daughter nucleus than it would be for an uncharged nucleus. The matrix element is greater and the transition rate enhanced, particularly at low electron energies. In β^+-decay the reverse is the case at low energies. In fact, at low energies the positron may be created at a point where its potential energy is greater than its kinetic energy at infinity; it therefore has to escape through a Coulomb barrier, the result is that positron decay rates are depressed, more so at large Z. These Coulomb effects have been calculated and tabulated. It is usual to lump them into a term $F(Z,E_e)$ which becomes a factor to be attached to equation (12.19).

Another effect hidden in assumption 1 is the contribution to the matrix

DEFINITIONS AND KEYWORDS

Fermi transitions β-decay in which the electron–antineutrino (or positron–neutrino) system is in a singlet total spin state.

Gamow–Teller transitions β-decay in which the electron–antineutrino (or positron–neutrino) system is in a triplet total spin state.

Allowed transitions β-decay transitions with nuclear $|\Delta j| = 0$ or 1 and no change in nuclear parity.

Forbidden transitions β-decay transitions with a nuclear spin change $|\Delta j| \geqslant 2$ or a change in nuclear parity.

First forbidden transitions β-decay transitions with a nuclear spin change $|\Delta j| = 0$, 1, or 2 and a change in nuclear parity.

Superallowed transitions β-decay transitions between members of a nuclear isotopic spin multiplet.

element from the initial and final nuclear states. Thus another factor $|M_{nuc}|^2$ may appear in equation 12.19, which is different from decay to decay and may vary over a very large range. The reasons for this are not hard to discern. Decays between mirror nuclei, for example ${}^7_4Be \rightarrow {}^7_3Li$ (Fig. 9.10) have the property that apart from one change (in this case $p \rightarrow n$), the nuclear wavefunction hardly changes, and we expect that there will be almost complete overlap of initial and final states and that M_{nuc} will be close to 1. Mirror nuclei are members of isotopic spin doublets (Section 10.11) and the same will be true for transitions between members of other nuclear isotopic spin multiplets: for example in the $A=6$ isotopic spin triplet, ${}^6He \rightarrow {}^6Li^*$ (0^+ excited state at 3.56 MeV) (see Fig. 9.11). These transitions are called **superallowed** and have transition rates enhanced above average. At the other extremity very little nuclear wavefunction overlap leads to a small nuclear matrix element. This is true of the forbidden transitions for which the degree of forbiddenness is closely related to the change in the angular momentum of the nucleus. Increasingly large changes mean decreasing nuclear matrix elements.

12.6 The Kurie plot

Verification of Fermi's theory for the β-decay spectrum is a difficult experimental task. For example, in the majority of devices used to measure electron momentum or energy, this can be done only after the electron has left the source. That will not happen without energy loss by ionization or by emission of X-rays in scattering (bremsstrahling process). Thus results can suffer from considerable errors, particularly at low electron energies. However, away from that region precision results have been obtained. These are frequently plotted as a Kurie plot.

Look at equation (12.18) and include the Coulomb factor $F(Z,E_e)$. If we write the electron intensity per unit momentum range as $I(P_e)$ we can express this equation as

$$\sqrt{\frac{I(P_e)}{P_e^2 K(Z,P_e)}} = E - E_e, \tag{12.20}$$

where $K(Z,P_e)$ is all the constants and the Coulomb dependence correction. Therefore a plot of the left-hand side of equation (12.20) against E_e will yield a straight line intercepting the E_e-axis at E. This allows a determination of E, if required. The line should be straight if the transition is allowed. Figure 12.10a shows the electron kinetic energy spectrum and Fig. 12.10b the Kurie plot for the decay ${}^3_1H \rightarrow {}^3_2He + e^- + \bar{\nu}_e$. Forbidden transitions may also be plotted as straight lines if $K(Z,P_e)$ contains terms correcting the distortion, due to angular momentum effects, of the spectrum from that expected in the allowed transitions.

12.7 The ft value and some approximations

Let us now include the nuclear matrix element M_{nuc} and integrate equation (12.9) to find the total transition rate and relate that to the half-life $t_{1/2}$ (Section 2.2):

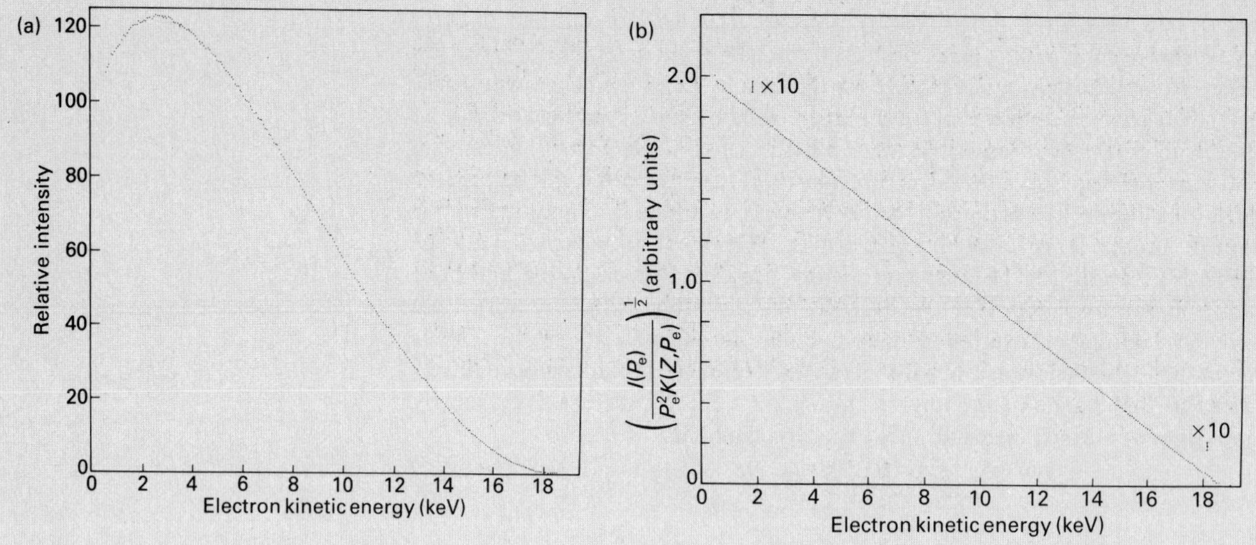

Fig. 12.10 (a) The kinetic energy spectrum of electrons emitted in the decay of tritium:

$${}^3_1\mathrm{H} \rightarrow {}^3_2\mathrm{He} + \mathrm{e}^- + \bar{\nu}_\mathrm{e}$$

as observed by implanting ${}^3_1\mathrm{H}$ nuclei into a silicon solid state detector. (b) The Kurie plot for the data of (a). The ordinate is the quantity

$$\left(\frac{I(P_\mathrm{e})}{P_\mathrm{e}^2 K(Z,P_\mathrm{e})}\right)^{1/2}$$

defined by the left-hand side of equation (12.20). Data obtained by J. J. Simpson of the University of Guelph and reproduced with his permission. The small vertical bars represent ten times the typical error at each end of the plot.

$$\omega = \frac{\ln 2}{t_{1/2}} = \int_{m_\mathrm{e}c^2}^{E} \frac{\mathrm{d}\omega}{\mathrm{d}E_\mathrm{e}} \mathrm{d}E_\mathrm{e}.$$

Therefore

$$\frac{\ln 2}{t_{1/2}} = \frac{G_\mathrm{F}^2 |M_\mathrm{nuc}|^2}{2\pi^3 \hbar^7 c^6} \int_{m_\mathrm{e}c^2}^{E} F(Z,E_\mathrm{e})(E_\mathrm{e}^2 - m_\mathrm{e}^2 c^4)^{1/2} E_\mathrm{e}(E - E_\mathrm{e})^2 \mathrm{d}E_\mathrm{e}$$

$$= \frac{G_\mathrm{f}^2 |M_\mathrm{nuc}|^2 m_\mathrm{e}^5 c^4}{2\pi^3 \hbar^7} \int_{m_\mathrm{e}c^2}^{E} F(Z,E_\mathrm{e}) \left(\frac{E_\mathrm{e}^2}{m_\mathrm{e}^2 c^4} - 1\right)^{1/2} \frac{E_\mathrm{e}(E - E_\mathrm{e})^2 \mathrm{d}E_\mathrm{e}}{m_\mathrm{e}^4 c^8}. \quad (12.21)$$

The integral looks difficult and since we do not know $F(Z,E_\mathrm{e})$ we cannot even attempt it. However, it has been numerically integrated and tabulated, and we put it equal to $f(Z,E)$. Then

$$f(Z,E)t_{1/2} = \frac{2\pi^3\hbar^7 \ln 2}{G_F^2 m_e^5 c^4 |M_{nuc}|^2}. \tag{12.22}$$

The l.h.s. of equation (12.22) is called the ft value of a decay: What is its importance? If G_F is a universal constant then ft is a direct measure of M_{nuc} and we could expect that superallowed transitions would have the same value of ft and that allowed transitions and transitions of different degrees of forbiddenness would group into well-defined values of ft. Unfortunately that does not happen. Figure 12.11 shows the distribution of some values of $\log_{10} ft$. The peak between 3 and 4 is due to the superallowed transitions. Otherwise the nuclear matrix element is too variable a factor to give well-separated groupings or always to allow firm decisions on the degree of forbiddenness.

Let us now look at an approximation: in equation (12.21), if we assume $F(Z,E_e)=1$ and $E_e \gg m_e c^2$ (outrageous!) and integrate from 0 to E, then the transition rate is given by

$$\omega = \frac{G_F^2 |M_{nuc}|^2}{2\pi^3\hbar^7 c^6} \frac{E^5}{30}. \tag{12.23}$$

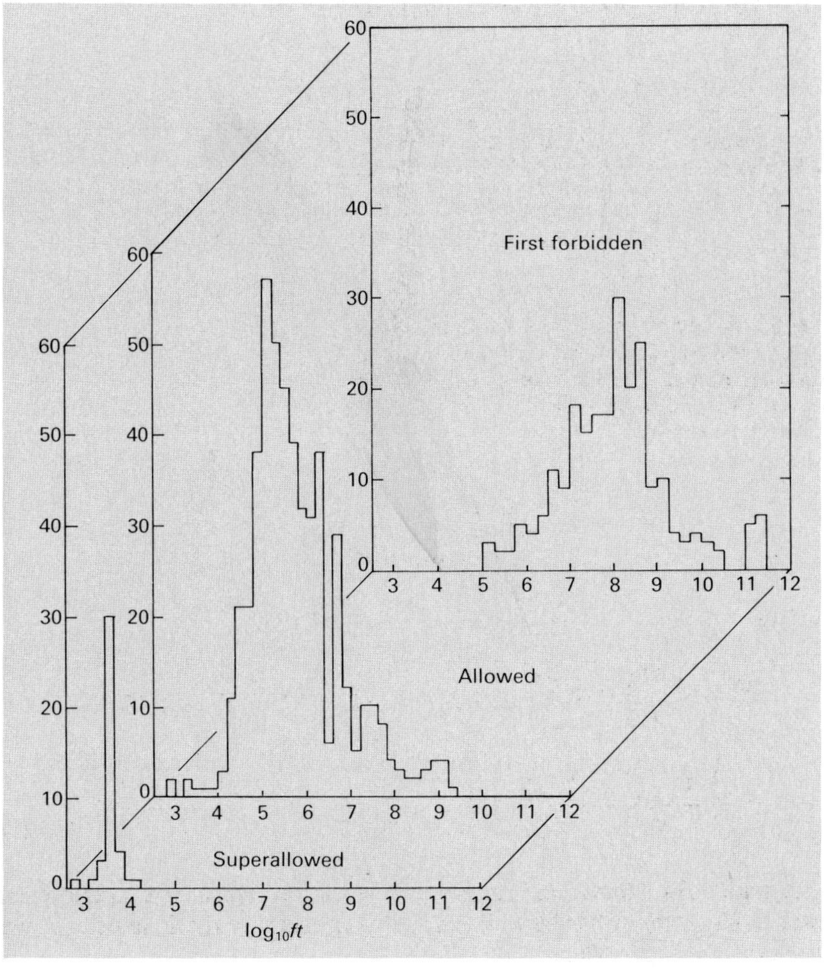

Fig. 12.11 A plot of the frequency distribution of the values of $\log_{10} ft$ for some superallowed, allowed and first forbidden β-decay transitions. It is clear that the ft value does not always allow on its own a clear decision on the class of the decay. This is because the nuclear part of the matrix element can vary greatly within each class.

We can improve this by replacing E either by $P_{max}c$, where P_{max} is the maximum momentum of the emitted electron, or by T_{max}, the maximum kinetic energy. These replacements have the virtue of agreeing with (12.23) when $E \gg m_e c^2$ and giving $\omega \to 0$ as P_{max} (or T_{max})$\to 0$, as must be the case. Fermi (1950) has given the analytic results of the integration over P_e that are applicable for all P_{max} but only so far as his assumption that $F(Z,E_e)=1$ is good.

With the replacement $E \to T_{max}$ in equation (12.23) the approximate dependence $\omega \propto T^5_{max}$ is now explained. Figure 12.12 shows a log–log plot of ω against T_{max} for a set of allowed and superallowed β-decays having a fifteen-decade range of ω: they provide a moderately good demonstration of the result found originally by Sargent.

We stop our theoretically inclined discussion at this point although this is clearly far from the end of the theory of β-decay. Refinements come from a proper relativistic treatment of the electrons, positrons and neutrinos by using Dirac's relativistic theory of fermions. In addition, the interaction operator has to be defined properly and experimental results used to decide which of several choices is correct. A description of that subject is beyond the scope of this book except that we can give the accepted theory a name—the V − A (V minus A) theory. The V means vector and A means axial vector and these words describe the nature of the fermion currents which are interacting. The theory is correct at low energies and it extends correctly to higher energies if the gauge bosons $W^{\pm}$, and Z^0 are included. Nuclear β-decay also involves a nuclear matrix element and

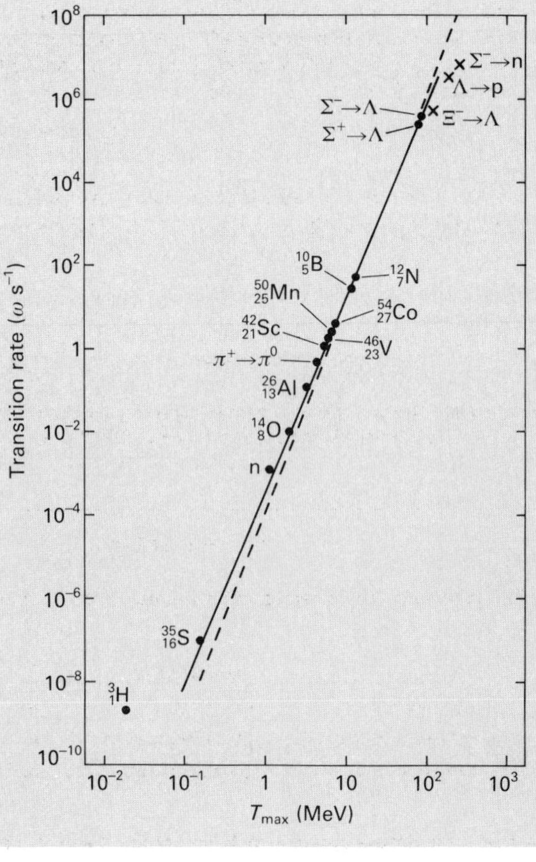

Fig. 12.12 A log–log plot of the transition rate ω against T_{max}, the maximum electron (or positron) kinetic energy of some allowed and superallowed β-decays. The continuous line has a slope of 4.67 where simple theory predicts 5 for $T_{max} \gg m_e c^2$. (That condition is not satisfied for most of the decays plotted.) The dotted line is the result for equation (12.23) with E replaced by T_{max}.

Three strangeness changing decays are also plotted (points ×); they have transition rates lower than expected on our simple theory.

a great deal of research has been done on understanding the connection with the theories of nuclear structure. In the case of weak interaction hadron decays, there exist analogous problems connected with the quark structure of the hadrons involved.

Using only the very simple theory that we have presented it is possible to make useful predictions about some particle decays. Since the strength of the Wqq and the W$l\nu$ vertices are quantitatively related, it is possible to analyse the decays in which a virtual W is emitted and becomes observable particles in order to predict branching fractions. Since the decay rate into three-body channels depends on the fifth power of the energy released in that decay, it is possible to predict some decay rates using the known rate for the decay of the μ^+ (equation (12.4)). These calculations are done for the decay of the τ-meson in Table 12.7.

Table 12.7 The decay of the τ^--meson.

The basic diagram is:

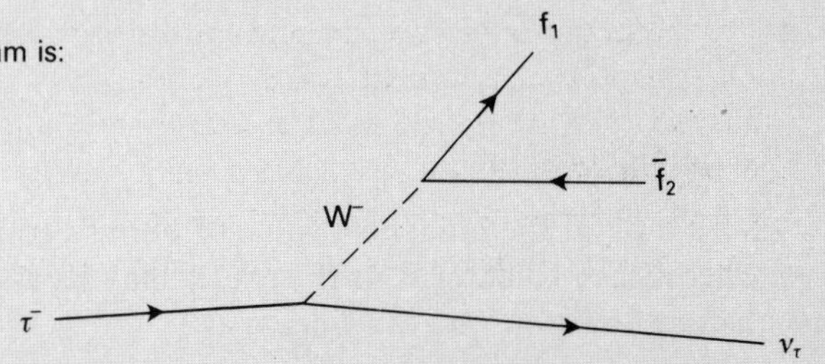

The fermions $f_1, \bar{f}_2$ can be $e^-\bar{\nu}_e$, $\mu^-\bar{\nu}_\mu$, or $d\bar{u}$ (×3 for colour). Decay to $\bar{c}s$ is forbidden by energy conservation. There will be a few percent branching to $\bar{u}s$ states which we can discount because of the almost equal decrease in branching to $\bar{u}d$. There are, therefore, five equal channels, and we expect the branching fraction to leptons to be about 40%. It is 35%.

The transition rate for

$$\mu^- \rightarrow e^- + \bar{\nu}_e + \nu_\mu$$

is $4.55 \times 10^5\ s^{-1}$. The decay rate for such leptonic decays is proportional to the fifth power of the energy release (see Section 12.7, equation (12.23)) so that

$$\omega(\tau^- \rightarrow e^- \bar{\nu}_e \nu_\tau) \simeq 4.55 \times 10^5 \left(\frac{m_\tau}{m_\mu}\right)^5 s^{-1}$$

$$= 6.23 \times 10^{11}\ s^{-1}.$$

Multiply by five for five decay channels and we find the τ^- mean life is expected to be 3.2×10^{-13} s. The observed mean life is $3.04 \pm 0.09 \times 10^{-13}$ s.

PROBLEMS

12.6 The following weak interaction decays have the Q-values and transition rates given:

${}^{14}_{8}O \rightarrow {}^{14}_{7}N^* + e^+ + \nu_e$,	$Q = 1.81$ MeV,	$\omega = 5.04 \times 10^{-3}\ s^{-1}$,
$\Sigma^+ \rightarrow \Lambda + e^+ + \nu_e$,	$Q = 73$ MeV,	$\omega = 2.53 \times 10^5\ s^{-1}$,
$\pi^+ \rightarrow \pi^0 + e^+ + \nu_e$,	$Q = 4.093$ MeV,	$\omega = 0.39\ s^{-1}$,
$n \rightarrow p + e^- + \bar{\nu}_e$,	$Q = 0.782$ MeV,	$\omega = 1.14 \times 10^{-3}\ s^{-1}$,

Show that these decays approximately conform to the rule that $\omega \propto T^5_{max}$.

12.7 Summarize the evidence that, in nuclear β-decay, a transmutation takes place between a neutron and a proton and that a neutrino (having spin-$\frac{1}{2}$ and zero rest mass) is emitted.
The following table gives the half-life $t_{1/2}$ for three β-transitions. Nuclear spins are shown in brackets and T_{max} is the maximum kinetic energy of the β-particle in megaelectronvolts.

	T_{max} (MeV)	$t_{1/2}$(s)
${}^6_2$He(0) → ${}^6_3$Li(1)	3.5	0.81
${}^3_1$H($\frac{1}{2}$) → ${}^3_2$He($\frac{1}{2}$)	0.018	4×10^8
${}^{10}_4$Be(0) → ${}^{10}_5$B(3)	0.55	5×10^{13}

Explain briefly the factors you could expect to be important in accounting for the magnitudes of these lifetimes.

(Adapted from the 1973 examination of the Final Honours School in Natural Science, Physics, University of Oxford.)

12.8 Estimate the mean life of D^+ meson given that the μ^+ mean life is 2.2×10^{-6} s. (Assume that the Wcs vertex has the same strength as the Weν_e vertex and that the Wcd vertex is negligibly weak.)

12.8 Fermi's coupling constant

Let us briefly take a look at the numerical value of Fermi's coupling constant, G_F, which we earlier met in Chapter 9, as the reader may recall. If we take our simple theory, then a transition rate for β-decay is

$$\omega = \frac{G_F^2 |M_{nuc}|^2}{2\pi^3 \hbar^7 c^6} \frac{T_{max}^5}{30}.$$

In Fig. 12.12 we have plotted ω against T_{max} on a logarithmic plot for a number of allowed decays extending over 15 orders of ω. The strong correlation is in approximate agreement with a fifth-power relation and gives us some confidence in the theory. The broken line is given by equation (12.23) replacing E by T_{max} and using

$$G_F = 8.962 \times 10^{-5}\ \text{MeV fm}^3.$$

Table 12.8 The Fermi coupling constant

$G_F = 8.962 \times 10^{-5}$ MeV fm^3
$= 1.436 \times 10^{-62}$ J m^3
$G_F/(\hbar c)^3 = 1.166 \times 10^{-5}$ GeV^{-2}

The deviations from the broken line are due to the fact that we have used the simple formula.

The most accurate value for G_F (Table 12.8) comes from the decay rate for μ^+-decay:

$$\mu^+ \to e^+ + \nu_e + \bar{\nu}_\mu,$$

which involves leptons alone. In the hadron decay sector the superallowed decays of spin-0 nuclei, of which there are nine good examples, yields a value about $2\frac{1}{2}$% less. This is the effect, that we have discussed in Section 12.4, of the fact that the $W^\pm$ bosons mix the quark generations: in Fig. 12.4 the ability of a W to change the flavour of a u quark has to be shared amongst the changes to d, s, or b quark. Thus the u↔d, u↔s, u↔b transition amplitudes are approximately in the ratio 0.975:0.222:0.004. The W does not mix the lepton generations, so we expect the coupling constant in ${}^{14}_8$O decay, for example, to be 0.975 of that in μ^+-decay. If we extend this to look at the β-decay of strange particles in which strangeness changes, for example

$$\Sigma^- \rightarrow n + e^- + \bar{\nu}_e,$$

then we expect the effective coupling constant at the Wsu vertex to be 0.222 G_F, not G_F. Three of these decays are shown in Fig. 12.11. Their transition rates are less than those expected for an unaltered G_F.

This property that W-bosons have of mixing the quark generations and of not mixing the lepton generations is remarkable. Why mix? And why mix in one case but not the other? These are unanswered but vital questions. We note that if there is no mixing, the lightest member of each generation is stable. In the lepton case these are the ν_e, ν_μ, and ν_τ and, without evidence, we assume they are stable. In the hadron case the confinement of quarks makes it uncertain in the first generation which would be stable, u or d; in the heavier generations the s and b (and of course $\bar{s}$ and $\bar{b}$) would be stable in the absence of mixing. The nature of our world if this were to be the case is difficult to imagine: K^+- and isolated K^--mesons would be stable, complex stable nuclei could contain Λ particles as well as protons and neutrons. In the latter case our chemistry would be very complex as the simplicity of the regularities of the masses of the periodic table could disappear and elements of a given Z could have isotopes of masses no longer near integer multiples of the hydrogen atomic mass.

12.9 Through the looking-glass

Until 1955 the assumption was made that parity was conserved in the weak interactions. This assumption was called into question when two decays of the K^+ meson were observed:

$$\theta^+ \rightarrow \pi^+ + \pi^0, \quad (12.24)$$

and

$$\tau^+ \rightarrow \pi^+ + \pi^+ + \pi^-. \quad (12.25)$$

At the time these decay modes were not associated with one particle (K^+) but two different particles at that time labelled θ^+,τ^+ (not the τ-fermion) with similar masses. The spinless nature of the π-mesons means that the final state of the decay 12.24 must have spin-parity (j^P) from the series 0^+, 1^-, 2^+, ... (see Problem 12.9) and that, if angular momentum and parity are conserved, the θ^+ will have the same spin-parity. A less simple analysis of the decay 12.25 showed that the possibilities for j^P for the so-called τ^+ were 0^-, 1^+, 1^-, 2^-, ... The common value of 1^- required an energy spectrum of the pions in the τ decay which was not observed. So there was a problem! Either θ and τ were different or parity was not conserved. We know now that only one parent, the K^+, is involved and it has $j^P = 0^-$. Thus the decay (12.24) violates parity conservation.

This is an example of how initial and final parities can be used to test for parity non-conservation. However, beware! The weak interaction is not forced to change the parity (as in θ^+ decay (12.24)) but is blind to it; that is, it does or does not change parity almost with indifference and in some transitions can give a final state which is a superposition of two states of opposite parity. For example, the decay $\Lambda \rightarrow p + \pi^-$ has a final state which is a superposition of a state with orbital angular momentum $l = 0$ and a state with $l = 1$; these two states have opposite parity. Just one word: the analysis of K^+-decays that revealed this phenomenon depended on conservation of angular momentum and so the possibility existed that the violation was with angular momentum and not parity. That unpalatable possibility disappeared as other properties due to parity non-conservation in the weak interaction were found. We wish to investigate one such property in addition to the direct observation of a change in parity from the initial to the final state.

When Alice went through the looking glass (Carroll, 1872) the time might have been about 4 o'clock of the afternoon of 4th November 1870 (Fig. 12.13). If she had glanced back at the looking-glass clock before starting her adventures and had read the time by the position of the hands she would have read 8 o'clock (that is, if the clock face had not changed into a smiling face!) If she had watched long enough she would have decided the clock was going backwards. But if you, the reader, think about the clock and its mirror image, the hands are rotating in the same sense about the line defined by the spindle carrying the hands (in this special case when the spindle is perpendicular to the mirror). The (axial) vector representing this rotation points along the direction around which the rotation is clockwise, that is, normally into the mirror on the real world side and normally away from the surface of the mirror in the looking-glass world. As seen from the real world these are seen as the same vector because they have the same direction and magnitude. The same applies to spin: if we replace the rotating clock hands by a spinning neutron, with its spin pointing into the mirror, then when we look at its mirror image we see a neutron spinning with its spin pointing away from the mirror surface and into the looking glass world. If this neutron decays to give an electron coming away from the mirror in the real

Fig. 12.13 Alice crawls into the mirror and emerges into the looking-glass world! The artist, John Tenniel, has changed some of the severe Victorian decoration to that appropriate to the lighter spirit of the looking-glass world. Two other details have received the correct transformation. Note that we assume Alice herself does not suffer the mirror transformation.

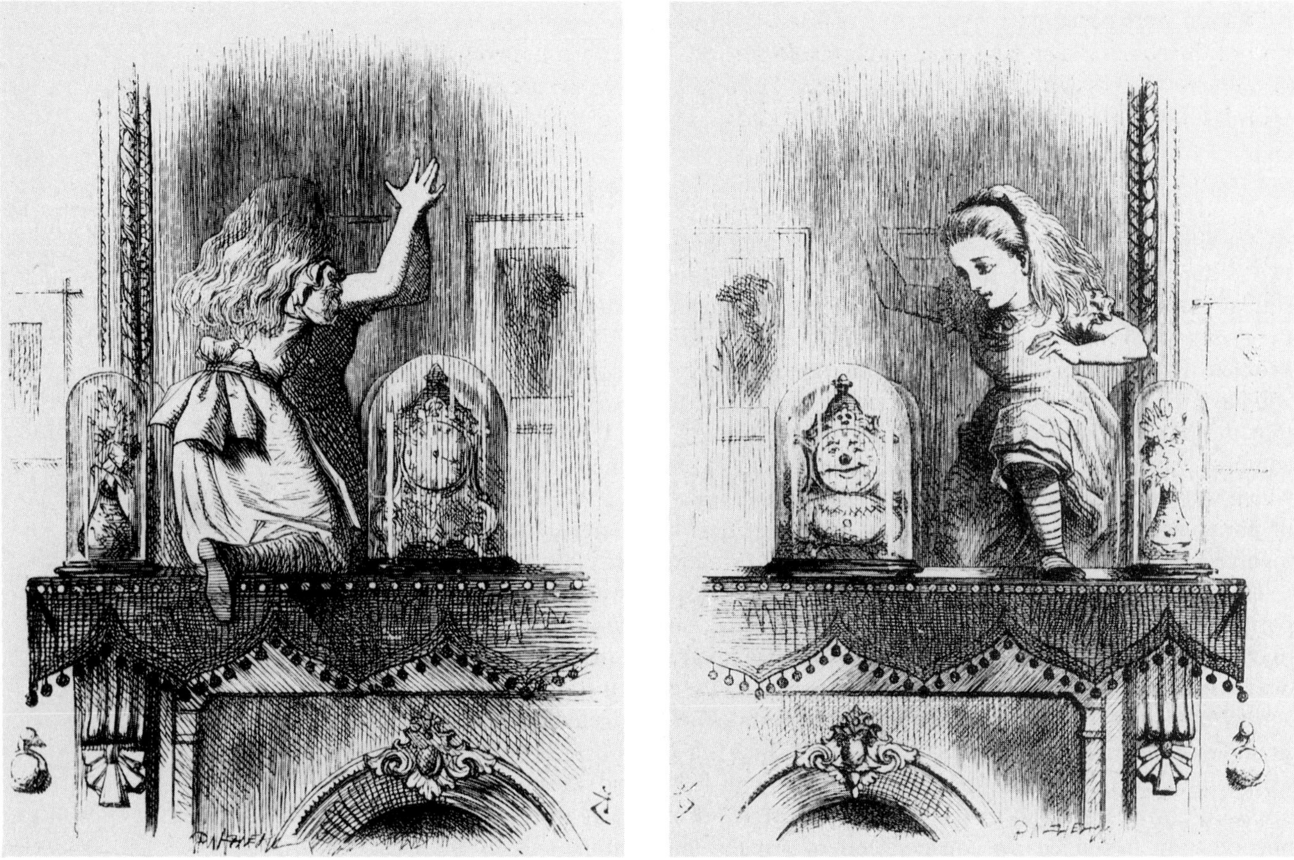

world, that electron emerges at near 180° to the direction of the spin. Looking into the mirror that electron is seen to be emitted from the mirror neutron away from the mirror surface, which is a direction near 0° with respect to that neutron's spin (see Fig. 12.14).

Where is this mirror world taking us? If parity is conserved, then a mirror image of a real world physical system is an equally likely situation in the real world. This is a statement of fact which we cannot prove here: however, it is a particular case of a general law which is briefly described in Table 13.3. Applying it to a spinning neutron which decays with a certain probability to give an electron at 180° to the spin vector then the mirror image, which decays to give an electron at 0°, is therefore equally probable. More generally, with respect to the spin direction the intensity of emitted electrons at angle θ equals that at $180° - \theta$. If, however, a decay is observed which does not have such a decay symmetry, that observation is unequivocal evidence for parity non-conservation in the decay.

One of the first tests of parity non-conservation completed was of this kind. The isotope $^{60}_{27}Co$ has nuclear spin 5 and decays to $^{60}_{28}Ni$ by β^--decay: a sample was spin polarized by application of a large magnetic field at liquid helium temperatures and the intensities of electrons emitted along the magnetic field and of those emitted against it were measured. There was a decay asymmetry which disappeared as the specimen warmed up and the $^{60}_{27}Co$ depolarized. The asymmetry was real and the conclusion had to be that parity was not conserved. Many other tests were made immediately after the time of that first test. All processes involving the weak interaction show effects of parity non-conservation in one way or another.

This discovery had a major effect in two ways. Firstly it provoked tests of most of the conservation laws, in all their manifestations, which had previously been unquestioned. Secondly it allowed a deeper investigation of the properties of the weak interaction than had previously been possible. All possibilities other

Fig. 12.14 The real and the looking-glass world seen by an observer watching a neutron spinning around an axis perpendicular to the mirror. The spin is represented by an arrow along the axis of rotation but, beware! This is not a vector but an axial-vector! To find how it would look in the mirror consider a disc with its plane perpendicular to the spin and rotating with the spin. The velocity of the edge of the disc is a vector and is represented by an arrow in the figure. It is straightforward to see how that velocity vector is seen in the mirror and from that how the mirror disc is rotating and then how the axial vector spin is pointing. Check this in the figure. If now the object neutron decays to emit an electron, momentum P_e, at about 150° (θ) with respect to its spin direction, then the mirror-image neutron decays to emit its electron at about 30° ($180° - \theta$) with respect to its spin direction.

Conservation of parity would require that the mirror image is an equally probable outcome of decay as is the object. Thus it follows that in neutron decay the intensity $I(\theta)$ of electrons emitted at angle θ to the spin should be the same as $I(180° - \theta)$; that is, the decay distribution should be symmetric about $\theta = 90°$. If $I(\theta) \neq I(180° - \theta)$, then the decay interaction violates parity conservation. For the neutron β-decay, the distribution is not symmetric and therefore parity is not conserved in its decay.

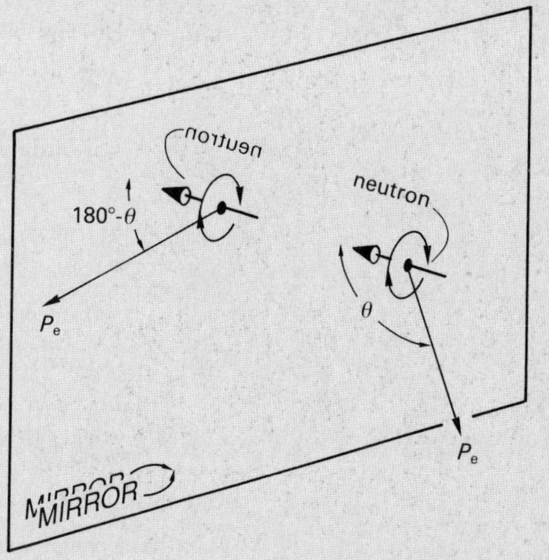

than the V–A (Section 12.7) nature of the four-fermion interaction of the Fermi theory were eliminated.

The investigation of the integrity of other conservation laws showed that the weak interaction is not the same for antiparticle as it is for particles. This meant that anti-$^{60}_{27}$Co (if it existed) would have a positron decay distribution asymmetry opposite to the decay asymmetry of $^{60}_{27}$Co. This effect is called violation of charge conjugation invariance! However, combine taking the mirror image (P = parity transformation) and changing particles into antiparticles (C = charge conjugation) and identical behaviour is restored. The anti-$^{60}_{27}$Co mirror image of a polarized $^{60}_{27}$Co object both have the same decay distribution with respect to the spin vector! This property is called CP invariance: The weak interaction has identical effects (it is invariant) for a system as for the mirror image, charge conjugated. Less than 10 years later it was found that even CP invariance was violated at the relative intensity level of about 5 in 10^6 in the decay properties of the K^0 and $\overline{K}^0$ mesons. But that is an exciting story which would take us far beyond the plan of this book (see Perkins, 1986). It is a story still developing because CP violation is far from being understood and it may turn out to have been of great significance in establishing the one-sidedness of the universe with respect to matter and antimatter (Section 14.13).

PROBLEMS

12.9 By 1955 the symbol θ^+ had been assigned to a particle which was observed to decay

$$\theta^+ \rightarrow \pi^+ + \pi^0,$$

and which had a mass of about 500 MeV/c^2. We now know that it was, in fact, the K^+-meson and one of its major decay modes. Classify the possible states of spin and parity of the final state and state the restrictions these classifications place on the spin and parity of the θ^+ under the various assumptions about the conservation laws that apply.

12.10 Explain qualitatively why all of the following processes can be attributed to the weak force despite the very different lifetimes involved:

(a) muon decay

$$\mu^+ \rightarrow e^+ + \nu_e + \bar{\nu}_\mu \qquad (t_{1/2} = 1.5 \times 10^{-6}\ \text{s},\ Q = 105\ \text{MeV});$$

(b) neutron decay

$$n \rightarrow p + e^- + \bar{\nu}_e \qquad (t_{1/2} = 607\ \text{s},\ Q = 0.782\ \text{MeV});$$

(c) atomic electron capture

$$^{7}_{4}\text{Be} + e^- \rightarrow {}^{7}_{3}\text{Li} + \nu_e \qquad (t_{1/2} = 4.6 \times 10^{6}\ \text{s},\ Q = 0.86\ \text{MeV});$$

(d) muon capture in muonic atoms

$$^{12}_{6}\text{C} + \mu^- \rightarrow {}^{12}_{5}\text{B} + \nu_\mu \qquad (t_{1/2} = 1.4 \times 10^{-6}\ \text{s},\ Q = 92\ \text{MeV}).$$

Describe an experiment which shows that parity is not conserved in weak interactions, explaining carefully why it shows that parity is not conserved.

The decay

$$^{16}_{8}\text{O}^*(j^P = 2^-) \rightarrow {}^{12}\text{C}(j^P = 0^+) + \alpha$$

has a partial width of order 10^{-10} eV. What can be inferred about the strong force?

(Adapted from the 1985 examination of the Final Honours School of Natural Science, Physics, University of Oxford.)

12.10 Neutrinos and the looking-glass

As the effects of parity non-conservation were explored, another physical effect became apparent. To investigate this let us suppose we have a mirror with an object which is a nucleus decaying to give an electron and an electron antineutrino, $\bar{\nu}_e$. We can imagine the antineutrino's momentum vector $\boldsymbol{P}_\nu$ and another vector $\boldsymbol{S}_\nu$ representing the spin of the $\bar{\nu}_e$ which we will align against $\boldsymbol{P}_\nu$ so that $\boldsymbol{S}_\nu.\boldsymbol{P}_\nu$ is negative. Now look at the mirror image: it is easy to see $\boldsymbol{P}_\nu$ but not so easy to see what has happened to $\boldsymbol{S}_\nu$, Fig. 12.15. If you realize $\boldsymbol{S}_\nu$ is really an axial vector and is best represented by a circulation around the spin axis (remember the spinning neutron in Fig. 12.14), then you will see that in the mirror image, $\boldsymbol{S}_\nu$ now points along $\boldsymbol{P}_\nu$ so that $\boldsymbol{S}_\nu.\boldsymbol{P}_\nu$ is now positive. The neutrino polarization with respect to its direction of motion has changed going from object to mirror image. If parity were conserved in the nuclear decay in which this neutrino is emitted, both these spin orientations would be equally likely and the average polarization of the emitted antineutrinos would be zero.

To proceed, we need a more concise vocabulary. The expectation value of $\boldsymbol{S}.\boldsymbol{P}/|\boldsymbol{S}|\,|\boldsymbol{P}|$ is called the **helicity**. If $\boldsymbol{S}.\boldsymbol{P}$ always has the greatest value possible for all the relevant fermions measured, then the helicity is $+1$: for the case of spin-$\frac{1}{2}$ fermions, this means that all measurements of the component of $\boldsymbol{S}$ along $\boldsymbol{P}$ would yield $+\frac{1}{2}\hbar$. Similarly if $\boldsymbol{S}.\boldsymbol{P}$ always has its most negative value the helicity is -1 and all measurements of $\boldsymbol{S}$ along $\boldsymbol{P}$ would yield $-\frac{1}{2}\hbar$. If the helicity is between $+1$ and -1, a single measurement of the component of $\boldsymbol{S}$ along $\boldsymbol{P}$ must yield either $+\frac{1}{2}\hbar$ or $-\frac{1}{2}\hbar$ but the average would be between these two values.

Comment *Usage of helicity*

Helicity is an expectation value and therefore is the average of $\boldsymbol{S}.\boldsymbol{P}/|\boldsymbol{S}|\,|\boldsymbol{P}|$ over many examples in a particular situation. For an individual spin-$\frac{1}{2}$ fermion $\boldsymbol{S}.\boldsymbol{P}/|\boldsymbol{P}|$ can only be $+\frac{1}{2}\hbar$ or $-\frac{1}{2}\hbar$. If the helicity is $+1(-1)$, every measurement will yield $+\frac{1}{2}\hbar(-\frac{1}{2}\hbar)$. This is probably the case for antineutrinos (neutrinos). Then individual neutrinos are often said to have helicity $+1(-1)$.

A very elegant experiment performed in 1958 by Goldhaber, Grodzins, and Sunyar (see Segrè, 1977) showed that the neutrinos emitted in electron capture have helicity -1. This is impossible unless parity is not conserved. By CP invariance it also means that antineutrinos are emitted with helicity $+1$. The evidence is that all this is true not only in β-decay and electron capture but wherever weak interactions produce ν and $\bar{\nu}$, and for all generations. The presently accepted theory of weak interactions, including the assumed roles of

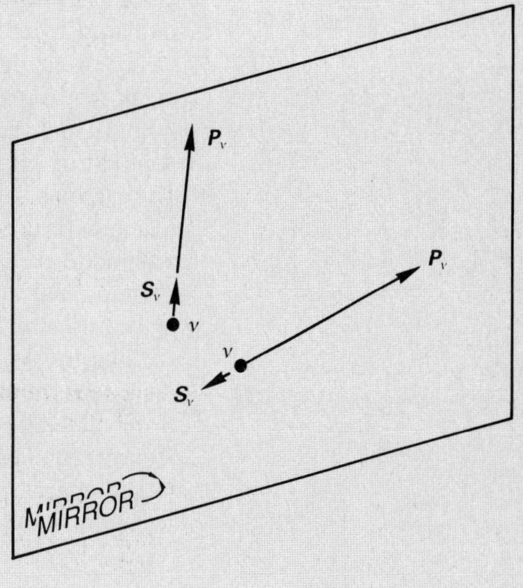

Fig. 12.15 Neutrinos and a mirror. In the foreground there is a neutrino with momentum $\boldsymbol{P}_\nu$ and helicity -1. Its image in the mirror will appear to be a neutrino with helicity $+1$. Thus the polarization of the neutrino with respect to the direction of motion changes from object to the mirror image. This result depends on how the spin of the object (neutrino) appears in the mirror. To make the correct change use the idea of the disc rotating with the spin, as in Fig. 12.14.

$W^{\pm}$ and Z^0, incorporates this result completely so that, if neutrinos are massless, they are always produced with helicities $+1$ and -1 for $\bar{\nu}$ and ν respectively. If they are not massless, then the average helicity will fall short of these extreme values. In fact the situation becomes that pertaining in the weak production of a fermion or antifermion: if the fermion velocity is v the helicity is $\pm v/c$; $-$ for fermions and $+$ for antifermions. Thus fermion (and antifermion) non-zero helicity is the normal occurrence in weak interactions: for example the positrons produced in the decay

$$\mu^+ \rightarrow e^+ + \nu_e + \bar{\nu}_\mu$$

have helicity very close to $+1$. If the fermion produced is a quark, then it must be found in a hadron and that complication changes a spin-$\frac{1}{2}$ baryon helicity from the expected $-v/c$. For example, the proton produced in the weak decay $\Lambda \rightarrow \pi^- + p$ has negative helicity, but the magnitude is not v/c. This negative helicity is certainly a consequence of the negative helicity of the u-quark produced in the weak s-quark to u-quark transition that must occur in this decay. In another approach, an analysis of the final state indicates that this helicity can only come about because the state of relative motion of pion and proton is a superpositron of two states of opposite parity. The Λ has positive parity and this confirms that explicit parity non-conservation is connected with the appearance of non-zero helicity.

The above description implies that the helicity properties are those that occur on production by decay. However, they also apply in collisions involving the weak interaction. Consider the example of charged current neutrino–quark scattering at very high energies (all velocities $\approx c$):

$$\bar{\nu}_\mu + u \rightarrow \mu^+ + d.$$

The μ^+ is produced with helicity $+1$, and d-quark with helicity -1. In the initial state, the antineutrino will only interact if it has helicity $+1$ and the u-quark if it has helicity -1; if these conditions are not met, the cross-section is zero (as long as the velocity of the u-quark can be taken to be that of light). These statements about helicity conditions in weak reactions change appropriately if fermions replaced by antifermions, and so on.

A left-handedly polarized photon has its spin angular momentum pointing against its direction of motion so it has helicity -1. Obviously a right-handed photon has helicity $+1$. However, circularly polarized photons cannot be emitted by unpolarized sources because parity is conserved in electromagnetic interactions. Have a look at Problem 12.13 to make sure you can do the mirror game with these situations. The idea of left and right is taken over into weak interactions: only helicity -1 fermions interact and so the jargon is that the fermion current that interacts weakly is left-handed. For antifermions the current that interacts weakly is right-handed (helicity $+1$). In fact, this must not be interpreted literally: the weak interaction can emit or absorb the 'wrong' helicity fermion but the transition rate is reduced by a factor $(1-v/c)/(1+v/c)$ from the value for the 'correct' helicity, where v is the velocity of the particle concerned. These remarks do not apply to non-zero spin hadrons because the other quarks present can dilute and change the helicity effects of the quark which is interacting weakly.

PROBLEMS

12.11 Show that, if right-handed (helicity +1) neutrinos are emitted in the β^--decay of the neutron, it is left-handed neutrinos that are emitted in β^+-decay of the anti-neutron.

12.12 Show that, if an unpolarized source emits electrons which are spin-polarized with helicity −1, then parity is not conserved in the decay.

12.13 The electromagnetic interaction conserves parity. Show that this means that an unpolarized source of γ-rays cannot emit circularly polarized photons.
An electric discharge is established in low-pressure helium gas in a magnetic field. Red light emitted along the direction of the magnetic field is found to consist of two lines with a small wavelength separation dependent on the strength of the magnetic field. These two lines are oppositely circularly polarized. Does this mean parity is not conserved in the transitions leading to the emission of these photons? Explain your answer.

12.11 Neutrino scattering

In Fig. 12.3 we outlined the first high-energy neutrino scattering experiment. The cross-sections for neutrino interactions with nucleons are very small but do increase with neutrino energy. Therefore it was the development and construction of proton accelerators at increasing energies and capable of producing high-energy neutrino beams that made that experiment possible and opened the way to the experimental study of neutrino charged and neutral current interactions with nucleons.

Let us look first at the cross-sections expected. In Table 12.9 we present a very simplified calculation of the cross-section for the reaction:

$$\bar{\nu}_e + p \rightarrow e^+ + n$$

at low energies $E_\nu \ll M_p c^2$. To do this we make some assumptions:

(1) that neutron and proton are very heavy and there is no neutron recoil,
(2) that we can neglect spins and angular momentum,
(3) that the four particles interact at a point, as in Fermi's simple theory,
(4) that we can neglect the effects of parity non-conservation.

The result for the total cross-section is

$$\sigma = \frac{4G_F^2 P_e E_e c}{\pi(\hbar c)^4}, \qquad (12.26)$$

where P_e and E_e are the momentum and total energy of the positron that is produced. The cross-section for antineutrinos of 2.3 MeV is $5.8 \times 10^{-18}\,\text{fm}^2$. This is tiny compared with strong interaction total cross-sections 3–5 fm^2. The smallness emphasizes the relative weakness of the weak interactions at these energies.

In Table 12.9, this calculation is extended to the other extreme where $E \gg M_p c^2$. We have to drop assumption (1) above and work in the centre-of-mass of the collision. Assumption (3) above implies that we are assuming that

Table 12.9 An estimate of the cross-section for the neutrino reaction

$$\bar{\nu}_e + \mathrm{p} \rightarrow \mathrm{e}^+ + \mathrm{n} + Q \; (= -1.804 \text{ MeV} = \mathrm{p} - (\mathrm{n} + \mathrm{e}) \text{ mass difference}).$$

We first do this at low energies and assume that the proton and neutron remain at rest. Let P_e be the electron momentum and E_ν the energy of the incident neutrino. Then from Table 12.7,

$$N = \frac{4\pi P_e^3 V}{3(2\pi\hbar)^3}.$$

Therefore

$$\frac{dN}{dE} = \frac{dN}{dE_\nu} = \frac{4\pi P_e^2 V}{(2\pi\hbar)^3} \frac{dP_e}{dE_\nu}.$$

The matrix element M for this transition is

$$M = \iiint_V \psi_n^* \psi_e^* G_F \psi_p \psi_\nu \, dV,$$

where ψ_n, ψ_e, ψ_p and ψ_ν are the wave functions of the particles involved. If these wave functions are plane waves and normalized one of each particle in a volume V, then the result of the integration over the volume V is

$$M = \frac{G_F}{V}.$$

Now the total energy E_e of the positron is $E_\nu - M_n c^2 + M_p c^2$, so that

$$P_e^2 c^2 = (E_\nu - M_n c^2 + M_p c^2)^2 - M_e^2 c^4.$$

Therefore

$$2P_e c^2 \frac{dP_e}{dE_\nu} = 2(E_\nu - M_p c^2) = 2E_e,$$

and hence the transition rate ω is given by

$$\omega = \frac{2\pi}{\hbar} \frac{G_F^2}{V} \frac{4\pi P_e E_e}{(2\pi\hbar)^3 c^2}.$$

Looking back at Table 12.7 we find that the prescription for finding the cross-section is to multiply ω by the normalization volume and divide by the sum of the colliding particles' velocities in the centre of mass: since we take the proton to be very heavy and at rest that sum is just c, the neutrino velocity. This gives

$$\sigma = \frac{G_F^2 (P_e c) E_e}{\pi(\hbar c)^4}$$

We have not used relativistic wavefunctions or the proper $V-A$ form of the interaction so that this oversimplified derivation misses a factor of 4 in the formula for the cross-section. Including this factor we calculate the numerical result for $E_\nu = 2.3$ MeV; this gives $E_e = 1.0$ MeV for which $P_e c = \sqrt{(E_e^2 - m_e^2 c^4)} = 0.8596$ MeV. We use $G_F = 8.962 \times 10^{-5}$ MeV fm^3 and $\hbar c = 197$ MeV fm. The result is

$$\sigma = 5.8 \times 10^{-18} \text{ fm}^2 = 5.8 \times 10^{-48} \text{ m}^2 \qquad \text{at } E_\nu = 2.3 \text{ MeV}.$$

This calculation can be extended to very high energies, in this case $E_\nu \gg M_p c^2$, where E_ν remains the neutrino energy in the laboratory. We have to assume that the proton and neutron are point-like in order to be able to think of the reaction mechanism as the interaction of four fermions at a point. In the centre of mass, the positron and neutron have equal and opposite momentum. Since the masses can now be neglected, the total centre of mass energy W is also equally shared. Therefore, for the positron in the centre of mass

$$P_e c = E_e = W/2 \quad \text{and} \quad \frac{dP_e}{dW} = \frac{1}{2c}.$$

Now

$$N = \frac{4\pi P_e^3 V}{3(2\pi\hbar)^3}.$$

Therefore

$$\frac{dN}{dE} = \frac{dN}{dW} = \frac{4\pi W^2 V}{8(2\pi\hbar c)^3}.$$

At such high energies $W^2 = 2E_\nu M_p c^2$ and we find for the transition rate ω:

$$\omega = \frac{2\pi}{\hbar} \frac{G_F^2}{V} \frac{\pi E_\nu M_p c^2}{(2\pi\hbar c)^3}.$$

To obtain the cross-section we must multiply ω by the normalization volume V and divide by the sum of the velocities of neutrino and proton in the centre of mass; that sum is $2c$. Then

$$\sigma = \frac{G_F^2 E_\nu M_p c^2}{8\pi(\hbar c)^4}.$$

This result again misses a factor of four and is not correct because the nucleon is not point-like and because the neutrino interacts with a quark, not with the whole nucleon at these energies. This is discussed in Section 12.11. We cannot calculate the neutrino–quark cross-section because we do not know what quark mass to put in place of the proton mass. And there are no free quarks! However, it is possible to derive formulae for the inelastic scattering cross-section for neutrinos by the assembly of quarks that is a nucleon in terms of the quark momentum distributions inside the nucleon: this subject is beyond the scope of this book and interested readers are referred to more advanced texts such as that by Halzen and Martin (1984).

the proton and neutron are point-like particles: this is incorrect and we shall return to this point shortly to discover what does happen. The result is:

$$\sigma = \frac{G_F^2 E_\nu M_p c^2}{2\pi(\hbar c)^4}. \tag{12.27}$$

At $E_\nu = 10$ GeV we have $\sigma = 8 \times 10^{-12}$ fm^2, a million-fold increase over the cross-section at 2.3 MeV. This is the increase in cross-section mentioned earlier. Look now at Problems 12.15 and 12.16 to see what cross-sections of these magnitudes mean.

Equation (12.27) reveals one of the problems of Fermi's simple theory. This cross-section increases indefinitely as the antineutrino energy increases. In fact, there is a theorem in quantum-mechanical scattering theory which states that the cross-section at a given energy in these circumstances cannot exceed

$4\pi\lambda\!\!\!^{-2}$ where $\lambda\!\!\!^{-}$ is the de Broglie wavelength, divided by 2π, of the incident antineutrino in the centre of mass. At very high energies this is, for a proton target, given by

$$\lambda\!\!\!^{-} = \frac{2\hbar}{\sqrt{(2E_\nu M_\mathrm{p})}},$$

and the not-to-be-exceeded cross-section is

$$\sigma = \frac{8\pi\hbar^2}{E_\nu M_\mathrm{P}}.$$

The increasing cross-section given by equation (12.27) meets this limit at a laboratory antineutrino energy of $E_\nu \simeq 1 \times 10^6$ GeV. Although this energy may be forever inaccessible, its existence highlighted one of the difficulties of Fermi's simple theory and led to early proposals about the existence of a vector boson as a carrier of the weak interaction (see also Section 9.13).

So what happens? The total cross-section for the reaction we are considering does increase but reaches a plateau at about $E_\nu = 1$ GeV because the momentum transfers involved are, except at scattering angles close to zero, too great for the soft structure of the proton to absorb with significant probability. Inelastic scattering, in the sense that the nucleon does not survive as the only hadron in the final state, becomes dominant in neutrino-nucleon interactions. The basic process is the interaction of the neutrino with one quark or one antiquark in the nucleon. For example:

$$\bar{\nu}_\mathrm{e} + \mathrm{u} \rightarrow \mathrm{e}^+ + \mathrm{d},$$

$$\nu_\mu + \mathrm{d} \rightarrow \mu^- + \mathrm{u}.$$

This is followed by fragmentation of this final quark and the spectator quarks from the proton into several observable final state hadrons. The quark appears to behave like a point-like particle and, in the simple Fermi theory, the cross-sections for neutrino–quark scattering therefore increase with energy as did the antineutrino–proton cross-section at low momentum transfers. Thus the cross-section for this inelastic scattering also increases with energy as long as the momentum transfers do not approach the value of $M_\mathrm{W}c^2$; when that happens the cross-sections become well-behaved in the sense of not violating the cross-section limit that we discussed above. Experiments have not yet been able to reach that kinematic regime. However, particularly important is the study of deep inelastic scattering of neutrinos by nucleons. (See Section 10.10 where we discussed the deep inelastic scattering of charged leptons.) Such neutrino scattering has been studied with both muon neutrinos and muon antineutrinos. The former produce μ^--, the latter μ^+-mesons: these are easy to detect and a measurement of their energy is straightforward. If the neutrino energy is known (usually with large error) from the cirumstances of its production, and the total energy of the final state hadrons can be measured, it is possible to obtain a useful amount of information about the kinematics of each event.

What is learnt from measurements of neutrino deep inelastic scattering? A Feynman diagram for the inelastic neutrino scattering process is shown in

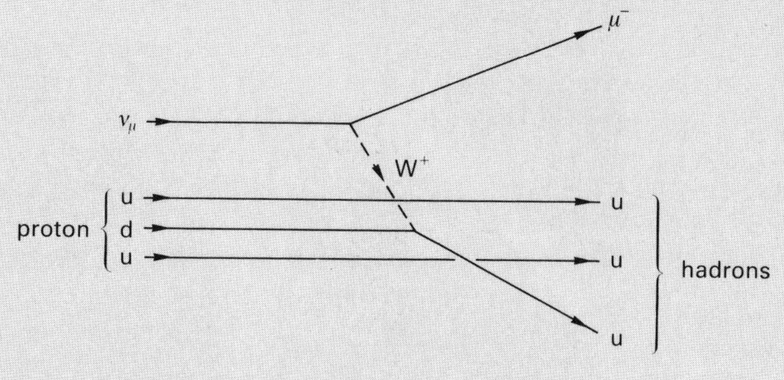

Fig. 12.16 One of the diagrams for the charged current, deep inelastic scattering of a muon neutrino by a proton:

$$\nu_\mu + p \to \mu^- + (\text{hadrons})^{++}.$$

This is a charged current interaction so that the reaction involves W exchange and there is a quark flavour change. As usual after such collisions the excited multiquark system produces a number of hadrons. Note that only the valence uud quarks of the proton are shown. However, scattering can also occur from one of the quarks or antiquarks of the sea (see Section 10.10).

The event of Fig. 12.17 is of this kind. In Fig. 12.18 the production of the charmed quark has involved the neutrino–quark reaction:

$$\nu_\mu + d \to \mu^- + c.$$

Neutral current scattering is also possible, in which case there is no quark flavour change and the scattered lepton is the neutrino.

Fig. 12.16 and the reader is invited to compare this with Fig. 10.18, which is the analogous diagram for charged lepton scattering. The information available from the latter is described briefly in Section 10.10, namely structure functions related to the momentum distribution of the charged partons (quarks) in the target. Neutrino scattering transfers a W^+ to the nucleon, which picks out a negatively charged quark or antiquark and thereby provides a measure of their momentum distribution. Conversely antineutrino scattering transfers a W^- which picks out a positively charged quark or antiquark and measures their distribution. Thus charged lepton and neutrino scattering can provide complementary information on the distribution of quarks and antiquarks in the nucleons. The relative weights of these distributions in the two cases confirm the assignment of charges made to the quarks. However, although these distributions are interesting in their own right, there appears to be at present no way of reliably calculating these distributions from QCD. In Section 10.10 we mentioned that the structure functions are almost scale invariant and the deviations are as expected in QCD: this statement includes those measured in neutrino scattering.

One type of detector used to investigate deep inelastic neutrino scattering was the large bubble chambers (see Section 10.2) filled with liquid hydrogen or deuterium, or a suitable heavier liquid. It thereby became possible to observe directly such events. Figures 12.17 and 12.18 show photographs of two such ν_μ

Fig. 12.17 A photograph of an event in which a ν_μ has undergone a charged current interaction with a nucleon of the liquid neon–hydrogen mixture of a large bubble chamber which was operated in a high-energy neutrino beam at the European Centre for Nuclear Research, Geneva, for some years. A magnetic field perpendicular to the object plane bends charged particle tracks into segments of circles or into helices. Each event has several charged particle secondaries including a μ^- which is indicated. In this event several charged secondary hadrons have been produced along with one or more π^0-mesons: the decay γ-rays from the latter have initiated electromagnetic showers (Section 11.4) and many electron–positron pairs may be distinguished.

The size of the chamber and the lengths of the tracks may be gauged by knowing that the inner circle, which is a part of chamber construction, has a radius of 1.1 m. The magnetic induction is 3.5 T. The camera view has a large depth of view along the cylindrical axis of the chamber; this property was obtained at the cost of some distortion, and some tracks which may be straight can appear curved.

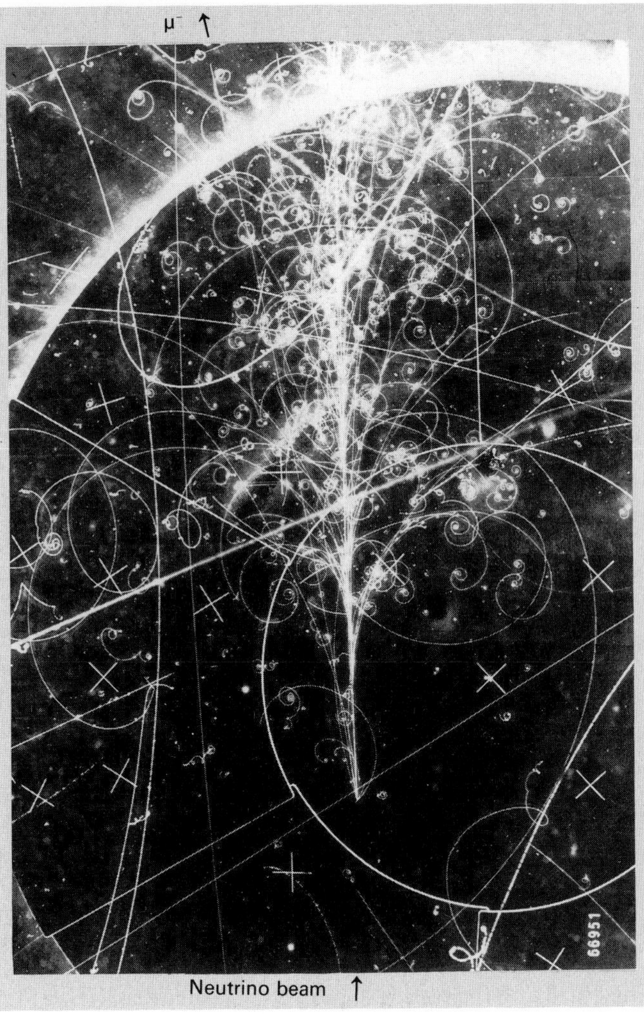

scattering events, each occurring on a bound nucleon or free proton in a bubble chamber filled with liquid neon–hydrogen mixture. One of these involves the production of a charmed meson and since there is a μ^- in the final state it must have involved the neutrino–quark scattering

$$\nu_\mu + \mathrm{d} \rightarrow \mu^- + \mathrm{c}.$$

This section has covered only the charged current interactions of neutrinos with nucleons. Neutral current neutrino interactions also occur but experimentally are more difficult to observe in a useful way; this is because the incident neutrino scatters but remains a neutrino and cannot be detected as a final state particle. Thus the kinematic information available from an event may be limited to a measurement of the energy transferred to the target.

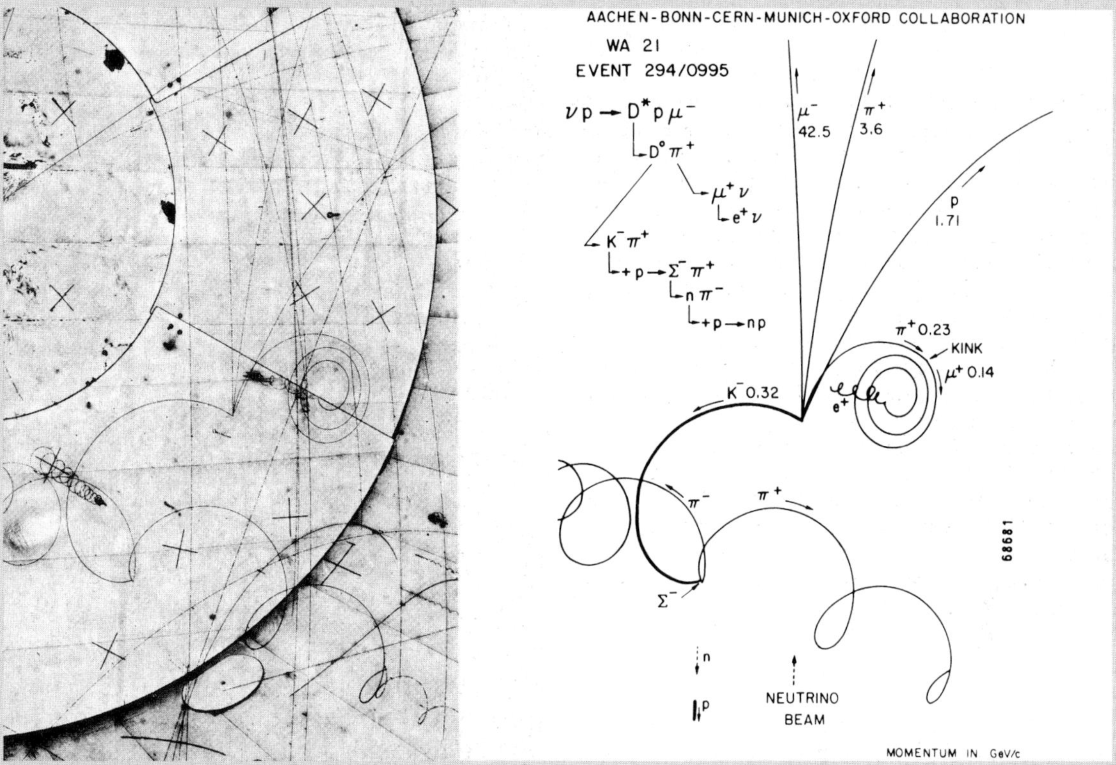

Fig. 12.18 A photograph taken in the same chamber as that of Fig. 12.17 but when the filling was liquid hydrogen. This event is remarkable: the event has been totally reconstructed and it is one in which a charmed meson has been produced and identified, a rare thing! A charged current muon neutrino interaction with a proton has occurred. The charmed meson is a $D^{*+}(c\bar{d}, j^P = 1^-$, 2010 MeV/$c^2$). Its decay sequence is given and the final visible, long-lived products are indicated in the figure. The D^* and daughter D^0 have mean lives of $< 3 \times 10^{-20}$ s and about 4×10^{-13} s respectively, so their paths are too short to be visible.

PROBLEMS

12.14 Beams of high-energy muon neutrinos can be obtained by generating intense beams of π^+-mesons and allowing them to decay while in flight. What fraction of the π^+-mesons in a beam of momentum 200 GeV/c will decay while travelling a distance of 300 m?

At the end of the decay path (an evacuated tunnel) the beam is a mixture of π^+-mesons, muons and neutrinos. What distinguishes these particles in their interactions with matter, and how is a neutrino beam free of contamination by π-mesons and muons obtained?

[π^+-meson mean-life: 2.6×10^{-8} s.]

(Adapted from the 1984 examination of the Final Honours School in Natural Science, Physics, University of Oxford.)

12.15 Antineutrinos of 2.3 MeV from the fission product decay in a reactor have a total cross-section with protons ($\bar{\nu}_e + p \rightarrow e^+ + n$) of 6×10^{-48} m^2. Calculate the mean free path of these neutrinos in water. Assume the antineutrinos are able to interact only with the free protons.

Estimate as well as you can the mean free path of an electron neutrino of 1.0 MeV in the material of a neutron star assuming nuclear density and the absence of protons and electrons.

$$\sigma(\nu_e + n \rightarrow p + e^-) = 10^{-47}\ \text{m}^2 \text{ at } 1.0\ \text{MeV}.$$

12.16 The muon neutrino–nucleon inelastic scattering cross-section is proportional to the neutrino energy. It is 2×10^{-43} m^2 at 1 GeV. What is the mean free path of 100 GeV ν_μ in steel assuming all the nucleons are potential targets? (Steel: density$=7.9\times10^3$ kg m^{-3}). A bubble chamber consisting of a right cylinder, cross-sectional area 1 m^2 contains 1000 kg of liquid propane. It is exposed to a beam of 10 GeV muon neutrinos which is incident normally on a circular face of the cylinder. The neutrinos arrive in bursts from the accelerator used to produce them (how?). The bubble chamber is expanded in synchronization with the burst and any neutrino interaction in the liquid that produces charged secondaries can be observed. How many neutrinos per burst are required to give an average of one interaction per 100 expansions?

12.17 Show that the event of Fig. 12.17 in which a charmed meson was produced is due to the neutrino–quark reaction

$$\nu_\mu + \mathrm{d} \rightarrow \mu^- + c,$$

and is unlikely to be due to the reaction

$$\nu_\mu + \mathrm{s} \rightarrow \mu^- + c,$$

where the s is from the sea of $\mathrm{q\bar{q}}$ pairs that a nucleon contains.

12.12 The neutrino mass

The measurements of neutrino mass can only yield upper limits and experiments are of such great difficulty that verification is necessary. These upper limits are given in Table 12.1. Let us look at the measurement of the $\bar{\nu}_e$ mass; consider the decay of the third isotope of hydrogen, namely tritium:

$${}^3_1\mathrm{H} \rightarrow {}^3_2\mathrm{He} + \mathrm{e}^- + \bar{\nu}_e + 18.6\,\mathrm{keV}.$$

Apart from a correction for nuclear recoil the maximum value of the electron kinetic energy is Q, unless the neutrino mass is non-zero, in which case the maximum value is less and the expected shape of the electron spectrum is changed near the new maximum as shown in Fig. 12.19. Prior to 1983 the upper limit on m_ν was about 35 eV: thus the change in spectrum shape, if it exists, affects less than the top 0.2% of the full range of electron kinetic energy from tritium decay, and an even smaller fraction of the total intensity of electrons emitted. These facts point to a very difficult experiment, and, in addition, one in which the interpretation is not as simple as the above discussion may suggest. Normally the tritium source has to be in the form of a tritium-bearing compound coated on to a suitable foil: the binding energy of the tritium atom and the state of the recoiling daughter ${}^3_2$He nucleus, singly charged ion or nucleus, can both alter the spectrum in the range of the last few tens of electron volts—just where the electron spectrum must be measured. Recent (1987) Russian experiments yielded values of $m_\nu = 14$ eV with large errors. Other experiments are under way.

The upper limits on the mass of muon and tau neutrinos are considerable: a brief note on how they were determined is included in Table 12.1.

On 23 February 1987, a supernova (SN1987A) was observed in the Large Magellanic Cloud at a distance known to be about 160 000 light years. All models of such supernova have the core of a star rich in iron collapsing gravitationally into a neutron star (or a black hole). In the former case about 10^{57} protons become neutrons emitting that number of neutrinos:

Fig. 12.19 This figure shows the change that occurs at the extremity of the β-decay electron kinetic energy spectrum from the case of $m_\nu=0$ to the case of $m_\nu \neq 0$. Since the electron neutrino mass is only a few electron-volts at the most, if it is not zero, then the actual part of the spectrum affected is a very small fraction of the whole. In addition this sketch indicates the spectrum and the change assuming in each case that the daughter atom is free and in its ground state. This is not necessarily the case. Thus examining the end of the spectrum is experimentally very difficult and plagued by uncertainties of interpretation.

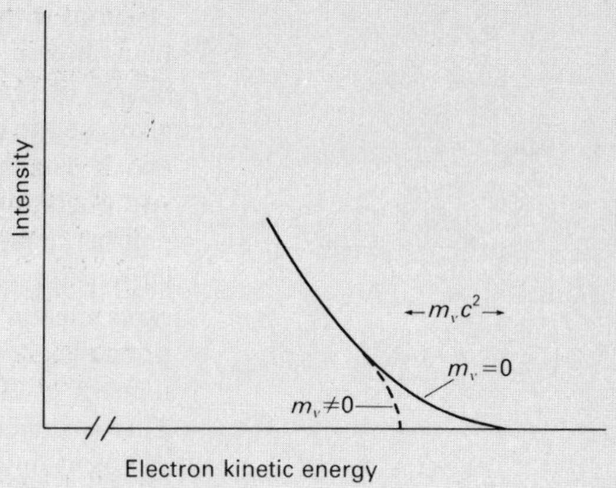

$$p + e^- \rightarrow n + \nu_e.$$

Neutrinos and antineutrinos are also produced at high-temperature ($>10^{10}$ K for which $kT \simeq 1$ MeV) by processes in which virtual Z^0-bosons are emitted and become $\nu_l \bar{\nu}_l$. All flavours, $l=$ e, μ, or τ, will be produced if the neutrino masses are small enough. These processes increase the number of neutrinos plus antineutrinos to about 10^{58}. It is believed that these processes are completed in a few seconds. The proximity of SN1987A meant that about 5×10^{13} neutrinos per square metre reached the Earth and in a period of about $12\frac{1}{2}$ s at 07.35 GMT on 23 February 1987, a total of 20 were detected in two operating experiments designed to search for proton decay. In fact 14 of this 20 were detected in two seconds. Both experiments contained a water detector, fiducial volumes of 2000 and 5000 tons. What reactions do the neutrinos cause? Predominantly

$$\bar{\nu}_e + p \rightarrow e^+ + n,$$

but elastic scattering also occurs

$$\nu_l(\bar{\nu}_l) + e^- \rightarrow \nu_l(\bar{\nu}_l) + e^-, \qquad l = e, \mu, \tau.$$

In both experiments recoiling e^+ or e^- moving through the water emitted Cerenkov light (see Fig. 13.4) which was detected by a large number of photomultipliers viewing the liquid. This permitted a crude measure of the energy of the recoiling electrons. (See Section 13.7 for a fuller description of these detectors.)

If the neutrinos have mass and are all emitted at one instant, the dispersion in their arrival at the Earth gives the mass if the energy is known. Unfortunately the production is not instantaneous and any assumptions about production are model dependent. There are also uncertainties about the time of arrival at one detector relative to the other. A reasonable calculation shows that the upper limit on the mass of neutrinos detected is about 6–10 eV. Have a go at Problem 12.19 to see the method of the calculation. (The accepted model of stellar core collapse is outlined in Section 14.9.)

There is another phenomenon which could reveal the existence of non-zero neutrino mass. If neutrinos do have mass and the mass eigenstates are not eigenstates of generation number, then mixing can occur. This means that a beam of neutrinos that starts a journey as ν_e, for example, will have a state vector which is a superposition of the three mass eigenstates, each of which develops with proper time as $\exp(im_k c^2 t/\hbar)$, $k=1,2,3$. (Of course, it will be more than three if there are more than three lepton generations.) Thus the relative phases of the mass eigenstates change with time, the superposition no longer gives a state of pure generation number and the ν_e becomes a mixture of ν_e, ν_μ, and ν_τ, changing with distance from the source; these changes are oscillatory. Although this mixing would represent a violation of the conservation of lepton number, it would have to be a relatively small effect and most neutrino-charged current interactions would conserve lepton number since we know that is the experimental situation on which the conservation law is based. Thus, in principle, it would be possible to determine the mixture by observing the number of electrons, muons or τ-mesons produced by that beam in a target. This is difficult because yields are low and unless the neutrino energies are sufficient only the electron neutrinos will interact to produce charged leptons. Several searches have been made for this mixing. For example, electron antineutrinos from a nuclear reactor can be detected (Section 12.2) by the production of positrons in interactions with protons: if the production rate varies with distance from the reactor in a manner which is different from that expected from the inverse square law, then mixing could be the cause. A similar experiment is possible with a high-energy beam of accelerator-produced muon neutrinos. All such searches have given negative results although not all regions of possible mass difference have been explored.

The possibility of neutrino oscillations is relevant to the solar neutrino problem which we discuss in Section 14.7.

PROBLEMS

12.18 Suppose that the neutrinos emitted by the supernova SN1987A were all generated and left the core in a period of one second. A large fraction of the twenty neutrinos detected 160 000 years later were all observed in a period of 2 s, with a mean energy of about 8 MeV. Make a rough estimate of an upper limit on the neutrino mass.

12.13 Another neutrino problem

In Section 12.2 we discussed the evidence that neutrino and antineutrino are different particles. If $\nu_e \neq \bar{\nu}_e$ they are labelled **Dirac neutrinos.** We shall now look at that evidence again and will find that it does not exclude $\nu_e = \bar{\nu}_e$, for which the label is **Majorana neutrino.**

In the situation that we have assumed in the earlier sections of this chapter, the antineutrino emitted in neutron decay is right handed (helicity $+1$). However, the neutrino which can cause

$$\nu_e + {}^{37}_{17}\text{Cl} \rightarrow {}^{37}_{18}\text{Ar} + e^-$$

is left-handed (helicity -1). Thus the absence of the conversion ${}^{37}_{17}\text{Cl} \rightarrow {}^{37}_{18}\text{Ar}$ caused by reactor neutrinos could be due to either

(1) that $\nu_e \neq \bar{\nu}_e$

or

(2) that the helicity does not match.

In the latter case, a Majorana neutrino is a possibility.

DEFINITIONS AND KEYWORDS

Dirac neutrino Neutrino which is not identical to its antineutrino.

Majorana neutrino Neutrino which is identical to its antineutrino.

There exist further possible developments in this story. In the next chapter we shall briefly look at what are unified theories of the particles: these are somewhat speculative at the present time but several predict that the neutrinos are Majorana particles with mass. The mass, as in the case of electrons and positrons, allows the absorption or emission of the wrong helicity. It follows that neutrinoless double β-decay would not be forbidden and could occur at what might be a very low rate. Experimental efforts are now being made to detect this process or at the least place very stringent upper limits on its transition rate for each of the several examples where it is energetically possible. If it is detected then we have Majorana neutrinos with mass, or . . .?

12.14 Conclusion

This chapter has covered some of the elementary aspects of the weak interaction. In doing this we have assumed that the roles of the $W^{\pm}$ and Z^0 are fully established and accepted: this is true only to the extent that the model (the electroweak theory) which describes their behaviour appears to explain correctly a large amount of low-energy data. However, although these particles have been produced and identified, an extensive program of experimental investigation of the properties of these gauge bosons as real, free particles has only just begun (1989). It is important to investigate whether their properties are as predicted: if they are, then some questions will remain about features of the electroweak theory, in particular on the origin of the mass of these bosons. If the properties are not as expected then the differences may provide us an opening in the search for answers to these questions.

References

Carroll, L. (1872). *Alice Through the Looking Glass* (1st edn). Macmillan, London.

Danby, G., Gaillard, J.-M., Goulianos, K., Lederman, L. M., Mistry, N. and Steinberger, J. (1962). *Physical Review Letters*, **9,** 36–44.

Davis, R. (1955). *Physical Review*, **97,** 766–9.

Davis, R. and Harmer, D. S. (1959). *Bulletin of the American Physical Society*, **4,** 217.

Fermi, E. (1950). *Nuclear Physics*. The University of Chicago Press.

Goldhaber, M., Grodzins, L., and Sunyar, A. W. (1958). *Physical Review*, **109,** 1015–7.

Halzen, F. and Martin, A. D. (1984). *Quarks and Leptons*. Wiley, New York.

Perkins, D. H. (1986). *Introduction to High Energy Physics* (3rd edn). Addison-Wesley.

Reines, F. and Cowan, C. L. (1959). *Physical Review*, **113,** 273–9.

Sargent, B. W. (1933). *Proceedings of the Royal Society*, London, **139,** 659–72.

Segrè, E. (1977). *Nuclei and Particles* (2nd edn). W. A. Benjamin, Reading, Massachusetts.

13

Particles: Summary and Outlook

13.1 The conservation laws

Throughout all of our previous chapters we have assumed that momentum, total energy and angular momentum are all conserved quantities. We have also assumed that electric charge is conserved—but, for example, we found that parity was not conserved in weak interactions. We must therefore draw up a table of quantities and label them as conserved or not conserved in various interactions (Table 13.1). The problem is that it is not a very interesting table because it has only five non-conservation entries, with three of those having a status which has not been clearly discussed in earlier chapters. To improve the situation we describe in more detail the meaning of a conservation law, where that meaning is understood.

The important connection is with the idea of a **transformation**. Suppose we have chosen to describe a system in an arbitrary set of coordinates ($\mathbf{r} = x, y, z$, for example). Then suppose we chose to move this system so that that part of it

Table 13.1 The conservations laws of particle physics: √ = conserved: x = not conserved; — = not relevant.

Quantity	Strong interaction	Electromagnetic interaction	Weak interaction
Momentum	√	√	√
Total energy	√	√	√
Angular momentum	√	√	√
Electric charge	√	√	√
Quark number*	√	√	√
Quark flavour	√	√	x
Lepton generation number*	—	√	√
Parity	√	√	x
Charge conjugation quantum number	√	√	x
Isotopic spin	√	x	x
Baryon number*	√	√	√

* May not be conserved if grand unified theories correctly predict the existence of leptoquarks (Section 13.6).

which starts at x, y, z is moved to $x+a, y, z$. We do not expect the physics of the system to change, that is, our expectation is that the behaviour of the system is **invariant** under transformations along the x-axis. Now there is a theorem in quantum mechanics which proves that if a system is invariant under a given (unitary) transformation then there is a conserved quantity. The proof of this theorem is outlined by Perkins (1986). In the case of invariance under a transformation along the x-axis the result is the conservation of the x-component of momentum. Such an invariance is normally called a **symmetry** of the system. We can now present Table 13.2, which shows transformations and the conserved quantities appropriate to each transformation, given that we assume the physics of the system is invariant under that transformation.

The observant reader will have noticed that the title of Table 13.2 is about **continuous transformations**. This implies that the transformation has the property that any one transformation can be made up of a sum of small transformations. That is clearly true of the entries in Table 13.2. But what is the opposite of a continuous transformation? It is called a **discrete transformation** and obviously it cannot be made up by a sum of small transformations.

An example of a discrete transformation is the parity transformation: change a system so that what was at **r** goes to −**r**. A few moments' thought shows that with whatever steps this transformation is performed, there must be one step which cannot be the sum of small transformations. In Table 13.3 we list some discrete transformations and the quantum numbers that are involved if the physics is invariant under each transformation. The reader will recall that in Section 12.9 we connected conservation of parity with invariance under the change of a system into its mirror image. Mirror imaging is equivalent to a parity transformation plus a rotation so that given conservation of angular momentum, invariance under mirror imaging means parity conservation. And, conversely, non-invariance as illustrated by neutron decay (Fig. 12.14) means parity is not conserved.

Charge conjugation means changing particles into antiparticles and vice versa. In Section 12.9 we indicated that the weak interaction is not invariant under this discrete transformation. Experimentally the evidence is that processes involving the strong and electromagnetic interactions are invariant. The corresponding quantity is called the charge conjugation quantum number: it is beyond the scope of this book to explore that subject but interested readers are invited to take up the challenge of Problem 13.1.

Time-reversal is also in Table 13.3. Suppose that we take a movie picture of a

DEFINITIONS AND KEYWORDS

Transformation A change of coordinate system. An active transformation is one in which the system described is moved with respect to the coordinates.

Invariant A system or a quantity is said to be invariant under a transformation if its behaviour does not alter, or its value change.

Symmetry A system which is invariant under a given transformation is said to possess a symmetry.

Continuous transformation A transformation which can be built from the summed effect of many small transformations.

Discrete transformation A transformation that cannot be built from a sum of many small transformations.

Table 13.2 The continuous transformations and their associated conservation laws.

Transformation	Conserved quantity
Displacement in positron	Momentum
Displacement in time	Energy
Rotation	Angular momentum

Note that the continuous transformations are associated with additive quantum numbers, additive in the sense that the contributions from various parts of a system add to give the total.

Table 13.3 Some discrete transformations and their conserved quantities if the physics is invariant.

Transformation	Description	Conserved quantity
Parity (P)	$\mathbf{r}\to-\mathbf{r}$	Parity*
Charge conjugation (C)	Particle ⇔ antiparticle	Charge conjugation quantum number*
Time reversal (T)	$t\to-t$	None†

* Note that these discrete transformations are associated with quantum numbers which can only be ±1 and are multiplicative.

† No conserved quantities, but gives relations between cross-sections.

process and then run it backwards, we see another physical process. Time-reversal invariance means that the reversed process is as equally probable as the forward process. This invariance allows the cross-section for an atomic or nuclear reaction to be related to the cross-section for the reversed reaction. By testing experimentally such relations it is possible to test the invariance: no direct evidence is found for non-invariance. (Time reversal is a discrete but not a unitary transformation and it does not have an associated conserved quantity.) However, there is indirect evidence for violation of time reversal invariance which is briefly mentioned at the end of Section 13.3.

An interesting case is that of the electric dipole moment of the neutron. The neutron contains three constituent quarks which interact through the gluon field. But they are also electrically charged and W exchange is possible between the u and d quarks. Thus all three interactions are affecting the structure in varying degrees. In Problem 13.2 the reader is invited to show that the electric dipole moment is zero unless there is violation of both parity invariance and time-reversal invariance: the weak interaction provides the first and thus the measurements of the electric dipole moment become a test of time-reversal invariance. There is no convincing evidence that it is not zero.

The conservation of charge and colour are related to a more subtle invariance. In Section 9.2 we discussed briefly the gauge invariance of electromagnetism: this invariance can be related to the conservation of electric charge. As we have mentioned in Section 9.11, such gauge invariances are now of vital importance in particle physics and we shall extend our discussion of their significance in Section 13.4.

13.2 Recognizing what is going on

This section heading means, given some information about a particle reaction, it should be possible to recognize the interaction responsible for setting the transition rate. Of course, in a given circumstance all interactions may be having an effect but we want to be able to point to the one which controls the rate: for example, we know that the baryon decay

$$\Lambda\to p+\pi^-,$$

involves the weak interaction since there is change in quark flavour (uds→udu + d$\bar{u}$). But the structure of the hadrons involved is determined by the strong interactions and that structure influences the decay rate. In addition, the final state proton and π-meson are interacting via the action of both the strong and electromagnetic interactions and this gives a non-trivial structure to the final state wavefunction, again influencing the transition rate. However, it is the weak interaction that is the critical link from the initial to final state, and that determines the magnitude of the mean life.

In fact we have already learnt some rules to recognize the weak interactions (Section 12.1) and the electromagnetic interactions (Section 11.10). We repeat them in Table 13.4 for completeness. We have not before examined the strong interaction with the idea of recognition: the main thing is that it conserves everything (Table 13.1) and the transition rates are large ($\lesssim 10^{20}\,\mathrm{s}^{-1}$) and the cross-sections are also typically large ($\lesssim 10^{-1}$ mb). There are two difficulties with the last two facts. First, the range of cross-sections and transition rates overlaps with those of the electromagnetic interactions. The second is that we should perhaps unravel primary parton–parton interactions involving QCD from the interactions involving hadrons which are composed of partons. The latter give total cross-sections governed by the size of the hadrons ($\sigma \sim 30$–40 mb) or by the existence of compound states. The former are due to the interaction between point-like particles and so have cross-sections that are scale invariant (Section 3.8) and vary with the reciprocal of the square of the centre-of-mass energy W. Thus consider the reactions

$$e^+ + e^- \rightarrow \mu^+ + \mu^- \tag{13.1}$$

$$u + \bar{u} \rightarrow d + \bar{d}. \tag{13.2}$$

Below $W \simeq 50$ GeV the first is dominated by a photon intermediate state and the cross-section is (Fig. 10.10)

$$\sigma_1 = \frac{4\pi\alpha^2(\hbar c)^2}{3W^2} \tag{13.3}$$

The second reaction is dominated by a gluon intermediate state and has a cross-section

$$\sigma_2 = \frac{2\pi\alpha_s^2(\hbar c)^2}{9W^2}, \tag{13.4}$$

where $\alpha_s = g_s^2/4\pi\hbar c$ is the QCD equivalent of the fine-structure constant $\alpha = e^2/4\pi\varepsilon_0\hbar c$ and g_s is the quark–gluon coupling constant. The difference in numerical factors between equations (13.3) and (13.4) is due to coefficients that come from the coupling of the colours of quarks and gluons. The coupling g_s varies slowly with W but at energies of order 10 GeV can be taken to be about 0.2. With $\alpha = (137)^{-1}$ we have $\sigma_2/\sigma_1 \simeq 120$. This factor is therefore a rough indicator of the effective relative strengths of the strong and electromagnetic interactions.

All the pointers which make it possible to recognize the controlling interaction are collected in Table 13.4. Reciprocal to recognizing what is going on is being able to guess, or better to estimate, or best to calculate a cross-section or a transition rate for a reaction or a decay (see Problem 13.3). We have given many examples in previous chapters and in Table 13.5 we briefly summarize the range of expectations for the case of elementary particle reactions.

Table 13.4 Recognizing interactions

The weak interaction If any of the following are satisfied the process is weak:

1. A neutrino or antineutrino is emitted or absorbed.
2. There is a change in quark flavour.
3. Parity conservation is violated or the system is not invariant under the parity transformation.
4. A mean life in seconds multiplied by the fifth power of the maximum momentum in MeV/*c* available to a decay product is of order 1 or greater.
5. A total cross-section in (fermi)^2 divided by the incident energy in MeV is of order 10^{-16} to 10^{-19} fm^2 MeV^{-1}.

The electromagnetic interaction If any of the following are satisfied the process is electromagnetic:

1. A real photon is emitted or absorbed.
2. A virtual photon is exchanged, as in quantum electrodynamic processes. Recognizing such an exchange is involved may not always be straightforward. The interaction of charged leptons is electromagnetic if not weak. If hadrons alone are involved the exchange of a virtual photon may be recognized by a change in total isotopic spin (Section 10.11) or by properties like those of Rutherford or inelastic Coulomb scattering where momentum transfers not near zero are strongly unfavoured.

Caveat: The electroweak theory shows that in certain physical regions the weak interaction and the electromagnetic are comparable. This will occur if the energies involved, or the momentum transfers ($\times c$), are comparable to the rest mass energy of the free $W^\pm$ and Z^0 bosons ($\sim$90 GeV). Thus the reaction $e^+ + e^- \rightarrow \mu^+ + \mu^-$ is dominated by photon exchange (see Fig. 9.7) at energies from threshold to the order of 10 GeV but at higher energies Z^0 exchange becomes important until at near 90 GeV the Z^0 is formed as a compound state and as a result dominates the reaction. At much higher energies the photon and Z^0 contribute comparably.

The strong interactions For collisions between hadrons or for the decay of a hadron to other hadrons the process will be strong if the following rules are satisfied:

1. All quantum numbers are conserved.
2. Transition rates are $> 10^{20}$ s^{-1} or total or partial cross-sections are $> 10^{-2}$ fm^2.

Table 13.5 This table is a partial attempt to show what to expect for a reaction cross-section or for a decay rate given the interaction responsible. As we have seen in many previous chapters, basically identical processes can vary in transition rate over very many orders of magnitude. Therefore, we restrict ourselves here to particle physics processes for which Coulomb and angular momentum barriers are not as important as they can be in nuclear processes.

Interaction	Cross-section (barns)	Decay transition rate (s^{-1})
Weak*	$(10^{-21} \rightarrow 10^{-18}) \times (E\,\text{MeV})$	$< (P\,\text{MeV})^5$
Electromagnetic	$10^{-7} \rightarrow 10^{-3}$	$10^{16} \rightarrow 10^{20}$
Strong	$10^{-5} \rightarrow 10^{-1}$	$10^{20} \rightarrow 10^{24}$

* E = incident particle energy, P = maximum momentum found in decay products. The weak interaction collision cross-section given is correct only for energies well below the rest mass energy of the $W^\pm$.

Reminder $1\ \text{b} = 10^{-24}\ \text{cm}^2 = 10^{-28}\ \text{m}^2 = 100\ \text{fm}^2$.

PROBLEMS

13.1 Consult Perkins (1986) or Gottfried and Weisskopf (1986) to explore the meaning of the charge-conjugation quantum number for neutral systems such as the π^0 meson and the $^3S_1$ and $^1S_0$ states of positronium. Show that the decays

$$\pi^0 \to \gamma + \gamma,$$
$$\mathrm{Ps}(1^1S_0) \to \gamma + \gamma,$$
$$\mathrm{Ps}(1^3S_1) \to \gamma + \gamma + \gamma,$$

conserve charge conjugation quantum number.

13.2 If the neutron has an electric dipole moment $\boldsymbol{p}$ it must be parallel (or opposite) to the spin

$$\boldsymbol{p} = \alpha \boldsymbol{S}$$

where α is a constant. The energy U of this dipole in an electric field $\boldsymbol{E}$ is

$$U = -\boldsymbol{p}.\boldsymbol{E} = -\alpha \boldsymbol{S}.\boldsymbol{E}$$

By considering what happens to $\boldsymbol{S}.\boldsymbol{E}$ under

(1) a parity transformation and

(2) time reversal

separately, show that α is zero unless the interactions responsible for the forces between the constituent quarks in the neutron contain an element of parity non-invariance and simultaneously of time-reversal non-invariance.

[The present value of the electric dipole moment of the neutron is $-0.3 \pm 0.5 \times 10^{-27}$ e m (e = charge on proton), consistent with zero.]

13.3 Indicate, with a brief explanation, whether the following reactions or decays can proceed through the strong, the electromagnetic, or the weak interaction. In each case estimate as well as you can the order of magnitude of the cross-section or transition rate, as appropriate.

(1) $\mathrm{n} + \mathrm{p} \to \mathrm{d} + \gamma$,

(2) $\Xi^- \to \Lambda + \pi$

(3) $\mathrm{p} + \mathrm{p} \to \pi^+ + \mathrm{d}$,

(4) $\mathrm{D}^+ \to \mathrm{K}^- + \pi^+ + \pi^+$,

(5) $\mathrm{K}^- + \mathrm{p} \to \mathrm{K}^0 + \mathrm{n}$,

(6) $\pi^- + \mathrm{p} \to \mathrm{K}^0 + \Lambda$,

(7) $\mathrm{e}^+ + \mathrm{e}^- \to \mathrm{W}^+ + \mathrm{W}^-$,

(8) $\mathrm{Z}^0 \to \mu^+ + \mu^-$,

(9) $\Omega^- \to \overline{\mathrm{K}}^0 + \mathrm{K}^-$.

(d here stands for deuteron, not d quark.)

13.3 *CP* violation

We can now use the language of invariance or violation of invariance on the experimental facts of the weak interaction. In Section 12.9 we essentially implied that the weak interactions violated invariance under the parity transformation (P) and under charge conjugation (C) separately. However, we also implied that it was invariant under the combined transformation CP so that, for example, the β-decay rates and asymmetries of $^{60}_{27}$Co are identical to those of a mirror image of anti-$^{60}_{27}$Co. All the evidence is that *CP*-invariance is not violated except in certain decays of the neutral K-mesons, K^0 and $\bar{K}^0$. That subject takes us outside the scope of the book and it suffices to say three things.

1. The two decay processes $K^0 \rightarrow \pi^+ + \pi^-$ and $\bar{K}^0 \rightarrow \pi^+ + \pi^-$ both have the property that the number of decays by this mode per unit time does not fall exponentially with time after the production of a sample of K^0 or of $\bar{K}^0$ mesons. This is the result of mixing of the K^0 and $\bar{K}^0$ by the weak interaction and of *CP*-invariance. In addition, the number of $\pi^+\pi^-$ decays at a range of times after production is slightly different between a sample which is initially K^0 and an equal number sample which is initially $\bar{K}^0$. This property is due to an interference term that has opposite sign for the two particles and is caused by a small *CP*-violating effect. The observation of the sign of the interference term, which can be done independently of the environment, labels the initial source as K^0 or $\bar{K}^0$ and allows the formulation of a recipe for distinguishing an environment of particles from one of antiparticles (see Problem 13.4). Other manifestations of *CP*-violation occur in the decay modes of the neutral K-mesons. (See Perkins 1986, for a fuller treatment of the K^0-$\bar{K}^0$ system.)
2. The strength of this *CP*-violation is given by the ratio of two decay amplitudes and that ratio is about 10^{-3}. However, the origin of this *CP*-violation is not known. There is some evidence that it is related to the ability of the weak interaction to cause quark flavour changes between quark generations (Fig. 12.5). This can only be the case if the truth quark (Table 10.6) exists and that has not been experimentally established. The experimental problem is that, apart from the K^0-$\bar{K}^0$ system, there may be no other accessible physical situations where experimental checks may be made. One possibility is the long-lived neutral mesons containing b- or $\bar{b}$-quarks. These are the 0^- ground states of $b\bar{d}$ and $\bar{b}d$, and of $b\bar{s}$ and $\bar{b}s$. They are stable to all but weak interactions and may provide in the future a laboratory for investigating systems analogous to K^0-$\bar{K}^0$. There may also be possibilities in the truth quark plus antiquark systems if they are ever observed. One difficulty throughout these heavy quark systems is that there are likely to be so many decay modes that the ones in which *CP*-violation is manifest may be too small a fraction of all to allow easy observation. In addition, the production of these states in conditions of low background and high yield may require a dedicated accelerator and a superb detection system. But, of course, these are challenges to physicists which may be met in the future.
3. The solar system and our local galaxy appear to be constituted from matter (baryons and electrons) and contain very little, if any, antimatter (anti-baryons and positrons). There is also some evidence that the same is true of other galaxies and clusters of galaxies. Thus the Universe appears to be not symmetric under charge conjugation. It is true that *CP*- and *C*-violation operating early in the history of the Universe are essential elements required for the generations of this asymmetry (unless the initial condition of the Universe was not symmetric). More on this in Section 14.13.

There is a theorem of quantum field theory based on impeccable assumptions which requires that if *CP*-invariance is violated, then so also is time-reversal invariance: thus the existence of *CP*-violation provides indirect evidence for violation of time reversal invariance. We cannot pursue this important subject (the *CPT* theorem and its consequences) here; interested readers are referred to Gottfried and Weisskopf (1986).

PROBLEM

13.4 You are planning to communicate with the inhabitants of a distant galaxy and wish to determine whether their environment is of particles by our reckoning, or of antiparticles. Investigate the statement that, although the weak interaction violates charge-conjugation invariance, there is no weak interaction experiment that can be described in words which permits a resolution of that question, other than experiments using the K^0–$\overline{K}^0$ system.

13.4 The standard model

The **standard model** is the name given to the present model that we have of the elementary particles and their interactions. The ingredients of this model are given in Table 13.6. The leptons and quarks we have met in earlier chapters: the lighter ones, e^-, u and d are the substance of ordinary matter. The gauge bosons are the particles responsible for the interactions between fermions: the gluons are coupled only to the quarks, whereas the massive electroweak bosons (Z^0, $W^\pm$) are coupled to all the fermions and the massless electroweak boson (γ) is coupled to all the electrically charged particles. The third entry is the Higgs bosons: we re-examine their role in the standard model by looking again at the importance of gauge invariance which is itself an essential element of the standard model. Much of this material is a reminder of subjects introduced in Sections 9.11–9.13 and we cover it again since our motivation has changed. Previously we introduced the gauge bosons because we wanted to use them in constructing simple models of the particles and their interactions. Now we wish to emphasize the apparent universal importance of gauge invariance.

Table 13.6 The particles and interactions of the standard model.

Particles

1. Spin-$\frac{1}{2}$ fermions and antifermions:
   (a) three generations of leptons (e^-, ν_e), (μ^-, ν_μ) and (τ^-, ν_τ),
   (b) three generations of quarks (d, u), (s, c) and (b, t),
   plus all their antiparticles.
2. Spin 1 gauge bosons:
   (a) one massless electroweak boson, the photon γ,
   (b) three massive electroweak bosons, W^+, W^-, and Z^0,
   (c) eight coloured gluons.
3. Spin-0 Higgs bosons.

Interactions

1. The electromagnetic coupling of the photon.
2. The interactions of the bosons $W^\pm$, Z^0.
3. The strong interactions of the gluons with gluons and with the quarks.

The electroweak theory unifies 1 and 2.

There is a theoretical basis which gives physicists confidence that the standard model is the most fundamental that it is possible to construct at our present level of understanding. This is in spite of the fact that quarks and gluons have not been observed directly as free, identifiable particles. Our belief in the existence of quarks is based on our knowledge of hadron spectroscopy, of deep inelastic lepton scattering by nucleons and of the production of hadrons in e^+e^- annihilation at high energies (see Table 10.11). There is also considerable experimental evidence for the existence of gluons, which is supported by the fact they are predicted in a theory of quarks that is required to have the property of **local gauge invariance**. In Section 9.2 we described the gauge invariance of the electromagnetic field. The construction of a quantum-mechanical theory of particles that is required to have local gauge invariance is impossible unless the particles are, in general, interacting with a field in a manner that is prescribed. In one simple version of such a local gauge invariance the field has all the properties of the electromagnetic field and the strength of the interaction is determined by a constant of the gauge transformation which we therefore identify as the charge of the particle. The theoretical physicists give this simple local gauge invariance the label $U(1)$. We cannot in this book explain the very respectable group-theoretical origins of this label except to say one thing: the integer in the parenthesis (1) tells us that there is one gauge boson associated with the field and thus we expect one kind of photon. Of course, the $U(1)$ local gauge invariance has been chosen because it has this result: however, the ideas set the pattern for what has to happen for other particles and interactions. In the case of quark–gluon interactions the theory that has, by far, the closest agreement with experiment is quantum chromodynamics (QCD) which has a local gauge invariance labelled $SU(3)$. There are now eight massless spin-1 gauge bosons (gluons), coupled to fermions (quarks) with a strength determined by the colour (Section 9.11 and 10.7) of the particles at a quark–quark–gluon vertex and a coupling constant. Colour is to $SU(3)$ what electric charge in units of the electronic charge is to $U(1)$. The actual strengths of the coupling, g_s for QCD and e for QED, are not given by the theory.

Requiring the fields to have local gauge invariance and thereby determining the nature of the interaction between them is the application of the **gauge principle**. It is the success and elegance of the gauge principle which gives physicists such confidence in the standard model of which it is an essential part.

The application of the gauge principle to the weak interaction is more complicated than in the case of the other interactions. The label is $SU(2)$ and that requires three massless spin-1 bosons. But the W^+, W^- and Z^0 have mass and the weak interactions do not conserve parity. The synthesis of the electromagnetic and weak interactions into the locally gauge invariant electroweak theory, first described in Section 9.13, requires a mechanism which gives mass to three vector bosons and leaves the photon massless. The mass-giving ingredient is called the Higgs mechanism: it requires the existence of one or more massive, spin-0, bosons, the so called Higgs bosons (see Section 9.13). The theoretical details of how the mass of the W and Z come about without destroying the underlying gauge invariance cannot be pursued here. (See however, Kane, 1987, for a reasonably elementary account.) This electroweak theory has three parameters: they are the Fermi coupling constant G_F, the electromagnetic coupling constant e, and the Weinberg angle, θ_W (see Table 9.8) which describes the mixing of the gauge fields to make the four observable bosons W^+, W^-, Z^0, and γ. The angle was measured in many low-energy processes involving the weak neutral and charged current interactions (see

Section 12.4). Then, given these three parameters, the masses of the Z^0 and $W^{\pm}$ bosons were predicted successfully before their discovery (see Section 9.13).

The mass of the Higgs boson (or bosons) is not given by the theory: there are limits but they are weakened by the assumptions that have to be made to obtain them. The Higgs mass could be as low as $4\,\mathrm{GeV}/c^2$ or greater than 1000 GeV/c^2. Searches for Higgs bosons are on the agenda for all accelerators capable of reaching centre-of-mass energies at which these particles might be produced.

Most experimental data are consistent with the standard model: the next test of its predictive power is coming when the properties of free W and Z bosons are investigated. The theory predicts lifetimes, decay branching ratios, decay asymmetries. These properties for the Z^0 will be measured at the large electron–positron collider (LEP) of the European Centre for Nuclear Research (CERN), Geneva, and at the linear collider (SLC) of the Stanford Linear Accelerator Center (SLAC), California. Later when the energy of e^+e^- collisions can be raised to above threshold for

$$e^+ + e^- \rightarrow W^+ + W^-$$

the properties of the $W^{\pm}$ will be investigated.

An examination of Table 13.6 and the record of experimental work shows that there are, apart from the Higgs bosons, two fermions that have not been observed (we do not count antifermions since their existence for an established fermion is certain and confirmed in most cases). The two missing are the truth (or top) quark and the τ-neutrinos. The reasons for assuming the existence of these particles is based on the pattern established in the first two generations but also on incompatibility of certain measurements and the standard model if it had to accommodate a b-quark without a generation partner, t, and a τ-meson without a τ-neutrino. For example, the weak interaction mixing of quark generations does not make sense if the b-quark is not a member of a pair in the third generation.

There is one property of gauge theories that is of great importance. We focus on the three coupling constants (e, g, and g_s) of the three gauge fields. Although called constants they are not constant and they vary with momentum transfer. They are then called **running coupling constants**: a contradiction in terms but now having a definite meaning for physicists. Let us consider what this means in the case of the charge on the electron: that charge has to be measured by the interaction with a photon. In Millikan's oil-drop experiment that photon is provided by a static electric field and very little momentum is transferred. In electron–electron scattering at very high energies in which the electrons scatter through large angles the momentum transfer (q) is very much larger. The wavelength associated $\hbar/q$ is small and the exchanged photon probes much closer to the point electron than did Millikan's experiment. The nature of quantum electrodynamics is such that the photon then sees a charge that is greater than that seen when q is near zero. Physically the picture is that the electron's field polarizes the vacuum with transient electron–positron pairs (for example: the diagram with a closed e^+e^- loop in Fig. 9.5). Since the polarization positron is closer than is the polarization electron to the electron, in any sphere centred on the electron there is an excess of polarization positive charge over polarization negative charge (see Fig. 13.1). The electron is therefore screened and its apparent charge is reduced. At large distances we measure -1.6×10^{-19} C. Getting closer, the screening decreases and the measured charge

DEFINITIONS AND KEYWORDS

The standard model The name given to the present model of elementary particles and their interactions.

Global gauge invariance Invariance under a gauge transformation that has one or more parameters which are constant and do not vary with the space–time coordinates.

Local gauge invariance Invariance under a gauge transformation that has one or more parameters that vary with the space-time co-ordinates.

The gauge principle The principle of requiring a field to possess a local gauge invariance which thereby imposes the existence of vector field interacting with the original field.

Running coupling constant The coupling constants of gauge theories turn out not to be constants. They vary with energy or momentum transfers and this phrase describes this situation.

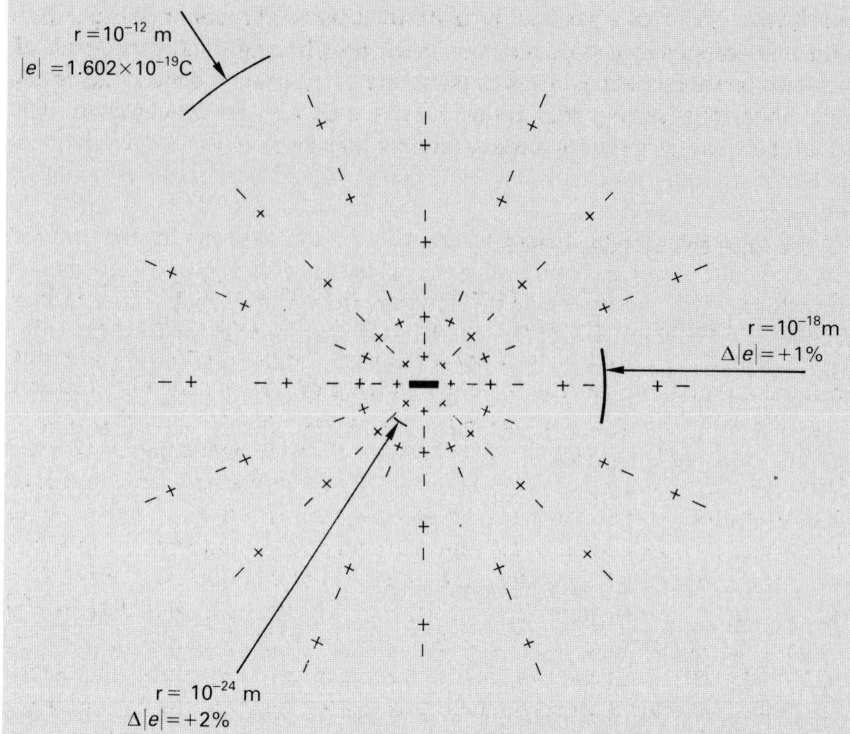

Fig. 13.1 Probing the charge on the electron at smaller and smaller distances. The bare electron charge polarises the vacuum by the production of virtual electron-positron pairs. The positron is, on the average, closer to the bare electron so that at any radius the sphere concentric with the electron encloses the bare electron charge plus a positive excess of polarization charge which screens the former. The effect is to reduce the field due to the electron from its value in the absence of polarization. As the distance of approach decreases the screening decreases and the apparent charge on the electron increases. The numbers on the figure are meant to show the slow rate of increase and are not correct in the sense that they have been estimated assuming, incorrectly, that only electron–positron pairs contribute to the screening.

increases. How close do we have to get in order to increase the apparent charge by 1%? Assuming that only electron and positron pairs are responsible for the screening, that distance is about 10^{-18} m, corresponding to momentum transfers of about 200 GeV/c. Such an increase is associated with a gauge theories of the $U(1)$ classification. The other gauge theories labelled $SU(2)$ and $SU(3)$ have the opposite behaviour: the fields become weaker as the interacting particles come closer to one another and the momentum transfer increases. In the case of quantum chromodynamics (QCD, Section 9.13) this property has been called asymptotic freedom, implying the particles experience ultimate freedom from interaction at infinitely large energies. At the other end of the energy scale, the QCD forces are very large, binding and confining quarks into hadrons.

If the strong QCD forces get weaker and the electromagnetic forces become stronger, there will be momentum transfer at which they become equal. A searching question is: do all the interactions of the gauge bosons, γ, $W^{\pm}$, Z and the eight gluons converge to one value? If they do it is in the region of momentum transfers, or what is equivalent, of masses, of the order of 4×10^{14} GeV. This mass is called the **unification mass**. This could be expected if all these interactions are all part of one interaction, having equal strengths in all places at very high energies, that is having a symmetry which makes the interactions equal at very high energies but which is broken at lower energies. We will take this point up again in Section 13.8.

13.5 Beyond the standard model

The standard model is not a complete theory. An analogy is Fermi's theory of the weak interaction before the electroweak unification. That theory was known to be incomplete in spite of its success because it was unrenormalizable and because it predicted some cross-sections growing without limit with increasing energy (Section 12.11). What makes us believe the standard model is incomplete? The first question could be, of course, what is a complete theory? We will sidestep that question by looking at some of the properties of the standard model that make us believe that it must be a part of a more comprehensive theory which may be able to relate some elements of the standard model which are presently not related. First the standard model does not predict the coupling constants (e, g, g_s, Sections 9.13, 12.4, 13.4, Tables 9.7, and 9.8) or the electroweak mixing angle (θ_W, Table 9.8, Section 9.13). Secondly it does not predict either the existence of three generations or the mixing of quark generations and the absence of mixing of lepton generations by the weak interaction. Thirdly, the masses of the fermions are parameters which have to be experimentally determined and cannot be predicted. The masses of the Z and W, although predicted, depend on the Higgs mechanism, and the fermion masses also depend on this mechanism. If the Higgs bosons are not found, then the problem of mass foreshadows a more fundamental physics about which we now have no concepts.

These difficulties are a challenge but do not devalue the standard model. It will remain a part of a more comprehensive theory. It is able to make quantitative predictions about the behaviour of particles at energies up to the mass of the W and Z (~ 100 GeV). These predictions are vital if experimental tests of the model are to be made with sufficient precision to be able to isolate the effects of physics beyond the standard model. We take a look at the kind of theory which attempts to predict this future physics in the next section. However, we must stress that these theories are very speculative.

13.6 Grand unified theories

Maxwell unified magnetism and electricity into classical electromagnetism. Einstein hoped to unify the latter with gravity but did not succeed. Glashow, Salam, and Weinberg unified electromagnetism with the weak interaction. Clearly this process of unification has further to go and theories attempting to do this are called **grand unified theories** (GUTs). We will indicate in a simple way how these are formulated and give an example of the simplest and the prediction that it made. The first idea is that the particles of any GUT should belong to a representation of a group. In Figs. 10.17 and 10.18 we showed the patterns of some of hadrons expected in the quark model: in fact these patterns are representations of the groups $SU(3)$ (Fig. 10.17) or $SU(4)$ (Fig. 10.18) given the assumption of 3 (uds) or 4 (udsc) fundamental quarks respectively. This use of group theory to help systematize the observed spectrum of hadrons is useful but in this case it is only phenomenological. However, physicists believe that the particles of the standard model do belong to a representation of a group yet to be discovered. Since that group is not known, builders of grand unified theories try various choices. Each choice has to accommodate the known particles of the standard model and will predict the existence of others. The latter happens because once the known fermions are accommodated in representations of a chosen group G, then the gauge principle and the group theory predict a set of massless gauge bosons capable by absorption or emission of changing one fermion into any other (for example, gluons change a quark of one colour into a quark of another colour in the $SU(3)$ theory of colour) or just changing the state of motion without changing the identity (for example, γ or Z^0). One particular motivation of the GUT approach is that at some unification energy the coupling of all these particles is specified in terms of one constant, g_G. Thus the theory starts with the concept of a high degree of symmetry. However, that symmetry is not observed at lower energies; some of the gauge bosons have mass and the coupling constants differ. The breaking of this gauge symmetry is called spontaneous symmetry breaking. It happens in many familiar physical situations: a liquid of a pure substance above its freezing point is symmetric in that it looks the same in all directions but once it has frozen the solid does not have this property because its crystal structure does not have it. In GUT building a sector of Higgs bosons is introduced to cause this symmetry breaking.

Gauge theories which have their symmetry broken will have gauge bosons of differing masses and coupling constants which are different in different sectors of the system: thus the coupling of photons to charged fermions is different from the coupling of gluons to coloured quarks.

The simplest possible GUT is labelled $SU(5)$. It has a place for all the fermions of one generation. The interaction is with gauge bosons associated with an $SU(5)$ local gauge invariance: that means 24 gauge bosons. Now we already have collected 12: γ, W^+, W^-, Z^0 and 8 coloured gluons. The group theory says that six others are each coloured one of red, green, or blue, with electric charge $-\frac{1}{3}$ (Y_R, Y_G, Y_B) or $-\frac{4}{3}|e|$ (X_R, X_G, X_B). The remaining six are the antiparticles of these six. They are all called **leptoquarks** or **diquark bosons**. The theory places particles such as e^- and d in the same representation of $SU(5)$, and a consequence is that such gauge bosons can transform a quark into antiquark or quark into a lepton. Some Feynman vertices for this are shown in Fig. 13.2. They are time ordered, so that new vertices may be found, as usual, by changing the topology of the three-arm vertex, or the directions of arrows.

The existence of these X- and Y-particles would have an immediate consequence that is illustrated in Fig. 13.3. The proton is no longer stable, decaying into mesons and a lepton. The properties of the leptoquarks is such that if B is baryon number and L is lepton number, then $B-L$ is conserved. However, to be of any use the theory must also predict a value for the proton mean lifetime, something we may be able to measure.

The argument which leads to a prediction for the lifetime depends on assigning a mass to the leptoquarks. Since $SU(5)$ applied in this way implies that there is a symmetry in the situation which makes all particle interactions equal, it is sensible to use the unification mass, 1.4×10^{14} MeV, for the masses of X and Y. This approach also fixes the $SU(5)$ coupling constant g_s at its 'unified' value estimated to be such that $g_s^2 \simeq \frac{1}{50}$. The contributions to the matrix element of the decay include a propagator for the X (or Y) which will be proportional to M_X^{-2}. This will appear squared in the transition rate. To obtain a dimensionally

DEFINITIONS AND KEYWORDS

Grand unified theories Theories with the objective of unifying the gauge interactions into one universal interaction.

Unification mass The energy in grand unified theories above which all the gauge interactions have the same strength.

Leptoquarks Gauge bosons of the grand unified theories which can transform a quark into an antiquark or a quark into a lepton (sometimes called **diquark bosons**).

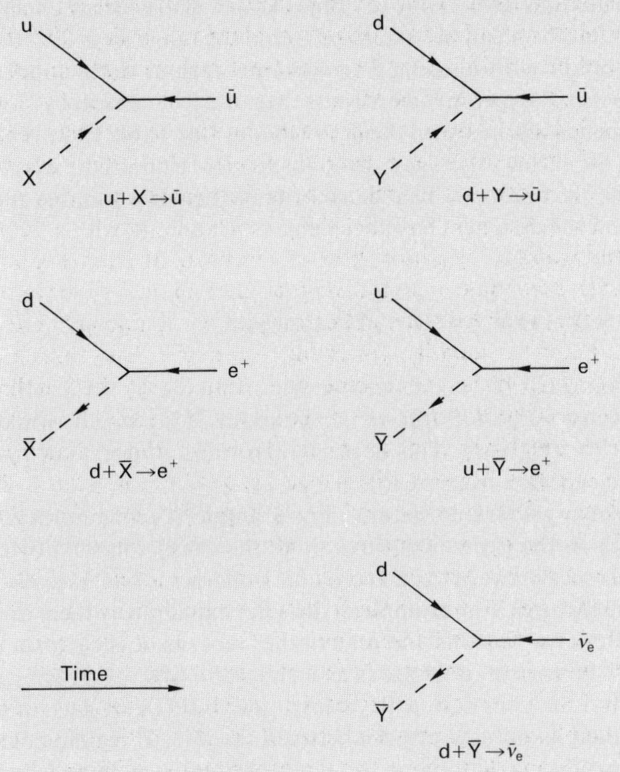

Fig. 13.2 Feynman vertices for the interaction of the predicted X (charge $-\frac{4}{3}$) and Y (charge $-\frac{1}{3}$) leptoquarks of $SU(5)$. Other diagrams of time-ordered processes may be made by changing the orientation of the arms and interpreting particles with arrows pointing back as antiparticles going forward in time, as usual. Each diagram above has a charge conjugate (reverse all arrows on each arm of a vertex). Colour labels have not been given: colour is conserved so that such labelling needs the understanding that $\mathrm{R} \to \overline{\mathrm{G}}\overline{\mathrm{B}}$, for example, is allowed.

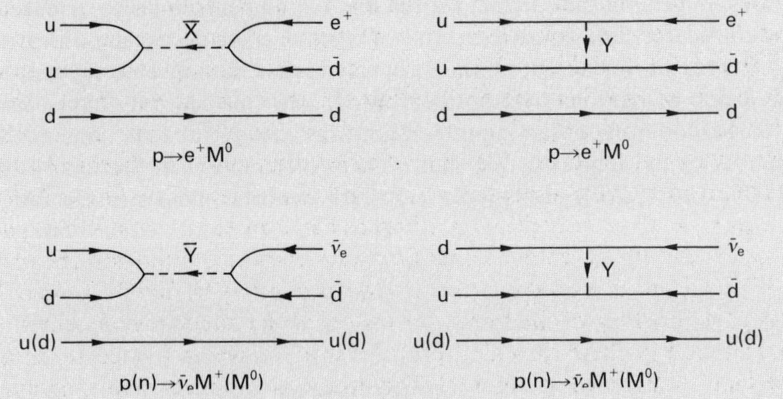

Fig. 13.3 Feynman diagrams for proton or bound neutron decay involving leptoquarks. The M^+ or M^0 represents a meson such as π^+ or π^0, or other states with the correct charge ($\rho, \omega, \ldots$). The usual fragmentation process suggests that additional mesons can be produced, within the limitations of energy and charge conservation.

correct formula we need five powers of an energy. The only one available is the proton rest mass (M_p) energy. Putting these things together gives a formula for the transition rate ω which looks like

$$\hbar\omega \sim \frac{g_s^2}{4\pi}\frac{M_p^5 c^2}{M_X^4}.$$

Then

$$\omega \sim 9 \times 10^{-37}\,\mathrm{s}^{-1},$$

or a mean life $\tau = \dfrac{1}{\omega} \sim 4 \times 10^{28}$ years.

More adequate calculations give about ten times longer but with an uncertainty which indicates that the mean life would be within the range from 2×10^{28} to 6×10^{30} years. The predicted branching fractions into various decay modes are important in assessing experimental results because the efficiency for detecting the decay depends markedly on the decay modes (see Table 13.7). The best limit comes from the Irvine–Michigan–Brookhaven (IMB) detector which will be described in the next section. This detector is particularly suitable for detecting $p\rightarrow e^{+}+\pi^{0}$ and the designers have been able to set a limit, with a 90% confidence level, of

$$\omega\times B(p\rightarrow\pi^{0}e^{+})<6.7\times10^{-33}\ (\text{years})^{-1}.$$

If the branching fraction B to $\pi^{0}e^{-}$ is one, this gives a mean life of $>1.5\times10^{32}$ years. The $SU(5)$ expected value for $B(p\rightarrow e^{+}\pi^{0})$ is about 35%, which would imply the lifetime is $>5\times10^{31}$ years. This is nearly ten times longer than the upper end of the range expected from $SU(5)$.

The $SU(5)$ GUT has many desirable features. For example, it accommodates the fact that the charge on the electron is three times the charge on the down quark and that the difference between the u and d quark charges is one. Equally, it has many undesirable features. For example, it does not explain why there are three generations of fermions. For this and many other reasons, it is clear that the $SU(5)$ GUT cannot be correct, unless it is as a subgroup of a larger gauge group. In addition, and in common with many other GUTs, it has one discouraging feature: there is no new physics between the $W^{\pm}$, Z regime near 100 GeV and the unification mass at about 10^{14} GeV. Nothing new in a twelve decade increase in energy!

There have been many attempts to construct GUTs based on higher symmetries. In general they predict proton and bound neutron decay, a feature which is considered desirable irrespective of whether these decays are observed or not. The reason for accepting this property is that the universe appears to have an excess of baryons over antibaryons. If true this can only have come about if a baryon non-conserving interaction was effective at some time in the early history of the Universe. We shall return to this point in Section 14.13. Most GUTs (not $SU(5)$) also predict non-zero neutrino masses in the range

Table 13.7 Possible GUT decay modes of the proton and neutron.

$$p\rightarrow e^{+}+M^{0},\qquad n\rightarrow e^{+}+M^{-},$$
$$p\rightarrow \mu^{+}+M^{0},\qquad n\rightarrow \mu^{+}+M^{-},$$
$$p\rightarrow \bar{\nu}+M^{+},\qquad n\rightarrow \bar{\nu}+M^{0}.$$

where

$M^{0}=\pi^{0}$, η, ρ^{0} or ω, or other totally neutral combinations of mesons,
$M^{\pm}=\pi^{\pm}$, $\rho^{\pm}$, $K^{\pm}$, or other singly charged combinations of mesons.

The $SU(5)$ predictions depend on assumptions about the formation of the hadronic part of the final state. Predictions for proton decay, give the modes $e^{+}\pi^{0}$:$e^{+}M^{0}$:$\bar{\nu}M^{+}$ branching ratios of order 35:50:15. For neutron decay modes $e^{+}\pi^{-}$:$e^{+}M^{-}$:$\bar{\nu}M^{0}$ the predictions are of order 60:20:20. Thus the $e\pi$ mode is important in both cases.

Major modes of subsequent decays:

$\pi^{+}\rightarrow\mu^{+}+\nu_{\mu}$,
(π^{-}, $K^{-}\rightarrow$ nuclear capture),
$\mu^{+}\rightarrow e^{+}+\bar{\nu}_{\mu}+\nu_{e}$,
$\pi^{0}\rightarrow\gamma+\gamma$,
$\eta\rightarrow\gamma+\gamma$, $\pi^{+}+\pi^{-}+\pi^{0}$, or $\pi^{0}+\pi^{0}+\pi^{0}$,
$\rho^{0}\rightarrow\pi^{+}+\pi^{-}$,
$\rho^{\pm}\rightarrow\pi^{\pm}+\pi^{0}$,
$\omega\rightarrow\pi^{+}+\pi^{-}+\pi^{0}$,
$K^{+}\rightarrow\mu^{+}+\nu_{\mu}$, $\pi^{+}+\pi^{0}$, ... (see Problem 2.3).

10 μeV to 10 eV. If correct, this could mean observable neutrino oscillations (Section 12.13), which would provide a useful experimental window on to GUTs.

Although the GUTs provide an attractive means of connecting the fundamental interactions into one, many parameters remain undetermined: the predictive power is rapidly lost and there is a huge increase in the number of elementary fields as the size of the symmetry group is increased beyond that of the unsuccessful $SU(5)$. Another difficulty arises in understanding the ratio of the unification mass (10^{14} GeV) to the $W^{\pm}$, Z masses (10^2 GeV) and there are related cosmological problems associated with the introduction of the Higgs fields. The absence of constraints on the properties of the Higgs sector make it appear to be unruly and to be completely lacking in aesthetic appeal. No alternative scheme exists which avoids that criticism and it is unfortunate that the grand unified schemes need a Higgs sector.

The mention of cosmological implications reminds us that gravity should be a part of a final unification scheme. No GUT so far proposed accommodates gravity: for that it is necessary to go to other theories: for example, to supersymmetry (Section 13.8).

13.7 Proton decay detectors

Whatever the status of grand unified theories, the detection of proton decay (including that of bound protons and neutrons) and the measurement of the decay properties, or the setting of a lower limit on its mean life, is an important project. The decay is one of the few ways of confronting GUTs and of indirectly exploring an energy region ($\sim 10^{14}$ GeV) probably forever inaccessible by direct means. These considerations motivated the construction of several large detectors designed to be sensitive to proton decay when the mean life was greater than 10^{30} years. Some users have reported upper limits for the partial transition rates into particular decay modes: most of these results are that such rates are less than 10^{-31} year^{-1}. However, two of the detectors received a bonus on 23 February 1987 when both detected neutrinos from a supernova in the Large Magellanic Cloud. In the next chapter we shall include a discussion of the nuclear and particle physics of supernovae (Section 14.9). Therefore we have prejudice in favour of describing the principle of these two detectors and why they were sensitive to neutrinos from the supernova. We apologize to the builders of other detectors for the neglect of their efforts.

The nucleon decay modes produce directly, or as secondary decay products, charged particles (Table 13.7). A convenient way of detecting such particles moving in a transparent medium is by the Cerenkov light that they emit. This process is described in Fig. 13.4 and it has some very useful properties:

1. The direction of the light is correlated with the direction of flight of the particle. (see Problem 13.5)
2. The number of photons depends on the path length of the particle.
3. The light is emitted instantly from the medium as the particle passes.

These properties ensure that the photons reaching the surface of a block of transparent material carry information by their timing, position and number on the direction and track length of a charged particle that caused the emission of these photons as it moved through the material. This is exemplified in Fig. 13.5.

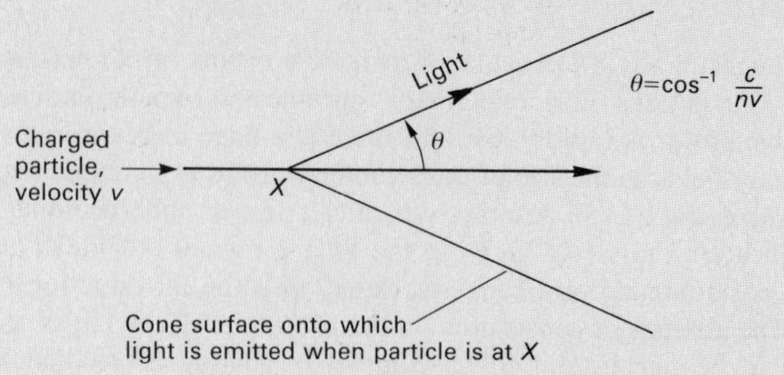

Fig. 13.4 The effect named after its discoverer, P. A. Cerenkov. A charged particle moving through an optically transparent medium of refractive index n causes the emission of light if the velocity v is greater than the velocity of light in that medium, c/n. This light is emitted at an angle θ, where $\cos\theta = c/nv$ (See Problem 13.5). The number of photons emitted per metre of path is about

$$\frac{2\pi\alpha}{c}\int \sin^2\theta \, \mathrm{d}f ,$$

where f is the frequency, and the integration will be over the frequency range of light transmission by the medium. For a relativistic electron this number is $2 \times 10^4\,\mathrm{m}^{-1}$ photons in a wavelength range 400–700 nm (visible light) in water ($n = 1.33$, $\theta = 41^\circ$).

A 10 MeV electron in water moves nearly 5 cm before coming to rest and, except for the last few millimetres, traverses its path at $v \approx c$. Therefore about 1000 visible photons are emitted. Allowing for overall efficiency, that is too few to be detected by the IMB detector but is detectable by the Kamiokande II detector.

Consider now the postulated decay

$$\mathrm{p} \rightarrow \mathrm{e}^+ + \pi^0 \quad (\pi^0 \rightarrow \gamma + \gamma)$$

The positron has a total energy of 459 MeV. The π^0 has 479 MeV which is not necessarily shared between the two photons equally. The positron is above the critical energy (Sections 11.3) of water, for example, so in that medium it is likely to radiate a photon which in turn Compton scatters or produces an electron–positron pair. The net effect is several electrons and positrons moving close to the direction of the original positron. Each γ-ray from the π^0 decay will also scatter or produce a pair (the radiation length in water is 0.36 m). The total effect is three small electromagnetic showers pointing back to the point of decay. The electrons and positrons emit Cerenkov light on a cone about their direction of motion. Three groups of visible photons can be detected at the water boundary. The number of photons in each group is closely connected to the energy of each shower, since each electron and positron is relativistic for almost all of its path. Given efficient visible photon detection at the boundary, it is possible to reconstruct with adequate precision the decay position and kinematics. Good resolution is essential to be able to discriminate against events like

$$\nu_e + \mathrm{p} \rightarrow \mathrm{n} + \mathrm{e}^+ + \pi^0$$

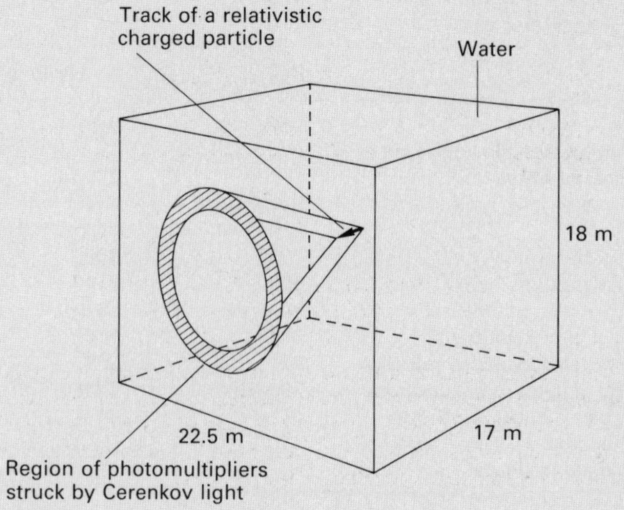

Fig. 13.5 A schematic of the arrangement of the IMB detector. It was, like the Kamiokande II detector, designed to detect nucleon decays in water. This water is purified to give the maximum achievable optical transmission. Cerenkov light is emitted by a moving charge particle onto the surface of cone coaxial with the direction of motion. This cone is sectioned by a wall or walls to give an ellipse or parts of ellipses. Each ellipse is diffused by the length of the track of the particle, by scattering of the particle and by the variation of refractive index with the wavelength of the emitted light. The walls are partially covered by photomultipliers each of which can detect a single photon striking its photocathods with an efficiency of about 15–30%. Thus a partial picture of the ellipse can be found. A determination of the timing (resolution about 2 ns) and the size of the signals add to give information on the position, length of the particle track, and hence, in particular, the energy it deposits in the water.

caused by neutrinos produced by cosmic rays. The properties of the π^0e^+ decay mode indicate why several groups chose to build proton decay detectors containing water and instrumented on their boundary with photomultipliers. The technical details of the two largest are given in Table 13.8.

The reader should have a look at Problem 13.6. This asks how these water Cerenkov detectors can be expected to respond to a couple of the other decay modes. The response is much less clear cut and makes the separation of real decays from candidates contaminated by events caused by neutrinos very uncertain.

We have already indicated that these devices are sensitive to neutrino events energetic enough to cause inelastic scattering on nucleons. However, the devices are also sensitive to the following reactions, which can be caused by neutrinos insufficiently energetic to cause inelastic scattering:

$$\nu + e^- \rightarrow \nu + e^-, \tag{13.5}$$

$$\bar{\nu}_e + p \rightarrow n + e^+. \tag{13.6}$$

The elastic scattering gives a recoil electron. The ν_e absorption gives a positron. Both these reactions can be detected if the e^- or e^+ has kinetic energy above the

Table 13.8 The water Cerenkov proton decay detectors operating at the time of SN1987A (23 February 1987).

	IMB	Kamiokande II
Location	USA	Japan
Depth underground in equivalent thickness of water	1570 m	2700 m
Shape	Rectangular $23 \times 18 \times 17$ m	Cylindrical $16\text{ m} \times 15.6$ m diam.
Mass of water (tons)	7000	3000
Fiducial mass (tons)	5000	2140
Diameter of photomultipliers	12.5 cm	50 cm
Number of photomultipliers	2048	1160
Surface covered by photomultipliers	2%	20%
Threshold of detection of electrons	20 MeV	5 MeV
Approximate number of photons detected from a 10 MeV electron stopping in the water	4*	44

* Too few to determine as belonging to a flash of Cerenkov light.

threshold of detection. Neutrinos from supernovae are expected to have energies in the range zero to 40–50 MeV with an average of about 12 MeV. The elastic scattering equation (13.5) does not transfer all the neutrino energy to the electron. The absorption reaction (13.6) gives an electron with almost all the neutrino energy. Thus a good fraction of events will be above threshold (5 MeV) of the Kamiokande II detector and a noticeable number above the threshold (20 MeV) of the IMB detector. In fact reaction (13.5) was probably not important in detecting neutrinos as reaction (13.6) has a very much larger cross-section. We shall discuss the interpretation of the data obtained by these detectors of the 1987 supernova in Section 14.10.

PROBLEMS

13.5 Use Huyghen's construction to show that Cerenkov radiation is emitted at an angle $\cos^{-1}(c/nv)$ to the direction of a charged particle moving with velocity v through a transparent medium of refractive index n.

13.6 Consider how a water Cerenkov detector for proton decay responds to the following two possible decay modes:

$$\text{(a)}\ \mathrm{p} \rightarrow \mu^+ + \mathrm{K}^0,$$
$$\text{(b)}\ \mathrm{p} \rightarrow \bar{\nu}_e + \pi^+.$$

List features of the detected light which could be used to distinguish these decays from other processes which might imitate them such as

$$\nu_\mu + \mathrm{p} \rightarrow \mu^+ + \mathrm{n} + \mathrm{K}^0,$$
$$\nu + \mathrm{p} \rightarrow \nu + \mathrm{n} + \pi^+,$$

respectively.

13.8 Theories of everything

Grand unified theories are not the only approach to finding a deeper understanding of the structure of matter than we possess at the present. They are all very speculative but are being actively pursued. The aim is ambitious: to understand all the interactions, the structure of matter and their relation to familiar space–time, including gravity. There may be many attacks on this problem but the following are prominent:

(1) composite models;

(2) supersymmetry;

(3) Kaluza–Klein theories;

(4) string models.

The first is almost self-explanatory: it seeks to explain the fundamental particles as composites of some fundamental entities. If this is the case then this level of structure may be beyond immediate access. How should it show up? If particles such as the quark or electron or muon were composite, then the reactions

$$e^+ + e^- \rightarrow \mu^+ + \mu^-,$$
$$e^+ + e^- \rightarrow q + \bar{q}$$

would have cross–sections which fell below their expected point-like cross-sections when the momentum transfer q at the fermion vertices reached a value such that $\hbar/q$ became comparable to the dimensions of the substructure. No such deviations have been detected up to 50 GeV e^+e^- collisions, so we believe that if such structure exists, it has a spatial size less than about 10^{-18} m.

Supersymmetry has many attractions. The prominent property of this theory is that every fermion has a supersymmetric boson partner, and vice versa. So quarks have squarks which are integer spin quarks. Photons have half-integer spin partners called photinos. It is easy to compose the names. The partner to a fermion is found by putting a prefix s in front of the regular fermion's name (electron to selectron, neutrino to sneutrino). The partner to a boson is found by adding a suffix ino to a root (gluon to gluino, graviton to gravitino). It looks like a game but it is serious because the theory has predictive power, can include gravity and may solve certain cosmological problems. Against that is the experimental situation: all efforts to find such particles have failed and give lower limits to masses which are quite high.

Kaluza–Klein theories are the development of an idea first proposed by Theodor Kaluza and published in 1921. Oscar Klein improved the formulation in 1926. The idea was to add a fifth dimension to the regular three-space plus one time: this fifth dimension is curled up and has no effect on usual physics. By correct procedure it is possible to define a field which in the usual four dimensions has components which satisfy the field equations of general relativity and other components which satisfy Maxwell's relations. Thus electromagnetism and general relativity are connected! However, the theory was far from satisfactory. Since the success of the electroweak unification, interest has revived in this approach and the one extra dimension is replaced by an integer number of dimensions, all of these extra dimensions spanning a 'compact' domain. Compact here means that each extra dimension is curled up and closed as is the longitude on the surface of a sphere. It is the legacy of Einstein's general theory of relativity, which treats gravity as a part of the fabric of space and time, that makes this approach appealing. The other interactions and the particles must also be part of that fabric and a space of dimension higher than three plus time may be the warp and weft from which the structure of matter is woven.

String theory is another attempt based on the idea that particles are not point-like but are strings. Particles can be closed or open strings. The theory has to be formulated in a space of possibly ten or more dimensions. The symmetries it possesses are broken when each of all but four of the dimensions is curled up on itself. The topology of this space determines the particles and their interactions. Of course, the known particles are used to eliminate topologies. Gravity has a place but, although the theory has many encouraging features, it is far from anything looking like the last word.

It is clear that these approaches are not isolated from one another. Multidimensional spaces are an important feature. The assumption of an overall symmetry with a broken version that we know forces the use of group theory to explore the consequences of that assumption and the use of its language to name the symmetries of the whole or of the parts into which the whole divides.

There is a feature of many theories which has had an impact on experimental physics. Apart from the search for supersymmetric partners to the known fermions and bosons, there have also been attempts to find other particles predicted. The searches for all that we mention have been unsuccessful in the sense of reproducible and confirmed results. Here is an incomplete catalogue:

1. Axions: Spin-0 particles of small mass which are an unavoidable feature of a mechanism proposed to eliminate a very undesirable feature of QCD.
2. Magnetic monopoles: A feature of most GUTs and normally very massive. They are of great interest since Dirac was able to prove that the existence of magnetic monopoles implied charge quantization.
3. Majorana neutrinos: The neutrino that is the same as its antiparticle (see Section 12.13).
4. Dyons: Magnetic monopoles with electric charge.
5. Weakly interacting massive particles: A generic name for particles which interact weakly and are massive. Presumed to have been produced in the early moments of the big bang (Section 14.13) and to be considered as candidates for the dark matter which is thought to be present in or around galaxies.

Gravity has been mentioned several times in connection with the theoretical attempts at 'theories of everything'. In GUTs the unification mass is expected to be at about 10^{15} GeV. However, that does not include gravity. If there is a unification mass that includes gravity it may be of order 10^{19} GeV, the energy corresponding to the Planck length which is $(G\hbar/c^3)^{1/2} = 1.6 \times 10^{-35}$ m, where G is the gravitational constant. This quantity is the only one with the dimension length that can be constructed from the gravitational constant and the fundamental constants.

Gravitons are the spin-2 particles which are expected to exist as the free quanta of the gravitational field. The formal quantization of this field has not yet been achieved, so that the consequences of gravity at very small distances are unknown and presumably will not be those expected from classical general relativity alone. Gravitons were not put in the class of objects described above because those were hypothetical, whereas physicists' faith in the final quantization of gravity makes gravitons unavoidable. The detection of gravity waves would be an important step in the experimental investigation of this subject. However, although a large amount of gravitational wave energy may exist (see Section 14.11) it is difficult to detect. The waves are very weakly coupled to matter so that it is necessary to detect small movements of massive pieces of material. This involves the measurement of very small displacements (10^{-15} m) in the presence of thermal noise and earth movements not related to gravitational radiation and at a frequency not known in advance. The simplest detector takes the form of two large masses, separated by a small distance. Gravitons have spin 2 so that the gravitational wave can excite a quadrupole vibration. For two identical masses that is a variation in separation. (This is not dipole because the two masses do not have opposite couplings to the field.) Thus a detection of a variation of separation is the objective. Other detectors have more elaborate systems. So far there has been no substantiated report of the successful detection of gravity waves. On the other hand the necessary sensitivity based on the expected intensity of such waves is only now being approached.

13.9 Open questions

Our discussion in the previous section about 'theories of everything' suggests that we summarize the problems that are facing physicists in the search for an understanding of matter. Those that are on the level of particle physics were mentioned in Section 13.5. The grand unified theories and others mentioned in Section 13.8 have not really solved any of those problems. For reference we have listed them in Table 13.9. Since each problem may overlap in its significance with others, this table does not claim to point out those which are more basic than others.

Behind these problems there are others which may be seen to be or are more basic.

1. Mass. The standard model requires one or more Higgs bosons to generate mass. If no evidence for the existence of such bosons can be found, beyond this role in a hypothetical mechanism, then understanding mass will be an outstanding problem.

Table 13.9 Particle physics waiting to be understood.

1. The pattern of leptons and quarks in each generation.
2. The existence of three (or more?) generations.
3. The violation of parity and charge conjugation invariance by the weak interactions.
4. The mixing of quark generations and the non-mixing of lepton generations by the weak interactions.
5. *CP* violation.
6. The masses of particles.
7. The pattern of gauge bosons.
8. The values of interaction coupling constants.
9. Higgs.
10. Gravity as another interaction.
    and ...

2. The quantization of gravity.
3. Quantum mechanics. This is a very successful system but nobody understands why it is the way it is.

References

Gottfried, K. and Weisskopf, V. F. (1986) *Concepts of Particle Physics*, Vols 1 and 2. Oxford University Press.

Kane, G. (1987) *Modern Elementary Particle Physics*. Addison-Wesley.

Perkins, D. H. (1986) *Introduction to High Energy Physics* (3rd edn). Addison-Wesley.

14

Nuclear and Particle Astrophysics

14.1 The expanding Universe

The astronomical evidence is that the Universe is expanding. The galaxies that we see from Earth have velocities that we can measure from the Doppler shift of the identifiable spectral lines. Apart from local irregularities that shift is towards the red and indicates that these galaxies are receding at velocities which increase linearly with distance. The constant of proportionality is called Hubble's parameter, named after the discoverer of the recession. Its value is uncertain but is about $1.8 \times 10^{-18}\,\mathrm{s}^{-1}$. This means that at a separation of 10^9 light years (9.46×10^{24} m, see Fig. 2.1) the velocity of recession is $1.5 \times 10^7\,\mathrm{m\,s}^{-1}$, one twentieth of the velocity of light. Incidentally, astronomers prefer the units which make Hubble's parameter about $50\,\mathrm{km\,s}^{-1}\,\mathrm{Mpc}^{-1}$, where a parsec (pc) is 3.26 light years!

In 1965 Penzias and Wilson (see Weinberg, 1977), discovered that the Universe is filled with black-body radiation with a temperature now measured to be 2.7 K. The interpretation of this discovery is that the Universe was created about 15×10^9 years ago in an explosion now called the **Big Bang** (see Fig. 14.6). The expansion is the recession from the explosion of the material that was produced, and the 2.7 K black body radiation is the cooled remnants of the electromagnetic radiation that existed at a certain time after the initial explosion.

Cosmologists are attempting to reconstruct the story but cannot yet do that back to time zero. In the forward direction we would like to know what is to happen to the Universe. As we examine this story it is convenient to find one time which for our purposes is a critical age dividing a forward story from a backward look. In this examination we shall be concerned primarily with the role of the physics of nuclei and elementary particles and will have to take on trust the physics which leads to an environment in which these are important matters. Our critical age is, therefore, the time at which the physics of complex nuclei began to play a role. That time is when the Universe was about 225 seconds old: the temperature was about 9×10^8 K ($kT = 77$ keV—see Table 14.1) and the density was about $2 \times 10^4\,\mathrm{kg\,m}^{-3}$. The Universe contained photons, neutrinos and antineutrinos, electrons and nucleons. Photons to neutrons to protons were in the ratio of about 10^{11}:13:87 and there were sufficient electrons to neutralize electrically the protons. These are the photons that, after scattering, absorption and re-emission, are destined to become the 2.7 K black body radiation 15 billion years later.

Table 14.1 To convert temperature T in Kelvin to kT in electron-volts

$kT\,(\mathrm{eV}) = 8.6 \times 10^{-5}\,T\,(\mathrm{K})$

The forward story takes us from Sections 14.2 to 14.12. In Section 14.13 we take the backward look to time near zero.

14.2 Big Bang nucleosynthesis

DEFINITIONS AND KEYWORDS

The Big Bang The singular event from which the present Universe is supposed to have evolved.

Nucleosynthesis The synthesis by nuclear reactions of complex nuclei from protons and neutrons.

We start at $t = 225$ s; some initial **nucleosynthesis** now takes place. It is associated with the big bang since nucleosynthesis will not be resumed until the stars form, a process not due to start for at least another 10^6 years.

The coexistence of neutrons and protons allows the nuclear reaction

$$\mathrm{n} + \mathrm{p} \rightarrow \mathrm{d} + \gamma + 2.22\ \mathrm{MeV} \tag{14.1}$$

and this must be the first step in the building of complex nuclei at this point of the story (d stands for deuteron not d quark). However, before $t = 225$ s the temperature was sufficiently high that the reverse reaction, deuteron photodisintegration, occurs almost immediately and the deuteron is destroyed before it can build into anything heavier. But after this time the deuteron survives long enough to begin significantly the reactions

$$\mathrm{p} + \mathrm{d} \rightarrow {}^3_2\mathrm{He} + \gamma + 5.49\ \mathrm{MeV}$$

and

$$\mathrm{n} + \mathrm{d} \rightarrow \mathrm{t} + \gamma + 6.26\ \mathrm{MeV}.$$

The nuclei ${}^3_2\mathrm{He}$ and t (${}^3_1\mathrm{H}$) are much more strongly bound than the deuteron and are not in danger of photodisintegration. Thus once the deuteron bottleneck is passed the following helium-producing reactions occur:

$$\begin{aligned}
\mathrm{t} + \mathrm{p} &\rightarrow \alpha + \gamma + 19.81\ \mathrm{MeV},\\
{}^3_2\mathrm{He} + \mathrm{n} &\rightarrow \alpha + \gamma + 20.58\ \mathrm{MeV},\\
\mathrm{t} + \mathrm{d} &\rightarrow \alpha + \mathrm{n} + 17.59\ \mathrm{MeV},\\
\mathrm{d} + \mathrm{d} &\rightarrow \alpha + \gamma + 23.85\ \mathrm{MeV}.
\end{aligned}$$

At this stage the density is too low to allow the reactions which jump across the gap caused by the absence of stable nuclei with $A = 5$ or 8. The necessary conditions will not occur until the stars form and age. The only nucleus heavier than helium which appears to survive the age of big bang nucleosynthesis is ${}^7_3\mathrm{Li}$ and then only in very small amounts. It is produced in the reaction

$${}^4_2\mathrm{He} + \mathrm{t} \rightarrow {}^7_3\mathrm{Li} + \gamma + 2.47\ \mathrm{MeV},$$

although inhibited by the Coulomb barrier. It is readily destroyed in

$${}^7_3\mathrm{Li} + \mathrm{p} \rightarrow \alpha + \alpha + 17.35\ \mathrm{MeV}.$$

The relative role of all these reactions at any instant is a function of the density, composition and temperature of the substance of the Universe at that time.

About 30 minutes later nucleosynthesis has ceased. The temperature is about 3×10^8 K ($kT = 26$ keV) and the density about 30 kg m^{-3}. The small amount of nuclear matter is approximately 76% by mass protons and 24% α-particles.

There are traces of deuterons and of $^3_2$He nuclei and possibly a tiny amount of $^7_3$Li. About 10^6 years later the temperature has fallen sufficiently (to about 2000 K) to allow the combination of electrons and nuclei into neutral atoms. Gravity is now free to play a part in the formation of galaxies and of stars within the galaxies. The number of photons to nucleons is about 10^9:1 and will remain so.

The next stage we have to consider is the processing of hydrogen and helium in the interior of stars. Although heavy nuclei are produced and that has been happening for over 10^{10} years, stellar nucleosynthesis has not altered significantly the gross composition of the Universe from that when its age was 30 minutes until the present epoch. The matter in the Universe now is still close to 24% helium, 76% hydrogen, by mass.

14.3 Stellar evolution

In this section we shall describe in outline the story of stellar evolution, with the particular intention of emphasizing the role of nuclear reactions. We assume a star starts life as a cool cloud of interstellar material, dominantly primordial hydrogen and helium. As this **protostar** contracts gravitationally it will heat up. This is a consequence of the **virial theorem** which, applied to a gravitationally bound system, states that the time averages of the internal kinetic energy $\overline{T}$ and of the gravitational potential energy $\overline{\Omega}$ for the system are related by

$$2\overline{T}+\overline{\Omega}=0. \qquad (14.2)$$

The total energy $E=\overline{T}+\overline{\Omega}$. If the system is bound, then E must be negative and the binding energy $B=-E$. Therefore

$$\mathrm{d}B=-\mathrm{d}(\overline{T}+\overline{\Omega})=\mathrm{d}\overline{T}.$$

Thus an increase in binding energy is accompanied by an increase in $\overline{T}$, that is an increase in temperature. As the star contracts, the increasing binding energy has to be radiated from the surface. At any time the star has a secular equilibrium in which the gravitational forces are balanced by the internal pressure (hydrostatic equilibrium) and the highest temperature is at the centre (T_c). This secular equilibrium changes slowly as the star radiates energy and contracts; what next happens depends on the mass. In the context of stellar evolution a convenient unit is the mass of the Sun, abbreviated to $M_\odot$ and equal to 1.99×10^{30} kg (see Table 14.2). If the mass of our protostar is less than approximately $0.1M_\odot$ the contraction stops because the soup of electrons, protons and α-particles at the centre can support this weight by its **electron degeneracy pressure**. The existence of this pressure may be understood by considering a box containing electrons: if the electron gas is degenerate, that is, electrons fill all the energy levels up to the Fermi energy, then compressing the box raises the Fermi energy and the total energy of the electrons. Thus work has to be done to compress the box and the electrons exert an opposing pressure; this is the electron degeneracy pressure. Thus this undeveloped star finally cools into a body of cold helium and hydrogen with a core containing wholly or partially degenerate electrons, depending on the mass.

Table 14.2 The solar constants

Mass	$M_\odot$	1.99×10^{30} kg
Radius	$R_\odot$	6.96×10^{8} m
Luminosity	$L_\odot$	3.83×10^{26} J s^{-1}
Mean density		1.41×10^{3} kg m^{-3}

For masses greater than 0.1 $M_\odot$, T_c will reach 10^7 K, at which point **hydrogen burning** starts. This has the effect

4 protons + 4 electrons → 1 alpha particle + 2 electrons + 2 neutrons + 26.7 MeV.

In this case we include the electrons in order to keep track of the charge because the stellar interior is a plasma; this means that, although fully or partially ionized, it is and remains electrically neutral averaged over any volume very large compared with atomic volumes. This burning stops the contraction and a new kind of secular equilibrium is established in which the losses by radiation are provided by nuclear energy, which thereby maintains the T_c and postpones further gravitational collapse. This situation continues for a considerable period. The Sun has been burning hydrogen for about 5×10^9 years and will continue to do so for another 5×10^9 years. More massive stars reach higher central densities and hydrogen burning is faster. When the hydrogen in the core of the star is exhausted, further gravitational contraction will occur. If the mass $> 0.25M_\odot$, T_c can reach 10^8 K and **helium burning** starts in the core. This is effectively

$$3\ \alpha\text{-particles} \rightarrow 1\ {}^{12}\text{C nucleus} + 7.27\ \text{MeV}.$$

Once this burning has started in the centre, hydrogen burning continues in a layer around the core.

This program of burning, contraction and burning of the products continues. For stars of mass $> 4M_\odot$ carbon burning ($T_c \simeq 6 \times 10^8$ K) follows helium exhaustion in the core, leading to a core of oxygen, neon and magnesium. Stars of mass $> 10M_\odot$ burn oxygen ($T_c \simeq 2 \times 10^9$ K) to silicon and then to nuclei close to iron ($T_c \simeq 4 \times 10^9$ K). This last stage of burning produces nuclei which have the maximum binding energy per nucleon (Section 4.5) so that nuclear reactions leading to heavier or lighter nuclei consume energy. Thus stars of this mass reach a stage of onion-like structure in which the iron core has reached the end of nuclear burning and successive layers outwards contain increasingly lighter elements which continue burning (Fig. 14.1). Burning in the silicon layer feeds a growth in the mass of the iron core until the electron degeneracy pressure of the latter is no longer able to support the pressure of the overlying layers and its own weight (see Section 14.9). At this point gravity takes over again and a catastrophic collapse occurs. We shall discuss the nuclear effects shortly: at this point it is sufficient to say that there is a huge release of gravitational energy, a part of which blows off the outer layers and returns all but about $1.4M_\odot$ of the star's mass into the interstellar medium. What remains is a **neutron star** or, in the case of more massive stars, gravity may really win with the formation of a **black hole**. Such an event is one type of **supernova**: the collapse occurs in about one second and more energy is liberated than the star has emitted in the whole of its previous life. For a few days the supernova outshines all the stars of its own galaxy, and there may be more than 10^{10} of those!

The stars that evade nuclear burning to iron none the less end as compact stars called **white dwarfs**. These consist of the ashes of the last burn and are saved from further collapse by the electron degeneracy pressure in the core. The partially burnt outer layers are blown off during the course of the star's progression to the state of white dwarf, where only the core remains. A neutron star which does not increase its mass by accretion of material is supported by the neutron degeneracy pressure and the short-range nuclear repulsive force.

One common feature to all these scenarios is the ejection of unburnt hydrogen and helium and many of the products of burning back into the interstellar medium. This material becomes the source of material for new

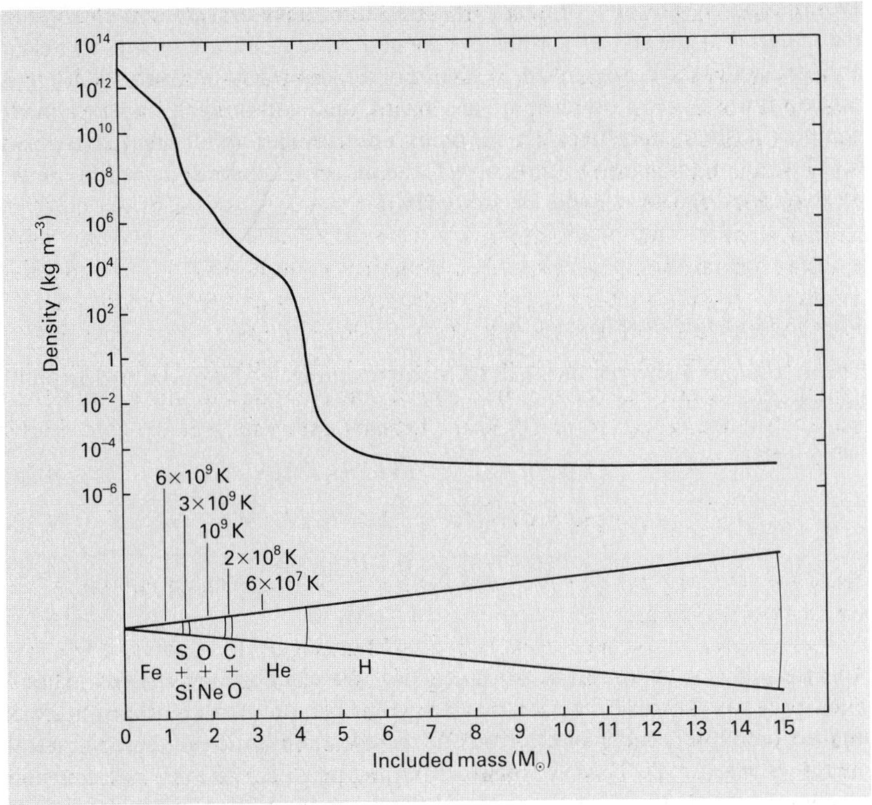

Fig. 14.1 The density profile as a function of included mass of a $15M_\odot$ star just before collapse. The figure includes a sector to the centre of the star showing the layers and the dominant elements in each layer. The average layer temperature is also given. Burning in each layer will be greatest where temperature and density is highest, that is just above its inner boundary. The core marked Fe is actually iron-like elements.

protostars and a new cycle of stellar contraction and burning. Each cycle leaves a compact star and gradually enriches with heavier elements the material partaking in the reprocessing. The material of which the Earth and its people are made has been through one cycle, and possibly more, of stellar nuclear processing.

This is not a book about astrophysics so we have simplified the story to a bare minimum. We shall, in the next section, discuss the nuclear reactions which occur at these various stages of stellar processing. The other physics of stellar

DEFINITIONS AND KEYWORDS

Protostar The material about to condense into a star.

Virial theorem The theorem which, applied to a system bound by gravitational attraction, relates the time average of its potential energy to the time average of its internal kinetic energy.

Electron degeneracy pressure Electron energy distributions are determined by the Fermi–Dirac statistics: the total energy of the distribution increases with decreasing volume, even at a temperature of 0 K, and this implies a pressure. This is the degeneracy pressure and it exists for all fermions, not only for electrons.

Hydrogen burning The processes by which a star converts hydrogen into helium.

Helium burning The processes by which a star converts helium into carbon.

Neutron star An object composed mostly of neutrons at nuclear density or greater. It has a mass in the range 1.0–1.5$M_\odot$ and a radius of the order of 10 km.

Black hole An object that has suffered complete gravitational collapse. This will have happened if its mass M is inside its Schwarzschild radius, $2GM/c^2$. (G is the gravitational constant.)

Supernova A stellar explosion in which the star undergoes rapid evolution involving the ejection of a large fraction of its mass. It may lead to complete dissolution of the star or to the generation either of a neutron star or of a black hole.

White dwarf A small star consisting of elements lighter than iron which has reached the stage where no further nuclear burning is possible.

evolution is an interplay of density, composition, temperature and entropy of the material of the star at various depths under the condition in which nuclear and gravitational energy must be transported outwards and gravity itself is pulling inwards. The building of models of the evolution of stars is a truly complex business and there remain many uncertainties about what is happening. For an introduction to this subject, the reader is referred to the first of the two volumes by Bowers and Deeming (1984).

14.4 Stellar nucleosynthesis 1

The first step is hydrogen burning: the main reactions leading from hydrogen to helium are

$$p + p \rightarrow d + e^{+} + \nu_e + 0.42\ \text{MeV}, \tag{14.3a}$$

$$p + d \rightarrow {}^{3}_{2}\text{He} + \gamma + 5.49\ \text{MeV}, \tag{14.3b}$$

$${}^{3}_{2}\text{He} + {}^{3}_{2}\text{He} \rightarrow p + p + \alpha + 12.86\ \text{MeV}, \tag{14.3c}$$

$$e^{+} + e^{-} \rightarrow \gamma + \gamma + 1.02\ \text{MeV}. \tag{14.3d}$$

This is the **pp chain**: a total of two of each of the reactions (a), (b) and (d) and one of (c) has the net effect of converting four protons to one helium nucleus and keeping the number of electrons to the correct atomic number. The total energy released is 26.72 MeV, of which an average of 0.26 MeV per neutrino escapes the star as two neutrinos. The photons produced in reactions (b) and (d) are absorbed and thereby deposit energy close to their production point. The reaction (a) is a weak interaction process, and has a very small cross-section $< 10^{-51}\ \text{m}^2$ ($10^{-21}\ \text{fm}^2$) at the energies corresponding to the temperatures of about 1.5×10^7 K in the appropriate part of a hydrogen burning star. This cross-section is too small to be measured and has to be calculated from weak interaction theory. The reactions (b) and (c) involve Coulomb barrier penetration but otherwise are electromagnetic (b) or strong (c) and have cross-sections very much larger than that for (a). Therefore it is reaction (a) which controls the rate of hydrogen burning. The result is that at densities of about $10^5\ \text{kg m}^{-3}$ and temperatures of about 10^7 K a proton has a mean life of about 10^{10} years. Once formed, the deuteron has a mean life of about 1 s and ${}^{3}_{2}\text{He}$ about 2×10^5 years. This is the situation in our own Sun: this very slow rate of hydrogen burning has given the Sun an age sufficiently long to permit the Earth to cool and to host the evolution of life to the level we know.

There are also three reaction chains which transform hydrogen to helium but require the initial presence of helium or carbon to act as a catalyst. The **helium catalysed chains** are shown in Table 14.3. Their importance relative to the pp chain depends on temperature and the amount of helium present. It is believed that in the Sun they account for about 8% of the total nuclear energy production. At higher temperatures this fraction would be greater. There is no problem in providing the helium catalyst as it is present in the primordial interstellar material and also as a product of the pp chain.

The carbon catalysed chain is called the **CNO chain**. The reactions involved are shown in Table 14.4. Essentially the catalysis closes the chain:

$${}^{12}_{6}\text{C} \rightarrow {}^{13}_{7}\text{N} \rightarrow {}^{13}_{6}\text{C} \rightarrow {}^{14}_{7}\text{N} \rightarrow {}^{15}_{7}\text{O} \rightarrow {}^{12}_{6}\text{C}.$$

Table 14.3 The nuclear reaction chains involved in the stellar burning of hydrogen.

pp chain

(a) $p+p \rightarrow d+e^{+}+\nu_e+0.42$ MeV

(b) $p+d \rightarrow {}^{3}_{2}He+\gamma+5.49$ MeV,

(c) ${}^{3}_{2}He+{}^{3}_{2}He \rightarrow \alpha+p+p+12.86$ MeV,

(d) $e^{-}+e^{+} \rightarrow \gamma+\gamma+1.02$ MeV.

Net effect of 2(a)+2(b)+(c)+2(d) is

$$4p+2e^{-} \rightarrow \alpha+2\nu_e+26.72 \text{ MeV}.$$

Average neutrino loss per α-particle produced=0.52 MeV.

α-particle catalysed chains: reactions (a) and (b) provide ${}^{3}_{2}He$ for

${}^{3}_{2}He+\alpha \rightarrow {}^{7}_{4}Be+\gamma+1.59$ MeV

${}^{7}_{4}Be+e^{-} \rightarrow {}^{7}_{3}Li+\nu_e+0.861$ MeV,	${}^{7}_{4}Be+p \rightarrow {}^{8}_{5}B+\gamma+0.135$ MeV,
${}^{7}_{3}Li+p \rightarrow \alpha+\alpha+17.35$ MeV.	${}^{8}_{5}B \rightarrow {}^{8}_{4}Be^{*}+e^{+}+\nu_e+14.02$ MeV,
	${}^{8}_{4}Be^{*} \rightarrow \alpha+\alpha+3.03$ MeV,
	$e^{-}+e^{+} \rightarrow \gamma+\gamma+1.02$ MeV.
Average neutrino loss=0.80 MeV	Average neutrino loss=7.2 MeV

Net effect is

$$p+{}^{3}_{2}He+e^{-} \rightarrow \alpha.$$

[${}^{8}_{4}Be^{*}$ here means the excited state of ${}^{8}_{4}Be$ at 2.94 MeV which is the daugher in 93% of the β^{+}-decays of ${}^{8}_{5}B$. The remaining 7% of the decays go to an excited state of ${}^{8}_{4}Be$ at 16.63 MeV.]

by three (p,γ) reactions and a couple of β^{+}-decays and a final (p,α) reaction. Thus one cycle again converts four protons to one α-particle. The cycle given is not the unique route from carbon to carbon: there are some branches involving ${}^{14}_{8}O$, ${}^{15}_{8}O$, ${}^{17}_{8}O$, ${}^{17}_{9}F$, ${}^{18}_{9}F$ and ${}^{19}_{9}F$. However, the route shown is the most important. The slowest reaction is the first ${}^{12}_{6}C(p,\gamma){}^{13}_{7}N$, but the reaction ${}^{14}_{7}N(p,\gamma){}^{15}_{8}O$ is also slow and the result is that there is a sizeable concentration of ${}^{14}_{7}N$ in hydrogen-burning regions involving the CNO cycle. This cycle is important in stars more massive than the Sun. There is the question of the origin of the ${}^{12}_{6}C$. In protostars entirely made of primordial hydrogen and helium there is no ${}^{12}_{6}C$, but once helium burning begins this nucleus is produced and, given adequate mixing, the CNO cycle may be able to contribute to hydrogen burning. However the

Table 14.4 The nuclear reactions of the CNO stellar cycle.

$p+{}^{12}_{6}C$	→	${}^{13}_{7}N+\gamma+1.94$ MeV,
${}^{13}_{7}N$	→	${}^{13}_{6}C+e^{+}+\nu_e+1.20$ MeV,
$p+{}^{13}_{6}C$	→	${}^{14}_{7}N+\gamma+7.55$ MeV,
$p+{}^{14}_{7}N$	→	${}^{15}_{8}O+\gamma+7.29$ MeV,
${}^{15}_{8}O$	→	${}^{15}_{7}N+e^{+}+\nu_e+1.74$ MeV,
$p+{}^{15}_{7}N$	→	${}^{12}_{6}C+\alpha+4.96$ MeV.

Net effect: 4 protons⇒1 α-particle+2 electron neutrinos. Average neutrino energy loss per α-particle produced=1.71 MeV. We have omitted the positron–electron annihilations that occur.

situation may be different for protostars formed out of interstellar material already processed through a star. If its history includes a long period of helium burning, material returned to the interstellar medium will be relatively rich in carbon, nitrogen and oxygen. New protostars formed from this material will have the ingredients necessary to allow the CNO cycle to flourish, given a high enough temperature.

Helium burning is the next stage. This requires not only high temperatures (10^8 K) but also high densities (10^8 kg m^{-3}). The reactions are

$$\alpha + \alpha \rightarrow {}^8_4\text{Be} - 0.09 \text{ MeV}, \tag{14.4a}$$

$$\alpha + {}^8_4\text{Be} \rightarrow {}^{12}_6\text{C} + 2\gamma + 7.37 \text{ MeV}. \tag{14.4b}$$

The interesting fact here is that ${}^8_4$Be has a mean life of about 1×10^{-16} s and it decays to two α-particles. Thus reaction (a) leads to an equilibrium concentration of ${}^8_4$Be, which is about 10^{-9} that of α-particles. This concentration is too small to give the observed burning rates unless something else favours reaction (b). This something is a resonance in reaction (b) at a centre-of-mass energy of 287 keV, leading to a compound state in ${}^{12}_6$C at an excitation of 7.65 MeV which can either return to $\alpha + {}^8_4$Be or, in two steps of photon emission, reach the ground state of ${}^{12}_6$C (see Fig. 14.2). The reaction

$${}^{12}_6\text{C} + \alpha \rightarrow {}^{16}_8\text{O} + \gamma + 7.16 \text{ MeV}$$

will also begin to occur during the helium-burning phase.

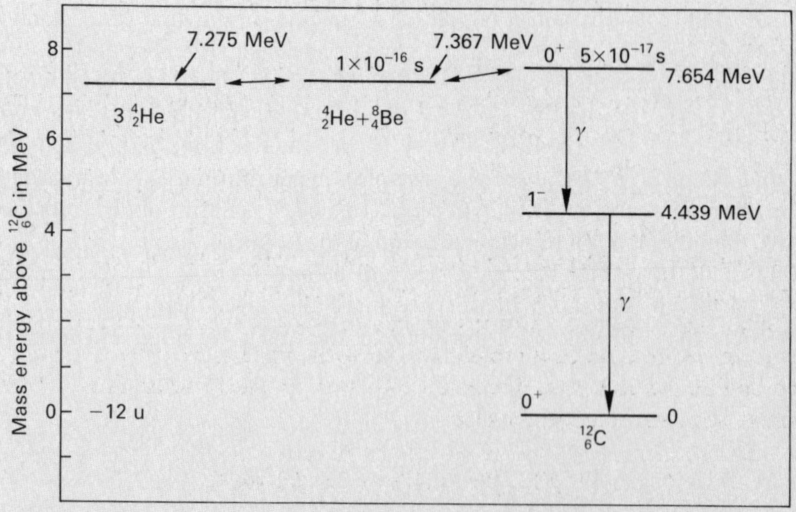

Fig. 14.2 The energy-level diagram of the nuclear levels with total $A = 12$ involved in the stellar burning of helium to produce ${}^{12}_6$C. The energy scale gives the mass or energy above the ground state of ${}^{12}_6$C. In a stellar helium-burning region there is a statistical equilibrium between the concentration of helium and the concentration of ${}^8_4$Be. The latter concentration is very low due to the high transition rate back to 2 ${}^4_2$He. (The mean life of ${}^8_4$Be is 1×10^{-16} s). The lifetime of the excited state of ${}^{12}_6$C at 7.654 MeV is 5×10^{-17} s and the branching fraction for the γ-ray decay mode is 4.2×10^{-4}. At any instant the concentration of ${}^{12}_6$C*(7.654) will also be very small but it can feed irreversibly the ${}^{12}_6$C ground state.

The next steps available in stellar nucleosynthesis are the burning of carbon and oxygen. The first requires temperatures of about 6×10^8 K, the latter about 2×10^9 K. These increasingly higher temperatures are, of course, the consequence of increasingly high Coulomb barriers to the reactions. The carbon reactions produce nuclei of $A=20$–24 and are listed in Table 14.5. The burning of oxygen produces nuclei in the mass range $A=24$–32: some of the reactions are also shown in Table 14.4. The dominant species produced are $^{28}_{14}Si$ and $^{32}_{16}S$.

The products of the carbon and oxygen burns are nuclei with $Z=10$ to 16 and the Coulomb barrier against reactions between these nuclei has now reached a level that the ignition temperature required is too high to allow continuation of the nuclear fusion style of burning applicable to carbon and oxygen burning. The reason is that the next temperature rise is sufficient to start the photodisintegration reactions

$$\gamma+{}^{28}_{14}\text{Si}\rightarrow{}^{27}_{13}\text{Al}+\text{p}-11.58\text{ MeV},$$

and

$$\gamma+{}^{28}_{14}\text{Si}\rightarrow{}^{24}_{12}\text{Mg}+\alpha-9.98\text{ MeV}.$$

This photodisintegration occurs only slowly so that the heavier elements remaining in the mix, such as silicon itself, can capture the photoproduced light particles and by successive stages build up higher mass nuclei. The lighter nuclei of the mix can also capture these light particles and build up towards silicon. Thus photodisintegration allows the redistribution of nucleons in favour of the build-up of nuclei beyond silicon: as we know the most strongly bound nuclei are in the region of iron and they are the product of this process, known as **silicon burning**.

We notice that in these processes of stellar nucleosynthesis there are some nuclei that are not produced. They are lithium, beryllium, and boron. Even if formed, they are easily broken down by proton-induced reactions at low temperatures and should not survive stellar processing and be returned to the interstellar medium. A possible production mechanism is cosmic ray bombardment of the interstellar gas. The cosmic radiation consists of charged nuclei, mostly protons, at energies of several hundred megaelectronvolts or greater. Their origin is not understood but they can cause nuclear reactions in the interstellar gas. Energetic protons or heavier particles striking nuclei of the interstellar medium can cause **spallation reactions**. In such reactions both or one of the colliding nuclei may break up into two or more parts: thus a 1 GeV proton striking carbon could cause

$$\text{p}+{}^{12}_{6}\text{C}\rightarrow{}^{9}_{4}\text{Be}+{}^{3}_{2}\text{He}+\text{p}-26.68\text{ MeV}.$$

It is obviously easy to invent reactions which produce the nuclei missing from the stellar product. This spallation process can account for the observed abundances of $^{6}_{3}Li$, $^{9}_{4}Be$, $^{10}_{5}B$, and $^{11}_{5}B$ but fails to account for the abundances of $^{7}_{3}Li$ and d. The amounts observed are possibly consistent with survival from the nucleosynthesis stage of the big bang. See Section 14.2.

Table 14.5 Some carbon- and oxygen-burning nuclear reactions

$^{12}_{6}C+{}^{12}_{6}C$	→	$^{20}_{10}Ne+\alpha+4.62$ MeV,
		$^{23}_{11}Na+p+2.24$ MeV,
		$^{23}_{12}Mg+n-2.61$ MeV,
		$^{24}_{12}Mg+\gamma+13.93$ MeV,
		$^{16}_{8}O+2\alpha-0.114$ MeV.
$^{16}_{8}O+{}^{16}_{8}O$	→	$^{24}_{12}Mg+2\alpha-0.393$ MeV,
		$^{28}_{14}Si+\alpha+9.59$ MeV,
		$^{31}_{15}P+p+7.68$ MeV,
		$^{31}_{16}S+n+1.46$ MeV,
		$^{32}_{16}S+\gamma+16.54$ MeV.

DEFINITIONS AND KEYWORDS

pp chain The principal chain of stellar nuclear reactions leading to the burning of hydrogen and the production of helium.

Helium catalysed chains The nuclear reactions in which the presence of helium catalyses the conversion of hydrogen to helium.

CNO chain The nuclear reactions in which the presence of carbon catalyses the conversion of hydrogen to helium.

Silicon burning The stellar nuclear reactions which produce iron-like nuclei from elements in the silicon region of the periodic table.

Spallation reactions Very energetic nuclear collisions in which a complex nucleus is disrupted into several lighter fragments.

14.5 Stellar nucleosynthesis 2

The increasing height of the Coulomb barrier ensures that it is the generation and capture of neutrons that plays the essential role in the synthesis of nuclei heavier than iron, although these processes are not important in the generation of energy by nuclear reactions in stars. Capture is radiative (Section 7.7) and may be followed by β-decay

$$(Z,A)+\mathrm{n}\rightarrow(Z,A+1)+\text{one or more } \gamma\text{-rays},$$
$$(Z,A+1)\rightarrow(Z+1,\ A+1)+\mathrm{e}^{-}+\nu_{\mathrm{e}}.$$

Successive neutron captures will build heavier nuclei. Suppose ω_{n} is the neutron capture rate and ω_{β} the beta decay rate, each one being an average value over many nuclei in a given circumstance, then two regimes may be defined:

(1) $\omega_{\mathrm{n}} \ll \omega_{\beta}$.

(2) $\omega_{\mathrm{n}} \gg \omega_{\beta}$.

DEFINITIONS AND KEYWORDS

s-process The process of stellar nucleosynthesis in which nuclei build to higher A capture of successive neutrons at intervals long compared to the mean life for β-decay.

r-process The process of stellar nucleosynthesis in which nuclei build to higher A by capture of successive neutrons at intervals short compared to all but the most energetic β-decays.

p-process The process of stellar nucleosynthesis in which proton-rich nuclei are produced by (p,γ) or (p,n) reactions.

The first defines the **s-process**, s for slow. The interval between neutron captures is much longer than for most β-decay processes and the build-up occurs along a route keeping close to the line of stability (Section 4.2). The second defines the **r-process**, r for rapid. Successive neutron captures occur before a significant number of β-decays can occur and nuclei are driven to neutron-rich regions far from the line of stability. At the neutron drip line (Section 4.9) further capture cannot occur. Thus more build-up is prevented unless a β^{-}-decay does occur. In the r-process the steps of nucleosynthesis therefore occur out to and just along the inside of this line. After the process is complete nuclei decay and populate the neutron-rich side of the line of stability.

It is possible to identify among the stable nuclei those most likely to have been made by the s-process and those by the r-process, although there is not necessarily a sharp distinction. However, among the stable nuclei there are some which cannot be made by either process. This can be seen by considering the mass curve for even-A nuclei. If there are two stable nuclei as in Fig. 5.7, the neutron capture processes produce isobars of this A (=100) on the neutron-rich side and the nucleus which is stable and has the lowest Z (=42) screens the production by decay of the nucleus $Z+2$ (=44). Such by-passed nuclei are probably produced by (p,n) or (p,γ) reactions. This is the **p-process**.

The s-process requires a source of neutrons and the reactions which are considered to be the strongest candidates are

$$^{22}_{10}\mathrm{Ne}+\alpha\rightarrow{}^{25}_{12}\mathrm{Mg}+\mathrm{n}-0.48\ \mathrm{MeV},$$

and

$$^{13}_{6}\mathrm{C}+\alpha\rightarrow{}^{16}_{8}\mathrm{O}+\mathrm{n}-0.91\ \mathrm{MeV}.$$

However, the circumstances in which these reactions can provide the necessary neutron flux and at the right place in the star are still a matter of investigation.

The s-process cannot make nuclei beyond $A=209$ because further neutron capture produces nuclei in the region of the periodic table where α-decay is fast and returns nuclei to $A=209$ or below. For example:

$$^{209}_{83}\text{Bi} + \text{n} \rightarrow {}^{210}_{83}\text{Bi} + \gamma,$$

$$\downarrow$$

$$^{210}_{83}\text{Bi} \rightarrow {}^{210}_{84}\text{Po} + \text{e}^- + \bar{\nu}_\text{e}, \qquad \tau = 7.2 \text{ days},$$

$$\downarrow$$

$$^{210}_{84}\text{Po} \rightarrow {}^{206}_{82}\text{Pb} + \alpha, \qquad \tau = 200 \text{ days}.$$

The natural occurrence of elements of A greater than 209 means that they are uniquely r-process products.

The source of neutrons for the r-process is clearly different from that for the s-process. The neutron flux must be very high ($\sim 10^{32}\ \text{m}^{-2}\ \text{s}^{-1}$) and need last for a short time only. This looks like explosive conditions and the best candidate is a supernova event. We shall see that it is likely that a supernova produces conditions in which a substantial part of the outer layer of the core becomes a mixture of iron, iron-like nuclei and neutrons in which the r-process occurs. This material may be ejected during the collapse of the rest of the core. The rapidity of the r-process means that the check at $A = 209$ does not operate and heavier nuclei are produced. However, the limit on increasing A by neutron capture is reached when the transition rates due to induced and spontaneous fission, which grow with A (Sections 5.5 and 7.12), destroy these heavy nuclei as fast as they are generated. The fission products are far down the periodic table and have to start again the journey of heavy element building. If this heavy material is ejected into the interstellar medium, the transuranic elements will have time to decay until they reach $^{235}_{92}\text{U}$, $^{238}_{92}\text{U}$, or $^{232}_{90}\text{Th}$. As we found in Section 5.4 these elements have lifetimes long enough to ensure their survival in significant quantities over 10^9 years. Have a look again at Problems 2.8, which now gives you the time since the terrestrial uranium was processed in a supernova.

14.6 Nucleosynthesis: summary

Figure 14.3 shows the abundances of the elements in the solar system relative to silicon as 10^6. The solar system includes the Sun, so we see immediately the 16 to 1 number ratio of hydrogen to helium typical of the end of big bang nucleosynthesis. The Sun has consumed only about 5% of its hydrogen and is still 5×10^9 years from the beginning of helium burning. Since spectroscopy reveals that it contains significant quantities of medium-A nuclei, some of the material from which it formed must have been through at least one stage of stellar processing. The planets are clearly the product of stellar processing, including the r-process.

Looking at other abundances we see the relative rarity of Li, Be, and B, which we have already mentioned and now understand as a consequence of the fact that they do not feature in helium to carbon burning. There are substantial abundances of $^{12}_{6}\text{C}$, $^{16}_{8}\text{O}$, and $^{20}_{10}\text{Ne}$, which can be understood in terms of their central role in nuclear burning and their even–even or doubly-closed shell structure: they have large binding energies and are hard to disintegrate once formed. The pairing term of the semi-empirical mass formula tells us that even-A nuclei are more strongly bound and that explains why they are more abundant than odd-A nuclei (Fig. 14.3). It is also implied that the even $N(=A-Z)$ nuclei will be the most abundant of the isotopes of a given element. There is a sudden fall in relative abundance beyond $Z = 20$. The nucleus $^{40}_{20}\text{Ca}$ is the last $Z = N$, even–even, doubly-closed shell nucleus and the next even–even

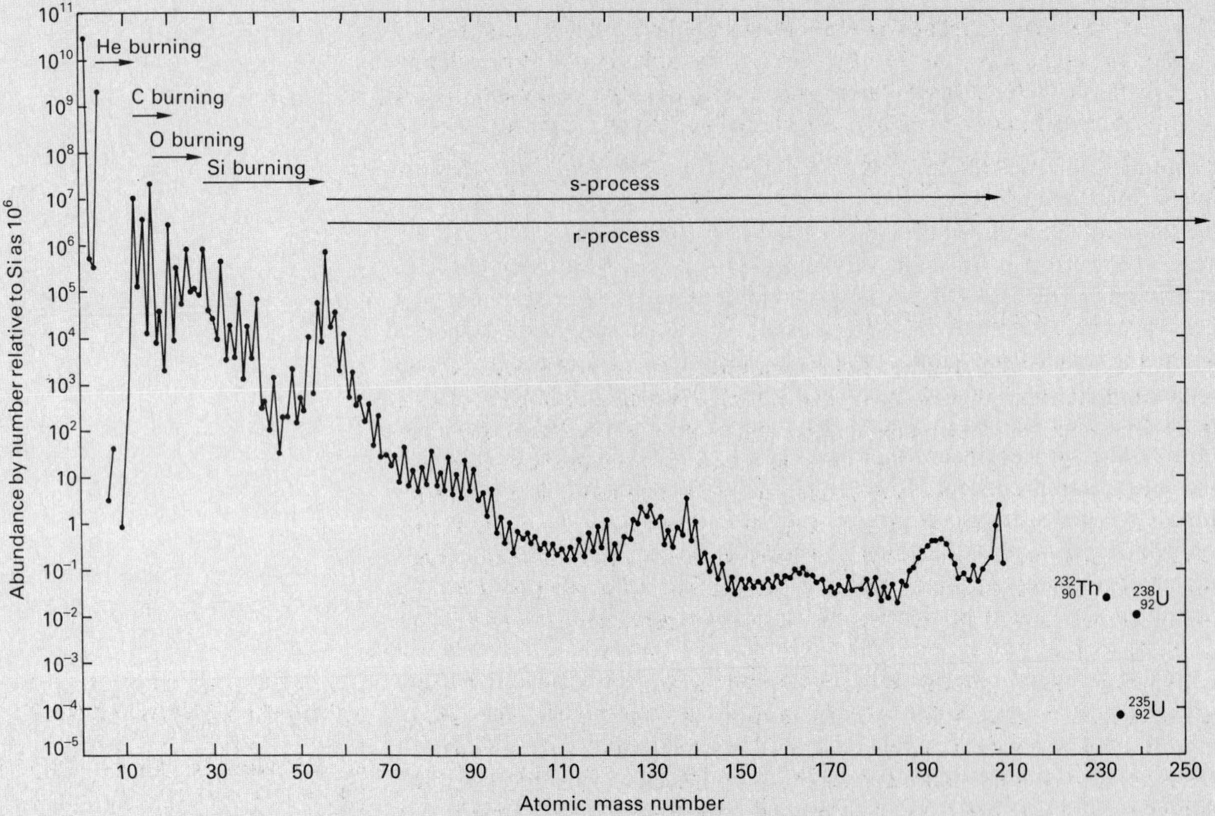

Fig. 14.3 The abundance by number of atoms relative to $^{28}_{14}Si + ^{29}_{14}Si + ^{30}_{14}Si$ as 10^6 in the solar system, as a function of the atomic mass number A. The relative H, He abundances are close to those existing after Big Bang nucleosynthesis. The obvious absences are those of $A = 5,8$, and $A > 209$ apart from $^{232}_{90}Th$, $^{234}_{92}U$ (not shown), $^{235}_{92}U$ and $^{238}_{92}U$. Large abundances occur for the even–even nuclei at $A = 12$, 16, 20, 24, 28, 32, and 40. The broad peak near $A = 56$ is the iron-like elements, noted as those with the greatest binding energy per nucleon in the periodic table. The abundance enhancing effect of magic numbers are also seen!

(1) at $A = 86$ to 90 due to $N = 50$,
(2) at $A = 114$ to 120 due to $Z = 50$,
(3) at $A = 138$ due to $N = 82$,
(4) at $A = 208$ due to $Z = 82$, $N = 126$.

In addition, the even-to-odd A abundance due to the pairing term effect on the binding energy is clearly visible.

stable nucleus, $^{42}_{20}Ca$, will be less strongly bound and more easily disintegrated by photons. The latter's abundance is less than 10^{-2} that of $^{40}_{20}Ca$.

Continuing up the periodic table we come to the high abundance of $^{56}_{26}Fe$: this is the most strongly bound nucleus, and element formation beyond this point costs energy. For this reason it is almost the last step in the nuclear burning program and we expect high abundance, certainly relative to heavier elements. Thereafter increasing A leads to steadily decreasing abundances with enhancements in regions where magic numbers occur. These are noted in the caption to Fig. 14.3. At $_{82}Pb$, there is another peak due to the doubly-closed shell $^{208}_{82}Pb$ and the other lead isotopes which are singly-closed shell nuclei: this structure makes lead the final resting point of the three naturally occurring radioactive series. In addition, $^{209}_{83}Bi$ (singly-closed) is the product of a series ($A = 2n + 1$) which is no longer to be found in nature but presumably was formed in the r-process and has decayed to bismuth (Section 5.4). Thus this region is doubly favoured by being the end of the s-process and the final depository in the decay of large-A r-process products.

Of course, theories of stellar nucleosynthesis must be tested quantitatively. The behaviour of stars is governed by so many variables that it is possible to model only parts of stellar interiors or histories and thus the predictions about stellar evolution and nucleosynthesis have limited validity. Such models that do exist make what are considered to be reasonably good predictions about relative nuclear abundances. These models have as a basis the story that we have described in this and the two previous sections.

14.7 Neutrinos in stellar evolution 1

In the hydrogen-burning stage of stellar evolution four protons become one α-particle and that must include two steps, in each of which a proton becomes a neutron by what is essentially either $p \to n + e^+ + \nu_e$ or its equivalent $e^- + p \to n + \nu_e$ (see Table 14.3). The electron neutrinos produced escape easily and immediately from the core of the star and contribute to the loss of energy from the star. For the pp chain the energy loss by neutrinos is an average of 0.52 MeV out of a total of 26.72 MeV generated with every α-particle produced. For the CNO cycle the loss is 1.71 MeV. The total effect for the Sun, allowing for all reactions, is that about 3% of the energy produced is removed by neutrinos (see Problem 14.1).

The remaining energy produced in the core of the Sun takes about 10^6 years to be transported to the surface. Fluctuations in the state of the core taking place with a characteristic time of less than 10^4 years would not therefore be visible as variations in the **luminosity** of the Sun, which is governed by surface conditions on a short time scale. However, the neutrinos escape immediately and are the one signal about the state of the core that is, in principle, accessible to observers. Table 14.6 gives some expected ν_e fluxes at the Earth. The problems of detecting these neutrinos are formidable: most have energies less than 1 MeV and interaction cross-sections which are very small (of order 10^{-47} m^2 and less, see Table 14.7): the interaction products are difficult to observe in the large detectors that must be employed. The first successful detection of solar neutrinos was made by R. Davis and collaborators (1978) using the reaction

$$\nu_e + {}^{37}_{17}\mathrm{Cl} \to {}^{37}_{18}\mathrm{Ar} + e^- - 0.81\ \mathrm{MeV}.$$

DEFINITIONS AND KEYWORDS

Luminosity The total energy flux from a star.

Auger effect The emission of atomic electrons as an atom de-excites after the creation of an inner shell vacancy by electron capture or by the photoelectric effect.

Table 14.6 Theoretically calculated electron neutrino fluxes at the Earth due to nuclear reactions in the Sun.

Reaction	$\nu_e\ \mathrm{m}^{-2}\,\mathrm{s}^{-1}$
$p + p \to e^+ + \nu_e + d$	6.1×10^{14}
${}^7_4\mathrm{Be} + e^- \to \nu_e + {}^7_3\mathrm{Li}$	3.4×10^{13}
${}^{14}_7\mathrm{N} \to {}^{14}_6\mathrm{C} + e^+ + \nu_e$	2.6×10^{12}
${}^{15}_8\mathrm{O} \to {}^{15}_7\mathrm{N} + e^+ + \nu_e$	1.8×10^{12}
${}^8_5\mathrm{B} \to {}^8_4\mathrm{Be} + e^+ + \nu_e$	3.2×10^{12}

These numbers are uncertain since they are sensitive to the assumptions of the solar model used.

Table 14.7 The approximate values of the theoretical total cross-sections for elastic scattering and reactions of neutrinos and antineutrinos with electrons, protons, neutrons and complex nuclei. We assume $m_\nu=0$ and $E_\nu > m_e c^2$, where E_ν is the neutrino energy.

Elastic scattering reactions	σ/E_ν (fm^2 MeV^{-1})
$\nu_e + e^- \to \nu_e + e^-$	9.4×10^{-19}
$\bar{\nu}_e + e^- \to \bar{\nu}_e + e^-$	3.9×10^{-19}
$\nu_l + e^- \to \nu_l + e^-$ ($l = \mu$ or τ)	1.6×10^{-19}
$\bar{\nu}_l + e^- \to \bar{\nu}_l + e^-$ ($l = \mu$ or τ)	1.3×10^{-19}

Coherent nuclear elastic scattering (all neutrinos and anti-neutrinos)	σ/E_ν^2 (fm^2 MeV^{-2})
$\nu + (Z,A) \to \nu + (Z,A)$	$8.8\, A^2 \times 10^{-20}$

Neutrino absorption	$\sigma/(E_\nu + Q)^2$ (fm^2 MeV^{-2})
$\bar{\nu}_e + p \to n + e^+ - 1.80$ MeV $\nu_e + n \to p + e^- + 0.78$ MeV	$\sim 6.7 \times 10^{-18}$

The ground state to ground state transition implied has a very small cross-section, but the transition to an excited state at 5.1 MeV of $^{37}_{18}$Ar (Fig. 14.4) is superallowed and has a relatively large cross-section (Table 14.8). The threshold is then 5.9 MeV and the production of $^{37}_{18}$Ar really only measures the flux neutrinos from the decay

$$^{8}_{5}\mathrm{B} \to {}^{8}_{4}\mathrm{Be}^* + e^+ + \nu_e + 14.02\ \mathrm{MeV},$$

which occurs in one of the α-particle catalysed reactions (Table 14.3). The electron neutrinos from this decay have an average energy of 7.2 MeV. However, this contributes only about 10^{-4} of the neutrino flux at the Earth, and the yield of $^{37}_{17}$Cl to $^{37}_{18}$Ar conversions will be correspondingly small.

The detector used by Davis (see Bahcall and Davis, 1976) was a tank containing 3.9×10^5 litres of the liquid, perchloroethylene (C_2Cl_4). The chlorine is 24.5% $^{37}_{17}$Cl. The ground state of $^{37}_{18}$A has a mean life of 49.4 days. The tank is flushed regularly with helium which carries the atoms of $^{37}_{18}$Ar to filters from which it is transferred to a detector. The detector is sensitive to the decay:

$$^{37}_{18}\mathrm{Ar} + e^- \to {}^{37}_{17}\mathrm{Cl} + \nu_e + 0.81\ \mathrm{MeV}, \qquad \tau = 49.4\ \text{days}.$$

See Fig. 14.4. The capture of the electron leaves a vacancy in the chlorine atomic K-shell which is refilled by rearrangement of the atomic electrons in the course of which an atomic electron of energy 2.8 keV can be emitted. (This atomic de-excitation by emitting one or more electrons is called the **Auger effect**.) This allows the detection of the $^{37}_{18}$Ar atoms with an efficiency of about 50%. The actual rate of production is about one $^{37}_{18}$Ar atom every two days! The whole equipment is installed about 1500 m underground to shield as far as possible

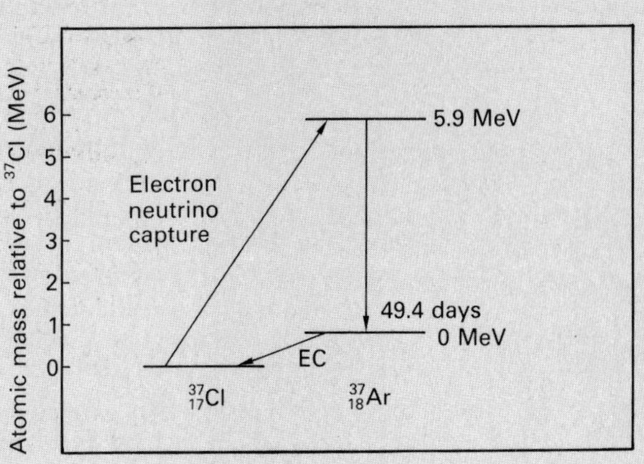

Fig 14.4 The reaction used by Davis and his collaborators in their solar neutrino detector. The dominant capture of ν_e from the Sun by the $^{37}_{17}$Cl is to the 5.9 MeV excited state of $^{37}_{18}$Ar, which decays rapidly to the ground state by γ-ray emission. This ground state has a mean life of 49.4 days and decays by electron capture to the ground state of $^{37}_{17}$Cl. (There are many other excited states in $^{37}_{18}$Ar not shown.) The $^{37}_{18}$Ar produced is detected by the Auger electrons emitted following electron capture by this nucleus.

Table 14.8 The average absorption cross-section of $^{37}_{17}$Cl for electron neutrinos from different reactions in the Sun.

Reaction	$\sigma(^{37}_{17}\text{Cl}+\nu_e \rightarrow e^- + ^{37}_{18}\text{Ar})$ m^2
$^{7}_{4}\text{Be}+e^- \rightarrow ^{7}_{3}\text{Li}+\nu_e$	2.4×10^{-50}
$^{13}_{7}\text{N} \rightarrow ^{13}_{6}\text{C}+e^+ +\nu_e$	1.6×10^{-50}
$^{15}_{8}\text{O} \rightarrow ^{15}_{7}\text{N}+e^+ +\nu_e$	6.6×10^{-50}
$^{8}_{5}\text{B} \rightarrow ^{8}_{4}\text{Be}+e^+ +\nu_e$	1.1×10^{-46}

from the effect of cosmic rays and sea level activity. Nonetheless to detect these few atoms against large residual backgrounds is a remarkable achievement. The rate of production of $^{37}_{18}$Ar is expressed in solar neutrino units (SNU): one SNU is 10^{-36} captures per second per target nucleus. The rate expected from current theories about the solar interior is 6 ± 1.5 SNU. Davis measures 2.2 ± 0.4 SNU. This result has stood the test of critical examination and repetition over several years.

This result is very important because it could mean one of several things:

1. The current models of the Sun's interior are incorrect and are overestimating the production of $^{8}_{5}$B.
2. The core of the Sun is not burning steadily but may suffer fluctuations or oscillations in temperature on a time scale of decades. The rate of production of $^{8}_{5}$B is very sensitive to temperature. The measurements of Davis could have caught the Sun's core in a quiet period.
3. Neutrinos may have a small mass. As we have mentioned in Section 12.12 this can lead to oscillations between the three neutrino generations:

$$\nu_e \leftrightarrow \nu_\mu \leftrightarrow \nu_\tau$$

Low energy ν_μ or ν_τ cannot cause easily detectable reactions and so it is possible that the flux of ν_e, which is detectable, is reduced at the Earth to a fraction of its value in the absence of oscillations.

Current research is aimed at discovering the cause of the discrepancy. Efforts are being made to improve stellar models. More important, efforts are being made to detect solar neutrinos by other methods because it is essential to confirm Davis's result and to attempt to detect neutrinos of lower energy. A long-term effort will also determine the time dependence of the neutrino detection rate.

The readers will obtain a better idea of the physical numbers involved by trying Problems 14.1–14.3.

PROBLEMS

14.1 Estimate the flux of pp chain solar neutrinos at the Earth given the solar luminosity $L_\odot$ (Table 14.2) and the fact that two neutrinos are produced in every conversion of four protons to one α-particle and that this conversion produces 26.72 MeV. Assume that the pp chain is solely responsible for the solar energy production. If 3% of the energy is lost this way what is the average neutrino energy?

14.2 The Davis detector contains 600 tonnes of C_2Cl_4 with the chlorine containing 24.47% by number of ${}^{37}_{17}$Cl. The cross-section for the reaction ${}^{37}_{17}\text{Cl}+\nu_e \to e^- + {}^{37}_{18}\text{Ar}$ averaged over the over-threshold part of the neutrino spectrum expected from the decay ${}^{8}_{5}\text{B} \to {}^{8}_{4}\text{B}+e^+ + \nu_e$ is 1×10^{-46} m^2. Given that the production of ${}^{37}_{18}$Ar is one every two days, calculate the capture rate in SNU and the flux of detectable neutrinos.

14.3 Investigate how the following reactions would be used in order to detect neutrinos:

1. ${}^{71}_{31}\text{Ga} + \nu_e \to {}^{71}_{32}\text{Ge} + e^- - 0.236\ \text{MeV}$
2. ${}^{115}_{49}\text{In} + \nu_e \to {}^{115}_{50}\text{Sn}^{**} + e^- - 0.120\ \text{MeV}$

$$\downarrow$$
$${}^{115}_{50}\text{Sn}^{*} + \gamma + 0.116\ \text{MeV}$$
$$\downarrow$$
$${}^{115}_{50}\text{Sn} + \gamma + 0.498\ \text{MeV}.$$

[For the gallium detector see Kirsten (1979) and for an indium detector see Raghavan (1976). The ** and * indicate excited states]

14.8 Neutrinos in stellar evolution 2

As stellar core temperatures rise above 10^8 K there is a rapidly increasing energy loss from the hot core by the production and escape of neutrino–antineutrino pairs. The production mechanisms are as follows.

1. The reaction

$$e^+ + e^- \to \nu_e + \bar{\nu}_e + 1.02\ \text{MeV},$$

which can occur by $W^\pm$ exchange or through a Z^0 intermediate state.

2. The reactions

$$e^+ + e^- \to \nu_l + \bar{\nu}_l, \qquad l = \mu \text{ or } \tau,$$

which occur through a Z^0 intermediate state.

3. The reactions

$$\gamma + e^\pm \to \nu_l + \bar{\nu}_l + e^\pm, \qquad l = \text{e}, \mu, \text{ or } \tau.$$

This is like the Compton scattering process $\gamma + e^- \rightarrow \gamma + e^-$ (Section 11.4) but instead of emitting a final state photon, a virtual Z^0 is emitted which becomes a $\nu\bar{\nu}$ pair.

4. The reactions

$$e^{\pm} + (Z,A) \rightarrow e^{\pm} + (Z,A) + \nu_l + \bar{\nu}_l, \quad l = e, \mu \text{ or } \tau.$$

This scattering of an electron or positron in the electric field of a nucleus (Z,A) in which a virtual Z^0 is emitted which in turn becomes a real neutrino–antineutrino pair.

Of course the positrons and electrons required for these reactions are present because of an equilibrium between these particles and the photons of the radiation field at these temperatures. Pairs are produced by photon materialization in the electric fields of either nuclei or electrons or positrons. Photons are produced as bremsstrahlung. (Try Problem 14.4.) All these processes are functions of the temperature and of the density, and become more important as nucleosynthesis reaches along the periodic table until, at the stage of silicon burning, the rate of neutrino energy loss can become equal to the rate of energy production in nuclear reactions. We shall refer to these reactions again so we summarize them in Table 14.9.

At this stage of evolution, where an iron-like core is growing, the pressure and temperature of this core are rising. As this happens the electrons present become degenerate and their Fermi energy increases to the point where electron capture can occur either by protons or by nuclei. For example,

$$p + e^- \rightarrow n + \nu_e - 0.782\ \text{MeV}.$$

This process is called **neutronization** and is also a source of neutrinos.

The escape of the neutrinos from the core will occur if the density does not become too high. (An example when neutrinos are trapped is described in the next section.) Thus at certain stages of stellar evolution a very rapid loss of energy from the core is possible which can strongly influence the evolution of a star.

DEFINITIONS AND KEYWORDS

Neutronization The process by which electron degeneracy forces electron capture by protons and nuclei so as to produce neutrons.

Table 14.9 Particle reactions producing neutrinos of all kinds in stellar interiors.

$e^+ + e^-$	$\rightarrow$	$\nu_l + \bar{\nu}_l,$	$l = e, \mu, \text{ or } \tau$
$\gamma + e^{\pm}$	$\rightarrow$	$e^{\pm} + \nu_l + \bar{\nu}_l,$	
$e^{\pm} + (Z,A)$	$\rightarrow$	$e^{\pm} + (Z,A) + \nu_l + \bar{\nu}_l,$	
$e^- + p$	$\rightarrow$	$n + \nu_e,$	
$e^+ + n$	$\rightarrow$	$p + \bar{\nu}_e.$	

PROBLEM

14.4 Draw the Feynman diagrams for the processes:

(a) $e^+ + e^- \rightarrow \nu_\mu + \bar{\nu}_\mu,$

(b) $e^+ + e^- \rightarrow \nu_e + \bar{\nu}_e,$

(c) $\gamma + e^{\pm} \rightarrow \nu_e + \bar{\nu}_e + e^{\pm},$

(d) $e^{\pm} + (Z,A) \rightarrow e^{\pm} + (Z,A) + \nu_e + \bar{\nu}_e.$

14.9 Supernova

In Section 14.3 we arrived at the point where we now believe that core collapse occurs. For stars of mass greater than $10M_{\odot}$ this collapse is the probable explanation of what are known as type II supernovae. There are several kinds of explosive stellar events but supernova are the most spectacular. Only six have been seen and recorded in our own galaxy in the last 1000 years and all those before the use of the telescope. Many others have been studied in nearby galaxies so that there is some information about their behaviour: the classification into two types may, however, be faulty and it is possible there is a spectrum of types with every supernova a unique event.

A supernova shows itself by a huge increase in luminosity of a previously undistinguished star. We have no visual sighting of the start of a supernova but within a day its light output can multiply by nearly 10^{11}, staying at that level for several days and then declining over a period of months. This increase in luminosity indicates a huge release of energy and the only possible source is gravitational potential energy or explosive nuclear burning.

Why does this collapse occur? During the previous evolution of the star the gravitational inward pull is balanced by the internal pressure. Nuclear reactions delay the contraction of the star. When nuclear burning ceases the core of the star can support gravity by means of the electron degeneracy pressure (Section 14.3). However, there is a limit to the mass that this pressure will support, known as the **Chandrasekhar mass**. It is 1.2–$2M_{\odot}$, depending on the ratio of electrons to baryons. Above this mass there is insufficient support and the star core must collapse. In a pre-supernova the possible scenario is that the iron core of a star of mass greater than $10M_{\odot}$ gradually grows in mass as the ashes of silicon burning in the layer above fall on to the iron core. When the core mass reaches the Chandrasekhar limit the system cannot adapt. As collapse occurs, released gravitational potential energy raises the temperature: but instead of raising the pressure, a large amount of energy is consumed in beginning the photodisintegration of iron to α-particles and finally to nucleons($>1.7\times10^{45}$ J for each 1 $M_{\odot}$ of iron dis-assembled) and neutronization ($>1.5\times10^{44}$ J for each $1M_{\odot}$ of protons and electrons converted to neutrons and neutrinos). These processes consume energy which might otherwise go to raising the temperature and pressure and, in the absence of increasing pressure, collapse begins in earnest.

The collapse separates the core from the surrounding layers and it is possible to consider the behaviour of the core in isolation. There remain many questions unanswered about the events in this collapse but the generally agreed picture is as follows. Collapse continues until densities approach nuclear densities ($\sim 2\times10^{17}$ kg m^{-3}), or greater, when the nuclei and nucleons of the core undergo a phase change to a state of degenerate nucleons. At this point the short-range nucleon–nucleon repulsive force stiffens the core, collapse stops and there is a bounce which sends a shock wave back out through the now infalling outer layers of the core. What remains is a hot neutron star. The gravitational potential energy released is of the order of 2–5×10^{46} J. A small part of this goes into the shock wave and ejecting the outer layers of the star ($\sim 10^{44}$ J). Table 14.10 summarizes the change of gross properties of the core from before to after collapse.

During the early stages of the contraction, neutronization leads to production in less than one second of a large flux of electron neutrinos, but as core densities pass 10^{15} kg m^{-3}, the neutrinos of all varieties produced in the core (see Section

Table 14.10 Stellar core properties just before collapse starts and after collapse.

	Pre-collapse	Post-collapse
Radius	~1000 km	~20 km
Mass	~1 $M_\odot$	~1 $M_\odot$
Temperature	10^8–10^9 K	~10^{11} K
Mean density	~5×10^{11} kg m^{-3}	~10^{17} kg m^{-3}
Composition	Iron-like elements	Mainly neutrons
Binding energy	~10^{44} J	~10^{46} J

14.8) have mean free paths less than the core radius, and decreasing, with neutrino elastic scattering from nuclei providing the major contributor to the cross-section (see Table 14.7 and the accompanying comment). Thus the neutrinos are trapped in the core and until contraction stops there is a rough equilibrium between neutrino production and absorption. When contraction of the core stops it starts to cool by neutrino emission; all three varieties of neutrino and their anti-neutrinos are produced, diffuse outwards and have their last collisions in an outer layer of the core called the **neutrino sphere**. The electron neutrinos ν_e emerge with an average energy about 10–15 MeV. For $\bar{\nu}_e$, the effective neutrino sphere is thicker because $\bar{\nu}_e$ cross-sections are smaller than those for ν_e and they leave with an energy reflecting the temperature of a deeper layer and have a mean energy 15–20 MeV. Similarly ν_μ, $\bar{\nu}_\mu$, ν_τ and $\bar{\nu}_\tau$ will have average energies of 20–30 MeV. In all a $1.4M_\odot$ core has about 3×10^{46} J of energy which escapes as neutrinos in a time of 3–10 s. What is left is a cool, degenerate neutron star.

Meanwhile what is happening to the outer layers of the star? Supernovae explosions are observed so that if our core story is correct there must be some mechanism by which about 10^{44} J of energy from core collapse is used to eject the outer layers and to give the spectacular optical display that is a supernova. There is an uncertainty about the mechanism. Outward push from the shock wave generated on core bounce is one suggestion. This shock wave develops in the core and has to fight its way out of the infalling outer layers of the core: there is some doubt as to whether it can do this successfully so as to reach and eject the outer layers. A second proposal is that the outgoing neutrino flux deposits energy and momentum by scattering in the outer layers and thereby causes their

Comment on neutrinos in the collapse of a supernova core

In a supernova core during collapse there is still a large fraction of iron-like nuclei and the coherent nuclear scattering determines the mean free path of neutrinos. Thus if the density is ρ and Z_A is the mass fraction of nuclei (Z,A), the partial mean free path λ for that nucleus is given by

$$\lambda = \frac{1}{n_A\sigma} = \frac{M_p}{\rho Z_A A E_\nu (8.8\times10^{50})}\ \mathrm{m},$$

where M_p is the proton mass, n_A is the number density and σ the total neutrino cross-section, both for nuclei (Z,A). At $\rho=10^{17}$ kg m^{-2}, $A=56$, $Z_A=0.5$, $E_\nu=10$ MeV, we have $\lambda=68$ m. These are the approximate, possible core conditions near the end of collapse when the radius is about 18 km. It is clear that the neutrinos are temporarily trapped and can only diffuse out of the core.

ejection. However, this mechanism may also be inadequate. This subject is now a matter of intense study.

It would be gratifying to be able to describe a neutron star and to give a clear picture of its structure. However, that would take us into the subject of the equation of state of matter at densities comparable to and greater than those of nuclei, applied to a system of great complexity. The subject is one of ongoing research and clarification and one which is outside the scope of this book. However, it is clear that a neutron star is not just a large nucleus of neutrons: it will contain electrons and protons, and the surface layers will probably contain nuclei. It has been suggested that the surface is actually solid. A neutron star is expected to be rotating and centrifugal forces at the equator limit the period of rotation to greater than about 1 ms. They are also expected to have intense magnetic fields, of order 10^8 T, due to a magnetic dipole moment: in general the magnetic axis need not coincide with the rotational axis so that a distant observer will see a rotating dipole field. If the neutron star is surrounded by a plasma, then the rotation of the dipole field will lead to the movement of charged particles along the magnetic field and that in turn will lead to radio-frequency emission along the direction of the magnetic poles or in a fan from the magnetic equator. To a distant observer the neutron star may look like a lighthouse with a beam which flashes into his radio telescope at regular intervals. The discovery in 1968 of objects, called **pulsars**, which are emitting short bursts of radio frequency at very regular intervals, gave life to the proposal that neutron stars existed. There is a pulsar at the centre of the Crab nebula which is believed to be the material ejected from the supernova of AD 1054. This and the properties of pulsars have given astrophysicists the confidence to identify these objects as neutron stars.

DEFINITIONS AND KEYWORDS

Chandrasekhar mass The largest mass a star may have and be supported by its electron degeneracy pressure.

Neutrino sphere That outer layer of a collapsed stellar core in which escaping neutrinos suffer their last collisions with core material.

Pulsars Objects emitting short radio pulses with a period (1–4000 ms) unique to each. The pulses have a time length which is a small fraction of the period.

The possibility of the production of a black hole instead of a neutron star is discussed briefly in Section 14.11.

Problems 14.5–14.9 will put some of the gigantic energies involved into perspective.

After the initial optical display, the luminosity of a supernova wanes and it may become optically invisible. However, at intermediate times the decline in luminosity is often exponential with a mean life identifiable with that of a radioactive material. Thus in the case of SN1987A, to be described in the next section, after about 120 days the decline in luminosity was that expected if the energy source was the β-decay of $^{56}_{27}$Co. This is also true of supernovae seen in other galaxies. The conclusion is that the luminous material is rich in material from the iron-like stellar core that has not been trapped in the collapse to a neutron star.

14.10 SN1987A

At 07.35 Greenwich Mean Time, 23 February 1987, two detectors looking for evidence of proton decay observed events which can be interpreted as interactions due to a flash of neutrinos from a supernova (SN1987A, A = for the first of 1987) which was visually spotted 18 hours later (see Fig. 14.5). The important thing is that this supernova, although not in our own galaxy, is in the Large Magellanic Cloud (LMC), a close satellite galaxy to our own, and therefore relatively close. This supernova has been the brightest and closest for nearly 400 years (the last was in AD 1667) and provided astronomers for the first time with an opportunity to study with modern techniques a close-by supernova with the unexpected bonus of the detected neutrinos.

The detectors operating at the time were those labelled IMB and Kamiokande II. They are described in Table 13.8. The properties of the Cerenkov effect and the principles of operation are described in Figs. 13.4 and 13.5. These detectors were designed to detect possible proton decays if these exist (Sections 13.6 and 13.7) and so are sensitive to positrons and electrons moving through the water that constitutes their sample of protons. An examination of Table 14.7 shows that, for a flux of neutrinos of all kinds of energy of order 10 MeV, the dominating reaction in water will be

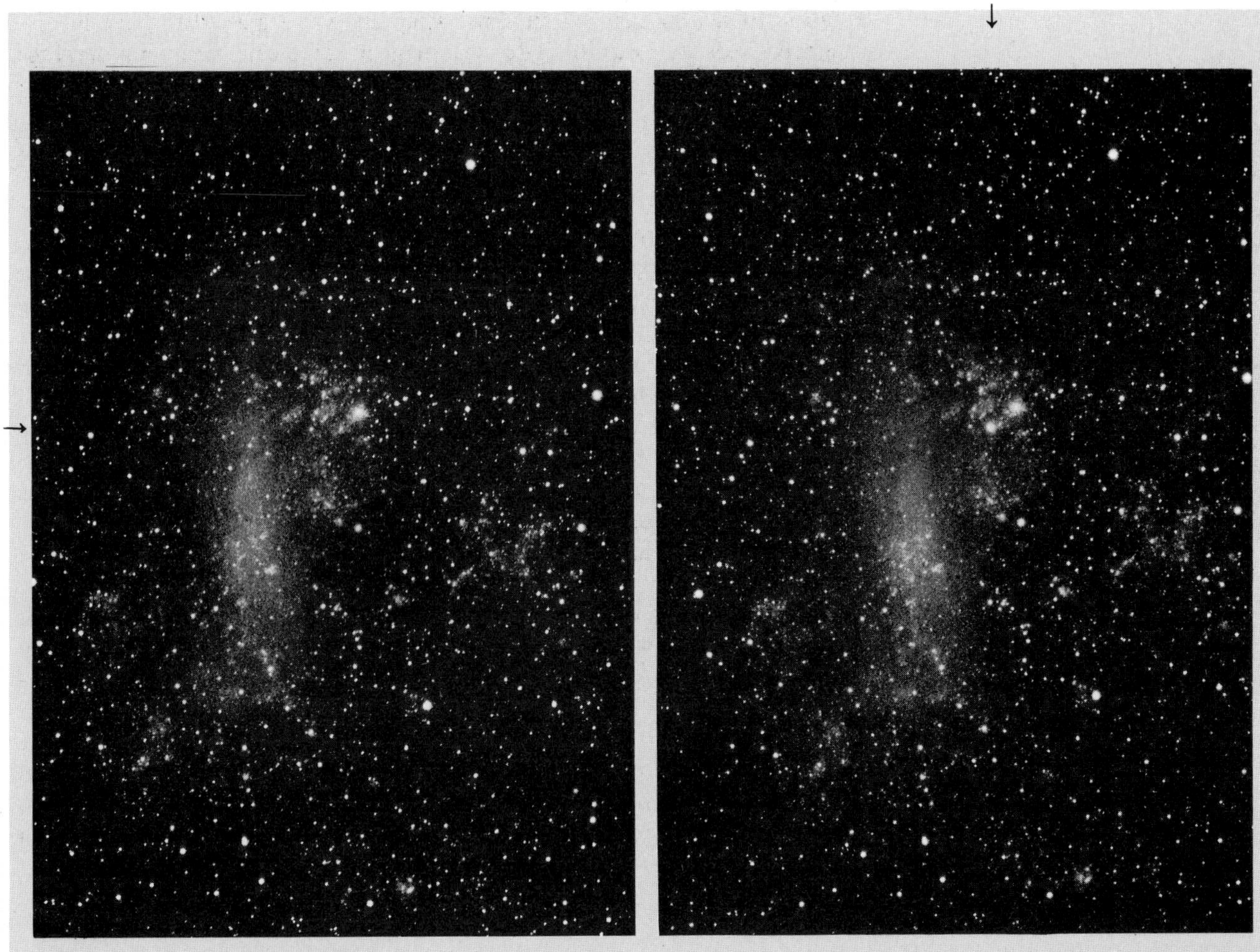

(a) (b)

Fig. 14.5 The Large Magellanic Cloud (a) before and (b) just after 23 February 1987. These photographs cover about four by three degrees of the sky near the South Pole. The general haze is caused by the large number of unresolved stars in this galaxy, the nearest galaxy neighbour to our own galaxy. In addition there are many nebulae in this galaxy; these glow with the light of recombining hydrogen and indicate regions active in the formation of new stars. The largest is the Tarantula Nebula which is visible right and up from the centre of (a). Nearby the Tarantula is another region of star formation called 30 Doradus which is the site of this 1987 supernova. In (b) the supernova is clearly visible.

Individual stars should appear as points of light but in the photo these points become diffraction discs in the limited aperture of the telescope used. In addition the photographic emulsion has limited resolution and the light from brighter stars spreads through the emulsion giving large images. The precursor to the supernova is lost in the unresolved haze of (a) but the supernova is so bright that its photographic image is as great as that of many stars much closer to us than is the Large Magellanic Cloud.

$$\bar{\nu}_e + p \rightarrow n + e^+ - 1.80 \text{ MeV}.$$

The neutrino-electron elastic scattering, allowing for five electrons per free proton in water, will contribute about 7% more detection efficiency for $\bar{\nu}_e$, and some for the ν_e and for the μ- and τ-neutrinos and antineutrinos. If we account only for the positron producing reaction we can use the detector results to make a rough estimate of the total ν flux at the Earth. This calculation is done in Table 14.12 using the results of the IMB and Kamiokande II events given in Table 14.11. We come up with about 5×10^{13} neutrinos per square metre. Calculating their energy we arrive at an astonishing 2×10^{46} J of neutrino energy of all kinds emitted from the core in about 10 s. This is about 300 times the entire energy output of the Sun in the last 5 billion years and is truly an astonishing amount. However, it points immediately to a gravitational source since this represents, for example, about 10% of the rest mass energy of the Sun. To get this 2×10^{46} J of energy in 10 s from nuclear burning would require the full conversion from hydrogen to iron of a mass of $12M_\odot$ in that time. No mechanism is known by which this could happen and turn the energy liberated into neutrinos. Thus the

Table 14.11 The Kamiokande II (KII) and Irvine–Michigan–Brookhaven (IMB) detector events caused by the electron neutrinos from the supernova SN1987A in the Large Magellanic Cloud (LMC).

Event	Time (ms)	Electron energy (MeV)	Angle with respect to direction from LMC (degrees)
KII			
1	0	20.0 ± 2.9	18 ± 18
2	107	13.5 ± 3.2	15 ± 27
3	303	7.5 ± 2.0	108 ± 32
4	324	9.2 ± 2.7	70 ± 30
5	507	12.8 ± 2.9	135 ± 23
6	1541	35.4 ± 8.0	32 ± 16
7	1728	21.0 ± 4.2	30 ± 18
8	1915	19.8 ± 3.2	38 ± 22
9	9219	8.6 ± 2.7	122 ± 30
10	10433	13.0 ± 2.6	49 ± 26
11	12439	8.9 ± 1.9	91 ± 39
	Total	169.7	
IMB			
1	0	38	74
2	420	37	52
3	650	40	56
4	1150	35	63
5	1570	29	40
6	2690	37	52
7	5010	20	39
8	5590	24	102
		(all ± 25%)	(all ± 15)
	Total	260	

Notes

1. The total electron energy must be increased by 1.80 MeV per event (positron mass plus neutron–proton mass difference) to get the total neutrino energy detected.
2. The angle with respect to the direction from the LMC is expected to be distributed isotropically since in the reaction $\nu_e + p \rightarrow n + e^+$, the target is heavy compared to the neutrino energy and the centre of mass of the collision is not moving appreciably.

gravitationally collapsing core models of supernova provide the only reasonable explanation for the observed neutrino events. Within that framework, the results are consistent with SN1987A giving a neutron star of mass about $1.4M_\odot$.

Table 14.12 The outline of a naïve calculation of the total neutrino energy emitted by SN1987A using the energy detected in the Kamiokande II (KII) and the Irvine–Michigan–Brookhaven (IMB) detectors (see Fig. 13.5).

The assumptions are as follows.

1. The neutrinos have Fermi–Dirac distribution of energy E_ν with zero chemical potential. Let $n(E_\nu)\mathrm{d}E_\nu$ be the number of $\bar{\nu}_e$ per square metre in energy range E_ν to $E_\nu+\mathrm{d}E_\nu$ at the Earth. Then

$$n(E_\nu)\mathrm{d}E_\nu = \frac{NE_\nu^2\mathrm{d}E_\nu}{\exp(E_\nu/kT)+1}$$

where N is a constant to be determined and we assume kT=4 MeV.

2. The neutrinos have zero mass.
3. The cross-section for $p+\bar{\nu}_e \to e^+ + n$ is $6.67\times10^{-48}\ E_\nu^2\ \mathrm{m}^2 = \sigma_0 E_\nu^2\ \mathrm{m}^2$, and this reaction is the sole one detected.
4. SN1987A is at a distance of 163 000 light years and emits all six varieties of neutrinos equally.
5 Both detectors are 100% efficient above their energy detection threshold, E_t.

		KII	IMB	
Mass of detector's fiducial volume protons	M	2.38×10^5	5.56×10^5	kg
Avogadro's number $\times 10^3$	$\mathscr{A}$	6.02×10^{26}		$(\mathrm{kg\ mole})^{-1}$
Threshold energy of detector	E_t	5	20	MeV
Total energy detected $\int_{E_t}^{\infty} n(E_\nu)E_\nu^3\sigma_0 M\mathscr{A}\,\mathrm{d}E_\nu$		189.5	274	MeV
Hence:				
Use this energy to find N	N	3.97×10^{11}	3.96×10^{11}	$\mathrm{MeV}^{-3}\,\mathrm{m}^{-2}$
Number flux of $\bar{\nu}_e$ at the Earth $\int_0^{\infty} n(E_\nu)\mathrm{d}E_\nu$		4.6×10^{13}	4.6×10^{13}	m^{-2}
Energy flux of $\bar{\nu}_e$ at the Earth $\int_0^{\infty} n(E_\nu)E_\nu\,\mathrm{d}E_\nu$		5.8×10^{14}	5.8×10^{14}	$\mathrm{MeV\,m}^{-2}$
Total neutrino energy output of SN1987A		1.7×10^{46}	1.7×10^{46}	J

It is an accident that both calculations arrive at the same value. More complete calculations arrive at a somewhat higher total neutrino energy production of $2–5\times10^{46}$ J, an average neutrino energy of 9–14 MeV and an effective source temperature with kT equal to 3–5 MeV. We have assumed 4 MeV for the latter and that gives a mean neutrino energy of 12.6 MeV. Note that we used the energy to find N: with the calculated N the expected number of events is 9 and 10 respectively. This is consistent with the observed numbers, 11 and 8, so a calculation using the number of events rather than the energy would have arrived at results comparable with those found here.

(The integrals over the Fermi–Dirac distribution were performed numerically.)

An exciting possibility is of a supernova in our own galaxy. Since the galaxy has a radius of about 33 000 light years, its centre is 5 times closer than SN1987A. Thus a supernova on our side of the centre could give perhaps 100 times as many neutrino events in a detector of the size of the IMB or Kamiokande II devices. With improved instrumentation, this would provide important information about the energy and time distributions of the neutrino flux produced in stellar core collapse. Unfortunately near-by supernovae are not common events.

14.11 Black hole formation

DEFINITIONS AND KEYWORDS

Schwarzschild radius See **Black hole**.

If a mass M is confined inside a radius $R_S = 2GM/c^2$, its **Schwarzschild radius**, then it is called a black hole because no light can leave. This is a result from general relativity. For $1M_\odot$ this radius is 3.97 km. In the course of stellar core collapse that we have described in the previous sections, it is possible that after the core has collapsed more material continues to fall on it and although a neutron star is formed it is too dense to support itself ($M > 2M_\odot$) and it collapses into a black hole. The equations of general relativity predict that all the mass is at $r = 0$ and the Schwarzschild radius defines a sphere from which no incoming material can ever return. If a black hole is formed in place of a neutron star, the timetable of neutrino emission is profoundly different. There may be a short (~ 0.3 s) flash of antineutrinos as neutronization of the outer layers of the core occurs. Neutrinos trapped in the contracting core have no time to escape before the core collapses into a black hole ($\sim 10^{-5}$ s). Thus the seconds-long neutrino burst observed from SN1987A would be absent.

The statement that all the mass is at $r = 0$ may require modification in the future when quantum mechanics is successfully wedded to general relativity. But, of course, the evidence for the existence of black holes is scanty. Several binary star systems have one invisible, massive (7–$15M_\odot$) partner. Since neutron stars cannot exceed about $2M_\odot$, these are black hole candidates. How these systems were formed is a matter of speculation and research.

Black holes appear to represent the final and complete, although long-delayed, victory for a remorseless gravity. However, black holes can evaporate! They are truly black in that they radiate black body radiation, albeit with a very low temperature ($\sim 10^{-6}$ K for a $1M_\odot$ hole). In addition, near R_S, electron–positron pairs can be formed and can quantum mechanically tunnel apart so that the energy of one particle increases at the expense of that of the other as it falls into the black hole, sufficiently to allow the former to escape. These processes will slowly dissipate the mass of an isolated black hole but only over an immensely long time: 10^{71} years for a $1M_\odot$ black hole. However, the lifetime is proportional to the initial mass cubed so that any very light ($< 10^{10}$ kg) black holes created in the big bang will have evaporated by now.

The description of black holes is the business of Einstein's general relativity. That theory predicts the existence of gravitational waves (see Sections 9.14 and 13.8) that will be emitted by any system with changing mass quadrupole moment. The weakness of the gravitational coupling constant (G) relative to the electromagnetic coupling constant (e) means that the emission of gravitational waves will not represent a significant mechanism of energy loss except under special conditions and that the waves will be hard to detect. Efforts have been made to detect such waves on the assumption that there must have been many catastrophic events which emitted copious gravitational radiation. However, there has so far been no confirmed detection of such radiation. There is some indirect evidence for the energy loss by this mechanism: there is a pulsar which is a member of a binary system where the companion is of comparable mass. The measured parameters of the system show that the pair are moving in an eccentric relative orbit (semi-major axis about 10^6 km) which is changing in period at a rate consistent with the expected loss of energy by gravitational radiation (Bowers and Deeming, 1984, Vol. I, p. 338).

PROBLEMS

14.5 Show that the potential energy due to gravitational attraction of a spherical mass M of uniform density and radius R is

$$-\frac{3GM^2}{5R},$$

where G is the gravitational constant.

14.6 Calculate the potential energy in joules for one solar mass ($1 M_\odot$) of material of uniform density and radius (a) 1 light year, (b) one solar radius ($R_\odot$ in Table 14.2), (c) 100 km, and (d) 10 km.

It is likely that the density is greater at the centre. How will this change the potential energy?

14.7 Suppose that the iron core of a pre-supernova star has a mass of $1.4\ M_\odot$ and a radius of 100 km and that it collapses to a uniform sphere of neutrons of radius 10 km. Assume that the virial theorem of the form of equation (14.2) holds. Calculate the energy consumed in neutronization and the number of electron neutrinos produced. Given that the remaining energy is radiated as neutrino-antineutrino pairs of all kinds of average energy 12+12 MeV, calculate the total number radiated.

14.8 Given that the supernova of Problem 14.7 is at a distance of 163 000 light years, calculate the total number of neutrinos of all types arriving at each square metre at the Earth. Also estimate the number of reactions

$$\bar{\nu}_e + p \rightarrow n + e^+$$

that will occur in 1000 tonnes of water. Assume that the cross-section is given by

$$\sigma = \frac{4P_e E_e G_F^2}{\pi \hbar^4 c^3},$$

where P_e and E_e are the positron momentum and energy respectively and G_F is the Fermi coupling constant. Assume only one-sixth of the neutrinos are electron antineutrinos.

14.9 Modify the nuclear semi-empirical mass formula by the addition of a term giving the mechanical potential energy due to gravitational attraction and apply it to a neutron star (Z=0). Find the smallest radius a neutron star can have assuming nuclear density throughout and the absence of any other constituents.

14.12 Now and the future

The previous sections have taken us through the evolution of the Universe from our watershed age at 225 s to the present time. That evolution was described with particular attention to the nuclear and particle physics aspects. What we see is that the Universe is a machine which is currently turning the primeval helium and hydrogen into compact stellar objects. These are either white dwarfs containing heavy elements, or neutron stars, or black holes. All this is happening in a Universe which is expanding. The question now is what will happen in the future? Will the Universe continue to expand and will the material in it become cold, compact stellar objects?

The expansion of the Universe can stop only if it contains sufficient mass for gravity to provide an adequate attractive force to slow down and reverse the expansion. The critical density above which expansion will be arrested is about $1 \times 10^{-26}\,\mathrm{kg\,m^{-3}}$. The observed average density of baryons, identified with the luminous matter of the Universe, is about $5 \times 10^{-28}\,\mathrm{kg\,m^{-3}}$ and on its own cannot reverse the expansion. However, the dynamics of galaxies and of clusters of galaxies indicates that there is matter present which is not luminous and which may bring the average density to near critical: this extra matter is called **dark matter**. It may be present in the form of many undeveloped stars (Section 14.1) or of already cold compact stellar objects. It could also be present in the form of

DEFINITIONS AND KEYWORDS

Dark matter Matter that may be present in the Universe but is not visible by being in a luminous state.

1 neutrinos with mass, or

2 unknown neutral particles with mass,

where in both cases the particles concerned were produced in the big bang. The presence of the neutrinos at 225 s has already been noted in Section 14.1. In Section 12.12 we discussed the determination of the mass of electron antineutrinos from the SN1987A event. This determination is not model independent but it is likely that an upper limit of 10 eV is realistic. The present number density of neutrinos of all types (and including antineutrinos) in the Universe is $10^8\,\mathrm{m^{-3}}$ so that, if all kinds have mass 10 eV, they give a density of about $1.9 \times 10^{-27}\,\mathrm{kg\,m^{-3}}$. Thus a background of neutrinos of mass 10 eV may not add enough to reach the critical mass.

The search is on for all kinds of dark matter. If insufficient is not found or inferred, our conclusion is that the Universe is set to expand forever, the galaxies becoming further apart, and gradually losing sight of each other. On the other hand, if the critical density is exceeded, then the expansion will slow down, stop and reverse. The expansion would be replaced by contraction and the reverse of the big bang, a big crunch, is inevitable. In that case perhaps the system will rebound in another big bang. A continuing cycle of big bang, expansion, contraction, big crunch is a possibility (Weinberg, 1977).

If, however, nucleon decay does occur then the future of matter in an ever expanding Universe will be one of gradual decay of the baryons in all stellar objects except black holes which decay by evaporation. Charge will be conserved so that in principle the positrons and electrons produced in the decay and evaporation will annihilate and all that will remain is a Universe of gravitons, photons and neutrinos, all becoming colder and colder.

14.13 The first 225 seconds

Our knowledge of particles and of the interactions between them gives us the basis upon which to construct models of the early history of the Universe. However, since we do not know what exists at masses greater than 100 GeV (temperature about 10^{15} K), we are limited in making conclusions about what happened to times later than that at which the temperature had this value. The gauge theories of the interactions do, however, give us one clue about the earlier times: as we indicated in Section 13.8 the interactions appear to converge to a common strength at an energy of about 10^{19} GeV (10^{32} K). Thus at a time ($\sim 10^{-43}$ s) when the temperature had this value all particles (gravitons, quarks, leptons, gauge bosons, Higgs bosons) were present and in statistical equilibrium. Of course if there are other particles to be discovered (lepto-quarks and who knows what), they must be included in this list. The statistical equilibrium means that each kind of particle had a production rate equal to the rate at which it suffered annihilation or absorption. As the Universe expanded the temperature fell. The fall from 10^{32} K reduced the gravitational coupling constant and the gravitons decoupled from the rest of the material and no longer played a role in the story. The isolated gravitons cooled as the Universe expanded and they now have a temperature of order 1 K. Two more similar decouplings occurred later.

In general, massive particles disappeared when the temperature fell to a value at which the photons from the black body radiation did not have sufficient energy to create a particle–antiparticle pair. The surviving particles decayed or annihilated. Thus at about 10^{15} K (86 GeV) the reaction $\gamma \rightarrow W^+ + W^-$ ceased and the remaining W^+ and W^- decayed if there were vacant states for their decay products or annihilated $W^+ + W^- \rightarrow 2\gamma$. At about 2×10^{12} K quarks and antiquarks formed into colourless hadrons. At 10^{12} K (86 MeV) the Universe consisted of electrons, positrons, and neutrinos of all kinds. There were also some π-mesons and μ-mesons but they were just about to disappear as the Ws disappeared earlier. There were also a relatively small number of protons and an excess of electrons over positrons that kept the total charge zero. The reason for this imbalance in favour of what we call matter over antimatter is a subject we shall discuss shortly. Thus all the conserved numbers, total charge, total quark number, total lepton number were all close to zero relative to the total number of particles. The elastic scattering of neutrinos of all kinds by electrons and positrons (Table 14.7) and their production (Table 14.9) kept them in thermal equilibrium with the matter. At 10^{11} K the remnant baryons, neutrons, and protons are kept in thermal equilibrium of number through the reactions

$$e^- + p \leftrightarrow n + \nu_e, \qquad (14.5)$$

and

$$e^+ + n \leftrightarrow p + \bar{\nu}_e. \qquad (14.6)$$

No complex nuclei were formed as the temperature was too high to allow deuterons to form (equation (14.1)). However, the fact that the neutron is heavier than the proton by 1.29 MeV means that below 10^{11} K (8.6 MeV) the neutron production cross-sections are significantly smaller than the proton production cross-sections. Thus the neutron–proton balance shifted from its high-temperature equilibrium of 50:50 to 25:75 at 10^{10} K. In addition, the other neutrino cross-sections had fallen along with the density and the neutrinos were almost decoupled from the remaining matter. The age of the Universe was 1 s. One other thing had happened: the temperature had fallen below the threshold for $\gamma \rightarrow e^+e^-$ (1.02 MeV $= 1.2 \times 10^{10}$ K), so that shortly afterwards electron–positron pair production ceased and the remaining positrons mopped up almost all the electrons in annihilation $e^+ + e^- \rightarrow 2\gamma$, thereby slowing down the temperature fall of the black body radiation.

About 13 s later neutron production had ceased, leaving a ratio of neutrons to protons of about 17:83. The black body temperature was 10^{10} K. The fall to 10^9 K was completed at about 180 s. During that interval 18% of the neutrons decayed ($\tau = 896$ s) changing the neutron to proton ratio to 14:86. At 225 seconds this ratio was about 13:87, the temperature had fallen below 10^9 K, deuterium production began and the Universe had arrived at the critical age from which we started in Section 14.2.

Note that the neutrinos decoupled at near 1 s but the decoupling of the photons of the black body radiation from the surviving electrons, protons, and the α-particles did not occur completely until the temperatures fell to about 10^4 K (0.86 eV) at an age of about 10^6 years. Until then Compton scattering and bremsstrahlung kept the photons in equilibrium with the electrons and nuclei. But, when the temperature fell well below the first excitation potentials of hydrogen and helium, the Universe became transparent to photons. Thereafter the black body radiation was free to cool independently of the matter.

Therefore the current catalogue of big bang residues is black body radiation at 2.7 K, neutrinos at a slightly lower temperature, gravitons at about 1 K, and a small amount of matter presumably tied up in galaxies. In the early stages of the Universe, the statistical equilibrium should have ensured equal numbers of particles and antiparticles. At later stages annihilation should have destroyed them in equal numbers unless there was some mechanism for separating into different regions matter (electrons and protons) from antimatter (positrons and antiprotons). There is no evidence that there are regions of antimatter in the Universe. The composition of cosmic radiation rules out antimatter in our own galaxy. In other galaxies there can be no direct evidence of their state except where two galaxies are having close encounters: in such cases there is no evidence of the huge energy release expected as matter from one annihilates with antimatter from the other. Therefore, why is there this asymmetry in the Universe?

This question has not been satisfactorily answered but the minimum conditions necessary have been identified. They are as follows.

1. Baryon, that is quark number, non-conserving interactions occur. We have seen in Section 13.6 that certain proposed unified theories allow that possibility (involving the hypothetical X and Y-particles) and, of course, the decay of the proton, if observed, would be an unequivocal evidence for such an interaction.
2. *CP* invariance must be violated. That we know does happen (Section 13.3).
3. Statistical thermal equilibrium does not exist at the time that conditions 1 and 2 were able to operate.

The ratio of the number of baryons in the universe to the number of photons is very small, 10^{-9}, so that we are looking for a tiny effect. Condition 2 allows the branching fraction for the decay $\bar{X} \rightarrow \bar{q}\bar{q}$ to be different from the for $X \rightarrow qq$. Statistical equilibrium would ensure no asymmetry but, of course, the expansion of the Universe means that we cannot have true equilibrium. Therefore the Universe could have emerged from the epoch when X and $\bar{X}$ were present with a tiny excess of q over $\bar{q}$. This time is guessed to be about 10^{-33} s. That tiny excess would be responsible for what may be 10^{12} galaxies of average mass $10^{11} M_{\odot}$, perhaps a total of 10^{80} nucleons in the Universe.

Fig. 14.6 The Big Bang: another view.

14.14 Conclusion

The story that we have told has one particularly strong message. It is that nuclear and particle physics are essential to understanding the history of the Universe and its contents. The future may well see an even closer symbiosis between particle physics and cosmology, particularly if the energies required to explore to greater particle masses cannot be achieved on Earth. In that case a study of the Big Bang and its relics as seen in astrophysical phenomena may be the sole means we shall have of research in that field of physics.

References

Bahcall, J. N. and Davis, R. (1976). *Science*, **191**, 264–7.

Bowers, R. L. and Deeming, T. (1984). *Astrophysics I and II*. Jones and Bartlett, Boston.

Davis, R., Evans, J. C. and Cleveland, B. T. (1978). *Conference Proceedings—Neutrino 1978*. Purdue University, Lafayette.

Kirsten, G. (1979). *Proceedings of the International Conference on Neutrino Physics and Astrophysics*, Vol. 2, p. 452 (eds. A. Haatuft and C. Jarlskog). University of Bergen, 1979.

Raghavan, R. S. (1976). *Physical Review Letters*, **37**, 259–62.

Weinberg, S. (1977). *The First Three Minutes*. Andre Deutsch, London.

Appendix A The Atomic Elements

Table A.1 The atomic elements by symbol, name and number

Chemical symbol	Name	Atomic number	Chemical symbol	Name	Atomic number	Chemical symbol	Name	Atomic number
Ac	Actinium	89	Gd	Gadolinium	64	Pm	Promethium	61
Ag	Silver	47	Ge	Germanium	32	Po	Polonium	84
Al	Aluminium	13	H	Hydrogen	1	Pr	Praeseodymium	59
Am	Americium	95	He	Helium	2	Pt	Platinum	78
Ar	Argon	18	Hf	Hafnium	72	Pu	Plutonium	94
As	Arsenic	33	Hg	Mercury	80	Ra	Radium	88
At	Astatine	85	Ho	Holmium	67	Rb	Rubidium	37
Au	Gold	79	I	Iodine	53	Re	Rhenium	75
B	Boron	5	In	Indium	49	Rh	Rhodium	45
Ba	Barium	56	Ir	Iridium	77	Rn	Radon	86
Be	Beryllium	97	K	Potassium	19	Ru	Ruthenium	44
Bi	Bismuth	83	Kr	Krypton	36	S	Sulphur	16
Bk	Berkelium	97	La	Lanthanum	57	Sb	Antimony	51
Br	Bromine	35	Li	Lithium	3	Sc	Scandium	21
C	Carbon	12	Lr	Lawrencium	103	Se	Selenium	34
Ca	Calcium	20	Lu	Lutetium	71	Si	Silicon	14
Cd	Cadmium	48	Md	Mendelevium	101	Sm	Samarium	62
Ce	Cerium	58	Mg	Magnesium	12	Sn	Tin	50
Cf	Californium	98	Mn	Manganese	25	Sr	Strontium	38
Cl	Chlorine	17	Mo	Molybdenum	42	Ta	Tantalum	73
Cm	Curium	96	N	Nitrogen	7	Tb	Terbium	65
Co	Cobalt	27	Na	Sodium	11	Tc	Technetium	43
Cr	Chromium	24	Nb	Niobium	41	Te	Tellurium	52
Cs	Cesium	55	Nd	Neodymium	60	Th	Thorium	90
Cu	Copper	29	Ne	Neon	10	Ti	Titanium	22
Dy	Dysprosium	66	Ni	Nickel	28	Tl	Thallium	81
Er	Erbium	68	No	Nobelium	102	Tm	Thulium	69
Es	Einsteinium	99	Np	Neptunium	93	U	Uranium	92
Eu	Europium	63	O	Oxygen	8	V	Vanadium	23
F	Fluorine	9	Os	Osmium	76	W	Tungsten	74
Fe	Iron	26	P	Phosphorus	15	Xe	Xenon	54
Fm	Fermium	100	Pa	Protoactinium	91	Y	Yttrium	39
Fr	Francium	87	Pb	Lead	82	Yb	Ytterbium	70
Ga	Gallium	31	Pd	Palladium	46	Zn	Zinc	30
						Zr	Zirconium	40

Appendix B Constants

Readers will need some numbers when doing problems. The atomic constants, α $\hbar$, c, and e may be obtained from Table 2.1, p.20. Atomic weights are available in Table 1.4, p.13. Nuclear binding energies for $A > 20$ may be calculated with sufficient precision using the semi-empirical mass formula in Table 4.1, p.60, or deduced from Fig. 4.6, p.61. Solar quantities are in Table 14.2, p.347 and Fermi's coupling constant is in Table 12.8, p.304. Particle masses may be found as follows:

Proton and neutron	Table 4.1, p.60
Pions and muon	Table 9.6, p.178
Other mesons	Tables 10.3 and 10.4, pp.198 and 200
Baryons	Table 10.5, p.202
W and Z bosons	Table 9.8, p.186
Quarks	Table 10.6, p.208
Leptons	Table 12.1, p.279

You may also need:

Gravitational constant, $G = 6.673 \times 10^{-11}\ \text{m}^3\,\text{kg}^{-1}\,\text{s}^{-2}$
Mean Earth–Sun distance $= 1.5 \times 10^8$ km.

Answers to Problems

1.1	(a) 5.3×10^4 fm^2; (b) 1.3×10^4 fm^2; (c) 5.6×10^3 fm^2.
1.2	5.6×10^5 fm^2 sr^{-1}.
1.3	Potential energy of α-particle at nuclear surface: Al, 10.4 MeV; Au, 32.5 MeV.
2.1	1.44 MeV; 22.7 MeV; 261 MeV.
2.2	9.9 ns.
2.3	8.1×10^{-13}.
2.4	5.1, 1.7, 0.45, 0.14, 0.26, 0.39×10^7 s^{-1}.
2.5	2.6×10^{15}; 3.4×10^{15}.
2.6	24.8 days.
2.7	1.9×10^9 years.
2.8	6.0×10^9 years.
2.9	1.5×10^8 years.
2.10	0.7; 4.3×10^9 years.
2.11	1760 years.
2.12	(1) 99.3 per cent; (2) 510 s^{-1}; (3) 4×10^{-5} m^{-2} s^{-1}.
2.13	326.
2.14	$Z = 3$; $A = 7$; 139.
2.15	0.0652; 0.06; 0.16.
2.16	9.9×10^{-4}.
3.3	$F(q^2) = (1 + q^2a^2/\hbar^2)^{-1}$.
3.4	57.5 MeV/c; 65.3 fm^2 sr^{-1}; 0.75.
3.5	-24.3 MeV.
3.6	(a) 284 fm; (b) 42.6 fm; $n = 2$; 2.63 MeV.
3.7	(a) 25 keV; (b) binding energy less by 499 eV.
4.2	154 MeV.
4.3	320 MeV; 11 MeV.
5.1	-1.9 MeV; $+3.1$ MeV.
5.3	3.7 MeV.
5.4	3.4 mg; 2.1×10^{-11} kg year^{-1}.
5.5	0.631 MeV/c^2; 22 MeV/c^2.
5.6	$^{237}_{93}$Np; $1.0:0:0.34:8.4 \times 10^{-5}$.
6.1	26.5 years.
6.2	1:0.37:0.036.
7.3	3.27 MeV; 17.6 MeV; 7.26 MeV; 20.8 MeV.
7.5	1.2.
7.6	12.4 MeV; Yes, at proton laboratory energy of 341 keV.
7.7	9.94 MeV.
7.10	2.5 MeV.
7.13	4.0×10^4 b.
8.4	0^+; $\frac{5}{2}^+$; $\frac{7}{2}^-$.
10.1	~ 12 m.
10.2	-534, 105, -381, -746, 28, -856, -1232 MeV.

10.3 f, h, and d.
10.5 31 mb.
10.7 1.6 m.
10.8 $3.7 \times 10^{-2}\,s^{-1}$; $0.12\,s^{-1}$.
10.9 0.28; $2.0 \times 10^{18}\,s^{-1}$, 2.1 per cent.
11.8 M4; $1.5 \times 10^{-3}\,s^{-1}$; M1; $1.4 \times 10^{19}\,s^{-1}$ (using the proton mass; however, how about using the mass and charge of a constituent quark?).
11.9 M1, $4.7 \times 10^{10}\,s^{-1}$.
M3, $5.9 \times 10^{-6}\,s^{-1}$.
E2, $7.2 \times 10^{10}\,s^{-1}$.
E1, $4.0 \times 10^{11}\,s^{-1}$.
M2, $7.8 \times 10^{7}\,s^{-1}$.
11.10 E4(?); E4/M4.
11.12 1.1×10^{-8} eV. (The convolution of two Lorentzian line shapes of equal width Γ is a Lorentzian of width 2Γ.)
11.13 1.1×10^{-7} eV; 3.2×10^{-8} eV.
11.14 $3.3 \times 10^{-4}\,mm\,s^{-1}$, down.
12.14 2.7 per cent.
12.16 3×10^{6} km; 2.5×10^{8}.
12.18 If 2 seconds represent the arrival time spread between 4 and 12 MeV neutrinos emitted at one instant, then the neutrino mass is 3.8 eV.
14.1 $6.3 \times 10^{14}\,\nu\,m^{-1}$; 0.4 MeV.
14.2 2.3 SNU; $2.3 \times 10^{10}\,\nu\,m^{2}\,s^{-1}$.
14.6 -1.7×10^{34} J; -2.3×10^{41} J; -1.6×10^{45} J; -1.6×10^{46} J.
14.7 2.4×10^{45} J; 7.7×10^{56}; 6.0×10^{57}.
14.8 $2.2 \times 10^{14}\,m^{-2}$; 2.

Index

Emboldened page numbers indicate an entry in a **Definitions and Keywords** panel